AT THE CROSSROADS OF ASTROPHYSICS AND COSMOLOGY: PERIOD–LUMINOSITY RELATIONS IN THE 2020S

IAU SYMPOSIUM 376

Uncrowded Cepheids observed with the *James Webb Space Telescope (JWST)* in the near-infrared. High-resolution *JWST* observations can test confusion-limited *Hubble Space telescope (HST)* observations that are used to measure the Hubble constant. *JWST* provides superior source separation to negate crowding noise resulting in a factor of > 2.5 reduction in the dispersion of the Cepheid period–luminosity relation in galaxies that host Type Ia supernovae (image shown for NGC 5584) and the geometric distance reference, NGC 4258, with the period–luminosity relation shown here. The agreement between Cepheid distance measures between *HST* and *JWST* is excellent and provides the strongest evidence to date that systematic errors in *HST* Cepheid measurements do not play a significant role in the present 'Hubble tension,' now at 5σ confidence. The source of the Hubble tension remains an outstanding mystery of potentially cosmological origin and warrants further study. (Credit: Adam Riess, Johns Hopkins University, USA)

IAU SYMPOSIUM PROCEEDINGS SERIES

INTERNATIONAL ASTRONOMICAL UNION

UNION ASTRONOMIQUE INTERNATIONALE

International Astronomical Union

IAU

AT THE CROSSROADS OF ASTROPHYSICS AND COSMOLOGY: PERIOD–LUMINOSITY RELATIONS IN THE 2020S

PROCEEDINGS OF THE 376th SYMPOSIUM OF THE INTERNATIONAL ASTRONOMICAL UNION
BUDAPEST, HUNGARY
17–21 April, 2023

Edited by

RICHARD DE GRIJS

Macquarie University, Sydney, Australia

PATRICIA A. WHITELOCK

South African Astronomical Observatory and University of Cape Town, South Africa

and

MÁRCIO CATELAN

Pontificia Universidad Católica de Chile, Santiago, Chile

CAMBRIDGE
UNIVERSITY PRESS

CAMBRIDGE UNIVERSITY PRESS
University Printing House, Cambridge CB2 8BS, United Kingdom
1 Liberty Plaza, Floor 20, New York, NY 10006, USA
10 Stamford Road, Oakleigh, Melbourne 3166, Australia

First published 2024

Printed in Great Britain by Henry Ling Limited, The Dorset Press, Dorchester, DT1 1HQ

Typeset in System LaTeX 2ε

A catalogue record for this book is available from the British Library of Congress Cataloguing in Publication data

ISBN 9781009353045 hardback
ISSN 1743-9213

Table of Contents

Preface

This volume includes contributions from IAU Symposium 376, *At the crossroads of astrophysics and cosmology: Period–luminosity relations in the 2020s.*

In the early 1900s, Henrietta Leavitt, in 1908—and later Adams & Joy and Shapley & Walton, both in 1927—showed that bright Cepheid variables were characterised by a narrow range in spectral type or, equivalently, temperature at a given period. This, in turn, led to the establishment of a tight period–luminosity relation (PLR): more luminous stars are expected to have longer pulsation periods. The spectral-type/temperature restriction of Cepheids naturally led to the realisation that pulsating variables only occur in a narrow "instability strip" in the Hertzsprung–Russell diagram.

This classical instability strip is host to a range of periodic variable stars, including classical Cepheids (δ Cephei stars), Mira variables at the top of the asymptotic giant branch (AGB), Type II Cepheids, and RR Lyrae stars. SX Phoenicis and δ Scuti variables (jointly known as dwarf Cepheids or ultrashort-period variables) as well as anomalous Cepheids, and ZZ Ceti, V777 Herculis and GW Virginis pulsating white dwarfs are also found in specific instability strips. Many of these stellar types obey specific period–mean density relations. Curiously, not only pulsating stars exhibit PLRs: well-defined and long-established relationships between orbital periods and luminosities also exist for contact binaries of W Ursae Majoris (EW) type. Crucial remaining open issues in this broad field include the metallicity dependence of the zero point of the Cepheid PLR, the possible presence of a break at a pulsation period around 10 days, and the effects of binarity and circumstellar envelopes. Intrinsic PLR widths may offer unique insights into the physical processes shaping these relations and the underlying physical properties of the contributing stars (stellar structure, atmospheric parameters, and pulsation properties).

A promising approach to reducing the current systematic uncertainties associated with the present-day expansion rate of the Universe—the well-known tension in the Hubble parameter—is by trying to achieve improved local calibrations of primary distance indicators and their derivatives, including calibration of the photometric zero point of the Cepheid PLR, at both optical and—potentially with much reduced scatter—infrared wavelengths (e.g., starting from existing *Spitzer Space Telescope* data), for instance through trigonometric-parallax measurements of carefully selected Cepheid samples by the European Space Agency's *Gaia* satellite. Improvements in the Cepheid and Mira distance scales will be achievable to a level of 3–4% or better, based on forthcoming mid-infrared observations with the *James Webb Space Telescope* (*JWST*). Perhaps the most powerful use of the variety of PLRs present for variable stars is the ability to cross-check results from any individual technique. The diverse stellar types exhibiting PLRs are drawn from distinct stellar populations, with their own age and metallicity distributions. Working together, these techniques can explore population-based systematics in distance determination to provide insights into potential calibration biases in the current Type Ia supernovae calibration.

After much discussion, we decided to frame IAU Symposium 376 by initially focusing on the discrepancy in the Hubble constant based on different approaches, including the "standard" Cepheid PLR. This choice led to many productive and engaging discussions, triggered by numerous inspiring talks that highlighted the latest developments. In essence, this theme followed on from the Spring school on the distance ladder that had been organised at Konkoly Observatory during the week prior to the conference week.

From the outset, it became clear that the *JWST* will soon become a game changer given the amazing data already generated by this new observatory. Thanks to its flawless launch and smooth initial operations, the community is now looking at an observatory that may last much longer than its nominal five-year operational lifetime. In turn, this offers an enormously expanded scope for studies of the distance ladder. Even at this early time, we are seeing significant improvements in spatial resolution with respect to that offered by the *Hubble Space Telescope* at similar (near-)infrared wavelengths. In fact, preliminary results imply that the scatter in the Cepheid PLR at 1.5 μm may be reduced by a factor of 2–3. Multiple participants presented very exciting results based on early *JWST* data.

Another eye-catching aspect brought up in many inspirational contributions is the phenomenal distance accuracy currently achievable from Cepheid PLRs. *Gaia* cluster Cepheids now show Large Magellanic Cloud-like scatter in their period–Wesenheit relations, while the distance to Messier 33 has now been established to better than 1.3%. This finally allows us to understand secondary effects, such as the impact of period fluctuations (e.g., in Messier 51) on distance estimates.

Similarly, tip of the red giant branch (TRGB) distance determination has come a long way over the past three or more decades and promises to deliver a powerful alternative approach to the standard PLR technique for Cepheids, pushing to ever greater distances. In addition, newly developed stellar population methods now allow us to apply the "J-AGB" method, exploiting the characteristics of carbon-rich stars.

Meanwhile, comparisons of distance estimates using different, independent methods are finally coming together. By pursuing minimisation of systematics and secondary variability effects, over the past decade the distance ladder has narrowed from a consideration of many complementary, independent techniques to a streamlined ladder with well-defined rungs, from Geometry to Cepheids, from Cepheids to SNe Ia, and from SNe Ia to redshifts. However, the field is now returning to the broader implications resulting from this much-improved ladder, with a greater focus on cross-checks with other methods. *Gaia*, *JWST*, and soon the Vera C. Rubin Observatory's Legacy Survey of Space and Time will be game changers out to nearby galaxies at distances as far as tens of megaparsecs.

Despite an early focus on the challenges posed by the extragalactic distance scale, a significant fraction of the conference was dedicated to the underlying physics governing stellar pulsation and variability, for different types of tracers. Among the classical Cepheids, renewed interest focuses on multimode variability and period fluctuations, from short-period overtones to ultralong-period variability. Current developments promise exciting opportunities to much better determine the stellar mass–luminosity relation, often based on painstaking, detailed work where every object counts. Perhaps even more importantly, we are now reaching observational regimes where we can finally get a good handle on metallicity effects in the context of Cepheid variability.

AGB stars are coming back into fashion, particularly because of improved calibration methods and new theoretical developments. Parallaxes from masers obtained with very long baseline interferometry (VLBI), at nearby distances anchored by *Gaia* parallaxes, are providing very useful constraints. These are important developments in the era of the new extremely large telescopes which are more infrared-sensitive than their smaller counterparts. Moreover, the Square Kilometre Array, the next-generation Very Large Array (ngVLA), the Event Horizon Telescope, and other major radio initiatives open up new areas of research involving AGB and semi-regular variables as useful tracers of the physical processes taking place in and distances to numerous nearby galaxies.

RR Lyrae stars remain the workhorse PLR tracers for older stellar populations given their pre-eminent importance for near-field cosmology and for understanding the formation history of the Galaxy, particularly when combined with dynamical data. *Gaia* has also triggered a revival in this field, complemented by many cutting-edge ground-based projects. A major focus at the present time is on the metallicity dependence of the RR Lyrae PLRs, so that their empirical basis now supports theoretical arguments.

The wealth of data available at the present time and their high quality now allow us to consider secondary effects that may play a role in this field, such as the light-curve shape, multi-mode pulsation properties, period changes, binarity, circumstellar envelopes, etc.

This conference has clearly shown that stellar pulsation physics has reached a high level of maturity, so that our attention is increasingly focused on more diverse pulsator types other than classical Cepheids and RR Lyrae, including anomalous and Type II Cepheids, δ Scuti stars, magnetically and chemically peculiar stars, scaling relations and PLRs pertaining to supergiant pulsators, contact binary systems, and many others.

It has become clear that the importance of *Gaia* for studies of variable stars cannot be overstated. It has already collected more than a trillion CCD measurements down to $G{\sim}21$ mag with uncertainties as small as 1 mmag. At the same time, the mission provides colours, astrometry, and spectroscopic time-series observations (for future release), thus making it the perfect tool for variability studies. At the present time, we already have access to *Gaia* observations of 271,000 RR Lyrae stars among some 10 million variables in Data Release 3 (DR3), with DR4 potentially reaching up to 100 million variable stars.

The key outstanding issue in this field, pertaining to PLRs in general, relates to the systematics affecting the current crop of *Gaia* parallaxes. We still need better ways to mitigate those effects, particularly as regards those objects that exhibit 'astrophysical confusion' in the sense of outflows, disks, etc. At present, it appears that the 'counter-corrections' required to derive accurate distances are often similar to the intrinsic, recommended corrections, so that more work is required to overcome those limitations. Many methods have been proposed (cluster membership, asteroseismology, binary systems with interferometric orbits, Miras with VLBI parallaxes, etc.), so it appears that the field is slowly moving to an acceptable solution in this area as well. The picture that is emerging is that there is not a single 'best' parallax correction that may or may not be magnitude-dependent, but we will have to resolve this issue using a sample-based approach.

Numerous contributions discussed the structure of the Milky Way, showing the significant synergies among different variability tracers, although one should keep in mind that different ages may trace different spatial components. Combined with dynamical measures, we now have a pretty good handle on the orbits of the Magellanic Clouds and their wake, on the warped and flaring young disk structure of the Milky Way, and even on the structure of our Galaxy on the other side of the Galactic Centre.

Theoretical advances are, of course, equally important as all those beautiful observational results. At this time, it seems that we are overwhelmed with high-quality data sets and theory is taking a bit of a back seat. Nevertheless, numerous teams are working on improving our theoretical understanding of the underlying pulsation physics, offering ever more detailed tests of models aided by the unmatched quality of observational data, including such models that focus on the boundaries of the classical instability strip. Major advances are seen in the context of understanding and modelling the effects of rotation and convection, where we are slowly moving from 1D to more realistic 3D models.

In view of all of these exciting developments, the future of this field looks extremely promising. Increasing data volumes are fundamentally changing our approaches, as we have already seen in the *Gaia* context, and this will only be accelerated when facilities such as the Rubin Observatory and its Legacy Survey of Space and Time come online. Likewise, the Roman Space Telescope/*WFIRST* will also contribute to more and larger data sets, while smaller, ground-based facilities will still be required to provide reference data, follow-up opportunities, and complementary temporal sampling.

Beyond photometry and light-curve analyses, spectroscopy represents the next frontier, and also here the future looks bright. Leading on from APOGEE and GALAH, the field is looking forward to exploiting such facilities as 4MOST, WEAVE, DESI,

SDSS-V, MOONS, and others, allowing us to start sampling the variable sky spectroscopically. In particular, radial velocity variability will be tackled with facilities like VELOCE (Cepheids) and, increasingly, *Gaia* (everything else?). It would be very helpful for the field as a whole to share analysis codes and improve accessibility to such tools so as to benefit a greater cross-section of the community.

Spectroscopic time-series observations clearly are the future of this field. Beyond the optical and near-infrared domains, X-ray variability, ultraviolet data, and radio approaches are increasingly hitting the forefront of the field, so we are warned to keep an eye out for new developments in those areas and an open mind as regards one's favourite wavelength range (even including gravitational waves?). Standardisation and cross-calibration remain a concern, but major efforts are undertaken to get this under control. We are eagerly looking forward to the era of the extremely large telescopes. And will artificial intelligence start to play an increasingly important role in our improved understanding? By the time that these developments have matured, a follow-up IAU Symposium will probably be warranted.

Richard de Grijs
Patricia A. Whitelock
Márcio Catelan

Editors

Richard de Grijs
Macquarie University, Sydney, Australia

Patricia A. Whitelock
South African Astronomical Observatory and University of Cape Town, South Africa

Márcio Catelan
Pontificia Universidad Católica de Chile, Santiago, Chile

Organising Committees

Scientific Organising Committee

Richard de Grijs (Chair), Macquarie University, Australia.
László Kiss (Co-chair), Konkoly Observatory, Hungary.
Rachael Beaton, Princeton University, USA.
Márcio Catelan, Pontificia Universidad Católica de Chile, Santiago, Chile.
Xiaodian Chen, National Astronomical Observatories, Chinese Academy of Sciences, Beijing, China.
Gisella Clementini, INAF, Osservatorio di Astrofisica e Scienza dello Spazio di Bologna, Italy.
Wolfgang Gieren, Universidad de Concepción, Chile.
Noriyuki Matsunaga, University of Tokyo, Japan.
Victoria Scowcroft, University of Bath, UK.
Róbert Szabó, Konkoly Observatory, Hungary.
Andrzej Udalski, University of Warsaw, Poland.
Patricia Whitelock, South African Astronomical Observatory and University of Cape Town, South Africa.

Local Organising Committee

Róbert Szabó (Chair), Konkoly Observatory, Hungary.
Adrienn Ádám (Co-chair), Konkoly Observatory, Hungary.
László Szabados (Co-chair), Konkoly Observatory, Hungary.
Attila Bódi, Konkoly Observatory, Hungary.
Susmita Das, Konkoly Observatory, Hungary.
Aliz Derekas, Eötvös University, Gothard Astrophysical Observatory, Hungary.
László Kiss, Konkoly Observatory, Hungary.
Adrienn Forró, Konkoly Observatory, Hungary.
Gábor Kovács, Konkoly Observatory, Hungary.
László Molnár, Konkoly Observatory, Hungary.
Emese Plachy, Konkoly Observatory, Hungary.
Dóra Tarczay-Nehéz, Konkoly Observatory, Hungary.
Zsófia Sódorné Bognár, Konkoly Observatory, Hungary.

List of Participants

Abdollahi, Mahdi
Ádám, Rozália
Afanasiev, Anton
Ahmad, Arief
Anderson, Richard
Andriantsaralaza, Miora
Čeki, Atila
Ardern, Steve
Arellano Ferro, Armando
Bódi, Attila
Baruch, John
Beaton, Rachael
Belokurov, Vasily
Benhida, Abdelmajid
Benkő, József
Bhardwaj, Anupam
Bienias, John
Bognár, Zsófia
Bora, Zsófia
Braga, Vittorio Francesco
Bras, Garance
Breuval, Louise
Catanzaro, Giovanni
Chen, Xiaodian
Chung, Chul
Clement, Christine
Clementini, Gisella
Colombo, Claudia
Cruz Reyes, Mauricio
Csoernyei, Geza
Dályа, Gergely
Das, Susmita
de Grijs, Richard
De Somma, Giulia
Deka, Mami
Derekas, Alíz
di Criscienzo, Marcella
Engels, Dieter
Espinoza-Arancibia, Felipe
Evans, Nancy
Eyer, Laurent
Fiorentino, Giuliana
Forró, Adrienn
Freedman, Wendy
Gallenne, Alexandre
Garca Hernández, Antonio
Gholami, Mahtab
Groenewegen, Martin
Hajdu, Gergely
Hintz, Eric
Hocdé, Vincent
Huang, Caroline D.
Iwanek, Patryk
Jeffery, Elizabeth
Jeon, Young-Beom
Jiang, Biwei
Jiménez-Esteban, Fran
Joyce, Meridith
Jurkovic, Monika
Kalup, Csilla
Kervella, Pierre
Khan, Saniya
Kiss, László
Kovács, Gábor
Kovács, Géza
Kurbah, Kerdaris
Latkovic, Olivera
Lee, Jae Woo
Lee, Myung Gyoon
Lengen, Bastian
Lewis, Megan
Liu, Niu
Marconi, Marcella
Martínez-Vázquez, Clara
Mateu, Cecilia
Medina Toledo, Gustavo
Menzies, John
Minniti, Javier
Molnár, László
Mondal, Prapti
Monelli, Matteo
Muraveva, Tatiana
Musella, Ilaria
Nakagawa, Akiharu
Narloch, Weronika
Netzel, Henryka
Olszewska, Justyna
Orosz, Gábor
Owens, Kayla
Paparó, Margit
Park, Seunghyun
Parto, Tahere
Paunzen, Ernst
Pawlak, Michał
Pilecki, Bogumił
Plachy, Emese
Ramezani, Tahereh
Ren, Yi
Ren, Fangzhou
Riess, Adam

Ripepi, Vincenzo
Sánchez-Benavente, Manuel
Sicignano, Teresa
Skowron, Dorota
Son, Junhyuk
Soszyński, Igor
Spetsieri, Zoi Tzogia
Storm, Jesper
Suárez, Juan Carlos
Surdha, Tashveena
Szabó, Róbert
Szabados, László
Tantalo, Maria
Taormina, Mónica
Tokarek, Jakub
Trabucchi, Michele
Trentin, Erasmo
Varga, Vázsony
Vázquez Ramió, Hector
Vinkó, József
Viviani, Giordano
Wang, Shu
Whitelock, Patricia A.
Wielgórski, Piotr
Zgirski, Bartłomiej
Zhang, Zehao

At the crossroads of astrophysics and cosmology: Period–luminosity relations in the 2020s
Proceedings IAU Symposium No. 376, 2024
R. de Grijs, P. A. Whitelock & M. Catelan, eds.
doi:10.1017/S1743921323003459

The Cepheid Extragalactic Distance Scale: Past, Present and Future

Wendy L. Freedman[1,2] **and Barry F. Madore**[1,3]

[1]Dept. of Astronomy & Astrophysics, University of Chicago, 5640 S. Ellis Ave., Chicago, IL 60637, USA. email: wfreedman@uchicago.edu

[2]Kavli Institute for Cosmological Physics, University of Chicago, 5640 S. Ellis Ave., Chicago, IL 60637, USA

[3]Carnegie Observatories, 813 Santa Barbara St., Pasadena CA 91101, USA

Abstract. Cepheids have been the cornerstone of the extragalactic distance scale for a century. With high-quality data, these luminous supergiants exhibit a small dispersion in their Leavitt (period–luminosity) relation, particularly at longer wavelengths, and few methods rival the precision possible with Cepheid distances. In these proceedings, we present an overview of major observational programs pertaining to the Cepheid extragalactic distance scale, its progress and remaining challenges. In addition, we present preliminary new results on Cepheids from the *James Webb Space Telescope* (*JWST*). The launch of *JWST* has opened a new chapter in the measurement of extragalactic distances and the Hubble constant. *JWST* offers a resolution three times that of the *Hubble Space Telescope* (*HST*) with nearly 10 times the sensitivity. It has been suggested that the discrepancy in the value of the Hubble constant based on Cepheids compared to that inferred from measurements of the cosmic microwave background requires new and additional physics beyond the standard cosmological model. *JWST* observations will be critical in reducing remaining systematics in the Cepheid measurements and for confirming if new physics is indeed required. Early *JWST* data for the galaxy, NGC 7250 show a decrease in scatter in the Cepheid Leavitt law by a factor of two relative to existing *HST* data and demonstrate that crowding/blending effects are a significant issue in a galaxy as close as 20 Mpc.

Keywords. Cepheids, *HST*, *JWST*, Extragalactic Distance Scale, Hubble constant

1. Introduction

No review of the Cepheid distance scale would be complete without recognition of the contribution by Henrietta Swan Leavitt. Leavitt's 1908 seminal paper (Leavitt 1908) on variable stars in the Large (LMC) and Small (SMC) Magellanic Clouds established the relationship between the period and luminosity of Cepheids. Subsequently, Leavitt and Pickering (Leavitt & Pickering 1912) published a period–luminosity relation for 25 Cepheids in the SMC, finding a nearly linear relationship between the logarithm of the luminosity and the logarithm of the period. Leavitt and Pickering recognized that the Magellanic Cloud Cepheids were likely at the same distance from Earth, "... their periods are apparently associated with their actual emission of light, as determined by their mass, density and surface brightness." These critical insights laid the foundation for the Cepheid distance scale, facilitating the measurement of distances to astronomical objects. Leavitt passed away in 1921, and sadly did not live to see the impact of her pioneering work. In 2008, a meeting was held at Harvard University to commemorate the 100-year anniversary of Leavitt's Cepheid discovery paper. An outcome of the meeting was the recommendation to name the Cepheid period–luminosity relation the 'Leavitt Law' (Freedman & Madore 2010).

Beginning in 1923, Edwin Hubble discovered Cepheids (Hubble 1925a,b, 1926) in the Local Group galaxies M33, M31, and NGC 6822, and estimated the distances to these galaxies employing Leavitt's period–luminosity relation. These and subsequent measurements culminated in his discovery of the relationship between galaxy distances and their recessional velocities (Hubble 1929), and set the foundation for modern cosmology with its implications for the expansion of the Universe.

Building upon a body of work characterizing the nature of variable stars in Milky Way clusters (Bailey 1902; Joy 1949; Sawyer Hogg 1955), Baade (Baade 1956; Baade & Swope 1963, 1965) unambiguously established the existence of two populations of Cepheid variables, resulting in a recalibration of Hubble's magnitude scale by 1.5 mag, or a factor of two in distance.

The next major advance in Cepheid studies came with Wisniewski & Johnson (1968)'s photoelectric *UBVRIJKL* observations of Milky Way Cepheids. Their high-precision photometry exhibited two major characteristics of Cepheid light curves: (1) the decrease in amplitude moving from the ultraviolet to the near-infrared, and (2) the shift in the relative phase of maximum light, also with increasing wavelength.

The 1970s through 1990s saw the 'factor-of-two' debate over the value of H_0. In a series of ten papers entitled "Steps Toward the Hubble Constant", published between 1974 and 1995, Allan Sandage and Gustav Tammann (Sandage & Tammann 2006) presented evidence generally favoring a value of $H_0 \sim 50$ km s^{-1} Mpc^{-1}. In contrast, de Vaucouleurs (1993) carried out his own calibration of the distance scale, finding values closer to 100 km s^{-1} Mpc^{-1}. An early detailed review of these programs was given by Hodge (1981). Hodge optimistically summarized the controversy: "Eventually, I assume, when sufficiently reliable measurements are possible ... there will be some number, such as H_0, that everyone will agree on, within some reasonable and understood limits." It would be another 20 years before the *Hubble Space Telescope* (*HST*) Key Project team (Freedman et al. 2001) measured a value of H_0 to 10% accuracy; a number that has stood the test of time.

The accurate measurement of Cepheid distances remains non-trivial. One of the main challenges is distinguishing Cepheids from fainter stars, both resolved and unresolved, that contribute to the overall background of the host galaxy. Avoiding/minimizing the effects of crowding and confusion is crucial for the successful identification and measurement of Cepheids in galaxies. A second challenge is the correction for the presence of dust, which both scatters and absorbs radiation. Cepheids are relatively young stars that have not had time to diffuse far from the locations where they were formed; hence they are found in the dusty disks of spiral (or irregular) galaxies. Third, the luminosity of a Cepheid may have a dependence on the abundance or metallicity of the star.

2. Advantages of the Near-Infrared

The recognition of the applicability of near-infrared observations of Galactic Cepheids to the distance scale did not occur immediately, despite the early near-infrared observations made by Wisniewski & Johnson (1968). The first application was by McGonegal and colleagues (McGonegal et al. 1982) (see Figure 1, left), who elucidated the advantages of the near-infrared, most specifically the decreased sensitivity to reddening and the decrease in the observed width of the Leavitt law, even with only single-epoch observations. The decrease in the width of the period–luminosity relation is primarily due to the decrease in sensitivity of the infrared surface brightness to the intrinsic temperature width of the instability strip. Near-infrared observations allowed for accurate distance determinations even with single, random-phase observations of Cepheids whose periods

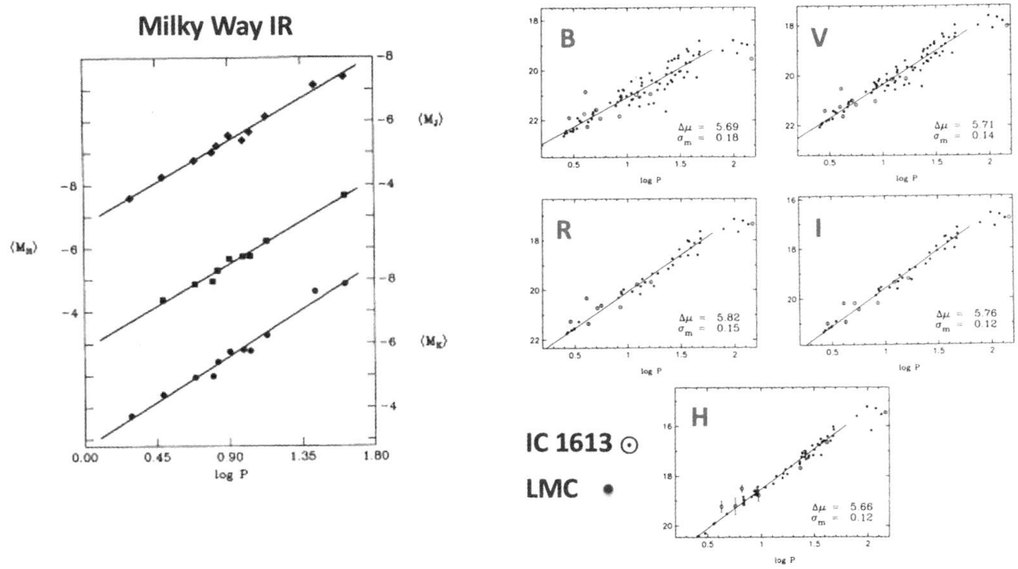

Figure 1. (left) Leavitt laws at *JHK* wavelengths (McGonegal et al. 1983) demonstrating the power of infrared photometry for the extragalactic distance scale. (right) Leavitt laws at *BVRI* wavelengths based on CCD photometry (Freedman 1988) (open circles); filled circles illustrate photoelectric photometry for LMC Cepheids.

had already been measured. These single-phase measurements were comparable in accuracy to complete time-averaged magnitudes derived from many nights of observations obtained at blue wavelengths.

3. The Era of CCDs and Infrared Arrays

With a wide wavelength sensitivity range, spanning from the B band (0.45 μm) to the I band (0.8 μm), CCD detectors allowed, for the first time, the possibility of high-resolution, panoramic, digital data for the extragalactic distance scale (Freedman 1984). With CCDs one could perform local sky subtraction and point-spread-function fitting to obtain accurate magnitudes and colors. In addition, CCDs enabled the handling of crowding and confusion errors at the one-square-arcsecond level. Furthermore, the large wavelength coverage of CCDs allowed the explicit and direct determination of reddening, by applying the interstellar extinction law as a function of wavelength, and eliminating the need to rely solely on foreground estimates and/or assuming negligible additional reddening (in the host galaxy), as many earlier studies had done. The approach, introduced by Freedman (1988) for determining total line-of-sight extinctions and true distance moduli of extragalactic Cepheids' multiwavelength data (e.g., see Figure 1, right) applied to single-epoch observations of Cepheids in IC 1613, M31, M33, and NGC 300, among others. In brief, the method involves determining differential apparent moduli by scaling them with respect to the corresponding LMC period–luminosity relations and assuming that the (observed) variation with inverse wavelength can be attributed to selective absorption.

3.1. *Metallicity Effects*

In addition to reddening, nailing down the metallicity dependence of Cepheids is also critical for application of the Leavitt law. Metallicity can affect the internal structure of a Cepheid: a higher metallicity results in a higher opacity, which can affect the temperature, radius, and luminosity of the star. Line blanketing, controlled by the abundance of metals in the atmospheres of Cepheids, can also result in a redistribution of light as a function of wavelength. The atmospheric effects are predicted to be greater at optical wavelengths than in the infrared.

The availability of CCDs also allowed the first empirical test of metallicity on the Cepheid period–luminosity relation. Freedman & Madore (1990) obtained *BVRI* CCD observations of Cepheids in M31 and estimated the mean reddenings and true distance moduli at three different radial positions (3, 10, and 20 kpc) in the disk of M31, for which a metallicity gradient of a factor of five had been measured from HII regions. They found evidence for a mild gradient of -0.2 ± 0.2 mag dex^{-1}. However, given the uncertainties, the value was also consistent with no measurable effect.

Twenty-five years later, there still remains uncertainty in the metallicity coefficient (Ripepi et al. 2020). A number of recent studies have found a range in the metallicity coefficient from -0.5 to 0 mag dex^{-1} (Udalski et al. 2001; Riess et al. 2016, 2021; Breuval et al. 2021; Ripepi et al. 2021, 2022). Generally, the effects of metallicity are found to be larger in the optical than in the near-infrared, although in contrast, the recent study by Breuval et al. (2022) finds nearly as large an effect in the near-infrared as in the optical.

4. The Hubble Space Telescope Key Project

The goal of the Hubble Key Project (Freedman et al. 2001) was to extend the Cepheid extragalactic distance scale to greater distances than possible from the ground, tie in to several independent methods for measuring distances beyond the Cepheids, and to ultimately measure the Hubble constant to an overall accuracy of 10%. To achieve these goals, the Key Project team used the *HST* to observe Cepheids in 18 galaxies spanning a range of distances. The observations included both wide-field and deep imaging in the F555W and F814W filters, with F555W (V band) allowing for the discovery, characterization, and measurement of the periods, amplitudes, and mean magnitudes of the Cepheids; and using the longer-wavelength F814W (I band) to form a color, thereby providing the means to correct for the effects of extinction and absorption by dust.

These measurements then enabled the calibration of secondary distance indicators (including Type Ia and II supernovae, the Tully–Fisher relation, surface brightness fluctuations, and the D_n–σ relation; see Figure 2. The final value of the Hubble constant was measured as $H_0 = 72 \pm 3$ (statistical) ± 7 (systematic) km s^{-1} Mpc^{-1}, with an estimated uncertainty of about 10%. This result provided early evidence for a Universe that is accelerating, consistent with the presence of dark energy, and consistent with the ages of globular clusters. And it resolved the longstanding factor-of-two discrepancy in the distance scale.

5. NICMOS and Ground-Based Near-Infrared Observations of Cepheids

5.1. *Early NICMOS Observations*

Following on the *HST* project, near-infrared observations of 14 nearby galaxies were obtained with *HST*/NICMOS (Macri et al. 2001). Distance moduli were obtained by combining single-epoch H-band data with the VI bands. These data provided support for the assumption of the universality of the standard (Milky Way) reddening law adopted by the Key Project (see Figure 3), by demonstrating the excellent correspondence between $E(V-H)$ and $E(V-I)$.

An additional goal of the NICMOS program was to undertake another test for the metallicity effect by observing two fields in the relatively face-on galaxy M101, a test

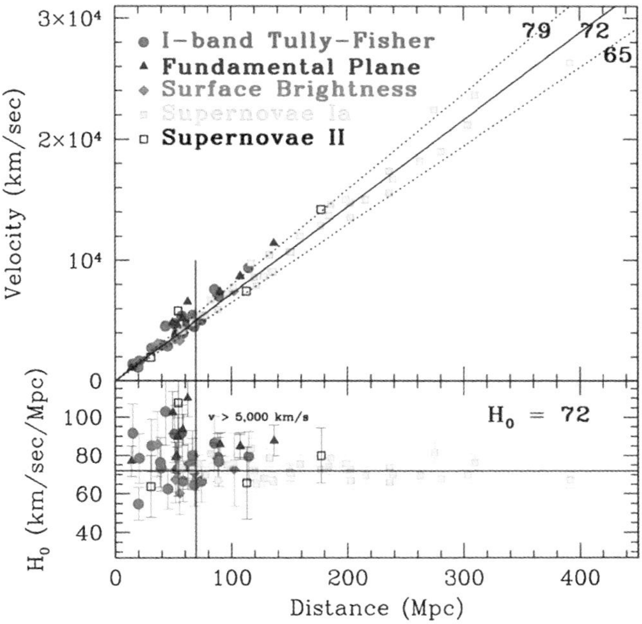

Figure 2. Hubble diagram showing the final results from the *HST* Key Project (Freedman et al. 2001). In the top panel velocity is plotted versus distance for the five secondary distance methods labeled in the plot. The solid line is for a slope of 72, flanked by dotted lines indicating a 10% uncertainty. The bottom panel shows the residuals for a value of $H_0 = 72 \pm 3$ (stat) ± 7 (sys) km s^{-1} Mpc^{-1}.

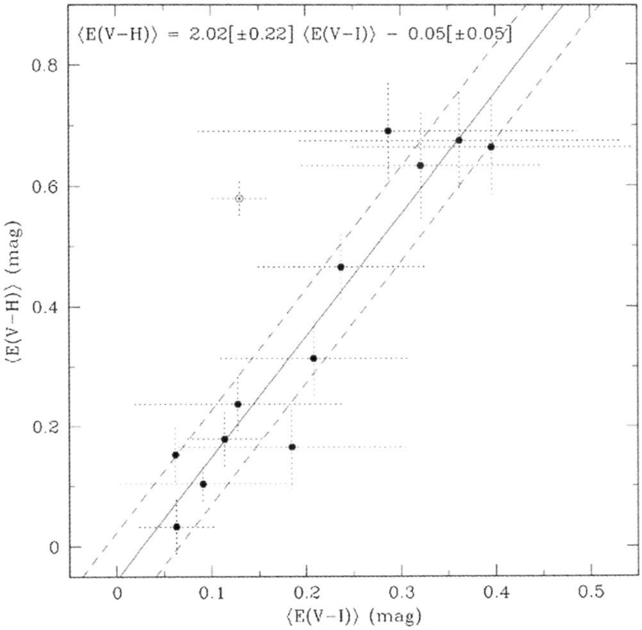

Figure 3. A comparison of reddening values measured via $E(V - H)$ with those from $E(V - I)$. The good agreement provides evidence that a standard extinction law is a reasonable approximation to correct for the dust in these galaxies.

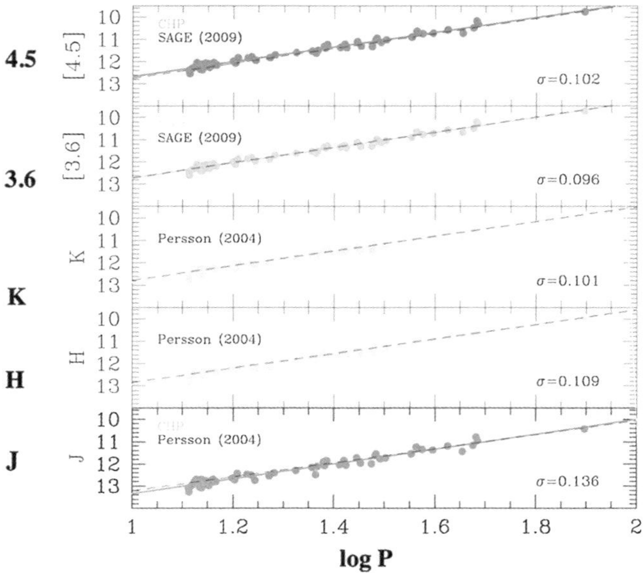

Figure 4. Leavitt laws for near-infrared (JHK) and mid-infrared (3.6 and 4.5 μm) for Cepheids in the LMC.

that had earlier been undertaken by the Key Project team. The inner and outer M101 fields were found to have a large difference in distance modulus (amounting to $+0.41 \pm 0.11$ mag at H and $+0.34 \pm 0.09$ mag at J, respectively). If interpreted in terms of a metallicity effect, this implies a dependence of order 0.6 mag dex^{-1}, much higher than theory predicts for the near-infrared. Simulations, as well as artificial star tests showed that crowding/blending effects were likely an issue for these data, explaining some (but not all) of this large difference.

These results provided a cautionary note regarding the difficulty of measuring Cepheids in the near-infrared in all but the nearest galaxies, and the need to avoid their inner (high-surface-brightness) regions. While the effects of dust are reduced at increasing wavelengths, the background contamination by red stars (red giants and even brighter asymptotic giants) increases substantially. Thus, the advantages of moving to the near-infrared to mitigate extinction effects were countered by the systematic disadvantages imposed by crowding and blending.

5.2. *Ground-Based Near-Infrared Observations of the LMC*

From the ground, an extensive and uniform near-infrared (JHK) study of 92 Cepheids in the LMC was undertaken by Persson and colleagues (Persson et al. 2004), with JHK sampling obtained at about 22 phase points for each star. The dispersion about the Leavitt law was found to be on the order of ± 0.1 mag for all three wavebands (see Figure 4). Fitting for a tilt to the LMC, the dispersions in the near-infrared period–luminosity–color relations drop to ~ 0.08 mag, showing again the immense power of the infrared in cases where crowding/blending effects were intentionally minimized and where the signal-to-noise is high. The high quality of these data is apparent when cross-comparing the positions of the individual Cepheids in these period–luminosity relations. The deviations from the mean period–luminosity relations are highly correlated across the various bandpasses, demonstrating intrinsic properties of the Cepheids and/or location within the LMC, that are not blurred out by photometric errors.

6. The Spitzer Space Telescope: Mid-infrared Observations of Cepheids

Using the IRAC camera on the *Spitzer Space Telescope* during its extended, warm mission, well-sampled mid-infrared light curves were obtained for 80 long-period LMC Cepheids (Scowcroft et al. 2011) falling in the period range $0.8 < \log(P) < 1.8$ [days]. These observations provided a well-measured slope of the period–luminosity relation based on the time-averaged 3.6 μm data. Time-averaged 3.6 μm data were also obtained for 10 high-metallicity Milky Way Cepheids with independently measured *HST* trigonometric parallaxes (Monson et al. 2012). The 3.6 and 4.5 μm Leavitt laws for the LMC can be seen in Figure 4. The 3.6 μm data resulted in a high-precision, reddening-corrected distance to the LMC of 18.477 ± 0.033 (systematic) mag. Freedman and collaborators (Freedman et al. 2012) applied the *Spitzer* calibration to the Key Project galaxy sample, yielding a value of $H_0 = 74.3$ km s^{-1} Mpc^{-1}, with a systematic uncertainty of $\pm$ 2.1 (systematic) km s^{-1} Mpc^{-1}. The *Spitzer* program reduced the systematic uncertainty in H_0 over that obtained by the *Hubble Space Telescope* Key Project by more than a factor of three.

7. The Araucaria Project

For 20 years, the Araucaria Project has been carrying out a program to observe a number of types of variable stars in nearby galaxies at both optical and near-infrared wavelengths (Pietrzyński et al. 2019). They have monitored Cepheid variables, RR Lyrae stars, and eclipsing binaries, as well as blue supergiants, tip of the red-giant branch (TRGB) stars, and carbon stars. The Araucaria Project team has measured Cepheid distances to galaxies in the Local Group (LMC, SMC, IC 1613, NGC 6822, M33), the Sculptor Group (NGC 247, NGC 300, NGC 7793), providing fundamental distances and tests of Cepheid metallicities (Udalski et al. 2001).

8. The SH0ES Program

The SH0ES (Supernovae, H_0, for the Equation of State) project, initially aimed at the study of dark energy, now has the objective of determining H_0 through the calibration of the extragalactic distance scale using Cepheid variables (Riess et al. 2009, 2012, 2016, 2022). Riess and colleagues report *HST* F555W, F814W, and F160W observations of Cepheid variables in host galaxies of Type Ia supernovae (SNe Ia) used to calibrate H_0. Reddening corrections are obtained using a small number of (2 to 4) random-phase observations in the F814W and F555W bands, with the distances coming mainly from ~11 observations taken in the F160W (H) band. The scatter in the F160W period–luminosity relations is generally of order 0.4–0.5 mag, i.e., a factor of four or so greater than the intrinsic dispersion observed in the uncrowded sample of Cepheids in the LMC. The zero-point calibration is set by geometric *Gaia* early Data Release 3 (EDR3) parallaxes, masers in the galaxy NGC 4258, and detached eclipsing binaries in the LMC. The most recent SH0ES paper quotes a 1% uncertainty with $H_0 = 73.04 \pm 1.04$ km s^{-1} Mpc^{-1}, based on a sample of 42 galaxies with distances in the range from 7 out to 80 Mpc.

9. An Independent External Check of Cepheid Distances: Tip of the Red Giant Branch

Few methods for measuring nearby distances rival the Cepheids in precision, accuracy, or numbers of galaxies for which the measurement can be made. A notable exception is the TRGB method (Lee et al. 1993; Mould & Sakai 2009; Hoyt 2023). The TRGB method also has a number of advantages relative to the Cepheid Leavitt law (Freedman et al.

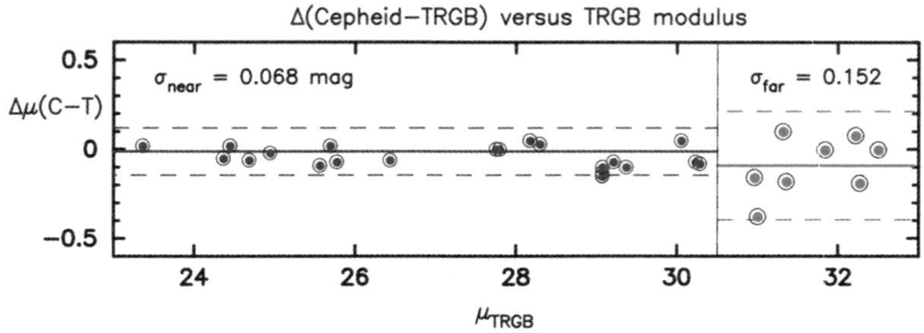

Figure 5. Comparison of published Cepheid and TRGB distances in the sense of (Cepheid − TRGB). The blue/red solid lines show the mean offset for galaxies with distance moduli closer/farther than 30.5 mag. The dashed lines indicate the 1σ scatter for the near and far samples, which are also labeled in the plot. The agreement between the Cepheid and TRGB distances is excellent, particularly for the nearby sample.

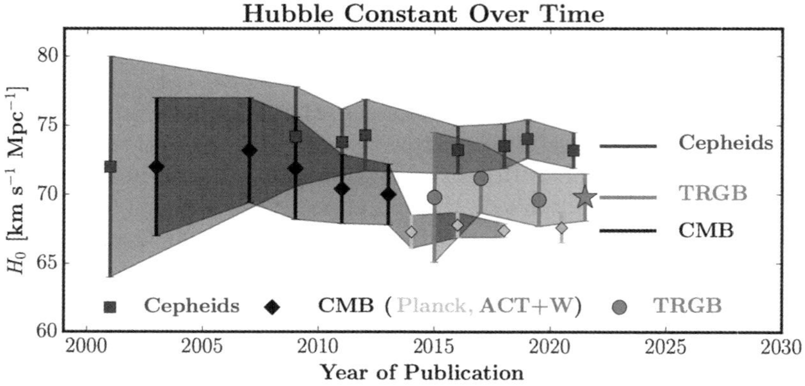

Figure 6. Measurements of H_0 from the *HST* Key Project through to today based on Cepheids (blue), the TRGB (red), and the CMB (*Planck*: yellow; ACT + WMAP: turquoise).

2019; Freedman 2021). It can be applied in the halos of galaxies where the effects of dust are generally negligible and where the surface brightness is about 100 times lower than in the disks, so that crowding/blending effects are not severe. It is a much easier practical issue to measure the luminosity of a halo TRGB star than it is to disentangle Cepheids from nearby neighbors in the disk. In addition, metallicity effects are much smaller for the TRGB (and can be directly calibrated), unlike in the case of Cepheids. Thus, beyond providing an independent calibration of H_0 (Freedman et al. 2019), the TRGB provides an excellent external constraint on the combined systematics of both methods.

In Figure 5 we show a comparison of distance moduli for Cepheids and the TRGB. The agreement between the two distance scales is excellent, with $\Delta\mu$(Cepheid − TRGB) $= -0.026 \pm 0.014$ mag. In this case, the zero point for the TRGB calibration is $M_I = -4.05$ mag. The scatter (expectedly) increases with increasing distance and photometric errors.

In Figure 6 we show an update to a recent comparison of the values of H_0 obtained from Cepheids (blue) and the TRGB (red; both tied into measurements of SNe Ia; Freedman 2021) and that inferred from observations of the cosmic microwave background (CMB; Planck Collaboration et al. 2020, black). The Cepheid and TRGB H_0 values agree to within 2σ of their quoted uncertainties; the TRGB measurements are in better agreement with the CMB value to within 1σ.

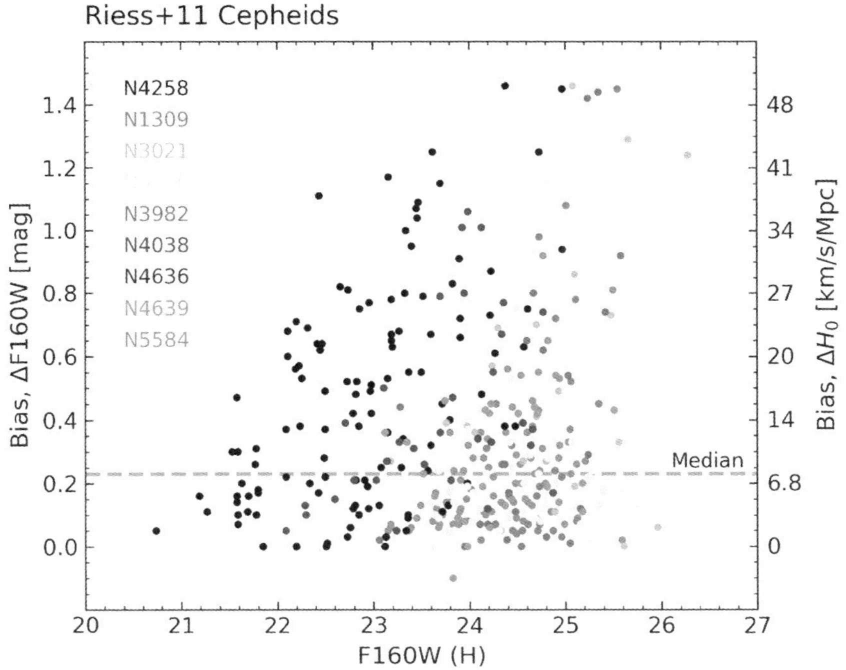

Figure 7. A measure of the bias in Cepheid photometry from Table 2 as published by Riess et al. (2012). The bias is determined from artificial star tests for each Cepheid's environment and is a statistical correction. The bias can be quite significant and ranges from 0 to 1.5 mag (a factor of up to four in luminosity). The median bias is 0.24 mag, corresponding to $\Delta H_0 = 7$ km s^{-1} Mpc^{-1}. The galaxies in this earlier study are amongst the closest in the more recent sample of 42 galaxies in the Riess et al. (2022) study.

10. Remaining Issues for the Cepheid Distance Scale

As previously noted, several challenges are still outstanding in measuring Cepheid distances, resulting from the fact that Cepheids are located in regions of star formation within the disks of galaxies, hence in dusty and crowded regions, particularly in their inner disks. The amount (and even the sign) and the wavelength dependence of the metallicity contributions to Cepheid luminosities, within and between galaxies, has yet to converge (Ripepi et al. 2020; Breuval et al. 2021; da Silva et al. 2022; Owens et al. 2023).

Both the *HST* Key Project and the SH0ES group approach the issue of crowding/blending of Cepheids with the use of artificial star tests. Figure 7 (courtesy of I. Jang) shows the F160W crowding-induced bias found for Cepheids observed in nine of the closest SNe Ia host galaxies (Riess et al. 2012). As can be seen, the bias can be quite large, reaching over 1 mag, with a median correction of $\sim$0.25 mag, which corresponds to a 10% difference in H_0 of >7 km s^{-1} Mpc^{-1} (as shown to the right of the plot). With the (unexplained) scatter in the F160W period–luminosity relation being as large as 0.4–0.5 mag, accurate crowding corrections are critical. And within that scatter, there remains the potential for a hidden systematic effect that may be difficult to identify and account for.

It is important to note that on average, for similar types of galaxies, crowding effects will become more severe with increasing distance. Sixty percent of the (Riess et al. 2022) sample of galaxies in which Cepheids have been discovered lie at greater distances than 20 Mpc, and 25% of the total sample lies beyond 40 Mpc. At a distance of 40 Mpc, four times the area will be contained within a given pixel. For the most distant SH0ES galaxy at 80 Mpc, 16 times the area will be covered. As the need for percent-level accuracy has grown in importance, and given the level of crowding for Cepheids in a galaxy at a distance of only 20 Mpc (see the *JWST* observations in the following section), it remains important to demonstrate that crowding effects do not produce a systematic bias in the photometry and, hence, the distance measurements for these more distant galaxies observed with *HST*.

11. The Future: Cepheids and JWST

While measurements of the Cepheid distance scale have improved since the time of the Key Project, great strides have been made in measurements of fluctuations in the temperature and polarization of the microwave background (Planck Collaboration et al. 2020). The current controversy over the value of H_0 now requires pushing the determination of local values to limits approaching 1% accuracy, in order to be of comparable accuracy to the CMB measurements. As we have seen, systematic errors have played, and continue to play, a dominant role, keeping more accurate measurement of H_0 a challenge.

Much like the 1970s and 1980s, when local distance measurement differences remained at an impasse, it is clear once again that improvements in technology will be required to settle the current differences, at the few-percent level, for the Cepheid extragalactic distance scale. We describe here a new program using the *James Webb Space Telescope* (*JWST*), aimed at addressing the current systematics in the *HST* measurements of Cepheid distances. The program is actually three-pronged and includes also the measurement of distances to nearby galaxies using not only Cepheids, but also TRGB stars and carbon/JAGB stars. In this review, however, we describe only the Cepheid portion of this program.

Because of the greater sensitivity of the surface brightness to temperature at bluer (optical) wavelengths, the amplitudes of Cepheid variables are larger than at longer wavelengths. Thus, its blue sensitivity and high spatial resolution (compared to the ground) made *HST* an ideal facility for the *discovery* of Cepheid variables (Freedman et al. 2001; Riess et al. 2016).

Two key features now make *JWST* the optimal telescope for addressing the *accuracy* of measurements of H_0: its red sensitivity and higher spatial resolution (compared to *HST*, and certainly far exceeding the spatial resolution and sensitivity of *Spitzer*). The extinction is significantly lower: for example, A_J is smaller by a factor of four relative to the visual extinction, A_V (Cardelli et al. 1989). For the F115W (J-band) filter on *JWST*/Near-Infrared Camera (NIRCAM; Rieke et al. 2023), the sampling resolution is four times that of *HST*/Wide-Field Camera 3 (WFC3) F160W (H band), with a FWHM of 0.04 arcsec versus 0.15 arcsec. With $\sim 4\times$ better resolution than *HST/WFC3*, crowding effects are reduced by more than an order of magnitude in flux. This is especially important because, in the near-infrared, contamination of the Cepheids by red giant and asymptotic giant branch stars exacerbates crowding effects relative to optical wavelengths.

As part of the Chicago Carnegie Hubble Program (CCHP), we have been awarded time in *JWST* Cycle 1 (JWST-GO-1995: PI W. L. Freedman, co-I B. F. Madore) for a program designed to address specifically known systematic effects: extinction and reddening by dust, metallicity effects, and crowding/blending of stellar images. We are obtaining NIRCAM J-band data and, simultaneously, F356W (mid-infrared) observations of 10 nearby SNe Ia-host galaxies with known Cepheids. For calibration purposes, we are also observing NGC 4258, a galaxy with a geometric distance derived from the motions of nuclear H_2O megamasers. In addition to the *JWST* results, our group is undertaking a

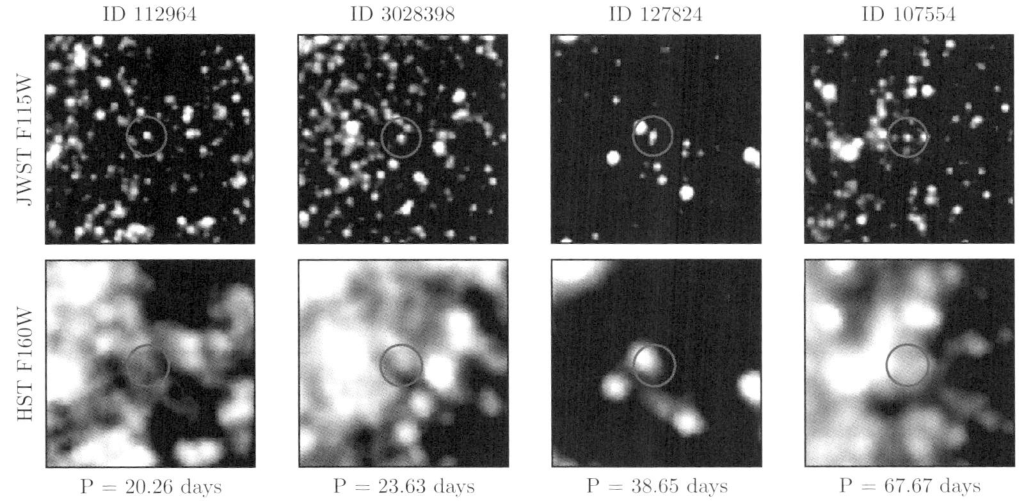

Figure 8. Four Cepheids in NGC 7250 discovered as part of the SH0ES project (bottom row: *HST* F160W/*H*-band exposures; top row: *JWST* F115W/*J* band). Each postage-stamp image is 2×2 arcsec on a side. The red circles are centered at the position of the Cepheid determined from the *HST* optical (F350LP white light) photometry. It is immediately evident that the crowding for these Cepheids is quite severe and the signal-to-noise ratio (SNR) for the *H*-band data tends to be low, ranging from 1 to 23. In contrast, the SNR for the *J*-band data ranges from 36 to 121. On average, the *JWST* data have almost an order of magnitude greater SNR, and a four times better angular resolution, allowing the Cepheids to be distinguished clearly from the background.

complete re-analysis (Owens et al. 2023) of the archival optical and near-infrared *HST* data obtained by the SH0ES group (Riess et al. 2016).

Here, we present some preliminary results from our program for the nearby galaxy NGC 7250, host to SN2013DY, located at a distance of about 20 Mpc. Figure 8 illustrates cut-out images of four Cepheids in NGC 7250 from Owens et al. (2023). Shown are images at F115W (from *JWST*; top panel) and F160W (from *HST*; bottom panel), which illustrate the superb resolution and the power of *JWST*. As can be seen in the *HST* data, many of the Cepheid candidates are crowded by nearby neighbors.

In Figure 9, we show the Leavitt laws for a sample of Cepheids in the LMC (Persson et al. 2004) and in NGC 7250 (Owens et al. 2023). The scatter in the *JWST* F115W data for NGC 7250 is a factor of two smaller than in the SHoES F160W data; i.e., the improved resolution and higher signal-to-noise ratio of the *JWST* data results in a lower variance (σ^2) for the F115W relation by almost a factor of four. This is all the more remarkable since the *J*-band data are single-phase observations only, while the *HST* observations have been corrected to mean light. The *HST* data exhibit more than three times the scatter of the *H*-band data for the LMC, the latter of which reflects the expected scatter for that band, as exemplified by the LMC data.

12. Summary

The accuracy of the Cepheid distance scale has continued to improve over the century during which it has been used to measure the distances to nearby galaxies and set the scale for the determination of H_0. Still, challenges remain in overcoming systematic uncertainties. Many of these challenges will be overcome with new capabilities provided by the *JWST*.

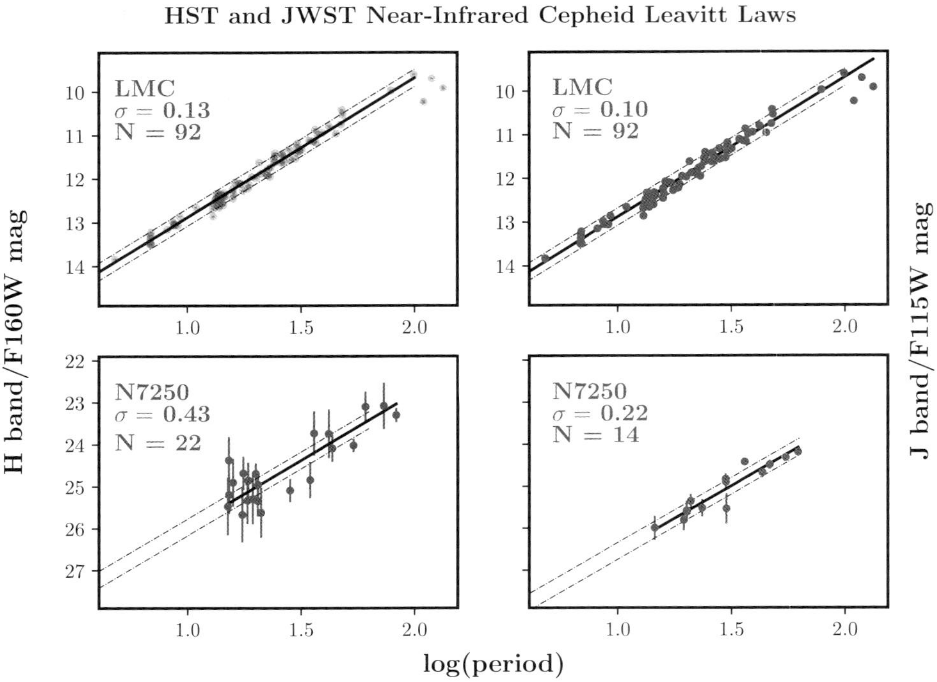

Figure 9. Period–luminosity relations for Cepheids in the LMC (top row) and NGC 7250 (bottom row). The left-hand panels are for ground-based H band for the LMC (top) and the similar F160W filter (from *HST*) for NGC 7250 (bottom). The right-hand panels are similar, but for ground-based J and the corresponding F115W filter (from *JWST*). The *JWST* data are plotted on an arbitrary magnitude scale. The scatter about the period–luminosity fit in each filter is labeled in each plot.

New *JWST* data for the nearby galaxy NGC 7250 already demonstrate that (1) many of the Cepheids observed with *HST/WFC3* are significantly crowded (and biased to brighter apparent magnitudes) by nearby neighbors. A re-analysis of the SH0ES optical data, then coupled with the new high-resolution and higher signal-to-noise *JWST* F115W data, leads to significantly reduced effects of crowding and smaller photometric uncertainties. (2) These improvements result in a factor of two lower scatter in the near-infrared Leavitt law for *JWST* F115W compared with *HST* F160W, even with single-epoch F115W *JWST* photometry.

The galaxies in our *JWST* CCHP program sample have all been selected to have with distances $\lesssim$20 Mpc, close enough to minimize crowding effects. As for the case of NGC 7250 presented here, these data will be combined with a re-analysis of the SH0ES *HST* optical data for the Cepheids. TRGB, carbon star, and Cepheid distances to the same sample of galaxies being observed as part of the CCHP will allow measurement of three independent distances to each of these galaxies. These new results already illustrate the power of *JWST* to improve the measurement of extragalactic distances, and specifically, to address remaining systematics in the determination of H_0.

Ultimately, these data will allow us to provide an answer to one of the most important problems in cosmology today: Is there new fundamental physics required beyond standard Λ Cold Dark Matter?

References

Baade, W. 1956, PASP, 68, 5; doi:10.1086/126870
Baade, W. & Swope, H. H. 1963, AJ, 68, 435; doi:10.1086/108996

Baade, W. & Swope, H. H. 1965, AJ, 70, 212; doi:10.1086/109717

Bailey, S. I. 1902, Annals of Harvard College Observatory, 38, 1

da Silva, R., Crestani, J., Bono, G., et al. 2022, A&A, 661, A104; doi:10.1051/0004-6361/202142957

Breuval, L., Kervella, P., Wielgórski, P., et al. 2021, ApJ, 913, 38; doi:10.3847/1538-4357/abf0ae

Breuval, L., Riess, A. G., Kervella, P., et al. 2022, ApJ, 939, 89; doi:10.3847/1538-4357/ac97e2

Cardelli, J. A., Clayton, G. C., & Mathis, J. S. 1989, ApJ, 345, 245; doi:10.1086/167900

de Vaucouleurs, G. 1993, ApJ, 415, 10; doi:10.1086/173138

Freedman, W. L. 1984, Bull. Am. Astron. Soc., 16, 888

Freedman, W. L. 1988, ApJ, 326, 691; doi:10.1086/166128

Freedman, W. L., Madore, B. F., Gibson, B. K., et al. 2001, ApJ, 553, 47; doi:10.1086/320638

Freedman, W. L., Madore, B. F., Scowcroft, V., et al. 2012, ApJ, 758, 24; doi:10.1088/0004-637X/758/1/24

Freedman, W. L., Madore, B. F., Hatt, D., et al. 2019, ApJ, 882, 34; doi:10.3847/1538-4357/ab2f73

Freedman, W. L. 2021, ApJ, 919, 16; doi:10.3847/1538-4357/ac0e95

Freedman, W. L. & Madore, B. F. 1990, ApJ, 365, 186; doi:10.1086/169469

Freedman, W. L. & Madore, B. F. 2010, ARA&A, 48, 673; doi:10.1146/annurev-astro-082708-101829

Hodge, P. W. 1981, ARA&A, 19, 357; doi:10.1146/annurev.aa.19.090181.002041

Sawyer Hogg, H. 1955, Publ. David Dunlap Obs., 2, 33

Hoyt, T. J. 2023, Nat. Astron., 7, 590; doi:10.1038/s41550-023-01913-1

Hubble, E. P. 1925a, The Observatory, 48, 139

Hubble, E. P. 1925b, ApJ, 62, 409; doi:10.1086/142943

Hubble, E. P. 1926, ApJ, 63, 236; doi:10.1086/142976

Hubble, E. 1929, PNAS, 15, 168; doi:10.1073/pnas.15.3.168

Joy, A. H. 1949, ApJ, 110, 105; doi:10.1086/145190

Leavitt, H. S. 1908, Annals of Harvard College Observatory, 60, 87

Leavitt, H. S. & Pickering, E. C. 1912, Harvard College Obs. Circ., 173

Lee, M. G., Freedman, W. L., & Madore, B. F. 1993, ApJ, 417, 553; doi:10.1086/173334

Macri, L. M., Calzetti, D., Freedman, W. L., et al. 2001, ApJ, 549, 721; doi:10.1086/319465

McGonegal, R., McAlary, C. W., Madore, B. F., et al. 1982, ApJL, 257, L33; doi:10.1086/183803

McGonegal, R., McAlary, C. W., McLaren, R. A., et al. 1983, ApJ, 269, 641; doi:10.1086/161071

Monson, A. J., Freedman, W. L., Madore, B. F., et al. 2012, ApJ, 759, 146; doi:10.1088/0004-637X/759/2/146

Mould, J. & Sakai, S. 2009, ApJ, 697, 996; doi:10.1088/0004-637X/697/2/996

Owens, K., et al. 2023, ApJ, submitted

Persson, S. E., Madore, B. F., Krzemiński, W., et al. 2004, AJ, 128, 2239; doi:10.1086/424934

Pietrzyński, G., Graczyk, D., Gallenne, A., et al. 2019, Nature, 567, 200; doi:10.1038/s41586-019-0999-4

Planck Collaboration, Aghanim, N., Akrami, Y., et al. 2020, A&A, 641, A6; doi:10.1051/0004-6361/201833910

Rieke, M. J., Kelly, D. M., Misselt, K., et al. 2023, PASP, 135, 028001; doi:10.1088/1538-3873/acac53

Riess, A. G., Macri, L., Casertano, S., et al. 2009, ApJ, 699, 539; doi:10.1088/0004-637X/699/1/539

Riess, A. G., Macri, L., Casertano, S., et al. 2012, ApJ, 752, 76; doi:10.1088/0004-637X/752/1/76

Riess, A. G., Macri, L. M., Hoffmann, S. L., et al. 2016, ApJ, 826, 56; doi:10.3847/0004-637X/826/1/56

Riess, A. G., Casertano, S., Yuan, W., et al. 2021, ApJL, 908, L6; doi:10.3847/2041-8213/abdbaf

Riess, A. G., Yuan, W., Macri, L. M., et al. 2022, ApJL, 934, L7; doi:10.3847/2041-8213/ac5c5b

Ripepi, V., Catanzaro, G., Molinaro, R., et al. 2020, A&A, 642, A230; doi:10.1051/0004-6361/202038714

Ripepi, V., Catanzaro, G., Molinaro, R., et al. 2021, MNRAS, 508, 4047; doi:10.1093/mnras/stab2460

Ripepi, V., Catanzaro, G., Clementini, G., et al. 2022, A&A, 659, A167; doi:10.1051/0004-6361/202142649

Sandage, A. & Tammann, G. A. 2006, ARA&A, 44, 93; doi:10.1146/annurev.astro.43.072103.150612

Scowcroft, V., Freedman, W. L., Madore, B. F., et al. 2011, ApJ 743, 76; doi:10.1088/0004-637X/743/1/76

Udalski, A., Wyrzykowski, L., Pietrzynski, G., et al. 2001, AcA, 51, 221

Wisniewski, W. Z. & Johnson, H. L. 1968, Commun. Lunar and Planetary Lab., 7, 57

Discussion

Question (Baruch): If we take SH0ES group paper (ApJL, 908, L6; 2021), it is possible to compare their use of the *Gaia* parallax data, $\pi_{\rm EDR3}$, with their own photometrically derived parallaxes, $\pi_{\rm R16}$. Let us take the data as it is in the paper with no alterations. For each Cepheid star, take the ratio of the two parallaxes, $\pi_{\rm EDR3}/\pi_{\rm R16}$, and calculate the mean of all the ratios. If the assumption is correct and the differences are due to errors in each of the measurements, then the mean should be approximately 1.00, within a standard deviation. It is 1.04, but what is the standard deviation? Calculation of the standard deviation, again using all their error data as presented in the paper, is 0.65%. Thus, the ratio of 1.04 is more than six standard deviations away from 1.000. This is the best possible evidence for absorption or grey extinction of Cepheid light within the Milky Way galaxy. The absorption coefficient is about 8% per kiloparsec and completely removes the Hubble tension. (Riess considers χ-squared fits, but the distributions are not normal and the measurements are related as squares, and inversely, so it is complicated, but the ratio above I find convincing for grey extinction.) If we look at your TRGB stars in the LMC, it is very difficult to match them with the SH0ES group with absorption until you consider dark matter and then the match is almost perfect. Dark matter must be the absorber. As a result of this, the absorption in the Galaxy for the density of dark matter in the solar neighbourhood is about 2% per kiloparsec. In intergalactic space, the absorption must be in the region of about 12% per Gigaparsec, removing the recent time acceleration in the expansion of the Universe. This totally depends upon the application of the absorption (grey extinction) found in the Milky Way to the TRGB stars in the LMC. What is your view of grey extinction for the Carnegie–Chicago data?

Answer: Thank you for the question. It appears that you have looked into this at some length for the Cepheids. No, I haven't considered the implications for the TRGB.

At the crossroads of astrophysics and cosmology: Period–luminosity relations in the 2020s
Proceedings IAU Symposium No. 376, 2024
R. de Grijs, P. A. Whitelock & M. Catelan, eds.
doi:10.1017/S1743921323003034

The Local Value of H_0

Adam G. Riess and Louise Breuval

Department of Physics and Astronomy, Johns Hopkins University, Baltimore, MD 21218, USA
Space Telescope Science Institute, 3700 San Martin Drive, Baltimore, MD 21218, USA.
email: ariess@stsci.edu

Abstract. We review the local determination of the Hubble constant, H_0, focusing on recent measurements of a distance ladder constructed from geometry, Cepheid variables and Type Ia supernovae (SNe Ia). We explain in some detail the components of the ladder: (1) geometry from Milky Way parallaxes, masers in NGC 4258 and detached eclipsing binaries in the Large Magellanic Cloud; (2) measurements of Cepheids with the *Hubble Space Telescope* (*HST*) in these anchors and in the hosts of 42 SNe Ia; and (3) SNe Ia in the Hubble flow. Great attention to negating systematic uncertainties through the use of differential measurements is reviewed. A wide array of tests are discussed. The measurements provide a strong indication of a discrepancy between the local measure of H_0 and its value predicted by Λ Cold Dark Matter theory, calibrated by the cosmic microwave background (*Planck*), a decade-long challenge known as the 'Hubble Tension'. We present new measurements with the *James Webb Space Telescope* of >320 Cepheids on both rungs of the distance ladder, in a SN Ia host and the geometric calibrator NGC 4258, showing good agreement with the same as measured with *HST*. This provides strong evidence that systematic errors in *HST* Cepheid photometry do not play a significant role in the present Hubble Tension. Future measurements are expected to refine the local determination of the Hubble constant.

Keywords. Hubble constant, Cepheids, supernovae, parallaxes

1. Why care about the Hubble Constant?

The Hubble constant, H_0, is a measure of the present expansion rate of the Universe and may be used to infer related quantities such as the age of the Universe, its fate, the distances to galaxies (with known redshifts), etc. Connecting the measured value of the Hubble constant to related properties of the Universe tests our fundamental understanding of our cosmic paradigm.

In the local Universe, $z \sim 0$, it is given by the constant of proportionalty between distances, D, and redshifts, z, in the relation $cz = H_0 D$. More generally, i.e., at $z > 0$ and $t < t_0$ (where t_0 refers to the present era) the relation between distance and redshift may evolve and so is more generally defined as

$$D = \frac{cz}{H_0}\left\{1 - \left[1 + \frac{q_0}{2}\right] z + \left[1 + q_0 + \frac{q_0^2}{2} - \frac{j_0}{6}\right] z^2 + O(z^3)\right\},$$

which follows from a Taylor expansion of the increasing scale factor, a:

$$a(t) = a_0 \left\{1 + H_0\,(t - t_0) - \frac{1}{2}\, q_0\, H_0^2\,(t - t_0)^2 + \frac{1}{3!}\, j_0\, H_0^3\,(t - t_0)^3 + O([t - t_0]^4)\right\},$$

with $H(t) = +(\mathrm{d}a/\mathrm{d}t)/a$ the expansion rate at time t, $q(t) = -(\mathrm{d}^2 a/\mathrm{d}t^2)\,[H(t)]^{-2}\,/a$ the deceleration parameter, and $j(t) = (\mathrm{d}^3 a/\mathrm{d}t^3)\,[H(t)]^{-3}\,/a$ the jerk parameter, and so on.

As provided, these relations ignore curvature but terms for a non-flat Universe can be included. Thus seen, the Hubble 'constant' is the present value of the expansion $H_0 = H(t_0)$ today. The higher-order derivatives of expansion, q_0 and j_0, can be directly determined from this relation, provided a set of distance and redshift measurements covering a wide range of redshifts is available (e.g., with Type Ia supernovae, SNe Ia, at $0 < z < 1$ free of cosmological model assumptions). This approach, following the definition, is the 'Direct' route, because it does not invoke a cosmological model and has negligible dependence on gravity.

The alternative route to knowledge of H_0 comes from predicting its present value from a fine calibration of the cosmological model, ('Vanilla') Λ cold dark matter (ΛCDM), measured from the cosmic microwave background (CMB). By invoking ΛCDM we are referring to a description of the Universe as composed of the simplest dark matter (i.e., non-interacting, non-decaying, from a particle with only gravitational interactions), the simplest dark energy (i.e., a 'Cosmological Constant'), atoms, photons, and neutrinos (three species), without spatial curvature, and *without any other cosmologically important features*. We can use the model as it would have looked at $z > 1000$ to predict the physical size of fluctuations in the primordial plasma and compare the fluctuation spectrum with the angular size as observed from the CMB. This comparison serves to calibrate the six free parameters in ΛCDM. Once calibrated, the model predicts that dramatic changes in the Universe will occur (matter dominated followed by vacuum energy dominated) and describes the expansion history, $H(z)$, from $z = 1000$ to $z = 0$, and hence the value of H_0. In the appendix we provide a detailed description of how, in practice, the value of H_0 is measured from the CMB. We encourage the reader to review it, as it may challenge the prior belief that this is simple! See Kamionkowski and Riess (2022) and Planck Collaboration et al. (2020) for details. As should be clear, ΛCDM is a *phenomenological* model with parameters that stand in for a physical description of the unknown nature of 96% of the present Universe. Thus, H_0 is uniquely suited to provide an 'end-to-end' test of ΛCDM and our understanding of the Universe. By comparing the direct and model-dependent routes, we can test the model. Figure 1 illustrates this test.

2. The Distance Ladder: From Geometry to Cepheids to Type Ia Supernovae

Because the indirect route offers better than percent-level precision, fully leveraging the comparison demands a percent-level local measurement. The first generation of such measurements from the *Hubble Space Telescope* (*HST*) Key Project (KP; see Freedman and Madore 2023, in this conference), foundational work of immense importance, reached 10% precision by 2001 (see also Sandage et al. 2006). To reach greater precision requires a considerable re-thinking of the approach and methods to reach percent-level while at the same time leveraging new geometric measurements (e.g., *Gaia*). In 2005 we started the SH0ES program to use new instruments on the *HST* to accomplish this. The SH0ES ladder is composed of three rungs: Geometry to Cepheids to Type Ia supernovae. The key and new elements of SH0ES are:

- Cancel flux calibration errors using *HST* for Cepheids (rung 1 and 2);
- Near-Infrared (NIR) for Cepheids to minimize dust;
- Best quality SN data, calibration (Pantheon+, Brout et al. 2022);
- Comprehensive error analysis, covariance, analysis variants;
- Public release of data, 10^7 data numbers, code to fit the data.

2.1. *Geometric Rung*

There are three geometric anchors of the distance ladder which rely on different systems and measurements and so are fully independent of each other. Milky Way Cepheid

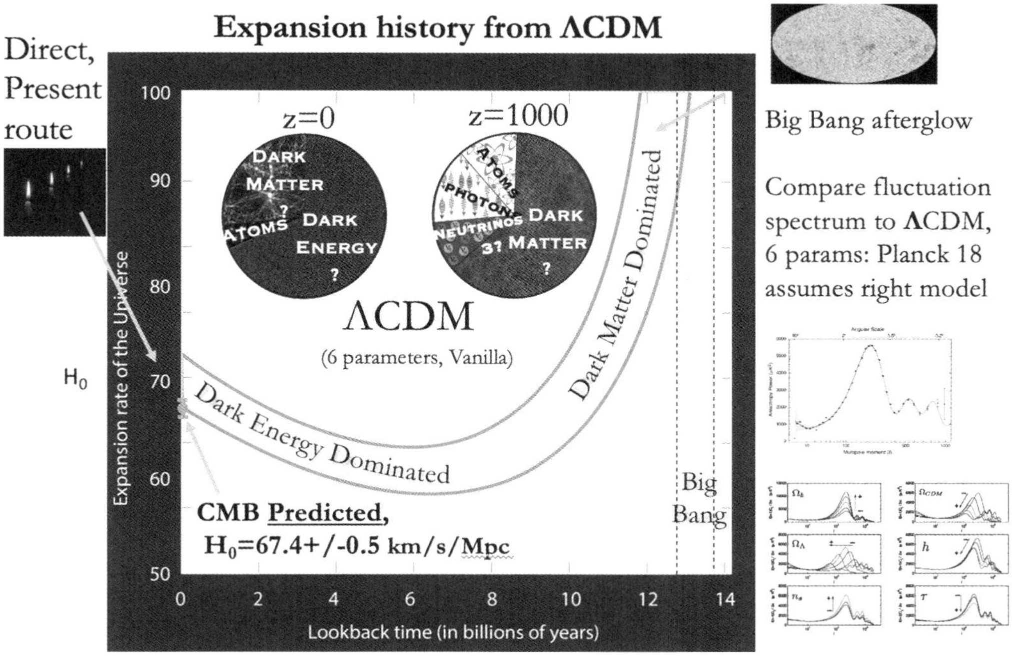

Figure 1. The expansion rate of the Universe can be predicted by the ΛCDM model with its parameters calibrated by the CMB or measured directly and locally from redshifts and distances.

parallaxes, thanks to *Gaia* (now in Data Release 3) provide $\sim$ 1% precision in the calibration of the Hubble constant. These are a tremendous advancement over past work. However, we could not leverage the geometric precision these afford without an equally precise photometric tie. The SH0ES project has spent a lot of effort to devise a spatial scanning method of photometric measurements with *HST* that demonstrates extreme precision and accuracy for direct calibration of Cepheids on the second rung (in SN Ia hosts). The scans were also used earlier to measure parallaxes (before *Gaia*). More recently, even greater precision and independence from *Gaia* systematics (the parallax offset term) has come from Cepheids in star clusters, where hundreds of stellar parallaxes may be averaged (Riess et al. 2022a). Figure 2 shows three different approaches to measuring Cepheid parallaxes, all in good agreement.

Another anchor comes from the exquisite detached-eclipsing binary measurements in the Large Magellanic Cloud (LMC; Pietrzyński et al. 2019) which reach 1.2% precision. Again, leveraging these requires direct measurements of Cepheids with *HST*. These measurements (Riess et al. 2019) yield both greater precision and accuracy, especially in the NIR, than those from the ground (see historic data in Freedman and Madore 2023, in this conference) with a dispersion of $\sim$0.07 mag. They also demonstrate the power of NIR data to mitigate dust and the use of colors to deredden for remaining dust. These are shown in Figure 3.

The third anchor is the water-maser host NGC 4258, with a 1.5% geometric distance (Reid et al. 2019). By collecting Cepheids in six fields, the SH0ES Team has collected 669 Cepheids in NGC 4258 (Yuan et al. 2022). The Cepheids in NGC 4258 also provide a strong test of the effect of the blotchy background on Cepheid photometry, because we can compare the Cepheid intercepts measured in the inner region (where crowding is high) versus in the outer region, where it is low. As shown in Figure 4, the agreement is

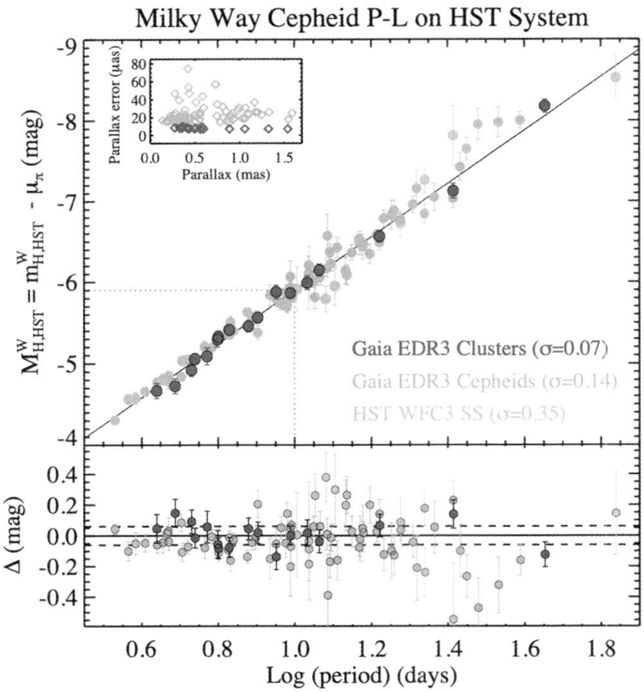

Figure 2. The Milky Way Cepheid period–luminosity relation in the *HST* NIR, reddening-free (Wesenheit) system as calibrated with three samples. Parallaxes from *HST* spatial scanning (Riess et al. 2018) for 8 Cepheids are in blue and yield 3% precision. The 68 points in gray (Riess et al. 2021) use *Gaia* early Data Release 3 (EDR3) parallaxes with simultaneous calibration of the parallax offset. The red points come from cluster Cepheids and do not require parallax offset calibration as they are measured in the range where *Gaia* is best calibrated (Riess et al. 2022a). These samples differ in their parallax precision (inset) leading to the low dispersion of $\sigma = 0.07$ mag for the cluster Cepheids.

good, at the ~0.05 mag level. Here we clarify some terminology. Freedman and Madore (2023), at this conference, referred to crowding as a bias in photometry. To be clear, this term (as shown in that contribution) is the measured difference in the background level between the true sky and the mean level of the (non-Cepheid) stars (measured with artificial stars). Thus, the true (Cepheid) background is both terms added together. Accounting for these results in *no bias*, but due to the variations in the latter, it does lead to greater statistical noise which is not 'unexplained' but, rather, included in the measurement errors. Multiple tests such as the one above show that the photometry is accurate and provided that one has access to more than ~25 Cepheids per host, the noise does not reach the level of the uncertainty in the SNe Ia the Cepheids calibrate.

2.2. *SNe Ia calibrations*

The next rung of the distance ladder is the consistent measurement of Cepheid variables in the hosts of recent SNe Ia. Cepheids can be measured effectively with *HST* up to $z \sim 0.01$ or $D < 40$ Mpc, a volume that produces a prototypical SN Ia with low reddening about once a year. Because each SN Ia yields a distance with ~6% precision, we need to collect many dozens in order to reach the target goal. The SH0ES team has now collected 42 SNe Ia in 37 hosts, an effort requiring about 1000 orbits of *HST*, resulting from more than a dozen Time Allocation Committee-selected proposals. These SNe necessarily

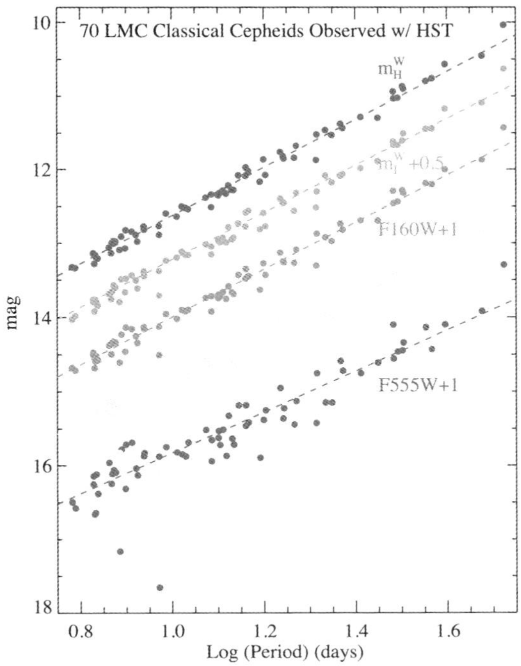

Figure 3. Period–mean magnitude relation for the 70 LMC Cepheids observed with *HST* using DASH mode.

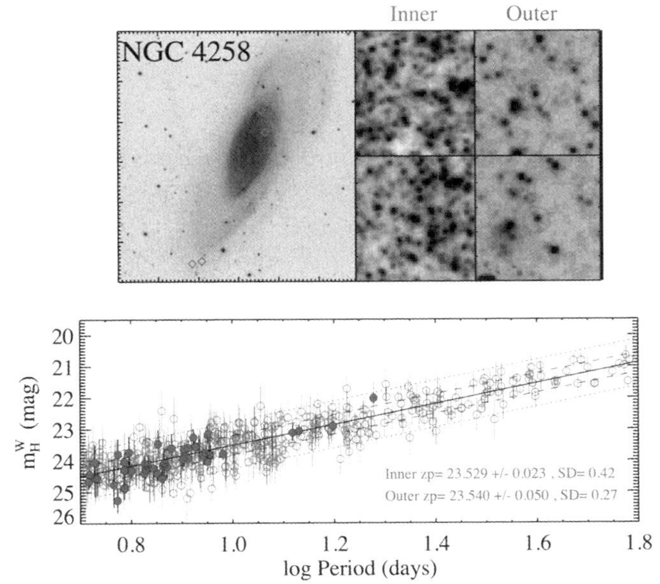

Figure 4. Comparison of Cepheids measured in a dense (inner) field (in red) and sparse (outer) field (in blue) of NGC 4258. Because these Cepheids are at the same distance, the comparison shows the accuracy of the background estimates, which differ in the mean over the same sampled range, $0.7 < \log P < 1.2$ [days], by 0.45 mag (relative to the Cepheids) yet yield a consistent intercept with $\sigma = 0.05$ mag. The difference in metallicity between the samples, ~0.08 dex, corresponds to a difference of 0.02 mag, smaller than the precision of this comparison.

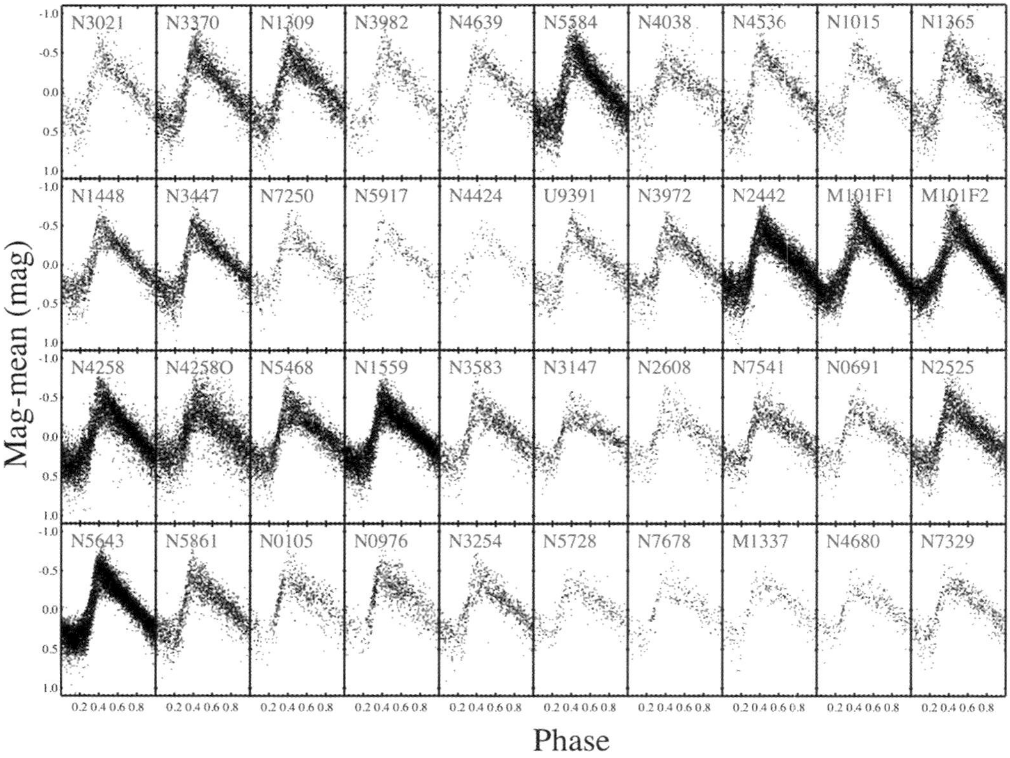

Figure 5. Composite visual (F555W) or white-light (F350LP) Cepheid light curves. Each *HST* Cepheid light curve with $10 < P < 80$ days is plotted after subtracting the mean magnitude and determining the phase of the observation.

come from the last four decades, the era of digital measurements and from many SNe surveys. The SNe measurements are standardized across many surveys by the Pantheon+ collaboration.

The Cepheids are found in these hosts primarily with a white light filter and then followed up in the optical (V and I bands) and the NIR (H band). It is important to note that the optical measurements are not 'random-phase' as claimed in the previous contribution (see Freedman and Madore 2023, in this issue); they are at known phases due to the simultanoues white light light curves and, thus, are used to measure the magnitudes at mean phase. Composite light curves of all Cepheids in all hosts are shown in Figure 5. These are used to identify Cepheids and determine their periods. The individual period–luminosity relations are shown in Figure 6. The number of Cepheids varies but has a mean of around 50 Cepheids per host.

Because this conference is focused on Cepheid variables, we can take a closer look at the extragalactic Cepheids and address the question if they are like those in the Milky Way. Naturally we would think so, but we can also bin their light curves by period and compare the changing shape to the well-known Hertzsprung progression. There is a rather striking match, as shown in Figure 7. The light curves at short periods are symmetric and as the periods grow, become more asymmetric and with higher amplitudes before becoming more shallow and sinusoidal-looking at long periods. There are also fine features like a bump migrating with phase due to a resonance between an overtone and fundamental oscillation.

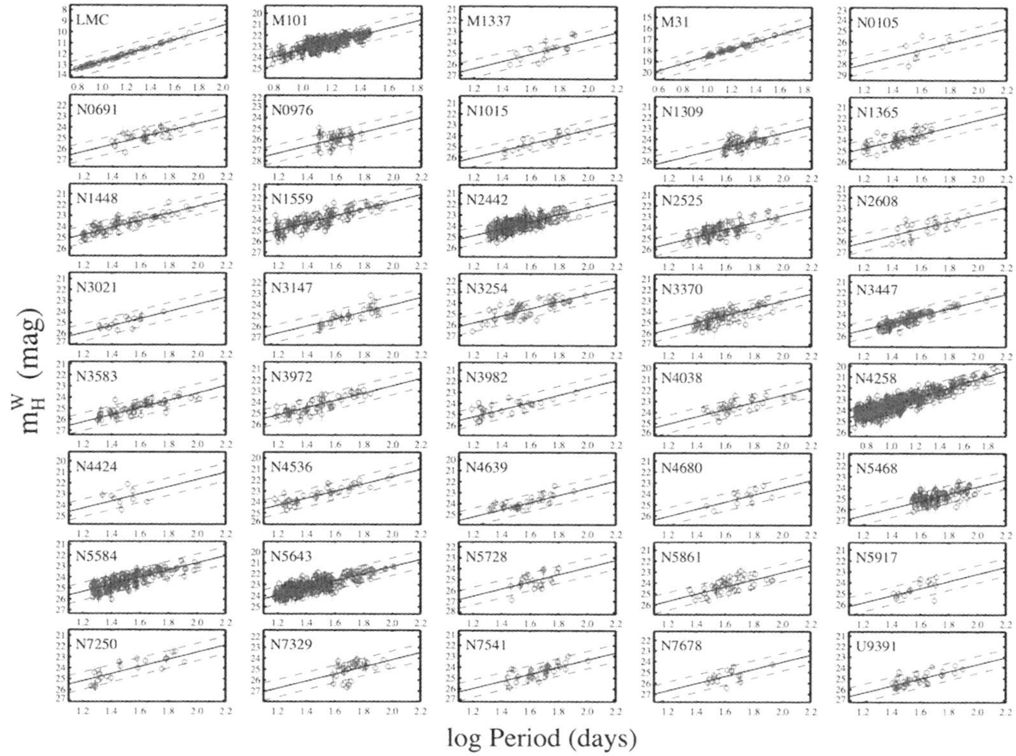

Figure 6. *HST* NIR Cepheid period–Wesenheit index relations. The Cepheid magnitudes are shown for 37 SNe Ia hosts, M31, and two of the three possible distance-scale anchors (LMC and NGC 4258). The uniformity of the photometry and metallicity reduces systematic errors along the distance ladder. A single slope is shown and used for the baseline, but we also allow for a break (two slopes) as well as limited period ranges in some analysis variants.

The final rung is the measurement of the Hubble flow from SNe Ia at moderate redshifts. We made these measurements from about 300 SNe Ia from the Pantheon+ sample at $0.023 < z < 0.15$. This redshift range was chosen to avoid the largest, local peculiar motions while avoiding the highest redshifts to avoid senstivity to higher-order terms, q_0 and j_0. We correct for peculiar velocities using mass maps and determine $q_0 = -0.55$ from high-redshift SNe Ia. The global fit of the distance ladder and H_0 is a large set of simultaneous, linear equations which can be optimized for five global parameters, including the fiducial luminosity of SNe Ia and Cepheids, two parameters specific to Cepheids, the slope of their period–luminosity relation and metallicity dependence, and the final parameter is ($5 \log H_0$). The entire data set as fitted is shown in Figure 8. The equations can be solved exactly by inverting a matrix, but we also use a brute-force, Markov Chain Monte Carlo approach to view the interdependency of parameters.

One of the more interesting parameters for this group is the Cepheid metallicity dependence. We find a value of -0.22 ± 0.04 mag dex^{-1}, which sits well in the middle range of recent findings. Figure 9 shows a summary of metallicity results: we see that the metallicity effect (γ) has now converged to a modest range of values with wide concurrence on a negative sign, with a good agreement between various measurements from independent teams and methods. Although a few studies diverge from this central value, Breuval et al. (2022) identified that these measures were affected by geometry effects in the Magellanic Clouds and by the use of an overestimated *Gaia* parallax zero-point in the Milky Way.

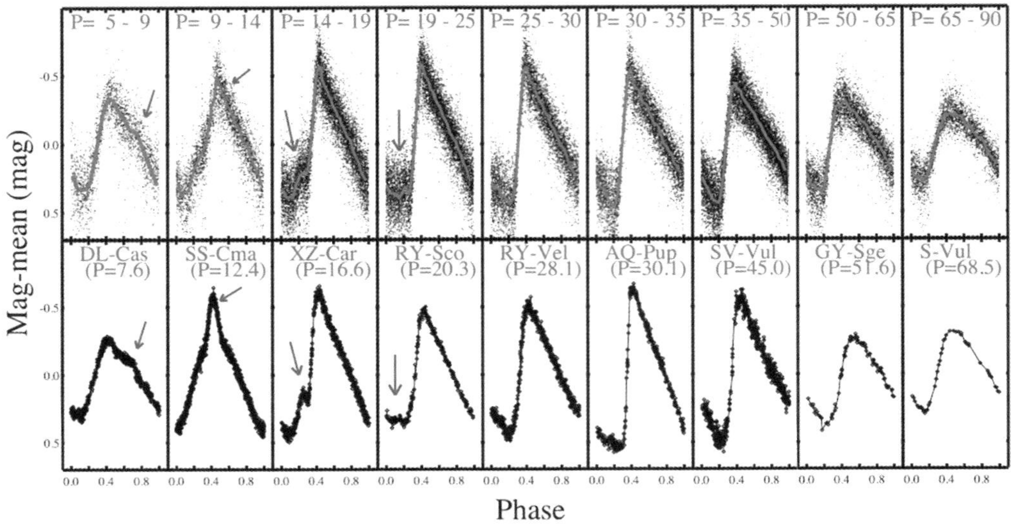

Figure 7. Composite visual (F555W) or white-light (F350LP) Cepheid light curves binned by period (top) and compared with individual Milky Way Cepheids near the middle period of the bin. The 'Hertzsprung progression' (the relation between the light-curve shape and the period) is apparent, including subtle features like the progression in phase of a resonance 'bump' between the second overtone and fundamental pulsation for $P < 20$ days. The red line is a cubic spline constrained by the averages of bins in phase.

Overall we find strong evidence, from dozens of calibrations in the literature, that metal-rich Cepheids are brighter than metal-poor ones by about -0.2 mag dex^{-1}. There is also agreement on the size and sign between the empirical measurements and theoretical modeling (presented at this meeting) of Wesenheit magnitudes (the only disagreement with the theory was on the monochromatic dependence) with Wesenheit magnitudes of greatest relevance for distance measurements. However, it is important to note that the value of the Hubble constant is not sensitive to this parameter, because the mean metallicity of the SNe Ia hosts is similar to the mean of the anchors. The anchors themselves have somewhat different metallicities, so the metallicity term is an important contributor to the consistency of the anchors.

The resulting determination of H_0 is 73.04 ± 1.04 km s^{-1} Mpc^{-1}, in good agreement with the mean of local determinations but inconsistent with the prediction of 67.4 ± 0.5 km s^{-1} Mpc^{-1} at the 5σ confidence level. This discrepancy is so famous, it has a name, the 'Hubble Tension', and it has been disturbing the cosmology community for a decade. It is certainly tempting to think the discrepancy may be telling us we are missing something in the cosmological model.

We also calculated 67 variants of the analysis varying many, many, many things. There are too many to discuss them all, but we select a subset of the most interesting and what value of H_0 they yield.

Variants
- Optical Cepheid data only (72.7)
- Different peculiar velocity map or none (73.1, 72.7)
- SN scatter monochromatic + mass step (73.5)
- No pre-2000 SNe (73.2)
- Closest half hosts (73.1)
- Most crowded half (73.4)

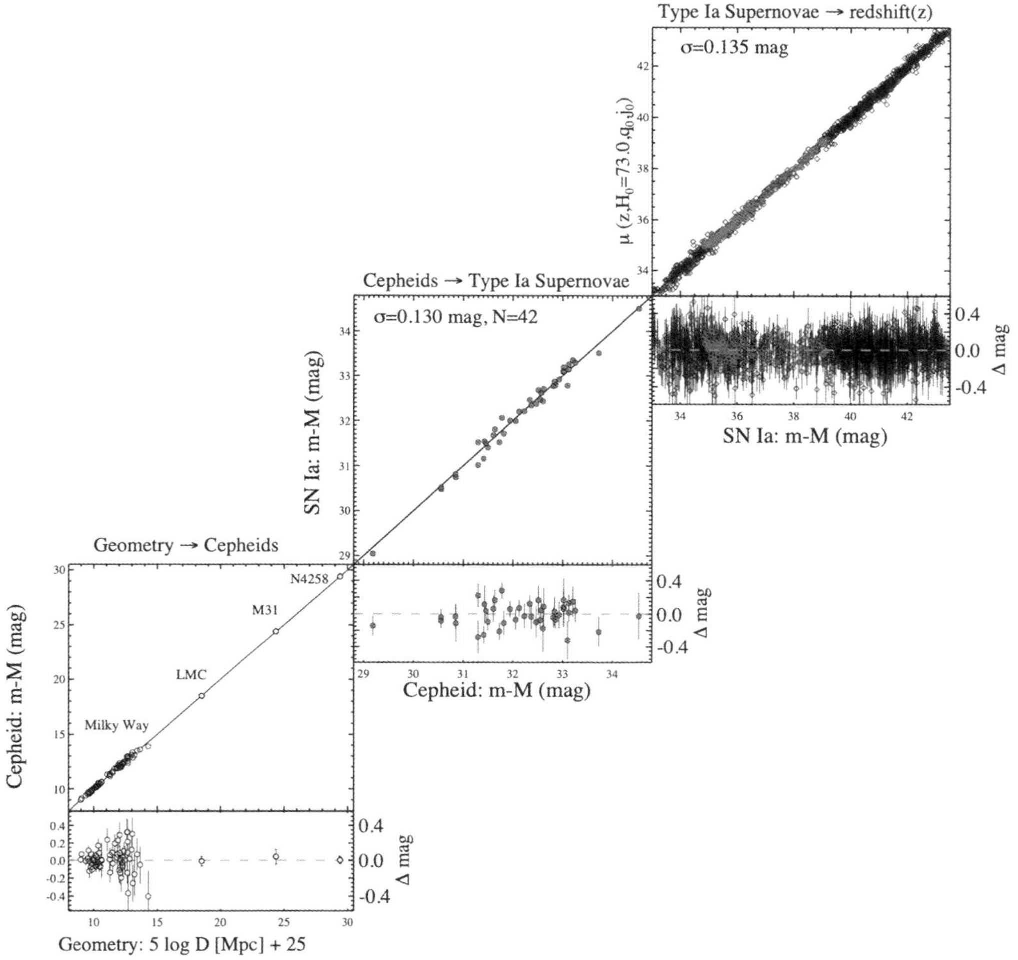

Figure 8. Complete distance ladder (Riess et al. 2022b). The simultaneous agreement of distance pairs: geometric and Cepheid-based (bottom left), Cepheid- and SN-based (middle), and SN- and redshift-based (top right) provides the measurement of H_0. For each step, geometric or calibrated distances on the abscissa serve to calibrate a relative distance indicator on the ordinate through the determination of M_B or H_0. Results shown are an approximation to the global fit, as discussed in the text. Red SN points are at $0.0233 < z < 0.15$, with the lower-redshift bound producing the appearance of asymmetric residuals when plotted against distance.

- Least crowded half (73.3)
- Skip 'local hole' $z > 0.06$ (73.4)
- All host types (73.3)
- Include tip of the red-giant branch (consistent) jointly (72.5)
- No metallicity term (73.5)
- Break in the period–luminosity relation at $P = 10$ days (72.7)
- No extinction correction (74.8)
- Individual host extinction law (73.9)
- Free parametric extinction law (73.3)
- Low $R_V = 2.5$ extinction law (73.2)
- Two of three anchors (73.0, 73.4, 73.2)
- No outlier rejection (73.4)

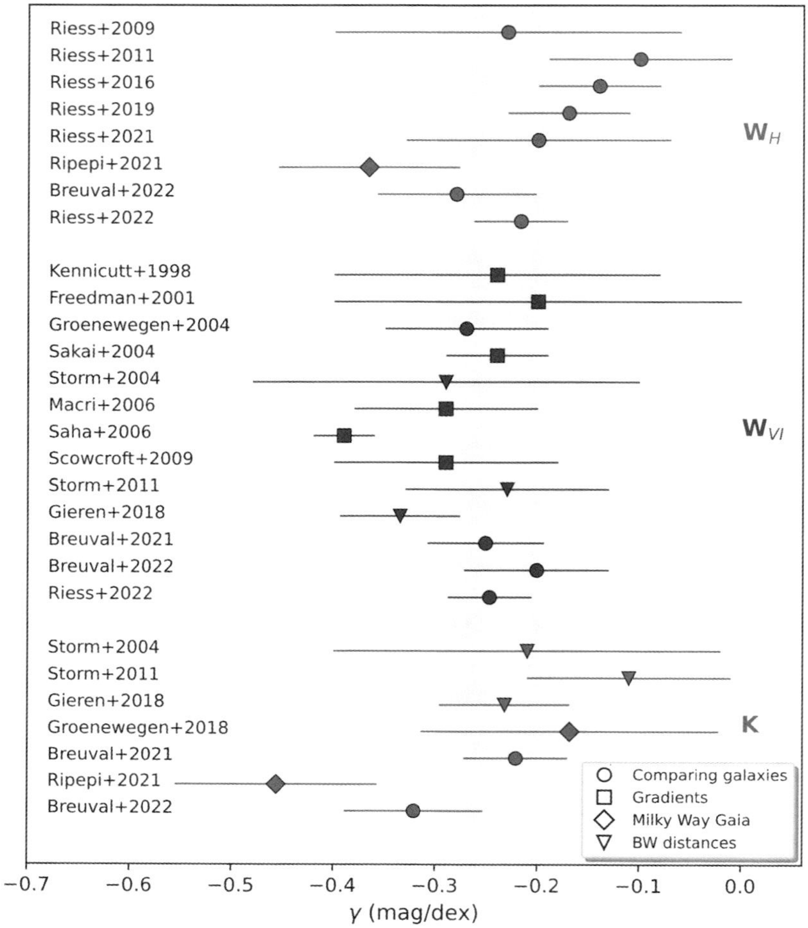

Figure 9. Recent Cepheid metallicity measurements. There is good consistency on the sign (negative) and scale, -0.1 to -0.3 mag dex^{-1}, for this term. It is important to know that this term has little effect on H_0 because the mean metallicity of Cepheids on the first and second rung is similar.

As shown, it is hard to move the value of the Hubble constant much below 72.5 or much above 73.5, which indicates that the baseline result is fairly robust. Although the SH0ES team measurement provides the highest precision, a consequence of having the largest sample of calibrated SNe Ia, the Hubble Tension is a broad finding that has lasted a decade and covers a wide range of different techniques, teams and methods. A recent summary of many measures is shown in Figure 10. This is a subset of recent measures but selected to be those which are the most cited, offer the most independence, and use the best source of data when multiple are available (e.g., *HST* versus ground).

3. What can we learn from *JWST*?

There are many tests of the distance ladder to discuss, but in the interest of space, we will focus on the most recent and novel from the *James Webb Space Telescope* (*JWST*). Observations with the *JWST* offer for the first time the resolution in the NIR (and mid-IR) to separate Cepheids from their stellar crowds. This greatly reduces the noise in the period–luminosity relations, whose source was the variations in background (i.e.,

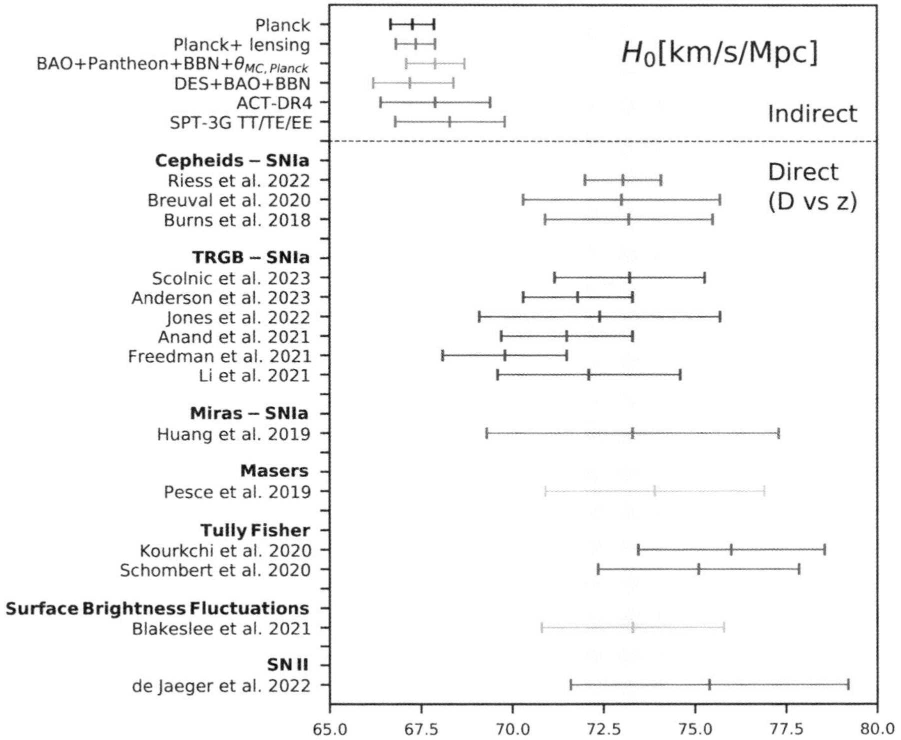

Figure 10. Recent measures and the Hubble Tension. The subset of recent measures was selected to be those which are the most cited, offer the most independence, and use the best source of data when multiple are available (e.g., *HST* versus ground).

fluctuations in the density of red giants) seen at *HST* NIR resolution. The previous contribution (see Freedman and Madore 2023, this volume) presented their group's *JWST* observations following up Cepheids discovered with *HST* in the SN Ia host NGC 7250. They showed 14 Cepheids with a dispersion of 0.22 mag at 1.15 μm, half the size of the dispersion seen with *HST* in the NIR (see also the wonderful contribution by Kayla Owens presenting this work). This is consistent with what we see in our *JWST* observations.

We observed >320 Cepheids in two hosts, NGC 5584 (the host of a SN Ia) and NGC 4258, the previously discussed calibrator. We observed the Cepheids in three filters centered at 0.9, 1.5 and 2.8 μm and across two epochs separated by 2–3 weeks. The differences in epochs allowed us to (1) confirm the astrometric identification of the Cepheids through their variability; (2) check the photometric stability of *JWST* over time by comparing non-variable stars; and (3) constrain the phase of the Cepheids (knowledge of which has dissipated over the years since they were discovered).

In Figure 11 we show the use of the epoch differences applied to the Cepheid phases. We found that the combination of two epochs (to reduce the random phase error) and the ability to constrain the phase reduced the dispersion from $\sim$0.43 mag to $\sim$0.17 mag, which is statistically equivalent to doubling the sample size. The results for both hosts and direct comparison with *HST* is shown in Figure 12. In Figure 13 we show a comparison of many postage stamps around Cepheids with *HST* and *JWST*. The contrast offered by *HST* is excellent in the optical, but becomes low, and the crowding high, in the NIR. This is the region where *JWST* wins by a large margin. The filters F160W for *HST* and F150W for *JWST* are nearly the same, which offers the best comparison.

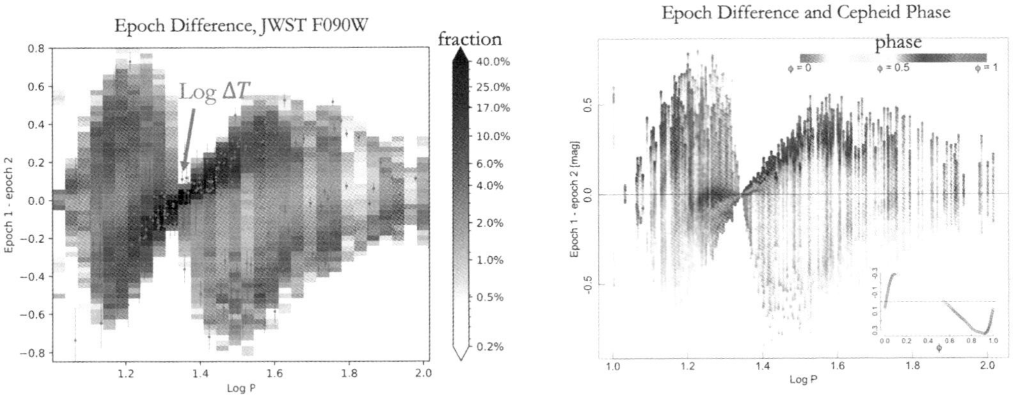

Figure 11. *JWST* two-epoch measurements of Cepheids in NGC 5584. (left) Epoch differences. (right) The expected differences which determine the phase.

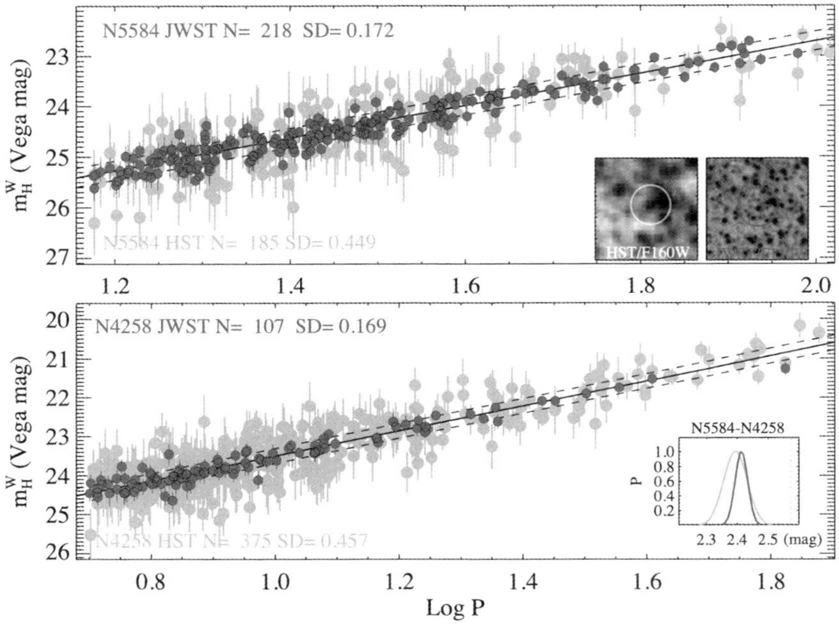

Figure 12. Comparison with the standard (SH0ES; Riess et al. 2022b) magnitude $W_{V,I}^{H}$ period–Wesenheit relation used to measure distances. The red points use *JWST* F150W, and the gray points are from *HST* F160W, including a small transformation, F150W−F160W $= 0.033 + 0.036[(V - I) - 1.0]$. The top panel is for NGC 5584, with the inset showing image stamps of the same Cepheid seen in the H band by each telescope. The bottom panel is for NGC 4258, with the inset showing the difference in distance moduli between NGC 5584 and NGC 4258 as measured with each telescope.

Because we have measured Cepheids with *JWST* at both rungs, a SN Ia host and NGC 4258, we can directly measure and compare the resulting calibration they provide for the distance ladder, a difference which is independent of uncertainties in the zero points of *JWST*, which cancel out. As shown in the inset of Figure 12, they provide a consistent difference (in magnitudes) between rungs as seen with *HST*. Because the Cepheid samples are large (>320 with *JWST* compared with >560 with *HST*) and this is purely a test of Cepheid measurements, the precision of this test is very high, 0.04

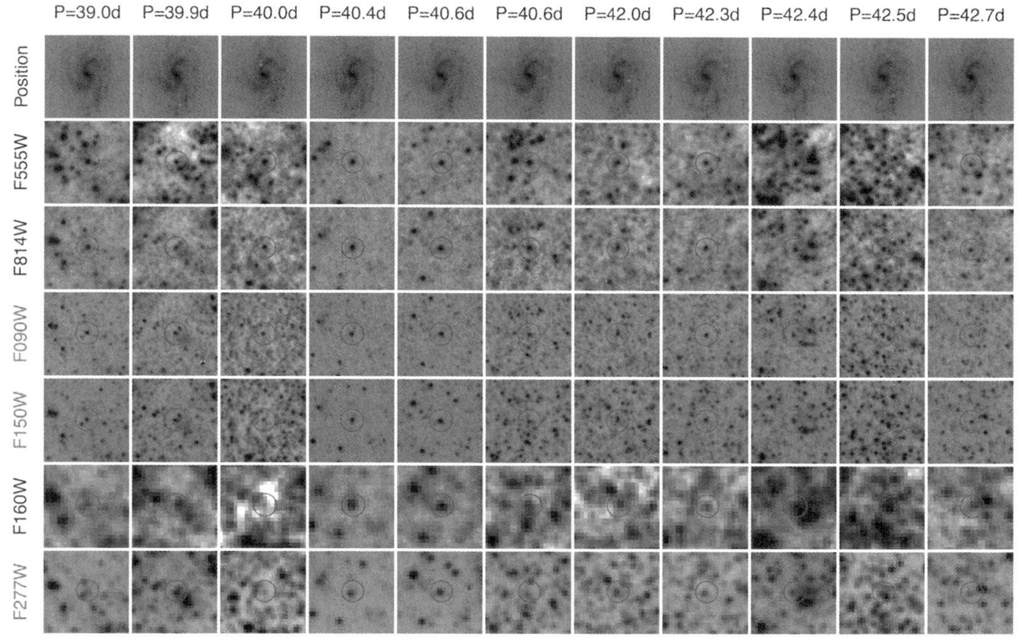

Figure 13. *HST* and *JWST* image stamps of NGC 4258 for all Cepheids with periods of 18–41 days. The top row indicates the location of each Cepheid. *HST* filters are labeled in black and *JWST* filters in magenta.

mag, which is many times the size of the 0.18 mag Hubble Tension. So, by stripping the noise and crowding from the Cepheid measurements with *HST*, these observations provide the strongest evidence yet that systematic errors in *HST* Cepheid photometry do not play a significant role in the present Hubble Tension (Riess et al. 2023). However, a more ambitious goal which will require more observations and the work of many is to improve the precision of the local measurement of H_0 to reach 1.0%.

So, what causes the Hubble Tension? We do not know. The facts are that it has lasted 10 years, it is statistically very significant and there is no precise (e.g., 3%) local measurement which is not higher than the CMB-based prediction, so it is pretty robust. We have undertaken a rather exhaustive set of tests and studies of systematics and there has been no indication of a problem on the measurement side. In this regard the *JWST* data are very impressive. One can speculate about an unknown systematic error, but the data have become very good, so concrete proposals for such systematics are inconsistent to date with the breadth of data and tests. It is also easy to speculate that it is something missing in ΛCDM but much harder to come up with a specific notion of what that is with sufficient rigor to test and yield a good fit to the cosmological data. This is a very hot topic in the theory community and we are seeing lots and lots of proposals, but nothing definitive or compelling (yet). We are optimistic. History has shown that basic errors are found by our community rapidly (months) and that discrepancies that last this long are telling us something interesting and thus worth the investment. Let's be curious.

(On another day the conference enacted a poll, which is a rather unscientific way to resolve the Hubble Tension but it is a snapshot of what people at the conference thought after a couple of days, and entertaining, and we attach the result as Figure 14. The polling was anonymous and used a smartphone app to allow the participants to vote. The three options voted on as possible cause of the Hubble Tension were a problem with the CMB analysis, a problem with ΛCDM or an issue with stellar physics. Since the conference was

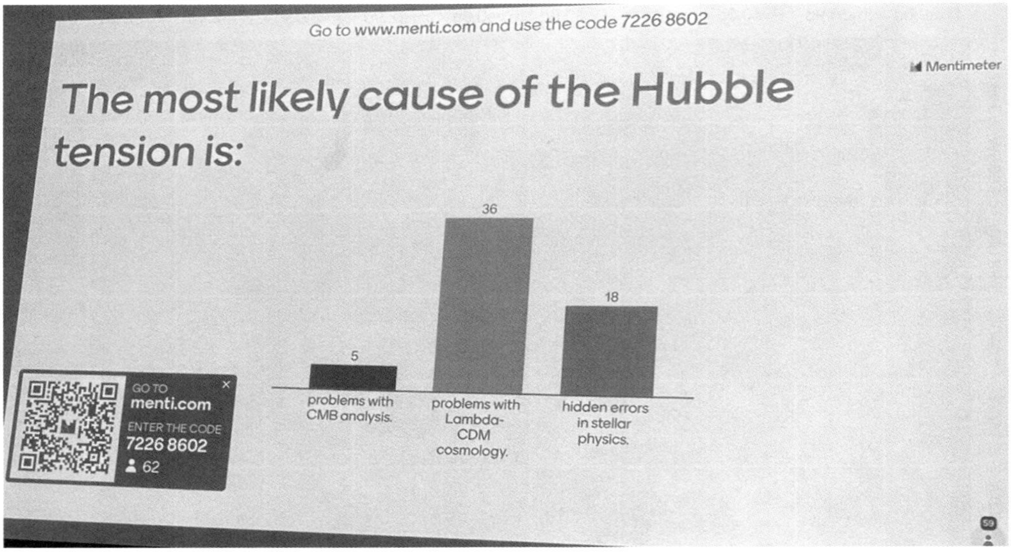

Figure 14. Meeting Poll.

a gathering of people studying stars and stellar physics, it was noteworthy that ~70% of attendees did not think that the local option of a problem with the physics of stars was the source of the Hubble Tension.)

We want to thank the organizers for a wonderful conference filled with discussions and goulash.

References

Breuval, L., Riess, A. G., Kervella, P. et al., 2022, ApJ 939, 89
Brout, D., Taylor, G., Scolnic, D., et al., 2022, ApJ 938, 111
Freedman, W. L. & Madore B. F., 2023, arXiv:2308.02474
Kamionkowski, M. & Riess, A. G., 2023, Annual Review of Nuclear and Particle Science, 73, 153
Pietrzyński, G., Graczyk, D., Gallenne, A., et al., 2019, Nature 567, 200
Planck Collaboration, Aghanim, N., et al., 2020, A&A 641, A6
Reid, M. J., Pesce, D. W. and Riess, A. G., 2019, ApJL 886, L27
Riess, A. G., Casertano, S., Yuan, W., et al., 2021, ApJ 855, 136
Riess, A. G., Casertano, S., Yuan, W., et al., 2019, ApJ 876, 85
Riess, A. G., Casertano, S., Yuan, W., et al., 2021, ApJL 908, L8
Riess, A. G., Breuval, L., Yuan, W., et al., 2022, ApJ 938, 36
Riess, A. G., Yuan, W., Macri, L. M., et al., 2022, ApJL 934, L7
Riess, A. G., Anand, G. S., Yuan, W., et al., 2023, ApJL, 956, L18
Sandage, A., Tammann, G. A., Saha, A., et al., 2006, ApJ 653, 843
Yuan, W., Macri, L. M., Riess, A. G. et al., 2022, ApJ 940, 64

Appendix A. From the CMB to H_0, a primer

The first and strongest peak we see in the power spectrum (expansion into spherical harmonics), l_s, determines the angle subtended by the sound horizon, the distance a fluctuation in the plasma will travel at the sound speed from the time of the Big Bang to the surface of last scattering (i.e., $z \sim 1000$). This scale results from the angular size of this feature, $\ell_s \simeq 2/\theta_s$. The angle subtended by the sound horizon is $\theta_s = r_s/D_A$, where D_A is the angular-diameter distance to the CMB surface of last scatter, and $r_s \sim c_s t_{dec}$

is the sound horizon. The angle is exceedingly well-measured by *Planck*: $\theta_s = (1.04109 \pm 0.00030) \times 10^{-2}$ rad.

We can calculate the physical size of the sound horizon from the physics of the early Universe as given by ΛCDM. It follows by integrating the sound speed $c_s(t)$ over time prior to recombination and is given by the integral,

$$r_s = \int_{z_{ls}}^{\infty} \frac{c_s(z)\,dz}{H(z)} = \frac{c}{\sqrt{3}H_{ls}} \int_{z_{ls}}^{\infty} \frac{dz}{[\rho(z)/\rho(z_{ls})]^{1/2}\,(1+R)^{1/2}}, \tag{1}$$

over redshift z. Here, $z_{ls} \simeq 1080$ is the redshift at which CMB photons last scatter, $c_s(z) = c\,[3(1+R)]^{-1/2}$ is the sound speed of the photon–baryon fluid, with $R = (3/4)(\omega_b/\omega_\gamma)/(1+z)$, and $\rho(z)$ is the total energy density at redshift z. Here, $\omega_b = \Omega_b h^2$ is the current physical baryon density, where $h \equiv H_0/(100\,\mathrm{km\,s^{-1}\,Mpc^{-1}})$ is a dimensionless Hubble constant. ω_b is a parameter constrained by higher-peak features in the CMB power spectrum (Silk damping at higher l and the relative heights of the even- and odd-numbered peaks with ω_m) and $\omega_\gamma = 2.47 \times 10^{-5}$ is the physical photon energy density given by COBE FIRAS. It is convenient to define the expansion rate at last scattering,

$$H_{ls} = 100\,\mathrm{km\,s^{-1}\,Mpc^{-1}}\,\omega_r^{1/2}(1+z_{ls})^2\sqrt{1 + \frac{\omega_m}{\omega_r}\frac{1}{1+z_{ls}}}, \tag{2}$$

where $\omega_m = \Omega_m h^2$ is the present physical matter density fit from the higher-peak structure in the CMB. In the standard cosmological model, the early-Universe energy density is $\rho(z) \propto \omega_m(1+z)^3 + \omega_r(1+z)^4$. The physical radiation density is

$$\omega_r = \left[1 + \frac{7}{8}N_{\mathrm{eff}}\left(\frac{4}{11}\right)^{4/3}\right]\omega_\gamma, \tag{3}$$

where the second term accounts for additional non-relativistic degrees of freedom. In the standard cosmological model, these include the three neutrino mass eigenstates, and $N_{\mathrm{eff}} = 3.06$.

The (comoving) angular-diameter distance to the surface of last scattering is then an integral,

$$D_A = \frac{c}{H_0}\int_0^{z_{ls}} \frac{dz}{[\rho(z)/\rho_0]^{1/2}}, \tag{4}$$

from recombination until the current time t_0, when the total energy density is ρ_0. The denominator here is $\rho(z)/\rho_0 = \Omega_m(1+z)^3 + (1-\Omega_m)(1+z)^{-3(1+w)}$ in the standard cosmological model, with a dark-energy equation-of-state parameter w. The cosmological constant corresponds to $w = -1$.

From $\theta_s = r_s/D_A$, we infer a Hubble constant,

$$H_0 = \sqrt{3}H_{ls}\theta_s \frac{\int_0^{z_{ls}} dz\ [\rho(z)/\rho_0]^{-1/2}}{\int_{z_{ls}}^{\infty} dz\ [\rho(z)/\rho(z_{ls})]^{-1/2}\,(1+R)^{-1/2}} \tag{5}$$

from the CMB.

By this route, H_0 depends on ω_b, ω_m, ω_γ, N_{eff} and also (to a lesser degree) on the other model parameters through degeneracies in the description of the fluctuation spectrum.

At the crossroads of astrophysics and cosmology: Period–luminosity relations in the 2020s
Proceedings IAU Symposium No. 376, 2024
R. de Grijs, P. A. Whitelock & M. Catelan, eds.
doi:10.1017/S1743921323002685

The Tip of the Red Giant Branch as a Cosmological Tool

Myung Gyoon Lee

Seoul National University, Department of Physics and Astronomy, SNUARC, Seoul, Republic of Korea. email: mglee@astro.snu.ac.kr

Abstract. The luminosity of the brightest stars (tip) of the red giant branch (TRGB) in the color–magnitude diagrams of old stars was used early on to introduce the 'multiple stellar populations' concept, in 1944, by Walter Baade. However, the precision and accuracy of the TRGB for distance estimation has not been known well for long. In the modern era, equipped with high spatial resolution imaging telescopes, the TRGB is considered an excellent standard candle for any type of resolved galaxies, thus representing a powerful probe for cosmology. The TRGB has several advantages over the classical Cepheids. I review how we can apply the TRGB in cosmology. Four science cases, from large to small scales, are presented: (1) the Hubble flow with Type Ia supernovae; (2) Virgo Cluster infall and dark matter; (3) dark galaxies; and (4) dark matter-free galaxies.

Keywords. Cosmology, Cosmic distance scale, Dark matter, Galaxies, Galaxy clusters

1. Introduction

Cepheids are great! Cepheids have become popular as the best standard candle for resolved galaxies. Period–luminosity relations of Cepheids (Leavitt's law) opened a new era of extragalactic astronomy and observational cosmology, one of the main topics of this symposium (see the reviews by Freedman and Riess in these proceedings).

However, Cepheids are found only in late-type galaxies, while the Universe is composed not only of late-type galaxies but also includes early-type galaxies; see Figure 1. We need a comparable standard candle that covers both late-type and early-type galaxies in the local Universe. That standard candle is the tip of the red-giant branch (TRGB) (Lee et al. 1993; Beaton et al. 2019; Madore et al. 2023, and references therein). Here, I introduce some examples of cosmological applications of the TRGB. I present some science cases, from large to small scales: (1) the Hubble flow with Type Ia Supernovae (SNe Ia); (2) Virgo Cluster infall and dark matter; (3) dark galaxies; and (4) dark matter-free galaxies.

2. What is the TRGB?

The TRGB is the brightest part of the red giant branch (RGB) in the color–magnitude diagrams (CMDs) of old stars in stellar systems such as galaxies or globular clusters. Theoretically, it corresponds to the core-helium flash stage in the evolution of low-mass stars in the Hertzsprung–Russell (HR) diagrams. Figure 2 (left) displays the conventional $M_V, (B-V)$ CMD of M55, a Milky Way globular cluster, and Figure 2 (right) shows the composite $M_I, (V-I)$ CMD of 46 Milky Way globular clusters (Freedman 2021). The $M_V, (B-V)$ CMD of M55 clearly shows the TRGB. The V-band magnitude of the TRGB varies significantly, depending on the metallicity. However, the TRGB's I-band magnitude, shown in Figure 2 (right), is much less metallicity-dependent so that it can be used as a precise standard candle. Photographic magnitudes of the TRGB

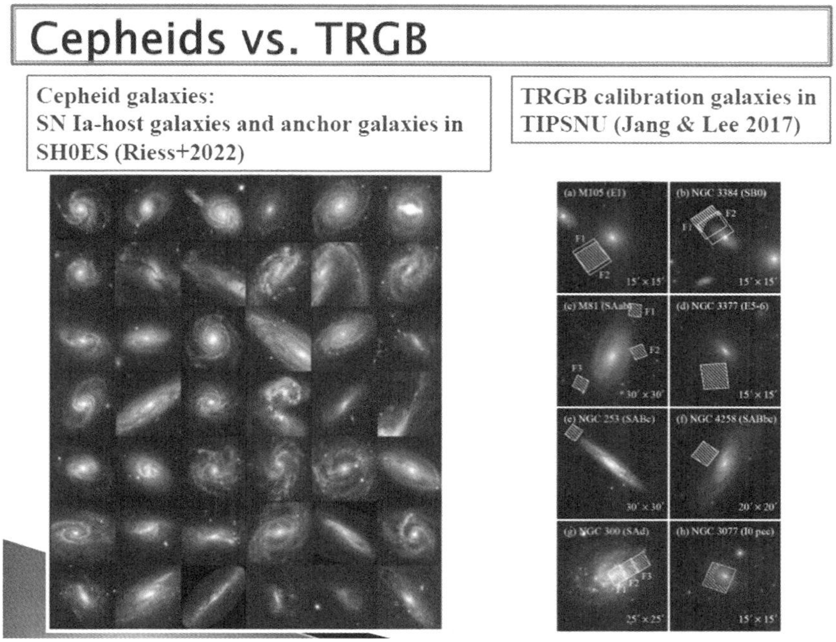

Figure 1. Morphological types of sample galaxies where, respectively, Cepheids (Riess et al. 2022, left) and the TRGB (Jang & Lee 2017a, right) have been used for distance measurements.

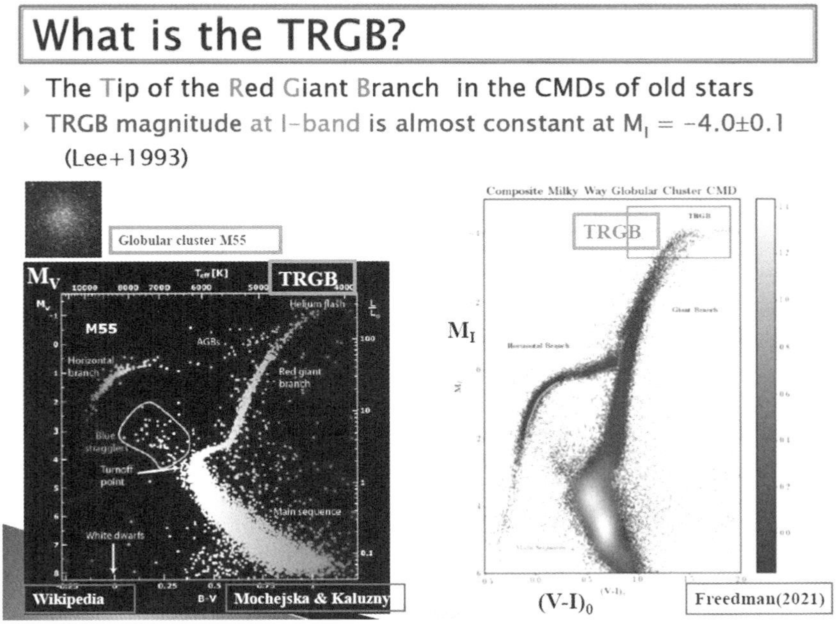

Figure 2. (left) $M_V, (B-V)$ CMD of the Milky Way globular cluster M55 (Mochejska and Kaluzny, via Wikipedia). (right) Composite $M_I, (V-I)$ CMD of 46 Milky Way globular clusters (Freedman 2021).

were used to introduce the concept of Populations I and II in M31 and its companion galaxies as early as 1944 by Walter Baade. Thereafter, the TRGB was sometimes used for distance estimation to nearby galaxies but its reliability has not been known well for long. Only in 1993 it was shown that the TRGB is an excellent standard candle, comparable to Cepheids (Lee et al. 1993; Lee 1993). The limit of the TRGB for distance measurement used to be about 3 Mpc based on ground-based observations, but that distance scale increased by a factor of 10 with the advent of the *Hubble Space Telescope* (*HST*), reaching about 30 Mpc today (Jang & Lee 2017a,b).

2.1. *TRGB vs. Cepheids*

Almost all galaxies in the local Universe were formed at early times. Not only typical elliptical and lenticular galaxies but also star-forming irregular/spiral galaxies host old stellar populations with ages older than a few giga-years. As such, the TRGB method can be applied to any type of resolved galaxies in the local Universe, while the Cepheid method can be applied only to the late-type, resolved galaxies. Old RGB stars are best visible in the halo of a galaxy, while Cepheids are located in star-forming regions in galaxy disks. Thus, the TRGB based on halo stars is almost free of internal extinction. In contrast, Cepheids suffer significantly from the internal extinction problem. To reduce this problem, Wesenheight relations (which are defined to be reddening-free) have been designed. In addition, the I-band TRGB magnitude varies little in the low-metallicity regime where most TRGB measurements are obtained. The TRGB magnitude is fainter than the Cepheid magnitudes in optical bands, but it becomes brighter in near- and mid-infrared bands, which cover the main window of the *James Webb Space Telescope* (*JWST*).

Today, the number of the galaxies with the TRGB distances has reached more than 500, which is about eight times larger than the number of Cepheid distances (Jang & Lee 2017b; Anand et al. 2021; Tully et al. 2023; Madore et al. 2023). This ratio is expected to increase significantly in the near future. This shows clearly that the TRGB is a more versatile tool for cosmology than the use of Cepheids.

2.2. *Methods for the TRGB Estimation*

Figure 3 shows an example galaxy for TRGB measurements, NGC 4592, a dwarf spiral galaxy located in front of the Virgo Cluster. We chose, as an example, this spiral galaxy rather than an early-type galaxy, because the CMDs of late-type galaxies are more complicated than those of early-type galaxies. Kim et al. (2020) used *HST*/Wide-Field Camera 3 (WFC3) F555W/F814W images of this galaxy from the archive. The middle panel shows a gray-scale map of the F814W image of this galaxy, the location of which is marked in the color map of the Subaru/HyperSuprime Cam images in the left panel. The right panel shows a F814W, (F555W − F184W) CMD of the measured stars in the off-disk region marked in the middle panel. The red dots in the middle panel denote the RGB star candidates selected from the CMD of the detected stars. Various stellar populations are visible: blue supergiants as well as bright main-sequence stars, red supergiants, RGB stars, and asymptotic giant-branch (AGB) stars. The most notable feature is the thick RGB. The stars in this RGB mainly belong to the old halo, while the other stars are mostly drawn from the young disk. The bright top of the RGB is distinguishable at F814W $\sim$ 25.8 mag, clearly showing the location of the TRGB.

The principle adopted for typical TRGB magnitude measurements is simple: we detect a significant jump in the luminosity functions (LFs) of the RGB star candidates selected from the CMDs of resolved stars, and we measure the magnitude of the jump. The jump is primarily due to the high number ratio of RGB and AGB stars in old stellar systems.

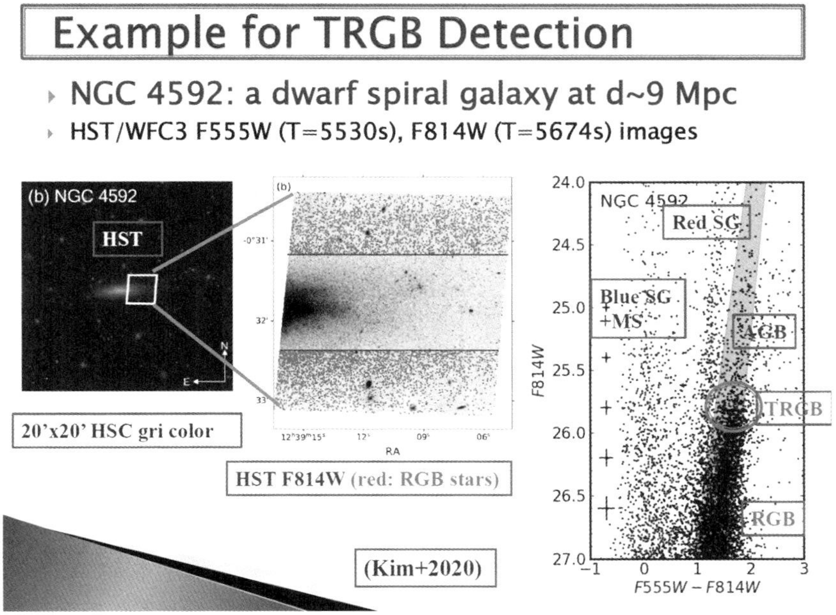

Figure 3. Example of a TRGB measurement: NGC 4592 (Kim et al. 2020). (left) Location of the *HST* field (white box) marked in pseudo-color Subaru/Hypersuprime Cam images of NGC 4592. (middle) Gray-scale map of the F814W image of the *HST* field. The outer regions in the halo of NGC 4592 were used for selection of the RGB star candidates (marked with red dots). (right) F814W, (F555W − F184W) CMD of the detected stars in the outer regions of the *HST* field. Various stellar populations are labeled. The red circle denotes the location of the TRGB. Plus signs on the left represent mean photometric errors for a given magnitude.

as shown in stellar evolution models. In optical bands, the I (or F814W) band is best for TRGB measurements, because the TRGB magnitude of low-metallicity RGB stars is almost constant in this band (Lee et al. 1993; Jang & Lee 2017a).

In the CMD of RGB stars spanning a large range in metallicity, the TRGB is represented by a line following the upper edge of the RGB. This upper edge (locus) of the RGB is almost horizontal in the I, $(V - I)$ CMDs. Therefore, we apply an edge detection algorithm to detect the TRGB in the LFs.

Typical methods to measure the TRGB are briefly summarized in Kim et al. (2020), as follows: (a) direct edge detection (DED) with a Sobel filter (Lee et al. 1993; Jang & Lee 2017a); (b) GLOESS (Gaussian windowed, locally weighted scatterplot smoothing) edge detection (GED) for smoothed LFs (Hatt et al. 2017); and (c) maximum-likelihood (ML) LF fitting with a model LF of RGB and AGB stars (Méndez et al. 2002; Makarov et al. 2006).

2.3. *Precision of the TRGB*

We check the precision of the TRGB using a comparison of measurements based on multiple methods. Figure 4 compares the TRGB magnitudes for NGC 4592 as measured using the three methods described above. The TRGB magnitudes from all three methods show excellent agreement at the level of ±0.01 mag. On this basis, Kim et al. (2020) derived a distance to this galaxy of $d = 9.07 \pm 0.27$ Mpc. The typical precision of TRGB magnitude measurements with good data is ±0.02–0.03 mag (corresponding to a distance error of 1–1.5%). The TRGB magnitude measurement error can be as large as ±0.1 mag for difficult cases (corresponding to a distance error of 5%).

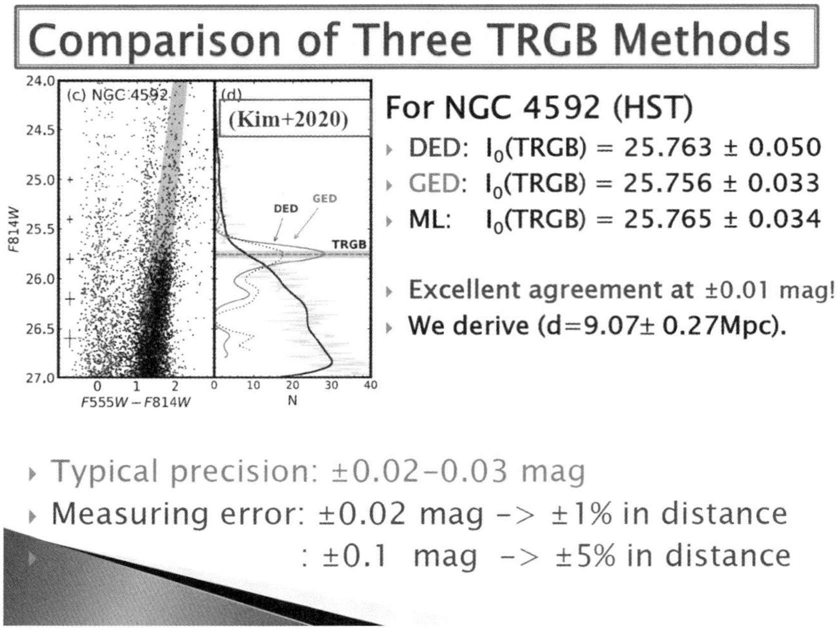

Figure 4. Comparison of TRGB magnitude measurements for NGC 4592 derived from three methods applied to the LF of the RGB stars (black line): direct edge detection (DED; blue line); GLOESS edge detection (GED; red line); and maximum-likelihood LF fitting with model LFs (ML) (Kim et al. 2020).

2.4. *Accuracy of the TRGB*

The accuracy of the TRGB depends mainly on the calibration error of the absolute magnitude of the TRGB. Calibration of the I-band magnitude of the TRGB was originally based on RR Lyrae (or the horizontal branch, HB; see Lee et al. 1993, and references therein). However, modern calibration of the TRGB (as well as for Cepheids) is based on geometric distance anchors (primary distance indicators), including detached eclipsing binaries in nearby galaxies (LMC, SMC), water megamasers in nearby galaxies (NGC 4258/M106), and parallaxes (*Gaia* stars in the Milky Way).

Figure 5 displays a comparison of TRGB loci in the M_I, $(V-I)$ diagram derived from observations (Lee et al. 1993; Rizzi et al. 2007; Madore et al. 2009; Jang & Lee 2017a; Freedman et al. 2020; Jang et al. 2021; Li et al. 2022; Soltis et al. 2021; Hoyt 2023) and theoretical models (Salaris & Cassisi 1997; Bellazzini 2008; Serenelli et al. 2017). Most calibrations show good agreement for the low-metallicity range ([Fe/H] < -1 dex). Note that the TRGB locus of Jang & Lee (2017a) was derived from a combination of eight nearby galaxies of diverse morphological types, covering a wide range of color. At this TRGB locus, the I-band TRGB magnitude is almost constant for the blue color range (i.e., for low metallicity, [Fe/H] < -1 dex), but it becomes fainter as the color increases (i.e., for higher metallicity). The blue region of the TRGB loci is mainly used for galaxy distance estimation, because most of the resolved stars in galaxy halos have low metallity ([Fe/H] < -1 dex).

In Figure 6, we display a zoomed-in view of the TRGB loci of two low-mass galaxies, the LMC and the SMC, recently derived by Hoyt (2023). It can be seen that the TRGB loci of these two galaxies are in excellent agreement with that derived from the combined sample of eight galaxies of Jang & Lee (2017b) (with updated zero points). Note that this latter TRGB locus exhibit a significant difference in the red color compared with the TRGB locus given by Rizzi et al. (2007).

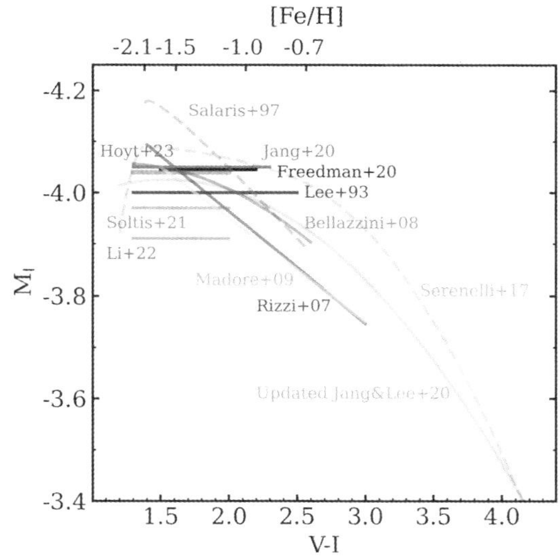

Figure 5. Comparison of the TRGB loci in the M_I, $(V - I)$ CMD derived from observations and theoretical models from the literature. Horizontal lines denote only the constant magnitudes for the low-metallicity range, which are generally used for distance estimation.

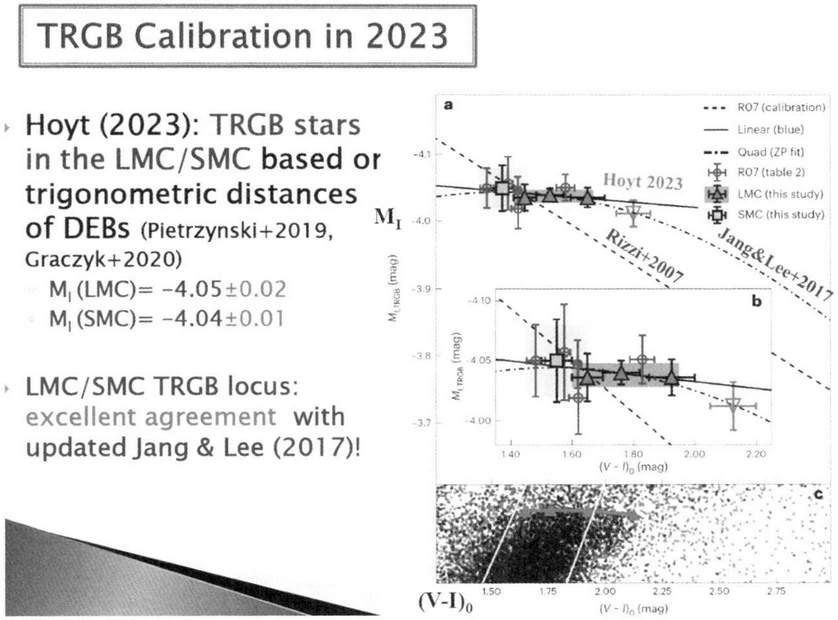

Figure 6. TRGB loci in the LMC (red symbols) and SMC (yellow symbols) in the M_I, $(V - I)_0$ CMD of Hoyt (2023) compared with those from Rizzi et al. (2007) (dashed line) and updated by Jang & Lee (2017a) (dot-dashed line). Note that the TRGB loci of the LMC and SMC are in excellent agreement with that of Jang & Lee (2017a) with updated zero points.

Figure 7 illustrates the absolute I-band magnitude of the TRGB as a function of year of publication. Since the earlier calibrations (Mould et al. 1983; Lee et al. 1993), numerous calibrations based on various methods have been presented. Calibration sources are diverse: Milky Way globular clusters, Local Group galaxies, as well as the LMC and SMC, NGC 4258, *Gaia* stars in the Milky Way, and theoretical models. Most of the new

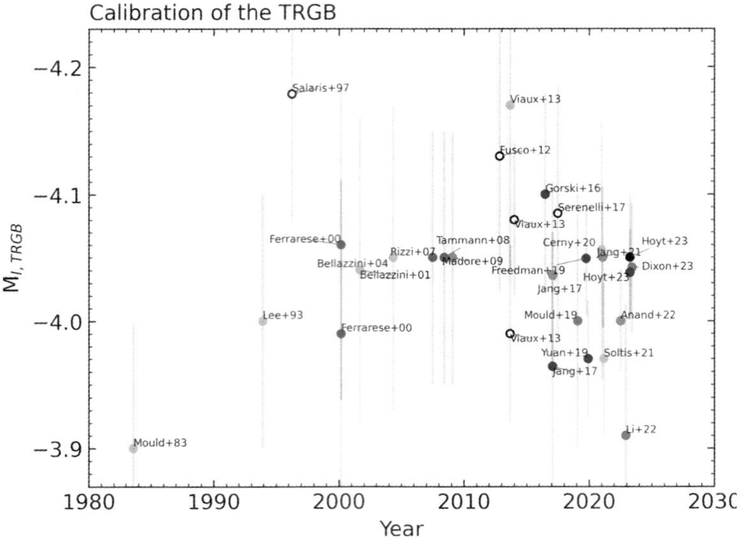

Figure 7. TRGB I-band magnitude calibration as a function of year of publication. Different symbols represent the calibration sources: yellow circles for Milky Way globular clusters, green circles for Local Group galaxies, blue circles for the LMC, black circles for the SMC, pink circles for NGC 4258, red circles for *Gaia* stars in the Milky Way, and open circles for theoretical models. (Prepared by In Sung Jang).

calibrations are consistent with the old value of Lee et al. (1993), $M_I = -4.0 \pm 0.1$ mag. However, the calibration errors are gradually becoming smaller, to the level of ± 0.05 mag, or even smaller. Note that Freedman (2021) derived $M_I = -4.049 \pm 0.015(\text{stat.}) \pm 0.035(\text{syst.})$ mag from a sample of four distance anchors: Milky Way globular clusters, LMC, SMC, and NGC 4258. This value is very similar to those of Jang et al. (2021) and Hoyt (2023). Interestingly, the two recent values derived from *Gaia* Data Release 3 of TRGB stars in the Milky Way (Li et al. 2022; Dixon et al. 2023) show a 0.1 mag difference. Although the Dixon et al. (2023)'s value is consistent with others discussed here, the Li et al. (2022) value is found at the faintest end.

In summary, it is reasonable to adopt the TRGB magnitude calibration for low-metallicity systems, $M_I = -4.05 \pm 0.05$ mag, as well as the TRGB locus of Jang & Lee (2017a), with the updated zero point, as shown in Figure 6. It is expected that the accuracy of the TRGB magnitude will become even better in the near future.

3. TRGB Science Cases

3.1. *The Hubble Flow with SNe Ia*

SNe Ia are a powerful standard(izable) candle for measuring the expansion history of the Universe. The first step in the SNe Ia cosmology is to calibrate the peak magnitude of SNe Ia using nearby SNe Ia host galaxies with known distances. Cepheids used to be the most popular tool for SNe Ia calibration. Note that the TRGB is another excellent candle for SNe Ia calibration and that SNe Ia are found not only in late-type but also in early-type galaxies. At present, the distance ladder to derive the value of the Hubble constant in supernova cosmology has become much simpler than in the past, as shown in Figure 8.

Figure 9 shows the evolution of the measured values of the local Hubble constant derived from Cepheid-calibrated SNe Ia since 2000 (blue symbols), in comparison with the values derived from extrapolation of the cosmic microwave background radiation (CMBR) measurements (black, yellow, and green symbols) (Freedman 2021). The value

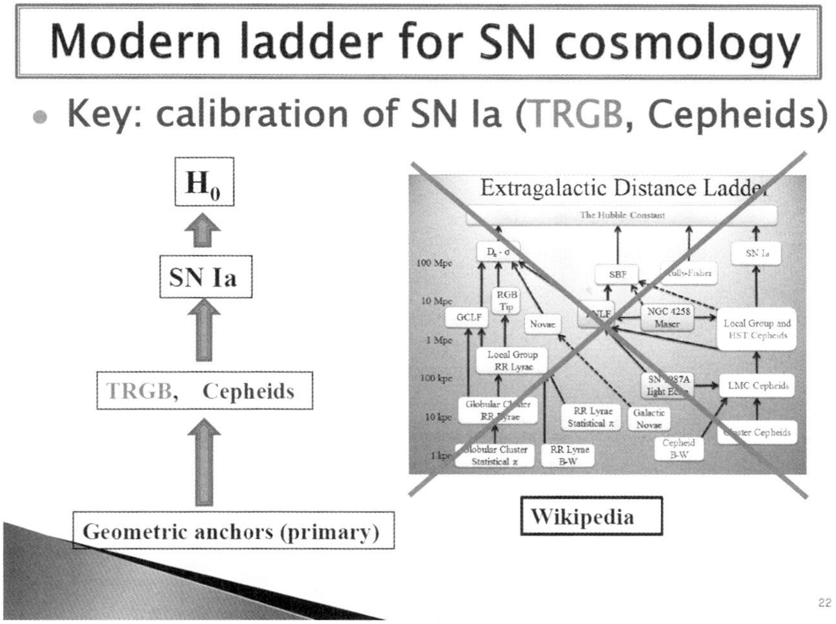

Figure 8. Modern distance ladder for supernova cosmology. Note that the TRGB and Cepheids are calibrated directly using geometric distance anchors. Currently, they are two major standard candles to calibrate SNe Ia.

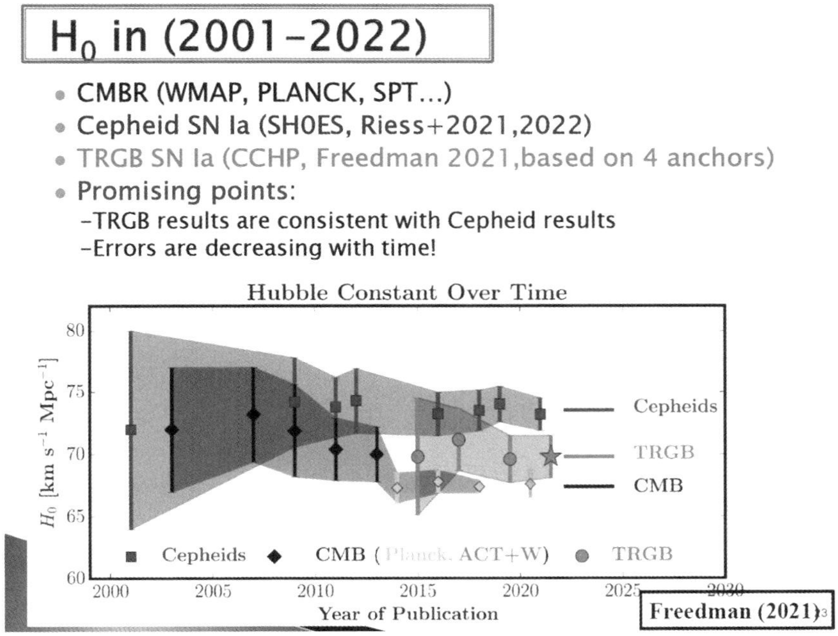

Figure 9. Comparison of local H_0 measurements derived from three methods: Cepheid-calibrated SNe Ia (blue symbols), TRGB-calibrated SNe Ia (red symbols), and the CMBR (black, yellow, and green symbols) (Freedman 2021).

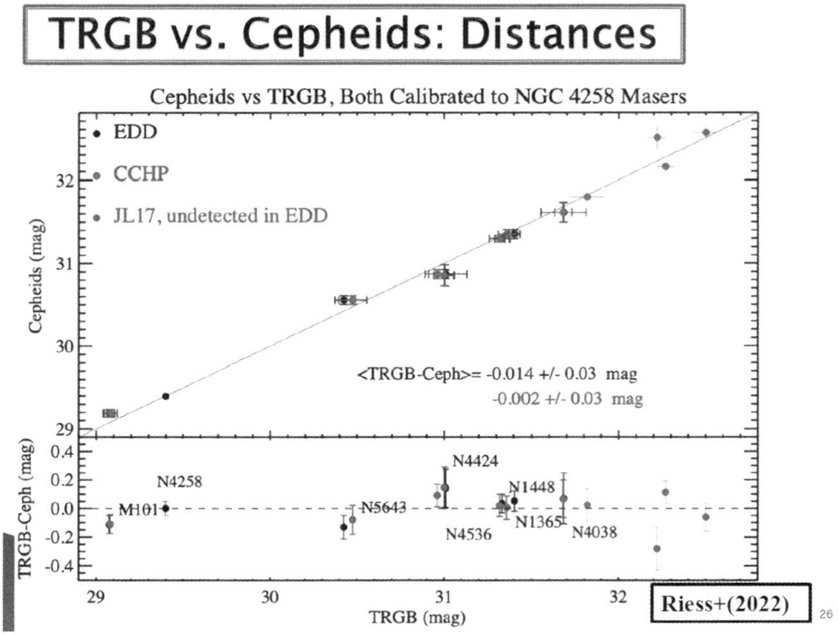

Figure 10. Comparison of TRGB and Cepheid distances to the same galaxies (Riess et al. 2022). The TRGB distances are from the CCHP (Freedman 2021, red symbols), EDD (Anand et al. 2021, black symbols), and TIPSNU for the most distant galaxies (Jang & Lee 2017b, blue symbols).

of the local Hubble constant derived from Cepheid-calibrated SNe Ia (Riess et al. 2021, 2022) is larger than the value extrapolated from the CMBR, at the level of 5–6σ. This problem is referred to as the 'Hubble tension'. It is one of the hottest topics in present-day cosmology (see Perivolaropoulos & Skara 2022, and references therein).

Figure 9 also shows recent measurements of the Hubble constant derived from TRGB-calibrated SNe Ia (Jang & Lee 2017b; Freedman et al. 2019; Freedman 2021), in red symbols. The TRGB-calibrated H_0 from Freedman (2021) is slightly larger than the CMBR value, while it is slightly smaller than the Cepheid-calibrated value (Riess et al. 2022). The differences are at the level of 2σ, showing little tension with either of the two measurements. Note also that the measurement errors from each method are, on average, decreasing as time passes. We expect that the errors will become even smaller in the near future.

Figure 10 displays a comparison of TRGB and Cepheid distances to the same galaxies given by Riess et al. (2022) (see also Anand et al. (2022)). The TRGB distances are from the Carnegie–Chicago Hubble Program (CCHP; Freedman 2021, red symbols), the Extragalactic Distance Database (EDD) for nearby galaxies (Anand et al. 2021, black symbols), and the Tip of the RGB for SN host galaxies in the Universe (TIPSNU) for the most distant galaxies (Jang & Lee 2017b, blue symbols). Note that Anand et al. (2021) provided a very useful catalog of TRGB distances in the EDD, but the galaxies in their catalog are located at $d < 20$ Mpc. TRGB distances to more distant galaxies are given by Jang & Lee (2017b). All these distances were calibrated with respect to the NGC 4258 maser distance for this comparison. The TRGB distances from these three different projects all show remarkable agreement with the Cepheid distances.

3.2. *Virgo Cluster Infall and Dark Matter*

While the motion of galaxies in low-density environments follow the Hubble flow, the motion of galaxies around galaxy clusters/groups is affected by the combination of the local gravity and the Hubble flow. Thus, the galaxies around galaxy clusters/groups are useful for estimating not only the local dark matter distribution but also the value of the Hubble constant. Assuming a spherical shell of radius R moving radially in a cluster with mass M, the motions of galaxies near the cluster in a Universe with Ω_Λ can be described as (Peirani & de Freitas Pacheco 2006, 2008):

$$\frac{\mathrm{d}^2 R}{\mathrm{d}t^2} = -\frac{GM}{R^2} + \Omega_\Lambda \mathrm{H}_0^2 R. \quad (1)$$

Solving this equation, we can derive H_0 and $M_0(R_0)$, where R_0 is the radius of the zero-velocity surface of the cluster and M_0 is the cluster mass within the zero-velocity radius. The clustercentric velocity V_C is given in terms of $M(R)$ and H_0, as (with the *Planck* value of $\Omega_\Lambda = 0.685$ and $n \sim 0.7$):

$$V_\mathrm{C}(R) = -\frac{0.940\mathrm{H}_0}{R^n}\left(\frac{GM}{\mathrm{H}_0^2}\right)^{(n+1)/3} + 1.335\mathrm{H}_0 R. \quad (2)$$

The cluster mass, M_0, is given in terms of R_0 and H_0 as:

$$M_0 = \frac{1.855}{G} \times \mathrm{H}_0^2 R_0^3 = 4.29 \times 10^{12} h_{100}^2 \left(\frac{R_0}{[\mathrm{Mpc}]}\right)^3 \mathrm{M}_\odot. \quad (3)$$

The Virgo Cluster, the nearest galaxy cluster, is an excellent laboratory where we can apply this method. In this regard, the galaxies located between the Virgo Cluster and the Milky Way play an important role (Karachentsev et al. 2018; Kashibadze & Karachentsev 2018; Kashibadze et al. 2020). It is expected that these galaxies are subject to infall into the Virgo Cluster center. The TRGB is the best standard candle to measure the distances to these infalling galaxies. Kim et al. (2020) used the kinematics of 33 galaxies with TRGB distances (McQuinn et al. 2014, 2017; Karachentsev et al. 2018; Kim et al. 2020; Tikhonov & Galazutdinova 2020) in front of the Virgo Cluster. Figure 11 shows the locations and line-of-sight velocities of these galaxies. Most are located in the northern region of Virgo, and only four galaxies are in the southern volume of the Virgo Cluster.

The virgocentric velocity of each infalling galaxy was derived from the observed line-of-sigh velocity and the separation angular distance with respect to the Virgo center (assuming a radial velocity of $v = 988$ km s^{-1} at a distance of 16.5 Mpc; Kashibadze et al. 2020). Following Kashibadze & Karachentsev (2018) and Kashibadze et al. (2020), Virgocentric velocities of infalling galaxies were derived for two extreme cases, i.e., a minor attractor model where the peculiar motion is negligible compared with the Hubble flow, and a major attractor model where the peculiar motion is comparable to the Hubble flow.

Figure 12 displays the Hubble diagram, showing the Virgocentric velocity versus Virgocentric distance of the infalling galaxies. Fitting the data with the above models and taking a mean of the two model results, Kim et al. (2020) derived a value of the local Hubble Constant, $\mathrm{H}_0 = 65.9 \pm 3.5(\mathrm{stat.}) \pm 2.4(\mathrm{syst.})$ km s^{-1} Mpc^{-1}. This value for H_0 is consistent with the value derived independently from the TRGB-calibrated SNe Ia, although its error is larger than the latter: see Figure 13.

The zero-velocity radius is derived as $R_0 = 6.76 \pm 0.35$ Mpc, and the corresponding mass is $M_0(< R_0) = (5.7 \pm 1.5) \times 10^{14}$ $\mathrm{M}_\odot$. The virial radius and mass of Virgo are $R_{200} = 1.7$ Mpc and $M_\mathrm{vir} = (4\text{–}6) \times 10^{14} \mathrm{M}_\odot$, respectively (Kashibadze et al. 2020). This confirms that the dark matter content of the Virgo Cluster is mostly located inside the

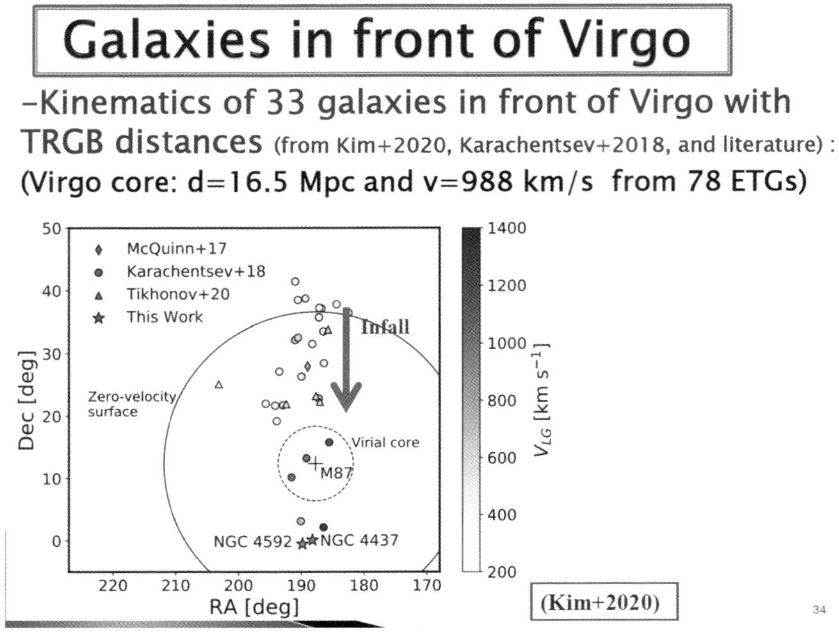

Figure 11. Locations of the galaxies in front of the Virgo Cluster (Kim et al. 2020). The color bar represents the line-of-sight velocity of each galaxy with respect to the center of the Local Group.

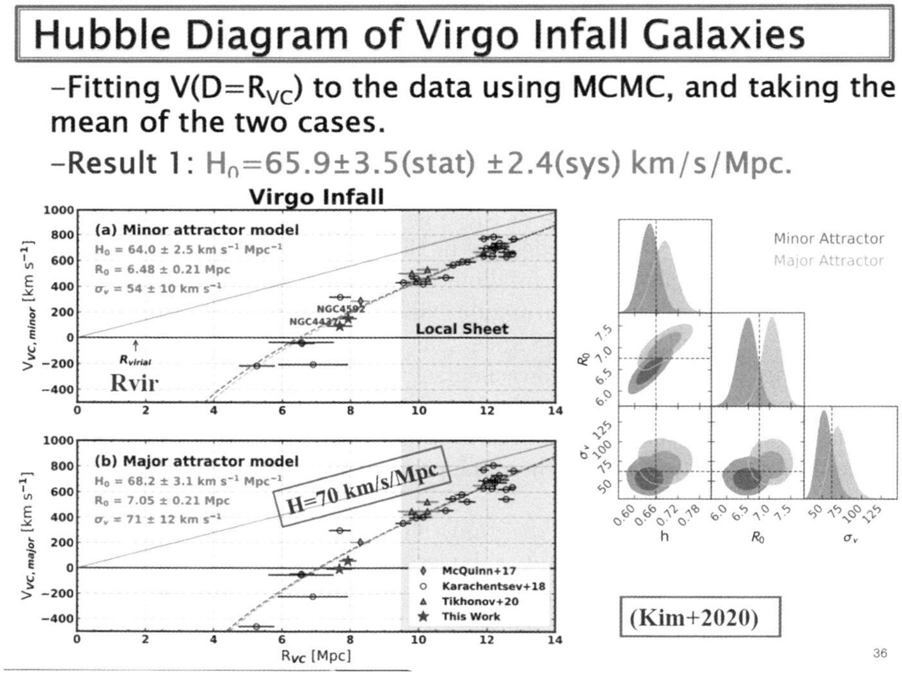

Figure 12. Hubble diagram of the Virgo infall galaxies showing Virgocentric velocity versus Virgocentric distance for the minor (top) and major (bottom) attractor models. The red dashed lines denote the model fits. The right panels show the posterior probability distributions for the fitted parameters, h, R_0, and σ_v. The blue and yellow colors represent the minor and major attractor models, respectively (Kim et al. 2020).

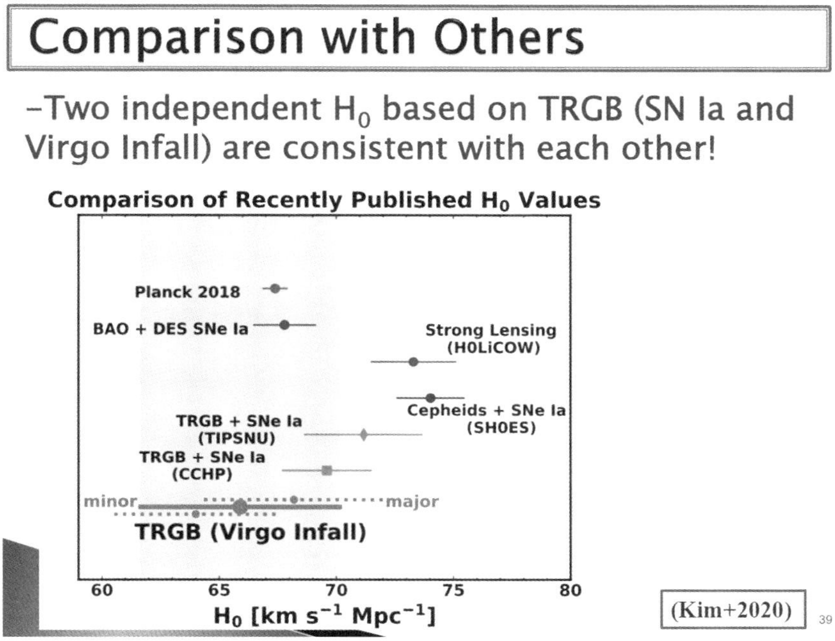

Figure 13. Comparison of the Hubble constants derived from the Virgo infall galaxies and other methods (Kim et al. 2020).

virial radius and that the fraction of dark matter between the virial radius and the zero-velocity radius is negligible (Karachentsev et al. 2018; Kashibadze et al. 2020).

In addition, the velocity dispersion of the infalling galaxies is estimated at $\sigma_0 = 62 \pm 14$ km s^{-1}. This is even smaller than the value given by Karachentsev et al. (2018), $\sigma_0 = (78\text{–}90)$ km s^{-1}. This confirms that the Hubble flow around the Virgo Cluster is cold (Kashibadze & Karachentsev 2018; Karachentsev et al. 2018).

3.3. *Dark Galaxies?*

Dark galaxies are isolated HI systems with barely visible optical counterparts. They are usually found in HI surveys (Cannon et al. 2015; Ball et al. 2018, and references therein). They can be either real galaxies or simply gas fragments in the Milky Way. Their origin is not clear. Some of them are associated with a low surface brightness (LSB) stellar system, which provides a critical clue to understand their origin. It is not easy to determine a distance to these LSB systems. If we know their distances, we can derive their physical properties and tell whether they are galaxies or gas clouds in the Milky Way. The TRGB is an excellent tool to measure the distances to these dark galaxies with LSB counterparts and investigate the nature of dark galaxies.

Coma P is an enigmatic example of an almost dark galaxy: an LSB optical counterpart is located at the location of AGC 229385, a large HI system with $v = 1348$ km s^{-1} (Janowiecki et al. 2015; Ball et al. 2018). Figure 14 shows an HI contour map and a color map based on *HST*/Advanced Camera for Surveys (ACS) F606W/F814W images of AGC 229385/Coma P (Ball et al. 2018).

TRGB distances to Coma P derived from the same *HST*/ACS images have been controversial, often showing large differences (Anand et al. 2018; Brunker et al. 2019; Tikhonov & Galazutdinova 2020). Figure 15 displays the CMDs of the detected stars in the *HST*/ACS F606W/F814W images of Coma P from different authors. Anand et al.

Figure 14. (left) HI map of AGC 229385/Coma P (red contours). (right) Color map of the *HST*/ACS images of Coma P. White contours represent the number density of detected stars in the *HST* images of Coma P, clearly showing the existence of a galaxy (Ball et al. 2018).

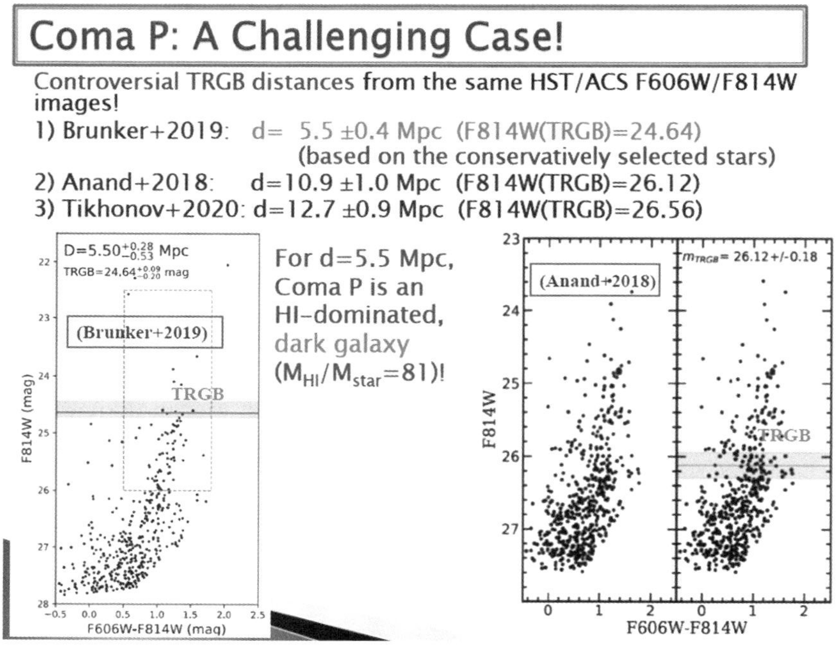

Figure 15. CMDs of the detected stars in the *HST*/ACS images of Coma P from Brunker et al. (2019) (left) and Anand et al. (2018). Note that the loci of the TRGB in the two CMDs are significantly different.

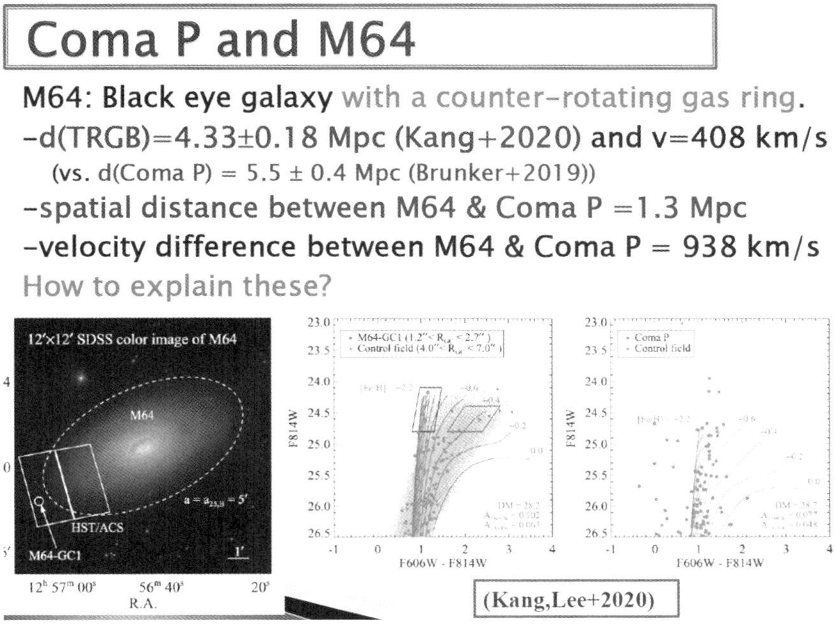

Figure 16. Relation between Coma P and M64. (left) SDSS color image of M64. The boxes denote the *HST*/ACS field. (middle) CMD of the detected stars in the *HST* images of M64 (gray dots). Magenta circles represent stars in the globular cluster M64-GC1 and blue circles are stars in the control field around the globular cluster. Green lines denote the PARSEC isochrones for an age of 12 Gyr and [Fe/H] $= -2.2$ to 0.0 dex (Bressan et al. 2012). (right) CMD of the detected stars in the *HST*/ACS images of Coma P (magenta circles). (Kang et al. 2020).

(2018) derived F814W(TRGB) $= 26.12$ mag and $d = 10.9 \pm 1.0$ Mpc. Then, selecting their sample stars more conservatively, Brunker et al. (2019) suggested that the TRGB magnitude is 1.5 mag brighter than the Anand et al. (2018) value, F814W(TRGB) $= 24.64$ mag, yielding a much shorter distance, $d = 5.5 \pm 0.4$ Mpc. Later, Tikhonov & Galazutdinova (2020) adopted the outer region of Coma P to measure the TRGB, obtaining an 0.4 mag fainter value than Anand et al. (2018), F814W(TRGB) $= 26.56$ mag (and $d = 12.7 \pm 0.9$ Mpc). Visual inspection of the CMDs in these references indicates that the Brunker et al. (2019) value may be the most persuasive. However, better *HST* data are needed to resolve this problem. These results show that Coma P is a difficult example for TRGB application.

For the distance to Coma P of Brunker et al. (2019), the ratio of the HI and stellar mass is derived as large as 81, showing that Coma P is indeed an HI-dominated dark galaxy. Coma P is located in a sparse field, and the nearest galaxy from Coma P in the sky is M64 (the Black Eye galaxy), an early-type SA spiral galaxy. Figure 16 displays the Sloan Digital Sky Survey (SDSS) color image of M64, where the position of one *HST*/ACS field is marked on the eastern side of M64. One interesting feature of M64 is the presence of a counter-rotating HI gas ring.

Kang et al. (2020) discovered one bright globular cluster, M64-GC1, in the *HST* field of M64 and showed that it is a metal-poor globular cluster in the halo of M64 based on the CMD of the detected globular cluster stars (see Figure 16, middle). They compared the CMDs of the resolved stars in M64-GC1 and Coma P (see Figure 16, right), showing that the mean metallicity of Coma P is [Fe/H] ≈ -0.9 dex, which is slightly higher than that of M64-GC1, [Fe/H] $= -1.5 \pm 0.2$ dex.

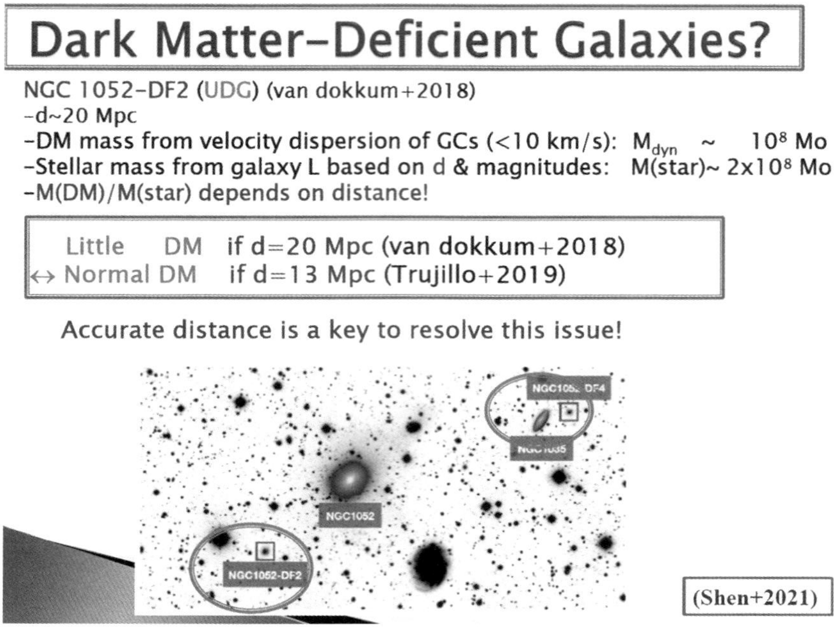

Figure 17. Locations of NGC 1052-DF2 and NGC 1052-DF4, both members of the NGC 1052 group (Shen et al. 2021).

Kang et al. (2020) derived a TRGB distance to M64 ($d = 4.44 \pm 0.18$ Mpc), showing that it is at the spatial distance of 1.3 Mpc from Coma P. However, the relative velocity of Coma P with respect to M64 is as large as $\Delta v = 938$ km s^{-1}. This can be explained in terms of a high-speed flyby scenario: a gas-rich massive progenitor of Coma P encountered M64 at high speed about one giga-year ago, and some of the gas removed from Coma P resulted in the counter-rotating gas ring in the outer disk of M64 (Brunker et al. 2019; Kang et al. 2020).

3.4. *Dark Matter-free Galaxies?*

The current paradigm for galaxy formation models is based on the assumption that the baryonic components of galaxies are embedded in much larger and more massive dark matter halos. Are there any galaxies whose dark matter fraction is significantly lower than for normal galaxies? If so, they must be large galaxies with very low surface brightness (also dubbed 'ultra-diffuse galaxies', UDGs). They are called dark matter-free (or dark matter-deficient) galaxies. NGC 1052-DF2 ([KKK2000]04), a UDG in the NGC 1052 group, is a strong candidate for such dark matter-free galaxies. Figure 17 shows the location of NGC 1052-DF2 in the NGC 1052 group (Shen et al. 2021).

van Dokkum et al. (2018) estimated the dynamical mass of NGC 1052-DF2, from the kinematics of its globular clusters, $M_{\rm dyn} \sim 10^8 {\rm M}_\odot$. They also derived a twice larger stellar mass for this galaxy, $M_* \sim 2 \times 10^8 {\rm M}_\odot$, adopting a distance of 20 Mpc. Based on these results, they suggested that NGC 1052-DF2 may be a dark matter-deficient galaxy. Later, Trujillo et al. (2019) derived a much closer distance to this galaxy, $d = 13$ Mpc, thus claiming that NGC 1052-DF2 is a normal galaxy. However, the *HST* images used for these studies are not deep enough to derive a reliable distance to this galaxy. Access to an accurate distance is key to resolving this issue. The TRGB is an excellent tool to measure the distance to these LSB galaxies and tell us whether or not they are dark matter-deficient.

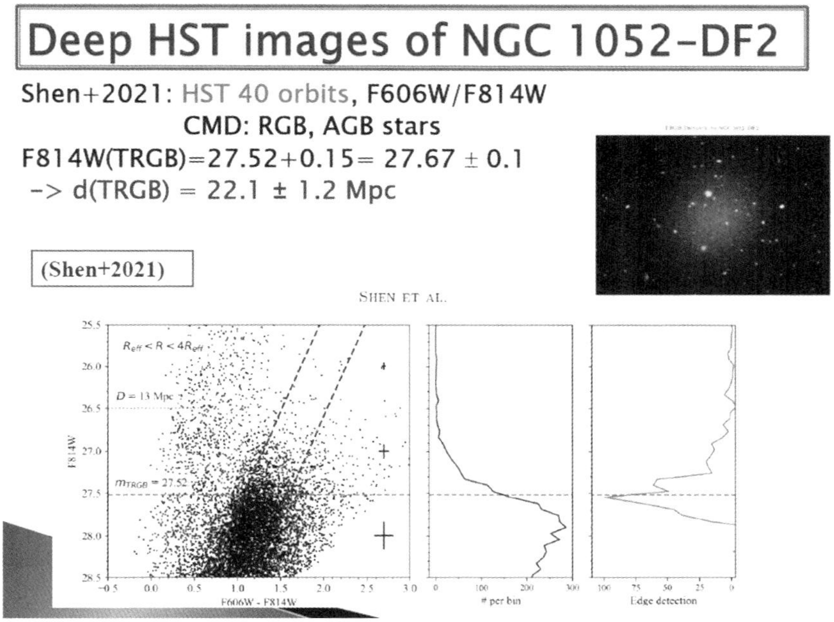

Figure 18. (top) Color map of the *HST* images of NGC 1052-DF2. (bottom) CMD, LF, and edge detector response of the detected stars in NGC 1052-DF2. The red dashed line denotes the locus of the TRGB (Shen et al. 2021).

Later, using much deeper *HST*/WFC3 F606W/F814W images, Shen et al. (2021) obtained much better photometry of the resolved stars in NGC 1052-DF2 and measured the TRGB magnitude and distance, F814W(TRGB) $= 27.67 \pm 0.1$ mag and $d = 22.1 \pm 1.2$. Figure 18 displays the CMD of the detected stars in NGC 1052-DF2, the LF of the selected RGB stars, and the corresponding edge detector response (Shen et al. 2021). The TRGB distance derived from the deeper *HST* images is consistent with the result of van Dokkum et al. (2018), showing that NGC 1052-DF2 is indeed a dark matter-deficient galaxy. Similarly, Danieli et al. (2020) derived a TRGB distance to NGC 1052-DF4, $d = 20. \pm 1.6$ Mpc, showing that it is another member of the NGC 1052 group and that it is also a dark matter-deficient galaxy. The presence of these dark matter-deficient galaxies can be explained by high-speed collisions of two gas-rich dwarfs or tidal stripping (Silk 2019; Lee et al. 2021; van Dokkum et al. 2022).

4. Summary and Outlook

The TRGB is the best standard candle for the cosmology of the local Universe. The accuracy of the TRGB continues to be improved. I have introduced four TRGB science cases, from large to small scales. However, they are only a small fraction of the possible science cases, and there are many more interesting cosmological projects one can do with the TRGB.

The *HST* has been a major workhorse for TRGB science since 1990. In 2022, the *JWST* started science observations, opening up a new era in near/mid-infrared high resolution imaging. The Vera Rubin Observatory/Legacy Survey of Space and Time will start operations soon from the ground, and the Nancy Roman Space Telescope will provide the wide-field, high-resolution space-borne imaging in a few years' time. Better spatial resolution, infrared coverage, and the wider fields of current and future telescopes will expand the TRGB reach from 30 Mpc to 100 Mpc, and also expand the TRGB sky coverage, filling a significant fraction of the sky. Let us enjoy TRGB cosmology in the forthcoming years!

Acknowledgements. Thanks to In Sung Jang for preparing some figures. This work was supported by a National Research Foundation grant funded by the Korean Government (NRF-2019R1A2C2084019).

References

Anand, G. S., Tully, R. B., Karachentsev, I. D., Makarov, D. I., Makarova, L., Rizzi, L., Shaya, Edward J. 2018, ApJL, 861, L6. doi:10.3847/2041-8213/aacc2b
Anand, G. S., et al. 2021, AJ, 162, 80. doi:10.3847/1538-3881/ac0440
Anand, G. S., Tully, R. B., Rizzi, L., Riess, A. G.; Yuan, W. 2022, ApJ, 932, 15. doi:10.3847/1538-4357/ac68df
Baade, W. 1944, ApJ, 100, 137. doi:10.1086/144650
Ball, C., et al. 2018, AJ, 155, 65. doi:10.3847/1538-3881/aaa156
Beaton, R. L., et al. 2019, Astronomical Distance Determination in the Space Age. Series: Space Sciences Series of ISSI, 89. doi:10.1007/978-94-024-1631-2_4
Bellazzini, M. 2008, Mem. Soc. Astron. It., 79, 440. doi:10.48550/arXiv.0711.2016
Bressan, A., Marigo, P., Girardi, L., Salasnich, Bernardo; Dal Cero, C;, Rubele, S., Nanni, A. 2012, MNRAS, 427, 127. doi:10.1111/j.1365-2966.2012.21948.x
Brunker, et al. 2019, AJ, 157, 76. doi:10.3847/1538-3881/aafb39
Cannon, J. M., et al. 2015, AJ, 149, 72. doi:10.1088/0004-6256/149/2/720049/198/1/6
Danieli, S., van Dokkum, P., Abraham, R., Conroy, C., Dolphin, A. E., Romanowsky, A. J. 2020, ApJL, 895, L4. doi:10.3847/2041-8213/ab8dc4
Dixon, M., Mould, J., Flynn, C., et al. 2023, MNRAS, 523, 2283. doi:10.1093/mnras/stad1500
Freedman, W. L., Hatt, D., et al. 2019, ApJ, 882, 34. doi:10.3847/1538-4357/ab2f73
Freedman, W. L., et al. 2020, ApJ, 891, 57. doi:10.3847/1538-4357/ab7339
Freedman, W. L. 2021, ApJ, 919, 16. doi:10.3847/1538-4357/ac0e95
Hatt, D., et al. 2017, ApJ, 845, 146. doi:10.3847/1538-4357/aa7f73
Hoyt, T. J. 2023, Nat. Astron., 7, 590. doi:10.1038/s41550-023-01913-1
Jang, I. S. & Lee, M. G. 2017a, ApJ, 835, 28. doi:10.3847/1538-4357/835/1/28
Jang, I. S. & Lee, M. G. 2017b, ApJ, 836, 74. doi:10.3847/1538-4357/836/1/74
Jang, I. S., et al. 2021, ApJ, 906, 125. doi:10.3847/1538-4357/abc8e9
Janowiecki, S., et al. 2015, ApJ, 801, 96
Kang, J., Kim, Y. J., Lee, M. G., Jang, I. S. 2020, ApJ, 897, 106. doi:10.3847/1538-4357/ab94ba
Karachentsev, I. D., Makarova, L. N., Tully, R. B., Rizzi, L., Shaya, E. J. 2018, ApJ, 858, 62. doi:10.3847/1538-4357/aabaf1
Kashibadze, O. G. & Karachentsev, I. D. 2018, A&A, 609, A11. doi:10.1051/0004-6361/201731645
Kashibadze, O. G., Karachentsev, I. D., & Karachentseva, V. E. 2020, A&A, 635, A135. doi:10.1051/0004-6361/201936172
Kim, Y. J., Kang, J., Lee, M. G., Jang, I. S. 2020, ApJ, 905, 104. doi:10.3847/1538-4357/abbd97
Lee, M. G. 1993, ApJ, 408, 409. doi:10.1086/172598
Lee, M. G., Freedman, W. L., & Madore, B. F. 1993, ApJ, 417, 553. doi:10.1086/173334
Lee, J., Shin, E.-. jin., & Kim, J.-h. 2021, ApJL, 917, L15. doi:10.3847/2041-8213/ac16e0
Li, S., Casertano, S., & Riess, A. G. 2022, ApJ, 939, 96. doi:10.3847/1538-4357/ac7559
Madore, B. F., Mager, V., & Freedman, W. L. 2009, ApJ, 690, 389. doi:10.1088/0004-637X/690/1/389
Madore, B. F., Freedman, W. L., Owens, K. A., Jang, I. S. 2023, AJ, 166, 2. doi:10.3847/1538-3881/acd3f3
Makarov, D., Makarova, L., Rizzi, L., Tully, R. Brent; Dolphin, A. E., Sakai, S., Shaya, E. J. 2006, AJ, 132, 2729. doi:10.1086/508925
McQuinn, K. B. W., et al. 2014, ApJ, 785, 3. doi:10.1088/0004-637X/785/1/3
McQuinn, K. B. W., Skillman, E. D., Dolphin, A. E., Berg, D., Kennicutt, R. 2017, AJ, 154, 51. doi:10.3847/1538-3881/aa7aad

Méndez, B., Davis, M., Moustakas, J., Newman, J., Madore, B. F., Freedman, W. L. 2002, AJ, 124, 213. doi:10.1086/341168

Mould, J. R., Kristian, J., & Da Costa, G. S. 1983, ApJ, 270, 471. doi:10.1086/161141

Mould, J., Clementini, G., & Da Costa, G. 2019, PASA, 36, e001. doi:10.1017/pasa.2018.46

Peirani, S., & de Freitas Pacheco, J. A. 2006, New Astron., 11, 325

Peirani, S., & de Freitas Pacheco, J. A. 2008, A&A, 488, 845

Perivolaropoulos, L. & Skara, F. 2022, New Astron. Rev., 95, 101659. doi:10.1016/j.newar.2022.101659

Riess, A. G., Casertano, S., Yuan, W., Bowers, J. B., Macri, L., Zinn, J. C., Scolnic, D. 2021, ApJL, 908, L6. doi:10.3847/2041-8213/abdbaf

Riess, A. G., et al. 2022, ApJL, 934, L7. doi:10.3847/2041-8213/ac5c5b

Rizzi, L., Tully, R. B., Makarov, D., Makarova, Lidia; Dolphin, A. E., Sakai, S., Shaya, E. J. 2007, ApJ, 661, 815. doi:10.1086/516566

Salaris, M. & Cassisi, S. 1997, MNRAS, 289, 406. doi:10.1093/mnras/289.2.406

Serenelli, A., Weiss, A., Cassisi, S., Salaris, M., Pietrinferni, A. 2017, A&A, 606, A33. doi:10.1051/0004-6361/201731004

Shen, Z., et al. 2021, ApJL, 914, L12. doi:10.3847/2041-8213/ac0335

Silk, J. 2019, MNRAS, 488, L24. doi:10.1093/mnrasl/slz090

Soltis, J., Casertano, S., & Riess, A. G. 2021, ApJL, 908, L5. doi:10.3847/2041-8213/abdbad

Tammann, G. A., Sandage, A., & Reindl, B. 2008, ApJ, 679, 52. doi:10.1086/529508

Tikhonov, N. A. & Galazutdinova, O. A. 2020, Astron. Lett., 45, 750. doi:10.1134/S1063773719110069

Trujillo, I., et al. 2019, MNRAS, 486, 1192. doi:10.1093/mnras/stz771

Tully, R. B., et al. 2023, ApJ, 944, 94. doi:10.3847/1538-4357/ac94d8

van Dokkum, P., et al. 2018, ApJL, 864, L18. doi:10.3847/2041-8213/aada4d

van Dokkum, P., et al. 2022, Nature, 605, 435. doi:10.1038/s41586-022-04665-6

At the crossroads of astrophysics and cosmology: Period–luminosity relations in the 2020s
Proceedings IAU Symposium No. 376, 2024
R. de Grijs, P. A. Whitelock & M. Catelan, eds.
doi:10.1017/S1743921323002557

Period–Luminosity Relations in the Local Group of Galaxies

Igor Soszyński

Astronomical Observatory, University of Warsaw, Al. Ujazdowskie 4, 00-478 Warszawa, Poland. email: soszynsk@astrouw.edu.pl

Abstract. Local Group galaxies, particularly the Large and Small Magellanic Clouds, have historically played and continue to play a unique role in studies of the period–luminosity (PL), period–luminosity–color (PLC), and period–Wesenheit (PW) relations, not just for pulsating stars. In recent years, significant efforts have been devoted to calibrate the PL, PLC, and PW relationships at different wavelengths, including studies of the influence of metallicity and non-linearities on the accuracy of measured distances. However, the PL diagram has many more astrophysical applications. It serves as a vital tool for classifying different types of pulsating stars and can even facilitate the discovery of new classes of variable stars. Moreover, it aids in distinguishing among various modes of pulsation, facilitates the identification of pulsating stars that are members of binary systems, and enables studies of the three-dimensional structures of neighboring galaxies. In this contribution, I present the latest results on the PL, PLC, and PW relations obeyed by various types of variable stars in Local Group galaxies – from δ Scuti stars to Mira variabless and from close binary systems to the mysterious long secondary periods exhibited by red giant and supergiant stars.

Keywords. Cepheids, RR Lyrae stars, delta Scuti stars, Miras, binaries, stars: distances, galaxies: structure, Local Group, Magellanic Clouds

1. Introduction

Period–luminosity (PL), period–luminosity–color (PLC), and period–Wesenheit index (PW) relationships for variable stars have found many astrophysical applications, of which the most well-known is measuring distances within and beyond our Galaxy. PL, PLC, and PW relations are exhibited by pulsating stars in the Cepheid instability strip (classical Cepheids, Type II Cepheids, anomalous Cepheids, RR Lyrae stars, and δ Scuti stars), long-period variables (including Miras and semiregular variables), close binary systems, and the mysterious long secondary period variables.

The first ever PL relation was discovered in 1908 by Henrietta Swan Leavitt who studied Cepheid variables in the Small Magellanic Cloud (SMC). The Magellanic Clouds are galaxies that are located close enough so that individual stars can be easily resolved by ground-based telescopes, but far enough away that stellar populations can be treated, to first approximation, as being located at the same distance. This allows us to obtain the apparent PL relation without measuring the distances to individual Cepheids, or even without knowing the distance to their host galaxy. To date, the Magellanic Clouds, and more broadly the Local Group of galaxies, remain among the most important sources of knowledge providers about the properties of variable stars, particularly about their PL, PLC, and PW relations.

The Large Magellanic Cloud (LMC) is the nearest galaxy containing a significant stellar population, which makes it indispensable for cross-calibrating various distance indicators.

This galaxy has a very favorable position relative to us. It is located nearly face-on, which translates into a compact distribution (±1.5 kpc) of distances across the face of the galaxy (e.g., Jacyszyn-Dobrzeniecka et al. 2016). The distance to the LMC (49.59 ± 0.54 kpc) is already known with an accuracy of about 1 percent, thanks to measurements made by the Araucaria project (Pietrzyński et al. 2019) using late-type detached eclipsing binary systems. Furthermore, interstellar reddening toward the Magellanic Clouds is very accurately known thanks to the reddening maps obtained by Skowron et al. (2021) based on red clump stars.

On the contrary, the SMC exhibits a significant elongation along the line of sight, resulting in a substantial dispersion of the PL relations in this galaxy. The distance to the SMC has been determined with an accuracy of better than 2 percent based on detached eclipsing binaries (Graczyk et al. 2020). However, the SMC is characterized by significantly lower metallicity than the LMC, which, among other things, allows us to study of the influence of chemical composition on the slope and intercept of the PL relations. Considering other large galaxies of the Local Group – M31 and M33 – extends the range of investigated metallicities to values higher than those commonly found in the Milky Way.

In this contribution, various aspects of the PL, PLC, and PW relations obeyed by different types of variable stars in the galaxies of the Local Group will be presented. This work focuses not only on classical Cepheids and not only on distance measurements but also demonstrates a wide range of applications of the PL relationships in astrophysics. Examples of the latest studies of neighboring galaxies that allow us to learn about the properties of variable stars will be presented, as well as examples of how the PL relationships allow us to learn about the structure of nearby galaxies. This article is divided into sections describing various types of variable stars that follow PL relations. All figures showing the PL and PW diagrams and example light curves of variable stars have been generated using photometric data gathered by the Optical Gravitational Lensing Experiment (OGLE) survey (Udalski et al. 2015a).

2. Classical Cepheids

The PL relation for classical Cepheids was, for the first time, mentioned by Leavitt (1908) in her paper describing the catalog of variable stars in the Magellanic Clouds. "It is worthy of notice that in Table IV the brighter variables have the longer periods" – this sentence describing the properties of 16 variables in the SMC is considered an announcement of the discovery of the PL relation for Cepheids, today known as the 'Leavitt Law'. The PL relation was first plotted on a graph in the paper by Leavitt & Pickering (1912). Since then, efforts have been made to calibrate the Leavitt Law to absolute magnitudes (e.g., Hertzsprung 1913; Shapley 1918) in order to use Cepheids as standard candles for measuring distances. A decade later, this led to the measurement of the distances to M31 and M33 (Hubble 1925), ultimately confirming the hypothesis of the Island Universe, and then to the discovery of the expansion of the Universe (Hubble 1929).

Over the next decades, the Leavitt Law was calibrated increasingly accurately, with the most significant revision of the distance scale made by Baade (1956). Until the 1990s, most PL diagrams for classical Cepheids showed only one relationship, populated by the fundamental-mode pulsators (e.g., Sandage & Tammann 1968; Madore 1982). The physical nature of so-called sinusoidal or s-Cepheids was subject of debate for years (e.g., Arp 1960; Ivanov & Nikolov 1976; Connolly 1980). A breakthrough came in the 1990s with the launch of large-scale sky surveys aimed primarily at searching for gravitational microlensing events, but their side products were huge photometric databases ideally suited for large-scale studies of variable stars. After several years of observations, the EROS (Beaulieu et al. 1995), MACHO (Alcock et al. 1995) and OGLE (Udalski et al.

1999) projects presented PL and PW diagrams for classical Cepheids in the Magellanic Clouds. Two parallel relationships were clearly visible, finally confirming that s-Cepheids are first-overtone pulsators.

The top panel of Figure 1 displays the modern PW diagram for classical Cepheids in the LMC. Four relationships can be distinguished, for stars pulsating in the fundamental and the first three overtone modes (the third overtone occurs only in multi-mode pulsators). This diagram was obtained based on long-term observations in the I and V bands carried out by the OGLE project – the only still active sky survey of those pioneering microlensing projects which started over 30 years ago. Throughout its long history, OGLE has evolved and currently, in the fourth phase of the project (OGLE-IV), regularly observes about two billion stars in the densest regions of the southern sky: the Galactic bulge and disk, the LMC and SMC, in total about 3600 square degrees across the sky. Since 1997, the survey has been conducted with the dedicated 1.3 m Warsaw Telescope located at Las Campanas Observatory in Chile.

Variable stars occupy a special position among the most important achievements of the OGLE project. The OGLE Collection of Variable Stars (OCVS) currently contains more than a million variable objects, mainly pulsating stars and eclipsing binaries. Among others, this catalog contains over 10,000 Cepheids of various types in the Magellanic Clouds (Soszyński et al. 2018, 2019b), which is a virtually complete list of such objects in these galaxies. The OCVS is characterized by a very high level of completeness and purity, since the final decision on the classification of each variable star is made by an experienced astronomer after visual inspection of the light curves. Therefore, the OCVS serves as a framework for various astrophysical studies, including many papers presented in these proceedings.

2.1. *Classical Cepheids as Standard Candles*

This presentation of the latest results pertaining to classical Cepheids will begin with their canonical role as distance indicators. Since the LMC is an anchor galaxy for the cosmic distance scale, the Leavitt Law in this galaxy has been studied over a wide range of wavelengths, from the optical to the mid-infrared domain (e.g., Ngeow et al. 2009; Macri et al. 2015; Chown et al. 2021).

One of the most remarkable calibrations of the Leavitt Law in recent years was presented by Riess et al. (2019). They used observations from the *Hubble Space Telescope* (*HST*) to study the PL and PW relations followed by 70 long-period Cepheids in the LMC. This was a significant achievement, because the same photometric system had been used earlier to measure distances to galaxies hosting Type Ia supernovae, resulting in a more precise determination of the Hubble constant, H_0. Calibrating LMC Cepheids with *HST* observations was a major research challenge due to the *HST* camera's small field of view ($2'$), implying that only a single Cepheid could typically be observed per image. The authors overcame this limitation by designing a new observation mode that used the *HST* gyroscopes for both slewing and guiding. Additionally, the *HST* has faced significant technical problems related to its gyroscopes, which made the observations even more challenging.

The Riess et al. (2019) calibration of the Leavitt Law in the *HST* passbands has already been employed to measure the distances to the Local Group galaxies M31 (Li et al. 2021) and M33 (Breuval et al. 2023) with unprecedented accuracy of 1.5% and 1.3%, respectively. Both galaxies are among the best sites for cross-calibrating various distance indicators, making them strong candidates as new anchors for the cosmic distance scale.

Achieving accurate calibrations of the PL relations for Cepheids, which is essential for measuring the Hubble constant with an accuracy of better than 1%, requires a

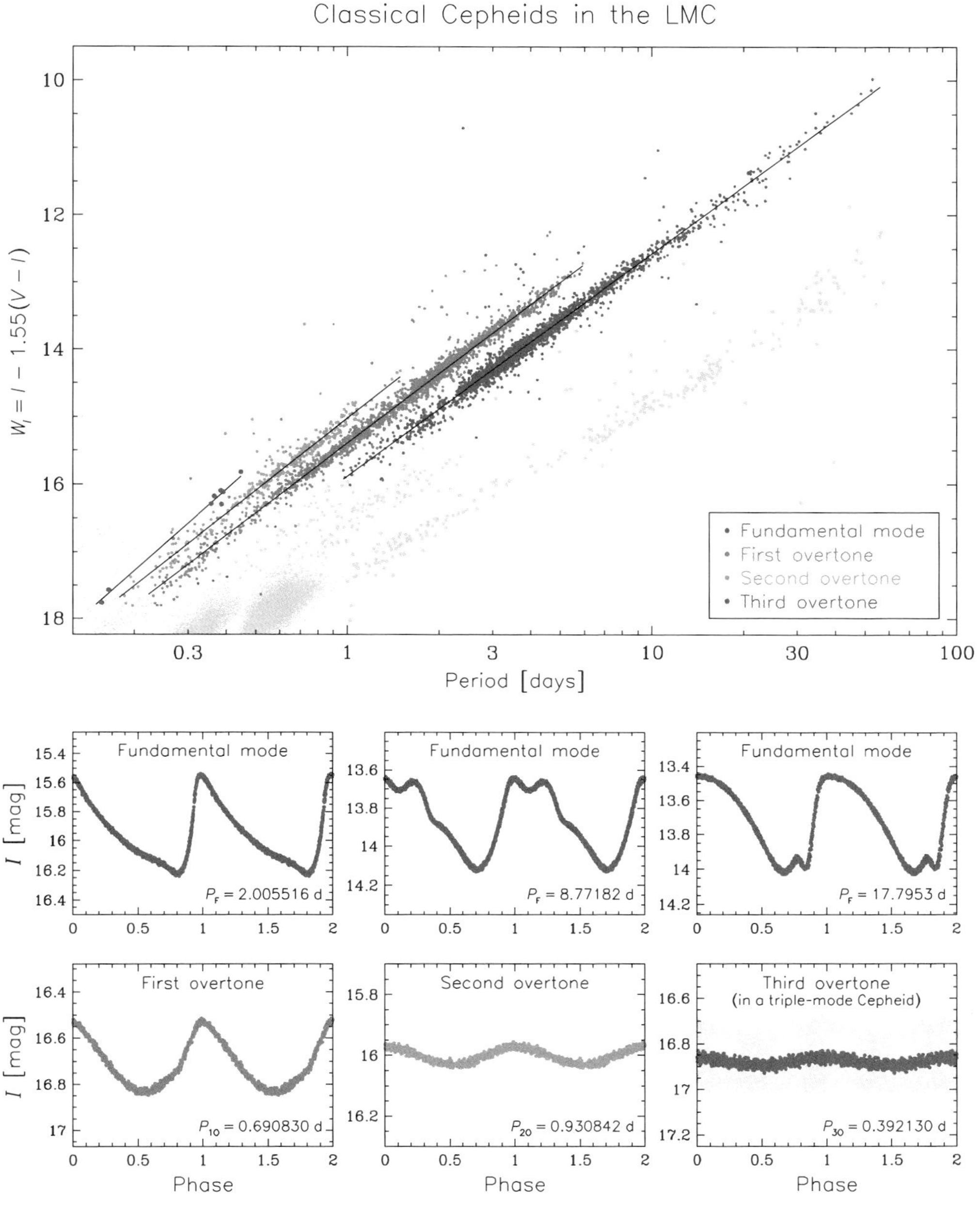

Figure 1. (top) Period–Wesenheit index diagram for classical Cepheids in the LMC (Soszyński et al. 2017, 2019b). The Wesenheit index is an extinction-free quantity defined as $W_I = I - 1.55(V - I)$, where the I- and V-band mean magnitudes of the Cepheids originate from the OGLE-IV survey (Udalski et al. 2015a). Blue, green, orange, and red dots mark the fundamental, first-, second-, and third-overtone pulsation modes, respectively, in both single-mode and multimode Cepheids. Gray points indicate other types of classical pulsators: Type II Cepheids, anomalous Cepheids, RR Lyrae stars, and δ Scuti stars. (bottom) Typical I-band light curves of classical Cepheids pulsating in different modes. The third overtone is only observed in double- and triple-mode Cepheids.

comprehensive understanding of potential systematic errors. These errors may arise from factors such as non-linearities in the PL relations or their dependence on metallicity.

The non-linearity of the Cepheid PL relation has been the subject of numerous studies over the past years (e.g., Kanbur et al. 2007; Ngeow et al. 2008; García-Varela et al. 2013). For instance, Bhardwaj et al. (2016) performed a statistical analysis of potential non-linearities in the PL, PW, and period–color relations obeyed by LMC classical Cepheids in optical and infrared bands. They applied various statistical tests to evaluate the significance of the non-linearities and found a small but statistically significant change in the slope of the PL and PW relations at $P = 10$ d for fundamental-mode Cepheids and at $P = 2.5$ d for first-overtone pulsators. The near-infrared period–W_{HK} relation, however, was found to be perfectly linear for both fundamental-mode and first-overtone Cepheids. Generally, the impact of possible non-linearities in the PL relations on extragalactic distance determinations and H_0 measurements was estimated to be marginal.

In addition, the chemical abundance of Cepheids has a significant contribution to the error budget of the Hubble constant. Therefore, in recent years, a lot of effort has been put into solving this problem (e.g., Wielgórski et al. 2017; Gieren et al. 2018; Groenewegen 2018; Ripepi et al. 2020; Breuval et al. 2021). The metallicity–luminosity relation for Cepheids is taken into account by including a corrective term, γ, in the PL relation:

$$M_\lambda = \alpha \log P + \beta + \gamma[\mathrm{Fe/H}]. \tag{1}$$

Various estimates of the γ values have been published in recent years. At present, most authors agree that this coefficient is negative, which means that metal-rich Cepheids are brighter than metal-poor ones. However, the dispersion in the measured values of the γ term is still significant, indicating that even after decades of research the metallicity effect on the Cepheid PL relation is not well-understood.

In their recent work, Breuval et al. (2022) compared the absolute PL relations for classical Cepheids in the Milky Way, LMC, and SMC, with the aim of deriving the metallicity effect in 10 filters covering the wavelength range of 0.5–4.5 μm. The authors took advantage of the fact that the iron abundance, [Fe/H], of young stellar populations in these three Local Group galaxies differs significantly, ranging from -0.75 to $+0.088$ dex. As a result, a uniform value of the γ term of about -0.28 mag dex^{-1} from optical to mid-infrared wavelengths was obtained.

2.2. *Cepheids in Binary Systems*

The PL relations for Cepheids in the Local Group are used for more than just calibrating the extragalactic distance scale. An original application of the PL diagram was proposed by Pilecki et al. (2021), who selected 41 classical Cepheids in the LMC lying at least 0.44 mag above the PL relation corresponding to their pulsation mode. In other words, they analyzed Cepheids that were at least 50% brighter than other Cepheids at the same pulsation period. The project aimed to extend the list of known classical Cepheids that are members of double-lined binaries beyond the eclipsing systems containing Cepheids discovered by the OGLE project (Pietrzyński et al. 2010; Udalski et al. 2015b).

Preliminary spectroscopic observations of 18 Cepheids from this overluminous sample revealed 16 double-lined binaries (Pilecki et al. 2021). Further observations led to the discovery of a Cepheid in a binary system with an orbital period of only 59 days, five times shorter than any such period measured before (Pilecki et al. 2022). Currently, 56 candidate Cepheids in the LMC, SMC, and Milky Way have been confirmed to reside in double-lined systems (B. Pilecki, private communication; see also these proceedings). The orbital motions of the Cepheids and their companions were detected in 32 of these systems. Moreover, eight of the confirmed systems are composed of two Cepheids. Such binary systems are excellent laboratories for testing evolutionary models of classical pulsators.

2.3. *Three-Dimensional Structure of Galaxies*

Studies of the three-dimensional geometry of Local Group galaxies, such as the LMC and SMC, are crucial to understand past interactions between star systems in the immediate vicinity of the Milky Way. Many investigations of the Magellanic System's spatial structure have been performed based on the PL relations of classical Cepheids (e.g., Inno et al. 2016; Jacyszyn-Dobrzeniecka et al. 2016; Ripepi et al. 2017; Deb et al. 2018). Here, I present the newest research by Ripepi et al. (2022), who used the OGLE Collection of Cepheids in the LMC (Soszyński et al. 2017) with time-series near-infrared photometry obtained by the VISTA Survey of the Magellanic Clouds (VMC; Cioni et al. 2011) to study the three-dimensional geometry of the young population in the LMC.

The near-infrared PL and PW relations obtained by Ripepi et al. (2022) are spectacularly narrow thanks to the deep VMC photometry and relatively good sampling of the light curves (typically 15–30 epochs). Thanks to the small dispersions in the PL relations, the distances to individual Cepheids were determined with a precision of about 1 kpc (2%). As a result, the authors presented accurate three-dimensional maps of the LMC, analyzed various substructures within this galaxy, measured the scale height of the disk, and derived the viewing angles of the LMC bar and disk. Additionally, Ripepi et al. (2022) reported the discovery of a new non-linearity in the PL relationship for first-overtone Cepheids with a break at $P = 0.58 \pm 0.1$ d, but the same nonlinearity (at $P = 0.5$ d) was earlier noticed by Soszyński et al. (2008a).

3. Type II Cepheids

Until the mid-20th century, astronomers were unaware of the existence of more than one type of Cepheid variable. The major revision of the distance scale made by Baade (1952) directly led the division of Cepheids into classical and Type II objects. The latter group is composed of low-mass stars belonging to the halo and old disk populations. Traditionally, Type II Cepheids are classified into BL Herculis (with pulsation periods in the range of 1–4 d), W Virginis (4–20 d), and RV Tauri stars (> 20 d). Additionally, Soszyński et al. (2008b) proposed a new subtype, dubbed peculiar W Virginis stars, which have light curves of different shapes and are bluer and brighter than ordinary W Virginis stars. A large fraction of peculiar W Virginis variables show signs of binarity, such as eclipsing or ellipsoidal modulation of their light curves. Therefore, it is probable that the pulsation of these stars is a consequence of their evolution in close binary systems.

3.1. *PL Relation for Type II Cepheids*

Type II Cepheids exhibit well-defined PL and PW relationships (Figure 2). These stars are 1.5–2 mag fainter than classical Cepheids at similar periods, but they are 1–4 mag brighter than RR Lyrae stars, which makes them useful distance indicators in environments where classical Cepheids are not present, such as globular clusters or dwarf spheroidal galaxies.

The most recent calibration of the PL relation for Type II Cepheids in the Milky Way and the LMC was published by Wielgórski et al. (2022). In the Milky Way, they used about 20 nearby BL Herculis and W Virginis stars along with their parallaxes from *Gaia* Data Release (DR) 3, while in the LMC, the sample of Type II Cepheids was

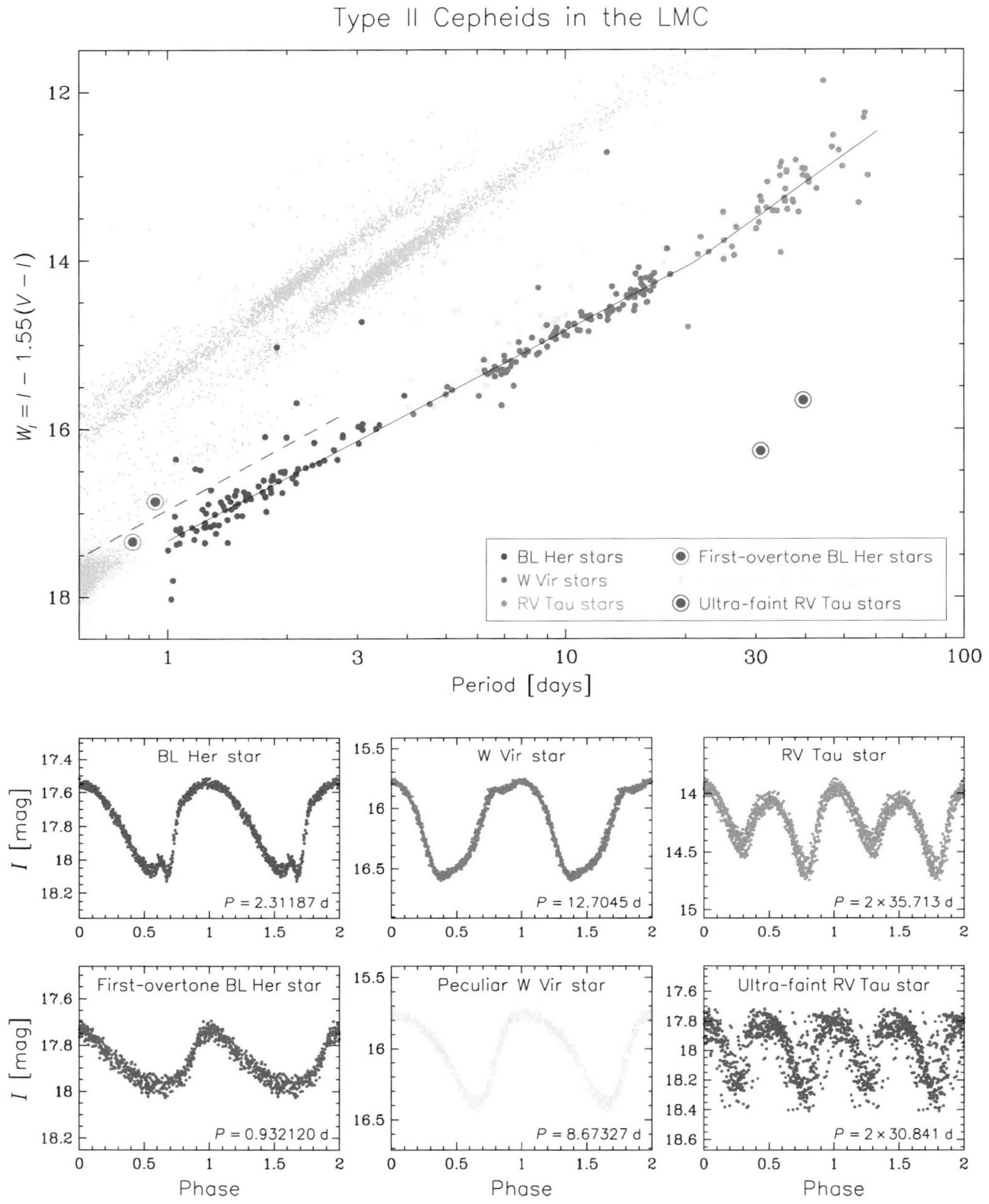

Figure 2. (top) PW diagram for Type II Cepheids in the LMC (Soszyński et al. 2018). Different types of classical pulsators are color coded: blue for fundamental-mode BL Herculis stars, green for W Virginis stars, orange for RV Tauri stars, purple for first-overtone BL Herculis stars (Soszyński et al. 2019a), cyan for peculiar W Virginis stars (Soszyński et al. 2008b), red for ultra-faint RV Tauri stars (Soszyński et al. 2018), and gray for other types of classical pulsators. (bottom) Example I-band light curves of different classes of Type II Cepheids.

taken from the OCVS (Soszyński et al. 2018). The slopes of the near-infrared relations in the Milky Way and LMC were found to be slightly different, at the level of 2σ. Additionally, Wielgórski et al. (2022) noticed a significant metallicity effect of about -0.2 mag dex^{-1}, indicating that metal-rich Type II Cepheids are intrinsically brighter than their metal-poor counterparts. Interestingly, the sign of this relation is opposite to that of RR Lyrae stars (e.g., Neeley et al. 2019), which may indicate that the common PL relation obeyed by RR Lyrae variables and Type II Cepheids in the LMC (Bhardwaj et al. 2017) is not a universal feature.

3.2. *First-Overtone Type II Cepheids*

For years, all Type II Cepheids were believed to be exclusively fundamental-mode pulsators, until Smolec et al. (2018) identified the first two cases of BL Herculis stars pulsating simultaneously in the fundamental and first-overtone modes. To date, a total of five double-mode Type II Cepheids have been identified in the Milky Way (Udalski et al. 2018; Soszyński et al. 2020). These stars form a very homogeneous class of pulsators, with their fundamental-mode periods ranging from 1.0 to 1.4 d and their period ratios ranging between 0.70 and 0.71.

Inspired by the discovery of multimode Type II Cepheids, Soszyński et al. (2019a) conducted a search for BL Herculis stars pulsating solely in the first overtone. The primary tool used for this task was the PL diagram for classical pulsating stars in the LMC. First-overtone Type II Cepheids should be located in close proximity to the purple dashed line in Figure 2, corresponding to a period ratio of 0.7 between the first overtone and the fundamental mode. Additionally, potential candidates for first-overtone BL Herculis stars should have light curves exhibiting features typical of overtone pulsators, in particular round minima. A search of the OGLE photometric database of the LMC revealed two stars fitting these criteria: OGLE-LMC-T2CEP-290 and OGLE-LMC-T2CEP-291. Their positions in the PW diagram are marked with purple circles in the top panel of Figure 2, while the bottom left panel of this figure shows the light curve of one of these stars (OGLE-LMC-T2CEP-290). Three other candidates for first-overtone Type II Cepheids were subsequently discovered in the Milky Way (Soszyński et al. 2020).

3.3. *Ultra-Faint RV Tauri Stars*

Another example of unique objects discovered in the PL diagram are two ultra-faint RV Tauri stars in the LMC (Soszyński et al. 2018). The light curves of OGLE-LMC-T2CEP-207 and OGLE-LMC-T2CEP-237 exhibit typical features of RV Tauri stars, including slight variations in pulsation period, long-period modulation of the mean brightness in the case of OGLE-LMC-T2CEP-207 (a characteristic of RVb stars), and small but detectable alternations of deeper and shallower minima. These stars also have a $(V - I)$ color index consistent with RV Tauri variables. However, they are roughly 2.5 mag fainter than other RV Tauri stars in the LMC (red circles in the top panel of Figure 2). A possible explanation of the nature of these ultra-faint RV Tauri stars is that they are post-red-giant-branch stars (Manick et al. 2018), as opposed to classical RV Tauri stars that are thought to be post-asymptotic-giant-branch stars.

4. Anomalous Cepheids

The term 'anomalous' Cepheids was coined by Zinn & Searle (1976) to describe a class of pulsating stars whose properties did not match those of classical or Type II Cepheids. Anomalous Cepheids are radially pulsating stars that burn helium in their cores and have low metal content, with masses between 1 and 2 $M_{\odot}$ (Bono et al. 1997; Marconi et al. 2004; Groenewegen & Jurkovic 2017). In the PL plane, they occupy the region between the relations of classical and Type II Cepheids (Figure 3).

Just 15 years ago, only a few dozen anomalous Cepheids were known, the vast majority found in nearby dwarf galaxies. Very few such stars were known in Galactic globular

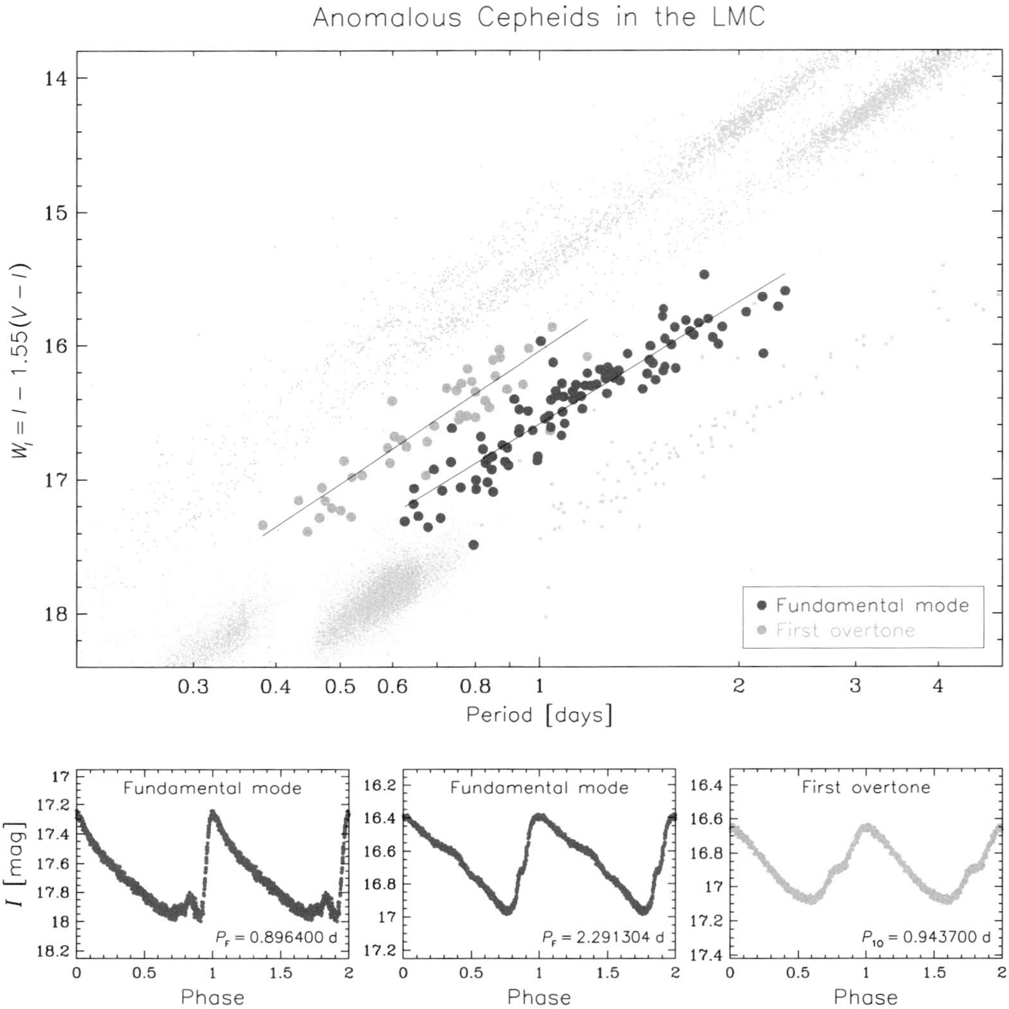

Figure 3. (top) Period–Wesenheit index diagram for anomalous Cepheids in the LMC (Soszyński et al. 2015, 2019b). Blue circles indicate fundamental-mode pulsators, while green symbols mark first-overtone Cepheids. Gray dots show the positions of classical Cepheids, Type II Cepheids, and RR Lyrae stars. (bottom) I-band light curves of typical anomalous Cepheids in the LMC.

clusters and only one anomalous Cepheid, XZ Ceti, was known in the field of the Milky Way (Szabados et al. 2007). However, this situation changed when the OGLE team first reported the discovery of 80 (Soszyński et al. 2008b) and later more than 260 anomalous Cepheids in the LMC and SMC (Soszyński et al. 2015, 2019b). The position of these stars in the PL diagram was the primary criterion that allowed us to separate anomalous Cepheids from classical Cepheids, Type II Cepheids, and RR Lyrae stars.

With this large and homogeneous sample of anomalous Cepheids in the Magellanic Clouds, Soszyński et al. (2015) developed a method for distinguishing them from other types of classical pulsating stars based on the light curve morphology, using the ϕ_{21} and ϕ_{31} parameters of the Fourier light curve decomposition. The period–Fourier coefficient diagram proved to be effective tools for identifying different types of Cepheids, both fundamental-mode and first-overtone ones. By applying this method,

Soszyński et al. (2015, 2020) detected approximately 120 anomalous Cepheids in the Milky Way. This way, the PL diagram for pulsating stars in the Magellanic Clouds enabled the identification of a significant number of anomalous Cepheids in our own Galaxy.

Ripepi et al. (2019) presented a similar application for the PL diagram for pulsating stars in the Magellanic Clouds, using different types of variables in these galaxies to improve the classification of Milky Way Cepheids included in the *Gaia* DR2 catalog of variable stars.

5. RR Lyrae Stars

RR Lyrae variables are the most abundant type of classical pulsators, with hundreds of thousands of stars of this type known across the Universe. Tens of thousands of them have been found in the galaxies of the Local Group (e.g., Soszyński et al. 2019b; Clementini et al. 2022; Nagarajan et al. 2022), including nearly 50,000 identified in the Magellanic System (Figure 4). RR Lyrae stars are old (> 10 Gyr) and low-mass ($< 0.85\ M_{\odot}$) stars that occupy the intersection of the instability strip with the horizontal branch. They are a perfect tool to trace the oldest stellar component and probe the structure of galactic halos. When using RR Lyrae stars as distance indicators, it is essential to consider the dependence of their brightness on metallicity. In other words, period–luminosity–metallicity (PLZ) relations should be used (e.g., Borissova et al. 2009; Muraveva et al. 2015; Bhardwaj et al. 2023).

The usefulness of the PLZ relations for RR Lyrae stars extends beyond measuring distances to star clusters and nearby galaxies. When distances are known from independent measurements, the PLZ relationship can be inverted to calculate the metallicities of individual RR Lyrae stars and trace the early chemical evolution of their host systems (e.g., Braga et al. 2016; Martínez-Vázquez et al. 2016). Moreover, when photometric observations in several optical and near-infrared bands are available, it is possible to simultaneously measure the distance, metallicity, and reddening of RR Lyrae stars (Bono et al. 2019). The versatile properties of RR Lyrae stars are commonly used to study the spatial structure, star formation history, and metallicity gradients in neighboring galaxies (e.g., Deb & Singh 2014; Jacyszyn-Dobrzeniecka et al. 2017; Muraveva et al. 2018; Ferguson & Strigari 2020).

In one of the most recent analyses, Cusano et al. (2021) studied the three-dimensional geometry of the LMC as traced by 29,000 RR Lyrae stars from the OGLE (Soszyński et al. 2016a, 2019b) and *Gaia* DR2 (Clementini et al. 2019) catalogs supplemented with near-infrared observations from the VMC survey. The LMC structure visible in old stars has a triaxial ellipsoid shape with axis ratios of 1:1.2:1.8 and no additional substructures.

6. δ Scuti Stars

δ Scuti variables are the faintest members of the family of classical pulsators. They are intermediate-mass stars (1.5–2.5 $M_{\odot}$) that pulsate in low-order radial or non-radial modes with periods of up to 0.3 d. In the Hertzsprung–Russell diagram, they lie close to the intersection of the Cepheid instability strip with the main sequence. δ Scuti pulsators consist of a mixture of stars in different evolutionary stages, including pre-main-sequence, main-sequence, and post-main-sequence stars, as well as blue stragglers (which are known as SX Phoenicis stars).

Due to the low intrinsic brightness of δ Scuti stars, most empirical efforts to measure their PL relations have relied on nearby objects with well-determined parallaxes (e.g.,

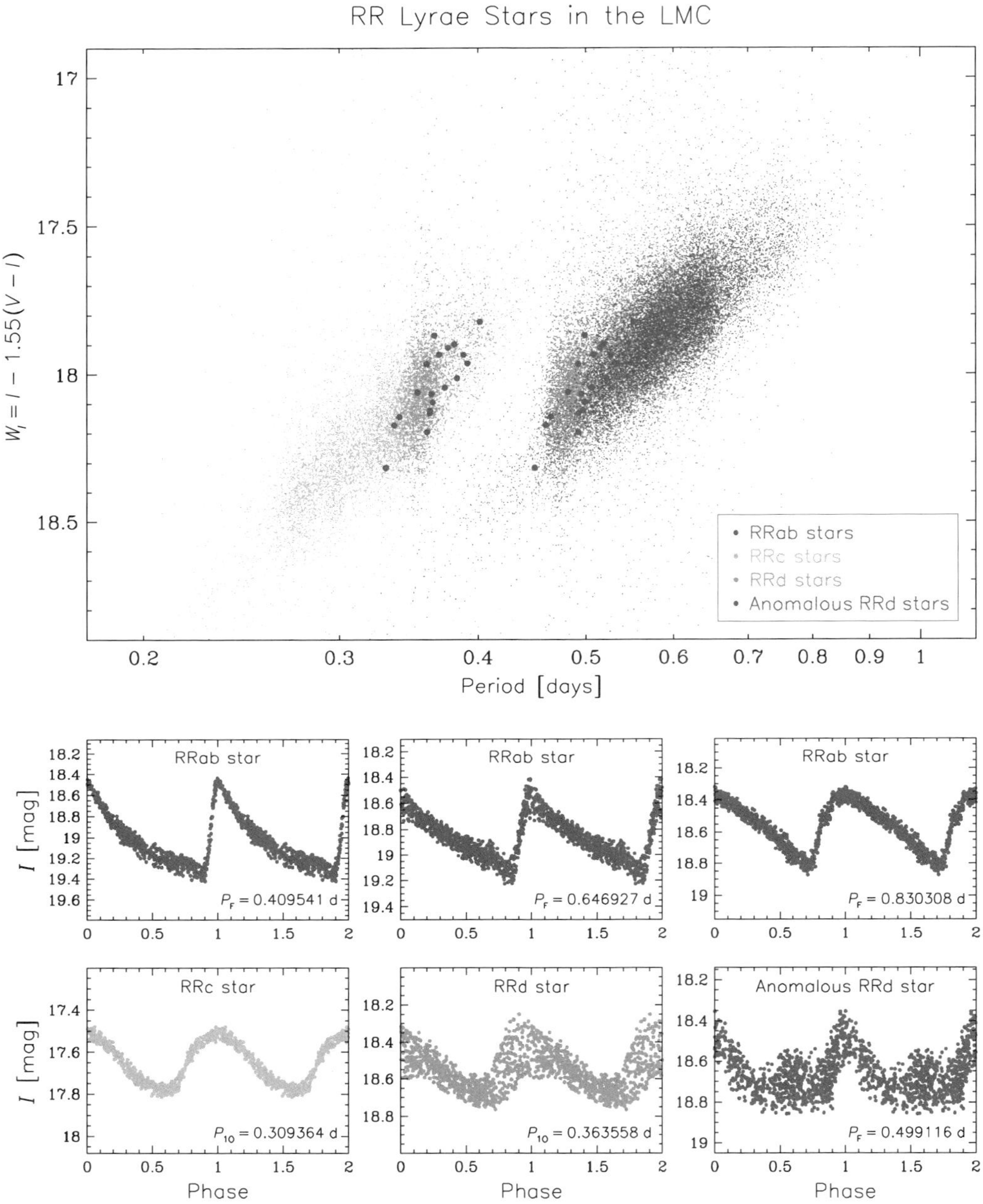

Figure 4. (top) Period–Wesenheit index diagram for RR Lyrae stars in the LMC (Soszyński et al. 2016a, 2019b). RRab stars are marked with blue points, RRc stars with green points, RRd stars with yellow points, and anomalous RRd stars (Soszyński et al. 2016b) with red circles. Each double-mode (RRd) star is represented by two points corresponding to the two pulsation periods. (bottom) Example I-band light curves of different types of RR Lyrae variables.

Ziaali et al. 2019; Jayasinghe et al. 2020; Barac et al. 2022) or on SX Phoenicis variables in Galactic globular clusters, whose distances have been independently determined using other distance indicators (e.g., Cohen & Sarajedini 2012; Ngeow et al. 2023). Until recently, the largest existing catalogs of extragalactic δ Scuti stars contained in total

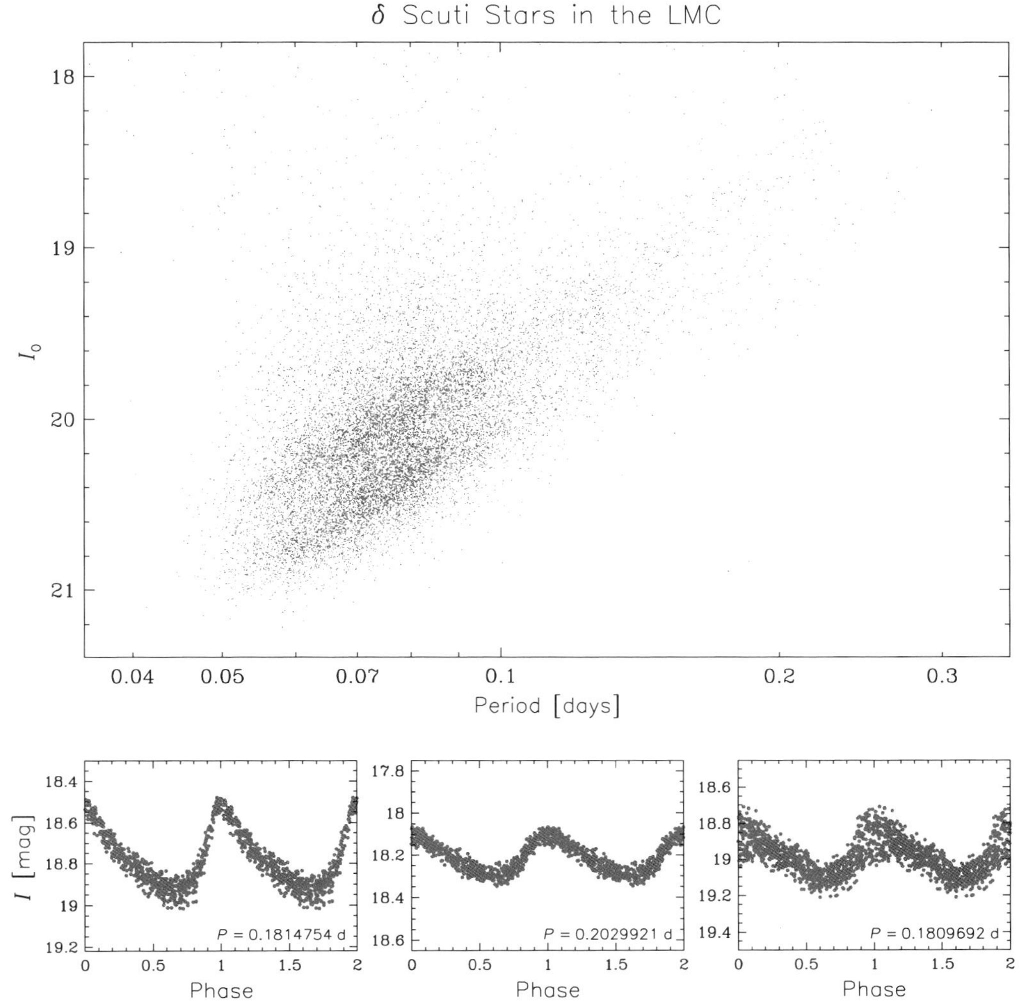

Figure 5. (top) I-band period–luminosity diagram for about 15,000 δ Scuti stars in the LMC (Soszyński et al. 2023). The magnitudes have been corrected for interstellar extinction using the reddening maps of Skowron et al. (2021). (bottom) Example I-band light curves of δ Scuti variables in the LMC.

several thousand objects in the LMC (Garg et al. 2010; Poleski et al. 2010) and SMC (Soszyński et al. 2022). This situation has been improved thanks to the recent publication of more than 15,000 δ Scuti stars in the LMC detected in the OGLE-IV database (Soszyński et al. 2023).

Before this publication, Martínez-Vázquez et al. (2022) used the available sources of extragalactic δ Scuti stars to investigate their distribution in the PL plane. The authors presented a single PL sequence that showed a significant change in slope at a pulsation period of approximately 0.09 d. However, analysis of the new OGLE-IV collection of δ Scuti pulsators in the LMC (Soszyński et al. 2023) does not confirm this conclusion. The extended sample is consistent with two linear PL relationships corresponding to the fundamental and first-overtone pulsation modes (Figure 5). It appears that the segmented PL relation reported by Martínez-Vázquez et al. (2022) was an illusion resulting from incompleteness of the available catalogs of extragalactic δ Scuti stars.

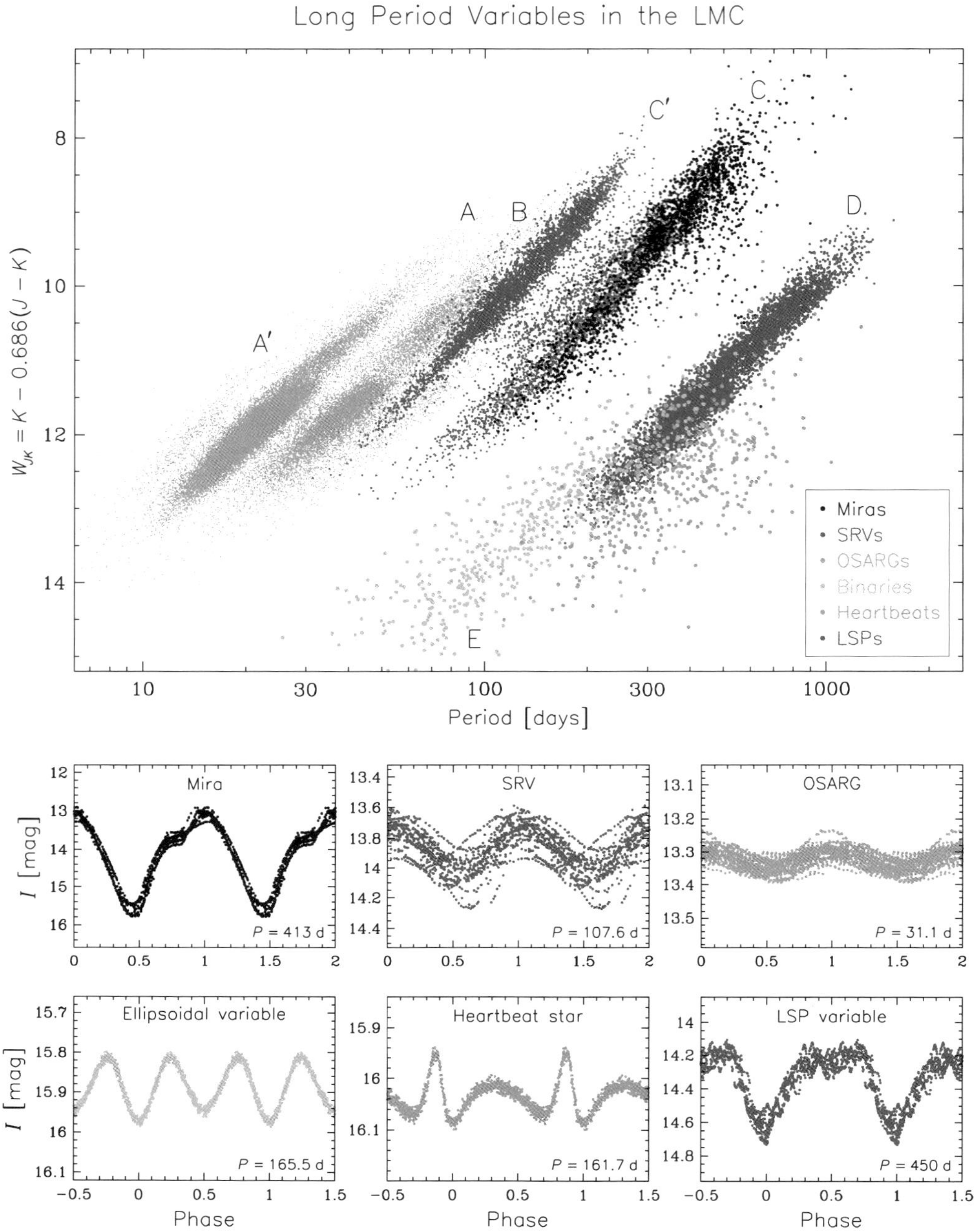

Figure 6. (top) Near-infrared period–Wesenheit index diagram for LPVs in the LMC (Soszyński et al. 2009). The near-infrared Wesenheit index is defined as $W_{JK} = K_s - 0.686 (J - K_s)$, where the J- and K_s-band magnitudes originate from the IRSF Catalog (Kato et al. 2007) or the 2MASS Catalog (Cutri et al. 2003). Different colors of the points indicate different types of LPVs: brown – Miras, red – SRVs, orange – OSARGs, green – eclipsing and ellipsoidal close binary systems containing a red giant as one of the components, lavender – heartbeat stars, and blue – LSP variables. (bottom) Typical I-band light curves of various types of LPVs.

7. Long-Period Variables

Long-period variables (LPVs) are pulsating stars belonging to either the red giant branch (RGB), asymptotic giant branch (AGB), or red supergiant branch. They are typically classified into three groups: Miras, semi-regular variables (SRVs), and OGLE small-amplitude red giants (OSARGs). The distribution of LPVs in the PL and PW planes is one of the primary tools for studying the physics and evolution of red giant stars.

Historically, the most important discoveries in the PL diagrams of LPVs have been made by studying variables in the LMC. Glass & Lloyd Evans (1981) were the first to demonstrate that Miras in the LMC follow a narrow PL relation in the infrared. Later, Wood & Sebo (1996) showed that SRVs follow the same PL sequence as Miras and additionally obey a second, parallel relation at shorter periods, which is today attributed to the first-overtone pulsation mode. The revolution in the PL diagram came with the advent of the microlensing sky surveys such as MACHO and OGLE, which published large samples of LPVs in the Magellanic Clouds. Wood et al. (1999) discovered five PL sequences for LPVs and labeled them with the letters A to E (with Miras lying along sequence C). Since then, the number of PL relationships for pulsating red giants has increased even more (e.g., Soszyński et al. 2007). Figure 6 depicts the contemporary near-infrared PW diagram for LPVs in the LMC, where the periods have been taken from the OCVS (Soszyński et al. 2009), while the near-infrared Wesenheit index was derived using photometric measurements from the IRSF (Kato et al. 2007) or 2MASS (Cutri et al. 2003) catalogs.

7.1. *Miras as Distance Indicators*

LPVs embody an interesting alternative to Cepheids as indicators of the extragalactic distance scale. Pulsating red giants are a numerous class of variable stars, they are bright at infrared wavelengths, and present in all types of galaxies. The vast majority of attempts to use the PL relations of LPVs to measure distances have used Mira variables (e.g., Yuan et al. 2018; Huang et al. 2020; Ou et al. 2023).

Iwanek et al. (2021) obtained the most precise PL relationships for Mira stars in mid-infrared bands using data from the *WISE* (Wright et al. 2010) and *Spitzer* (Werner et al. 2004) space telescopes. These PL sequences proved useful in determining the distances to individual oxygen- and carbon-rich Miras with an accuracy of 5% and 12%, respectively. In a subsequent study, Iwanek et al. (2023) applied these relations to the OGLE-IV collection of Mira stars (Iwanek et al. 2022) to map the central regions of the Milky Way.

The PL relations for LPVs are typically studied in infrared bands, since the light of Miras is affected by extinction from circumstellar dust that usually accompanies large-amplitude pulsations. However, Bhardwaj et al. (2019) analyzed the OGLE I- and V-band light curves of Miras in the Magellanic Clouds and found that the PLC relation in optical bands can also be quite narrow, similar to that at infrared wavelengths, particularly when the maximum brightness is used instead of the mean luminosity of individual stars.

7.2. *PL Relations for Red Supergiants*

Red supergiant stars also follow the PL relations, making them promising extragalactic distance tracers, especially since they are several magnitudes brighter than the tip of the RGB (TRGB). Ren et al. (2019) analyzed time-series photometry of red supergiants in M31 and M33 collected by the intermediate Palomar Transient Factory (iPTF) survey.

In both galaxies, red supergiant stars formed two distinct PL relationships corresponding to the fundamental and first-overtone modes.

Interestingly, the relative number of the fundamental-mode and first-overtone red supergiants in M31 and M33 differed. Extending this study to other Local Group galaxies (the Milky Way, LMC, and SMC) revealed that this was not a coincidence. There is a strong correlation between metallicity and the relative number of red supergiants pulsating in different modes. In the metal-poorest SMC, all supergiants are first-overtone pulsators, but the fraction of fundamental-mode variables increases with increasing metallicity, reaching more than 80% in the most metal-rich M31. Ren et al. (2019) suggested that this correlation may be related to convection since high metallicity leads to less effective radiative transfer due to higher opacity, and consequently, stronger convection is required, which then affects the preferred pulsation mode.

7.3. *Potential of SRVs and OSARGs as Distance Indicators*

Trabucchi et al. (2021) tested the potential of SRVs as standard candles. They found that some SRVs exhibit the same PL relationship as Miras and can therefore be used successfully to measure distances. However, the untapped potential of the PL sequences followed by OSARG variables is worth noting. Although analysis of the pulsation periods of these stars requires a larger number of observations than for Mira variables due to their multi-period nature and smaller amplitudes, OSARGs have other advantages. They are two orders of magnitude more numerous than Miras, and their light curves are less affected by circumstellar matter due to their reduced mass loss. Identification of pulsation modes in OSARGs would be an ideal task for artificial intelligence algorithms.

An interesting application of OSARGs was recently presented by Anderson et al. (2023), who reported absolute calibration of the TRGB distance scale with an unprecedented accuracy of 1.4%. The authors used OSARG variables in the LMC, carefully selected in the PL diagram by the OGLE team (Soszyński et al. 2009). This method produced more accurate calibration of the TRGB compared to the calibration based on the general RGB population. Anderson et al. (2023) used their calibration to derive the Hubble constant and obtained better than 1σ agreement with the Cepheids-based H_0 value.

8. Close Binary Systems

Since the pioneering works of Rucinski (1994) and Rucinski & Duerbeck (1997), it has been established that contact binaries follow the PL and PLC relations due to the correlation between the size of the components and the radius of the orbit (e.g., Chen et al. 2016, 2018; Ngeow et al. 2021). Pawlak (2016) used a sample of contact and nearly-contact massive binary systems in the LMC to calibrate their PLC relationship. One of the conclusions of this study was that massive contact binaries follow a different PLC relation than low-mass W UMa stars (e.g., Rucinski 2004), indicating that the PLC relation for the entire population of contact binaries is non-linear.

In turn, Pawlak et al. (2014) investigated close binaries (eclipsing systems and ellipsoidal variables) with a red giant star as one of the components that form the PL relation known as sequence E (Wood et al. 1999). While this sequence is generally wider than other PL relations followed by LPVs (Figure 6), the dispersion in sequence E can be significantly reduced by considering ellipsoidal variables with larger amplitudes. Pawlak et al. (2014) distinguished three PL sequences, designated E1–E3, of which sequence E2 consists of systems with two red giant stars, sequence E3 is composed of giant–dwarf binaries, and sequence E1 is a mixture of the previous two types of binary systems.

Lavender dots in Figure 6 represent ellipsoidal variables that do not satisfy the PL relation for close binary systems containing red giants (sequence E). These are so-called heartbeat stars, which are ellipsoidal variables with eccentric orbits (Soszyński et al. 2004). The term 'heartbeat' stars was coined by Thompson et al. (2012), since the light curves of these variables resemble electrocardiogram signatures. The orbital periods of heartbeat stars are naturally longer than those of similar binary systems in circular orbits. The greater the eccentricity of the system's orbit, the larger the displacement of the star relative to sequence E (Wrona et al. 2022a). Recently, Wrona et al. (2022b) published a catalog of nearly 1000 heartbeat stars found in the OGLE photometric database. Tidally excited oscillations have been detected in some of these stars.

9. Long Secondary Periods in Red Giant Stars

The final type of variable stars discussed in this article is the most mysterious one. Long secondary periods (LSPs) occur in at least 30% of LPVs (red giant and supergiant stars) and form a PL relationship designated by Wood as sequence D (blue points in Figure 6). LSPs are roughly ten times longer than the pulsation periods of the same stars and, contrary to their name, are often associated with larger amplitudes of light variations than stellar pulsation (for an example light curve, see the bottom right panel of Figure 6). The LSP phenomenon has been observed for many decades (e.g., O'Connell 1933; Payne-Gaposchkin 1954; Houk 1963), but until recently its origin was completely unknown. Possible explanations of the LSP phenomenon that have been considered in the literature include radial or non-radial stellar pulsations, binarity, episodic dust ejections, turnover of giant convective cells, and magnetic activity.

The first clue that LSPs are related to binarity was provided by the PL diagram, where sequence D partially overlaps and is a continuation of sequence E, populated by close binary systems (Soszyński et al. 2004). However, radial velocity measurements (e.g., Hinkle et al. 2002; Wood et al. 2004) suggested that the hypothetical companion of an LSP red giant should be a brown dwarf or very-low-mass stellar object, which was inconsistent with the so-called 'brown dwarf desert' – an observed deficit of brown dwarf companions to main-sequence stars (Grether & Lineweaver 2006; Nicholls et al. 2009).

A strong argument in favor of the binary origin of the LSP phenomenon was provided by Soszyński et al. (2021), who detected secondary eclipses in their mid-infrared light curves (Figure 7). It is interesting that the secondary eclipses are not visible in the optical range. According to the binary model, an extended cloud of dust is formed around the red giant's low-mass companion, regularly obscuring the main component of the system causing the primary eclipses that are visible in the optical and infrared wavelengths. When the cloud surrounding the brown dwarf passes behind the red giant, the secondary eclipse is visible only in the infrared light curve. To reconcile the brown dwarf desert with a large fraction (at least 30%) of substellar objects accompanying red giants, it should be assumed that brown dwarfs are former planets that have accreted material lost by their parent stars via stellar winds. If this hypothesis were true, it would mean that LSP variables can be used as tracers of planetary systems in the Milky Way and other galaxies. Using the PL relation of LSP stars, it would even be possible to study the three-dimensional distribution of former planetary systems in various stellar environments.

10. Summary

Studies of the PL, PLC, and PW relationships in nearby galaxies are of crucial importance for the calibration of the distance ladder and, therefore, for the confirmation of the Hubble tension. The diverse chemical compositions, different spatial structures, and star

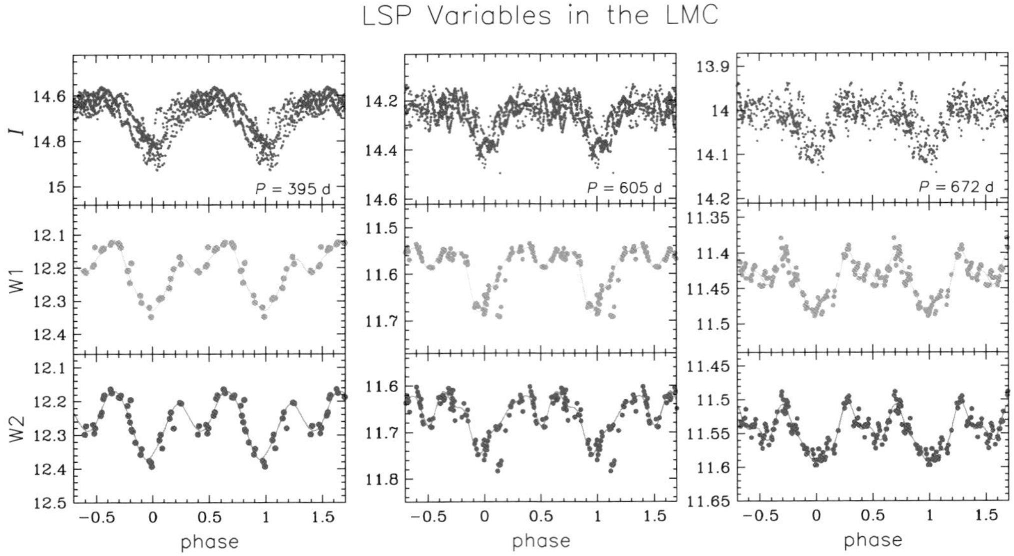

Figure 7. Folded light curves of three LSP variables in the LMC. The top panels show OGLE *I*-band (0.8 μm); middle panels, *WISE W*1 (3.4 μm); and bottom panels, *WISE W*2 (4.6 μm) photometry. The primary minima visible in all wavebands (at phases 0 and 1) are caused by eclipses of the red giant by an orbiting cloud of dust. The secondary minima, visible only in the mid-infrared light curves (at phase 0.5), occur when the cloud is hidden behind the red giant.

formation histories observed in Local Group galaxies provide us with valuable opportunities to explore various aspects, including the influence of metallicity and the linearity of the PL relations on Hubble constant measurements. This will be particularly significant in the face of current and upcoming large telescopes, like the *James Webb Space Telescope* and the Vera C. Rubin Observatory.

However, the PL relations in our closest galactic neighborhood serve a broader purpose, beyond distance measurement. In this work, we have presented examples of applications of the PL, PLC, and PW diagrams to classify variable stars, to discover new types of variables, to elucidate the origins of their light changes, to identify pulsation modes conclusively, to search for binary systems hosting pulsators, and to investigate the three-dimensional geometry of galaxies.

Acknowledgements. This work has been funded by the National Science Centre, Poland, through grant 2022/45/B/ST9/00243. For the purpose of Open Access, the author has applied a CC-BY public copyright licence to any Author Accepted Manuscript (AAM) version arising from this submission.

References

Alcock, C., et al. 1995, AJ, 109, 1653
Anderson, R. I., Koblischke, N. W., & Eyer, L. 2023, arXiv:2303.04790
Arp, H. 1960, AJ, 65, 404
Baade, W. 1954, Trans. IAU, 8, 397
Baade, W. 1956, PASP, 68, 5
Barac, N., Bedding, T. R., Murphy, S. J., & Hey D. R 2022, MNRAS, 516, 2080
Beaulieu, J. P., et al. 1995, A&A, 303, 137
Bhardwaj, A., Kanbur, S. M., Macri, L. M., Singh H. P., Ngeow C.-C., & Ishida E. E. O 2016, MNRAS, 457, 1644
Bhardwaj, A., Macri, L. M., Rejkuba, M., Kanbur S. M., Ngeow C.-C., & Singh H. P. 2017, AJ, 153, 154

Bhardwaj, A., et al. 2019, ApJ, 884, 20
Bhardwaj, A., et al. 2023, ApJL, 944, L51
Bono, G., Caputo, F., Santolamazza, P., Cassisi S., & Piersimoni A. 1997, AJ, 113, 2209
Bono, G., et al. 2019, ApJ, 870, 115
Borissova, J., Rejkuba, M., Minniti, D., Catelan M., & Ivanov V. D. 2009, A&A, 502, 505
Braga, V. F., et al. 2016, AJ, 152, 170
Breuval, L., et al. 2021, ApJ, 913, 38
Breuval, L., Riess, A. G., Kervella, P., Anderson R. I., & Romaniello M. 2022, ApJ, 939, 89
Breuval, L., et al. 2023, ApJ, 951, 118
Chen, X., de Grijs, R., & Deng, L. 2016, ApJ, 832, 138
Chen, X., Deng, L., de Grijs, R., Wang S., & Feng Y. 2018, ApJ, 859, 140
Chown, A. H., Scowcroft, V., & Wuyts, S. 2021, MNRAS, 500, 817
Cioni, M.-R. L., et al. 2011, A&A, 527, A116
Clementini, G., et al. 2019, A&A, 622, A60
Clementini, G., et al. 2023, A&A, 674, A18
Cohen, R. E. & Sarajedini, A. 2012, MNRAS, 419, 342
Connolly, L. P. 1980, PASP, 92, 165
Cusano, F., et al. 2021, MNRAS, 504, 1
Cutri, R. M., et al. 2003, The 2MASS All Sky Point Source Catalog, IPAC, Pasadena
Deb, S. & Singh, H. P. 2014, MNRAS, 438, 2440
Deb, S., Ngeow, C.-C., Kanbur, S. M., Singh H. P., Wysocki D., & Kumar S. 2018, MNRAS, 478, 2526
Ferguson, P. S. & Strigari, L. E. 2020, MNRAS, 495, 4124
García-Varela, A., Sabogal, B. E., & Ramírez-Tannus, M. C. 2013, MNRAS, 431, 2278
Garg, A., et al. 2010, AJ, 140, 328
Gieren, W., et al. 2018, A&A, 620, A99
Glass, I. S. & Evans, T. L. 1981, Nature, 291, 303
Graczyk, D., et al. 2020, ApJ, 904, 13
Grether, D. & Lineweaver, C. H. 2006, ApJ, 640, 1051
Groenewegen, M. A. T. & Jurkovic, M. I. 2017, A&A, 603, A70
Groenewegen, M. A. T. 2018, A&A, 619, A8
Hertzsprung, E. 1913, Astron. Nachr., 196, 201
Hinkle, K. H., Lebzelter, T., Joyce, R. R., & Fekel F. C. 2002, AJ, 123, 1002
Houk, N. 1963, AJ, 68, 253
Huang, C. D., et al. 2020, ApJ, 889, 5
Hubble, E. P. 1925, Pop. Astron., 33, 252
Hubble, E. 1929, PNAS, 15, 168
Inno, L., et al. 2016, ApJ, 832, 176
Ivanov, G. R. & Nikolov, N. S. 1976, Astrophys. Lett., 17, 115
Iwanek, P., Soszyński, I., & Kozłowski, S. 2021, ApJ, 919, 99
Iwanek, P., et al. 2022, ApJS, 260, 46
Iwanek, P., et al. 2023, ApJS, 264, 20
Jacyszyn-Dobrzeniecka, A. M., et al. 2016, Acta Astron., 66, 149
Jacyszyn-Dobrzeniecka, A. M., et al. 2017, Acta Astron., 67, 1
Jayasinghe, T., et al. 2020, MNRAS, 493, 4186
Kanbur, S. M., Ngeow, C., Nanthakumar, A., & Stevens R. 2007, PASP, 119, 512
Kato, D., et al. 2007, PASJ, 59, 615
Leavitt, H. S. 1908, Ann. Harvard Coll. Obs., 60, 87
Leavitt, H. S. & Pickering, E. C. 1912, Harvard Coll. Obs. Circ., 173
Li, S., Riess, A. G., Busch, M. P., Casertano S., Macri L. M., & Yuan W. 2021, ApJ, 920, 84
Macri, L. M., Ngeow, C.-C., Kanbur, S. M., Mahzooni S., & Smitka M. T. 2015, AJ, 149, 117
Madore, B. F. 1982, ApJ, 253, 575
Manick, R., Van Winckel, H., Kamath, D., Sekaran S., & Kolenberg K. 2018, A&A, 618, A21
Marconi, M., Fiorentino, G., & Caputo, F. 2004, A&A, 417, 1101

Martínez-Vázquez, C. E., et al. 2016, MNRAS, 461, L41
Martínez-Vázquez, C. E., Salinas, R., Vivas, A. K., & Catelan M. 2022, ApJL, 940, L25
Muraveva, T., et al. 2015, ApJ, 807, 127
Muraveva, T., et al. 2018, MNRAS, 473, 3131
Nagarajan, P., Weisz, D. R., & El-Badry, K. 2022, ApJ, 932, 19
Neeley, J. R., et al. 2019, MNRAS, 490, 4254
Ngeow, C., Kanbur, S. M., & Nanthakumar, A. 2008, A&A, 477, 621
Ngeow, C.-C., Kanbur, S. M., Neilson, H. R., Nanthakumar A., & Buonaccorsi J. 2009, ApJ, 693, 691
Ngeow, C.-C., et al. 2021, AJ, 162, 63
Ngeow, C.-C., Bhardwaj, A., Graham, M. J., Healy B. F., Laher R. R., Riddle R., & Wold A. 2023, AJ, 165, 190
Nicholls, C. P., Wood, P. R., Cioni, M.-R. L., & Soszyński I. 2009, MNRAS, 399, 2063
O'Connell, D. J. K. 1933, Harvard Coll. Obs. Bull., 893, 19
Ou, J.-Y., Ngeow, C.-C., Bhardwaj, A., Graham M. J., Laher R. R., Masci F. J., & Riddle R. 2023, AJ, 165, 137
Pawlak, M., et al. 2014, Acta Astron., 64, 293
Pawlak, M. 2016, MNRAS, 457, 4323
Payne-Gaposchkin, C. 1954, Ann. Harvard Coll. Obs., 113, 189
Pietrzyński, G., et al. 2010, Nature, 468, 542
Pietrzyński, G., et al. 2019, Nature, 567, 200
Pilecki, B., Pietrzyński, G., Anderson, R. I., Gieren, W., Taormina, M., Narloch, W., Evans, N. R., & Storm, J. 2021, ApJ, 910, 118
Pilecki, B., et al. 2022, ApJL, 940, L48
Poleski, R., et al. 2010, Acta Astron., 60, 1
Ren, Y., Jiang, B.-W., Yang, M., & Gao J. 2019, ApJS, 241, 35
Riess, A. G., Casertano, S., Yuan, W., Macri L. M., & Scolnic D. 2019, ApJ, 876, 85
Ripepi, V., et al. 2017, MNRAS, 472, 888
Ripepi, V., Molinaro, R., Musella, I., Marconi M., Leccia S., & Eyer L. 2019, A&A, 625, A14
Ripepi, V., et al. 2020, A&A, 642, A230
Ripepi, V., et al. 2022, MNRAS, 512, 563
Rucinski, S. M. 1994, PASP, 106, 462
Rucinski, S. M. & Duerbeck, H. W. 1997, PASP, 109, 1340
Rucinski, S. M. 2004, New Astron. Rev., 48, 703
Sandage, A. & Tammann, G. A. 1968, ApJ, 151, 531
Shapley, H. 1918, ApJ, 48, 89
Skowron, D. M., et al. 2021, ApJS, 252, 23
Smolec, R., Moskalik, P., Plachy, E., Soszyński I., & Udalski A. 2018, MNRAS, 481, 3724
Soszyński, I., et al. 2004, Acta Astron., 54, 347
Soszyński, I., et al. 2007, Acta Astron., 57, 201
Soszyński, I., et al. 2008a, Acta Astron., 58, 163
Soszyński, I., et al. 2008b, Acta Astron., 58, 293
Soszyński, I., et al. 2009, Acta Astron., 59, 239
Soszyński, I., et al. 2015, Acta Astron., 65, 233
Soszyński, I., et al. 2016a, Acta Astron., 66, 131
Soszyński, I., et al. 2016b, MNRAS, 463, 1332
Soszyński, I., et al. 2017, Acta Astron., 67, 103
Soszyński, I., et al. 2018, Acta Astron., 68, 89
Soszyński, I., Smolec, R., Udalski, A., & Pietrukowicz P. 2019a, ApJ, 873, 43
Soszyński, I., et al. 2019b, Acta Astron., 69, 87
Soszyński, I., et al. 2020, Acta Astron., 70, 101
Soszyński, I., et al. 2021, ApJL, 911, L22
Soszyński, I., et al. 2022, Acta Astron., 72, 245
Soszyński, I., et al. 2023, Acta Astron., 73, 105

Szabados, L., Kiss, L. L., & Derekas, A. 2007, A&A, 461, 613
Thompson, S. E., et al. 2012, ApJ, 753, 86
Trabucchi, M., Mowlavi, N., & Lebzelter, T. 2021, A&A, 656, A66.
Udalski, A., Soszyński, I., Szymański, M., Kubiak M., Pietrzyński G., Woźniak P., & Żebruń K. 1999, Acta Astron., 49, 223
Udalski, A., Szymański, M. K., & Szymański, G. 2015a, Acta Astron., 65, 1
Udalski, A., et al. 2015b, Acta Astron., 65, 341
Udalski, A., et al. 2018, Acta Astron., 68, 315
Werner, M. W., et al. 2004, ApJS, 154, 1
Wielgórski, P., et al. 2017, ApJ, 842, 116
Wielgórski, P., et al. 2022, ApJ, 927, 89
Wood, P. R. & Sebo, K. M. 1996, MNRAS, 282, 958
Wood, P. R., et al. 1999, Proc. IAU Symp., 191, 151
Wood, P. R., Olivier, E. A., & Kawaler, S. D. 2004, ApJ, 604, 800
Wright, E. L., et al. 2010, AJ, 140, 1868
Wrona, M., Kołaczek-Szymański, P. A., Ratajczak, M., & Kozowski, S. 2022a, ApJ, 928, 135
Wrona, M., et al. 2022b, ApJS, 259, 16
Yuan, W., Macri, L. M., Javadi, A., Lin Z., & Huang J. Z. 2018, AJ, 156, 112
Ziaali, E., Bedding, T. R., Murphy, S. J., Van Reeth T., & Hey D. R. 2019, MNRAS, 486, 4348
Zinn, R. & Searle, L. 1976, ApJ, 209, 734

Discussion

Question (Martínez-Vázquez): I have a question regarding the PLR in δ Scuti stars. Did you see the same two sequences in the Galactic bulge δ Scuti stars as you showed for the LMC?

Answer: I have not studied it thoroughly. At first glance, there is no double PL relation in the Galactic bulge, which is understandable owing to the large depth of the bulge along the line of sight, which significantly broadens the PL relations.

Question (Braga): You showed us that the I-band PLR of δ Scuti stars does not show any break in the slope, as did the V-band PL relation of Martínez-Vázquez. Since OGLE also has V-band data, can you tell us what the PL(V) looks like in OGLE?

Answer: In the V band, the PL diagram for δ Scuti stars in the LMC looks essentially the same as in the I band (there are two linear relationships), only the scatter is larger.

Question (Braga): You showed ultra-faint RV Tau light curves, and they were folded at twice the period, as usually done for RV Tau stars, with alternating deep/shallow minima. However, those two light curves did not look like they showed alternating minima. Did you use twice the period for all RV Tau stars for your PLRs?

Answer: Good question. During my talk, I presented pre-whitened light curves of these two stars (corrected for period changes and long-term brightness modulations). To see the original light curve, please check Figure 6 in the paper of Soszyński et al. (2018). Alternating deeper and shallower minima are not always obvious in RV Tauri (or related) stars. In the case of OGLE-LMC-T2CEP-237, the alternations are visible, while for OGLE-LMC-T2CEP-207 we can see long-term changes of the mean magnitudes, as is observed in RV Tauri-type b (RVb) stars.

At the crossroads of astrophysics and cosmology: Period–luminosity relations in the 2020s
Proceedings IAU Symposium No. 376, 2024
R. de Grijs, P. A. Whitelock & M. Catelan, eds.
doi:10.1017/S1743921323003228

Theoretical Stellar Pulsation Physics

M. Marconi, G. De Somma, R. Molinaro and I. Musella

INAF-Osservatorio Astronomico di Capodimonte. email: marcella.marconi@inaf.it

Abstract. Pulsating stars play a crucial role in the calibration of the cosmic distance scale as well as in tracing the properties of the associated stellar populations. In the era of large observational surveys and precise astrometric missions, it is crucial to rely on accurate stellar pulsation models able to predict the observed behaviors for different physical assumptions. Indeed, the relations currently used in the literature to derive individual and mean distances of mainly radially pulsating stars such as Cepheids and RR Lyrae are well physically understood, but are also known to depend on a number of often unknown parameters. Recent extensive sets of stellar pulsation models developed by various authors show how variations in the physical assumptions can affect the theoretical prediction of the instability strip boundaries, the morphology and amplitude of light and radial velocity curves, and the consequent Period-Luminosity, Period-Luminosity-Color and Period-Wesenheit relations. These aspects are discussed in the framework of current open problems in the field of classical pulsating stars.

Keywords. pulsating stars, distance scale, stellar pulsation, stellar parameters

1. Introduction

Pulsating stars play a very important role in stellar astrophysics and not only there. They are easily identified thanks to their variability. Moreover, their main pulsation observables, namely the period and the amplitude of the magnitude variation, are unaffected by distance and reddening uncertainties. On the other hand, the acoustic nature of the oscillation implies a fundamental relation connecting the oscillation period to the mean stellar density, which in turn implies relations among the period, the luminosity, the mass, and the effective temperature, with a non-negligible dependence on chemical composition. This is the fundamental relation between pulsation and evolutionary stellar parameters (Bono et al. 2000a; De Somma et al. 2021) that is the basis of the well known Period-Luminosity (PL) relation obeyed by several classes of pulsating stars and, in turn, of their role as standard candles. At the same time the relation between the pulsation period and evolutionary properties allows us to test stellar evolution predictions through independent pulsational constraints. In this review, we discuss the physics on the basis of stellar pulsation and of PL relations (in Section 2). In Section 3 we illustrate the state of the art of the modeling of pulsating stars, whereas in Sections 4 and 5 we still discuss some recent results for Cepheids and RR Lyrae, respectively. The Conclusions and some future developments are finally summarized in Section 6.

2. Stellar pulsation physics

Since the pioneering investigations by Eddington (1918), Zhevakin (1963) and Cox & Hodson (1978), the physical basis of stellar pulsation has been progressively clarified. The evidence that all the "classical" pulsating stars lie within the same almost vertical region in the color-magnitude diagram (the instability strip) suggested the efficiency of driving pulsation mechanisms associated with the surface parameters and, in particular, with the effective temperature of the stars. This driving phenomenon was identified with a sort of valve mechanism that favoured an absorption of energy during the contraction phase and the release of this energy excess during the subsequent expansion phase in the form of mechanical work. This valve effect was early associated with variations of the stellar opacity (the so-called κ mechanism) and of the adiabatic exponents (the so-called γ mechanism) in the ionization regions of the most abundant elements of stellar envelopes, namely H, He and He^+ (Zhevakin 1963). It is then crucial to adopt sound physical ingredients, in particular accurate opacity tables and equation of state, to properly model pulsating stars.

2.1. *The physical basis of the instability strip*

As the driving pulsation mechanisms are associated with variations of the opacity and adiabatic exponents in the ionization regions of the most abundant chemical elements in the stellar envelope, in order to be efficient, they require that the effective temperature of the star not be too high. Indeed, for each stellar mass and luminosity level, only when the effective temperature is below a critical value, the ionization regions are deep enough in the stellar envelope to allow pulsation to occur, because the driving zones prevail over the damping ones. This is the explanation for the existence of a blue edge of the instability strip for each selected pulsation mode. On the other hand, the existence of a red edge is due to the increasing efficiency of convection as the effective temperature decreases. In fact, the convective phenomenon tends to quench pulsation, by making the κ and γ mechanisms less and less active. The effect of the coupling between pulsation and convection has been verified by investigating how the predicted light and radial velocity curves, as well as the shape and the width of the instability strip vary, increasing the efficiency of super-adiabatic convection. Several authors have shown that the pulsation amplitudes decrease, the blue edge tends to become mildly redder and the red edge significantly bluer (see e.g. Di Criscienzo et al. 2004; Fiorentino et al. 2007; Paxton et al. 2019; De Somma et al. 2020, 2021). These are all effects of the increased opacity and convective efficiency in the stellar envelope.

2.2. *The Physics of PL relations*

The combination of the period-mean density relation with the Stefan-Boltzmann law implies a relation connecting the period to the mass, the luminosity and the effective temperature of the star, for each selected chemical composition. This is the so-called van Albada & Baker relation, derived for the first time by these authors for RR Lyrae stars on the basis of their linear non-adiabatic pulsation models (van Albada & Baker 1971).

In the case of RR Lyrae stars, no PL relation exists in the optical bands, as these pulsating stars belong to the Horizontal Branch (HB) phase (see Catelan et al. 2004), with colors and periods varying at a roughly fixed (but metallicity-dependent) magnitude level. However, due to the dependence of Near-Infrared (NIR) bolometric corrections to the effective temperature for RR Lyrae models, accurate NIR (metal-dependent) PL relations are predicted, confirming observations (since Longmore et al. 1990). On the other hand, in the case of Classical Cepheids a mass-luminosity (-chemical composition) relation is predicted by stellar evolution models (see e.g. Chiosi et al. 1993; Bono et al. 2000a, and references therein), so that the mass can be removed in the previous van Albada & Baker relation, obtaining a Period-Luminosity-Temperature or a Period-Luminosity-Color (PLC) relation, that is foreseen to be obeyed by each individual Cepheid. The traditionally used Period-Luminosity (PL) relation of Classical Cepheids is the projection of the PLC relation onto the PL plane and is statistically derived as the average period at each fixed luminosity level. On this basis, reliable PL relations can be derived only if a statistically significant sample of pulsators is available and reflect the topology of the instability strip, in particular in the optical bands. As also shown by the observations (see e.g. Freedman & Madore 2010, and references therein), the intrinsic dispersion of PL relations decreases from the optical to the NIR bands; the metallicity and nonlinearity effects are predicted to do the same (see e.g. Caputo et al. 2000; Marconi et al. 2005). There are two possible approaches to reduce the PL systematics: i) to work in the NIR or Mid Infrared (MIR) filters (e.g. with the *James Webb Space Telescope*); ii) to adopt Period-Wesenheit (PW) relations (since Madore 1982) that are reddening-free formulations of PL relations, as they include a color term whose coefficient is exactly the ratio between the total and the selective extinction. The PW relations have the advantage of being insensitive to reddening and of partially correcting for the color extension of the strip. At the same time, they have the disadvantages of differing from true PLC relations and of relying on the assumption of an extinction law (e.g. Cardelli et al. 1989).

3. The state of the art of the modelling of pulsating stars

Several efforts have been made in the last few decades to model the observed properties of Cepheids and RR Lyrae among the various classes of pulsating stars. Recent theoretical studies of Cepheids and RR Lyrae are mainly based on the following three approaches: i) Linear non-adiabatic hydrodynamical models; ii) Nonlinear convective hydrodynamical 1D models; iii) Multidimensional hydrodynamical simulations. In linear non-adiabatic hydrodynamical models, non-adiabatic effects related to the driving mechanisms are taken into account but the hydrodynamic equations describing the pulsation phenomenon are linearized. In this scenario, only the periods and the instability strip edges can be estimated. Among the most recent results obtained from this kind of pulsation modeling, we can quote the prediction of Cepheid instability strip boundaries and period-age relations as a function of both metallicity and the efficiency of rotation, by Anderson et al. (2016). More recently, interesting results for the prediction of double mode RR Lyrae periods and period ratios have been obtained by Kovacs & Karamiqucham (2021) on the basis of their extensive sets of linear nonadiabatic models. In nonlinear convective hydrodynamical 1D models the hydrodynamic equations describing the pulsation phenomenon are not linearized, so that not only the periods and the instability strip edges but also the pulsation amplitudes (full amplitude variation of all the relevant quantities along the pulsation cycle) can be predicted. Several authors developed nonlinear convective pulsation models of Cepheids and RR Lyrae (Gehmeyr 1992; Bono & Stellingwerf 1994; Yecho et al. 1997; Kolláth et al. 2002; Bono et al. 1999; Szabó et al. 2004; Smolec & Moskalik 2008; Paxton et al. 2019). The bolometric light curves are then usually transformed into a variety of photometric filters through the adoption of model atmospheres (e.g. Castelli & Kurucz 2003), so that mean magnitudes and colors can also be derived. These are used, together with the pulsation periods to obtain theoretical PL, PLC and PW relations. The derived accurate topology of the instability strip with the blue and, in particular, the red edge location driven by the coupling between pulsation and convection allows us to derive synthetic PL relations.These are obtained by populating the region between the predicted boundaries with an assumed mass distribution (see e.g. Caputo et al. 2000; Fiorentino et al. 2007, for details). Figure 1 shows synthetic PL relations in the *HST* filters F555W, F184W and F160W as derived on the basis of the nonlinear convective pulsation models by De Somma et al. (2022) with metallicity varying from $Z = 0.004$ to $Z = 0.03$ (see labels). In multidimensional hydrodynamic simulations, the 1D assumption is released and the pulsation-convection coupling is treated in 2 or 3 dimensions (see e.g. Mundprecht et al. 2013, 2015, for a 2D approach to pulsation). These simulations provide more realistic predictions and useful

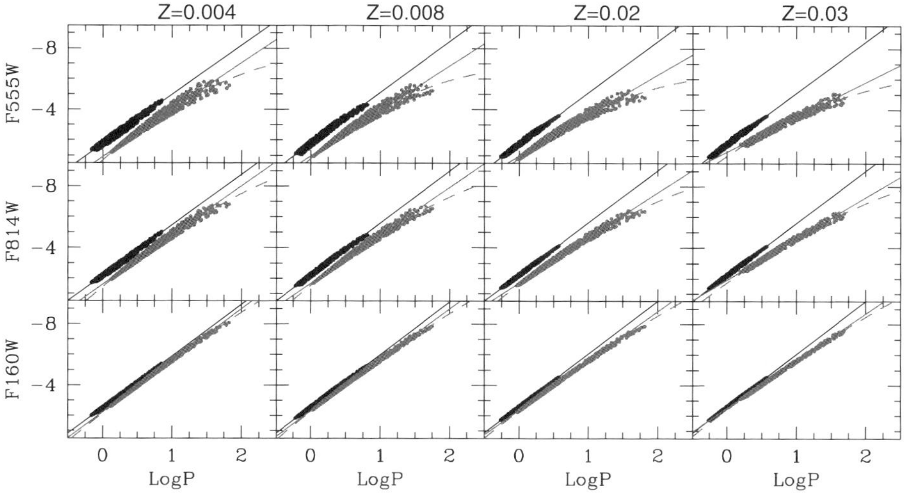

Figure 1. An example of synthetic PL relations for F-mode (red symbols) and FO-mode (black symbols) Cepheids based on nonlinear convective pulsation models for the labelled metallicities.

guidelines for developing descriptions of convection to be applied in 1D modeling (see also Deupree 2021, and reference therein)

4. Recent results for Classical Cepheids

The modelling of Classical Cepheids through non linear convective pulsation codes provides useful information for understanding the role of these pulsators as distance indicators and relatively young stellar popoulsation tracers. The recent nonlinear convective pulsation model results obtained through the radial stellar pulsation (RSP) module in the MESA code (Paxton et al. 2019), provided a new atlas of light and pulsation velocity curves, as well as period-amplitude diagrams, for both Cepheids and RR Lyrae, investigating the dependence of metallicity at fixed convective model parameters and of various model assumptions at fixed metallicity. Extensive model grids of Classical Cepheids have also been computed using an updated version of the original Stellingwerf's code (see Bono & Stellingwerf 1994; Bono et al. 1999, for details) for various assumptions about the efficiency of super-adiabatic convection, the mass-luminosity (ML) relation and the metal content (see De Somma et al. 2020, 2022, for details). In the quoted hydrocodes, the full amplitude behaviour of Cepheid models is obtained by integrating the pulsation equations together with those describing heat transfer and the coupling between pulsation and convection till the stability of the pulsation cycle. Then accurate predictions for the instability strip boundaries, and the variations of all relevant quantities along the pulsation cycle, can be derived for each assumed pulsation mode and combination of stellar parameters.

4.1. *Classical Cepheids as distance indicators*

Figure 2 shows the predicted instability strip as a function of the assumed chemical composition. Confirming previous results by Bono et al. (1999), we notice that the theoretical instability strip gets redder as the metallicity increases. This trend directly affects the predicted PL relations, especially in the optical bands, in the sense that more metal-rich Cepheids are expected to be fainter at a fixed period. On the other hand, the predicted dependence of the zero point of PW relations on metallicity is in the opposite

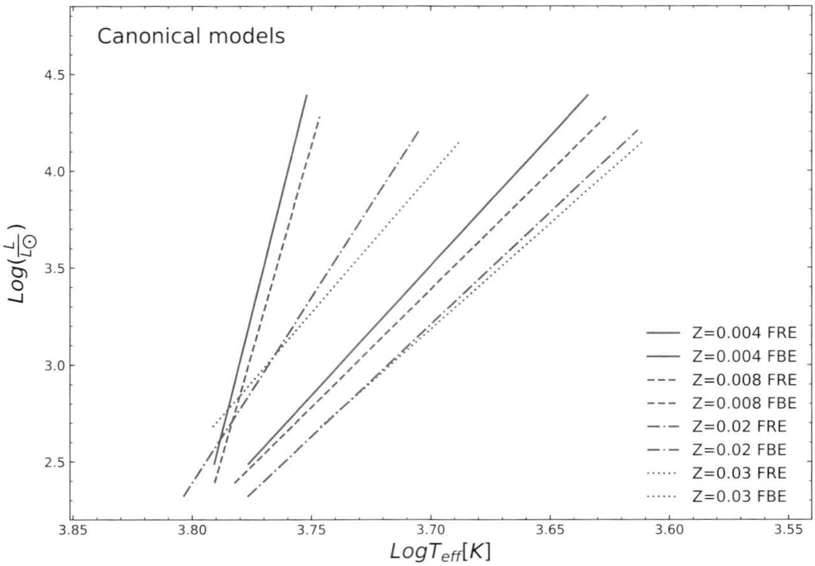

Figure 2. The predicted instability strip as a function of the assumed metallicity (see labels).

sense. Indeed fixing the period and the color, metal-rich Cepheids are predicted to be brighter (see e.g. Tables 19-22 in De Somma et al. 2022). In particular, in the *Gaia* filters, metal-dependent PW relations point towards a metallicity effect on the zero point varying from ~ -0.1 dex to ~ -0.2 dex for the Fundamental (F)-mode relations and from ~ -0.1 dex to ~ -0.3 dex for the First Overtone (FO)-mode relations. By applying these PWZ relations to a sample of Galactic Cepheids in the *Gaia* Data Release 3 (DR3) with available spectroscopic metal abundance measurements (see De Somma et al. 2022, for details), accurate theoretical parallaxes can be obtained. These results were directly compared with *Gaia* DR3 values. The agreement was found to be good, pointing towards a zero point parallax offset in agreement with Riess et al. (2021) results but dependent on the assumed convective efficiency and ML relation. In particular, if the effect of the assumed mixing length parameter is almost negligible, the effect of a brighter ML relation by 0.2 dex implies that the corresponding PWZ relations predict shorter distances by about 5 %. In principle a percentage variation in the inferred distance scale directly reflect in an equivalent variation in the resulting Hubble constant, thanks to the Hubble law. This point is interesting in view of improving our understanding of the so called Hubble constant tension, between early universe measurements and determinations and the values inferred from the extragalactic distance scale (see e.g. Verde et al. 2019; Riess et al. 2022, for detail).

4.2. *Classical Cepheids as age indicators*

Thanks to the relations between the stellar age and mass and between the stellar mass and luminosity, combined with the PL relation, accurate period-age (PA) and period-age-color (PAC) relations can be derived for Classical Cepheids, combining stellar evolution predictions for intermediate mass evolutionary tracks crossing the modeled instability strip (see e.g. Bono et al. 2005; Marconi et al. 2006; Anderson et al. 2016; De Somma et al. 2020, and references therein) with the Period-Mass-Luminosity-Temperature relations based on pulsation models. Using an extensive set of stellar evolution and nonlinear convective pulsation models for several chemical compositions, De Somma et al. (2021)

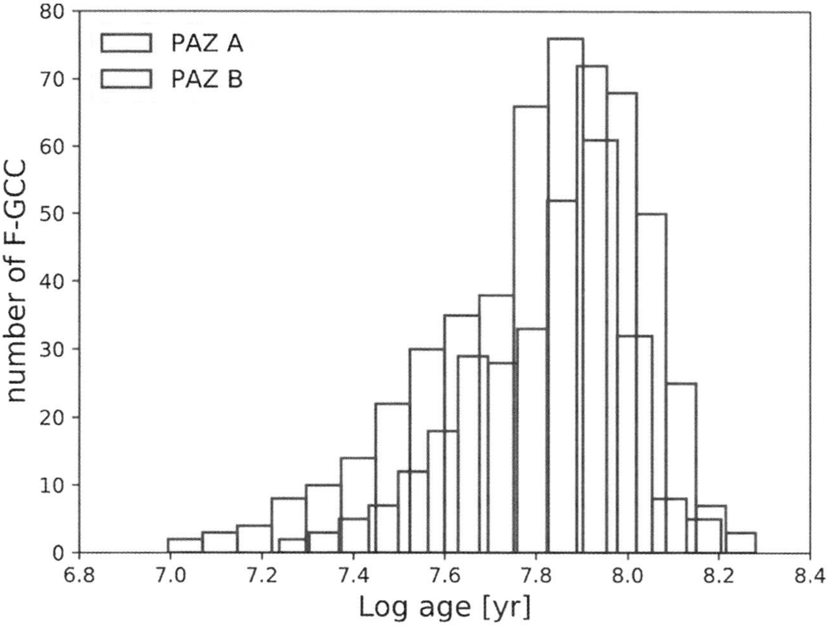

Figure 3. Inferred age distribution obtained by applying the canonical (A) and non-canonical (B) PAZ relations to the quoted *Gaia* F-mode Galactic Cepheid sample. Adapted from De Somma et al. (2021).

derived accurate metal-dependent PA (PAZ) and PAC (PACZ) relations (the latter in the *Gaia* filters). These relations were applied to the same sample of Galactic Cepheids in *Gaia* DR3 mentioned above and the theoretical age distribution was inferred for both F-mode and FO-mode pulsators (see De Somma et al. 2021, for details). The effect of the assumed efficiency of super-adiabatic convection is negligible, whereas a moderate overluminosity with respect to the canonical ML relation, implies that moderately older ages are predicted from PAZ and PACZ relations (see De Somma et al. 2021, for details). For example, in Figure 3 we show the inferred age distribution obtained by applying the canonical (A) and non-canonical (B) PAZ relations to the quoted *Gaia* F-mode Galactic Cepheid sample.

4.3. *The model fitting of Cepheid light and radial velocity curves*

One of the advantages of nonlinear convective hydrodynamic codes is the capability to predict the variation of luminosity, radius, effective temperature, gravity and radial velocity along a pulsation cycle. These theoretical variation curves can be directly compared with their observational counterparts. The first application of this model fitting technique was performed by Wood et al. (1997) for a Cepheid in the Large Magellanic Clouds. Several applications followed both in the Milky Way (see e.g. Natale et al. 2008; Gaia Collaboration et al. 2017) and in the Magellanic Clouds (see e.g. Bono et al. 2002; Keller & Wood 2002, 2006; Marconi et al. 2013a,b, 2017; Ragosta et al. 2019). In Figure 4 an example of model fitting of the observed light and radial velocity curves is shown for a few Galactic Cepheids in the *Gaia* DR3 database. Theoretical light curves were fitted to the observational ones, by shifting the former along the magnitude axis, in order to match the mean observational magnitude and by searching for the best-fit model matching the pulsation periods, amplitude and light curve morphology. On the other hand, theoretical pulsational velocity curves were fitted to observational radial velocities, by shifting the

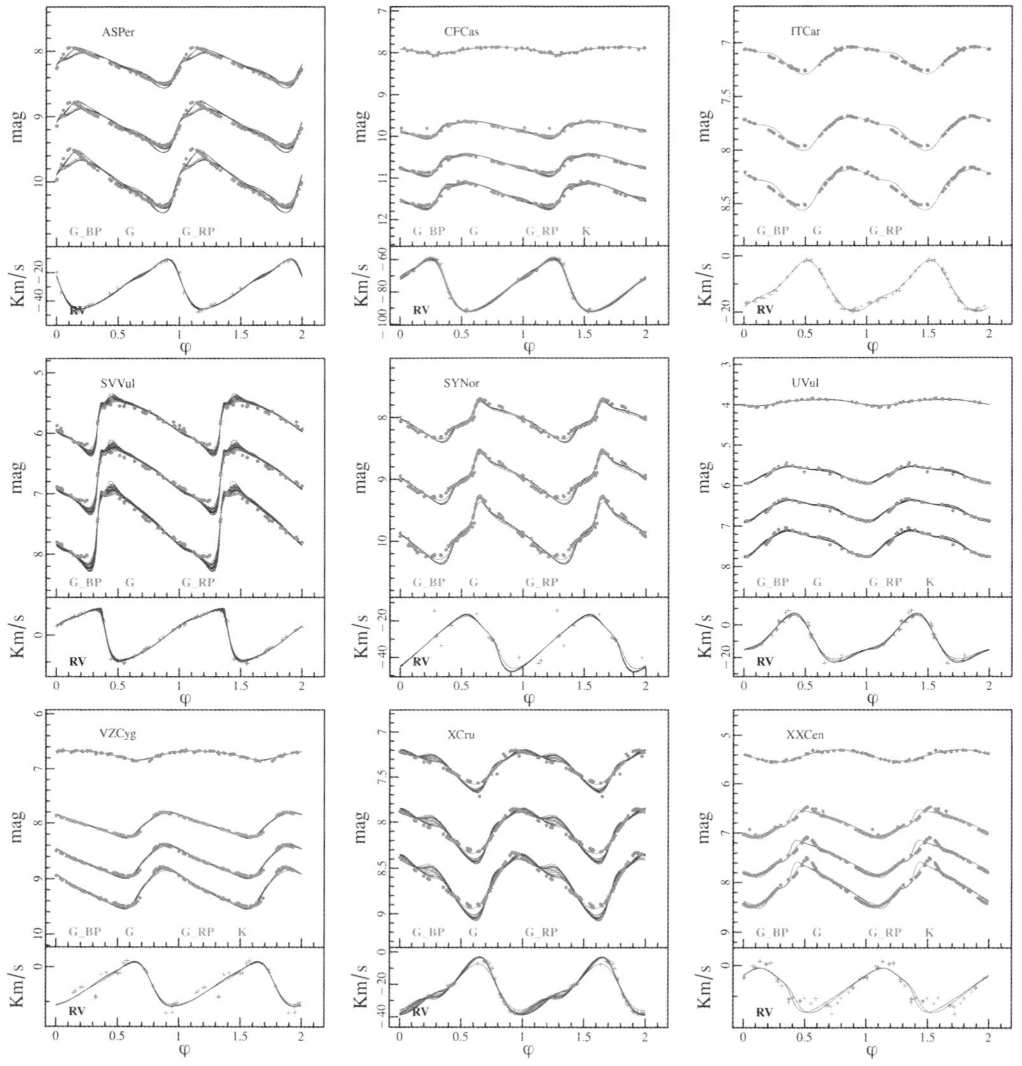

Figure 4. Model fitting of the observed light and radial velocity curves is shown for a few Galactic Cepheids in the *Gaia* DR3 database.

former along the velocity axis, in order to fit the mean observational velocity, and by stretching (through the p-factor) them in order to match the observational peak-to-peak radial velocity amplitude. A set of 100 bootstrap simulations was performed to compute the errors on the fitted parameters. In particular, the observational light and radial velocity curves were resampled (allowing data point repetitions) and the fitting procedure was repeated for every simulation. A statistical analysis of the obtained parameters of the fit allowed us to give an estimate of their errors. This investigation is part of a more complete analysis involving a few tens of Galactic Cepheids in the *Gaia* database (Molinaro et al. in prep).

From the model fitting, the individual distances and the intrinsic stellar parameters can be derived. In Figure 5 a comparison between the theoretical parallaxes inferred from the model fitting shown above, and the *Gaia* DR3 values, is displayed for the investigated stars. The good agreement obtained seems to support the predictive capability

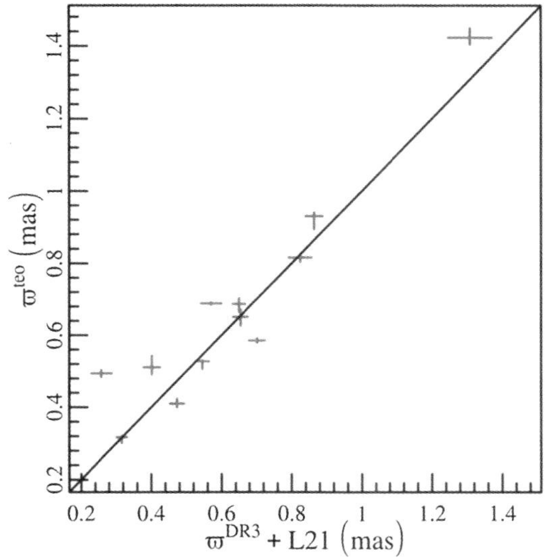

Figure 5. Comparison between the theoretical parallaxes inferred from the model fitting method aapplied to a subset of the considered *Gaia* Cepheids and the corresponding *Gaia* DR3 values.

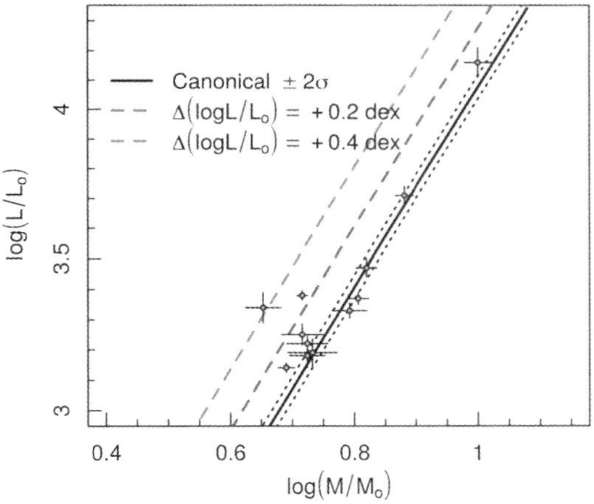

Figure 6. The inferred ML relation obtained from the model fitting of the investigated Cepheids, compared with the canonical (black solid line), moderately (red dashed line) and fully (cyan dashed line) non-canonical ML relations.

of current theoretical scenario. Similarly, in Figure 6, we show the inferred ML relation obtained from the model fitting method compared with the canonical (black solid line), moderately (red dashed line) and fully (cyan dashed line) non-canonical ML relations. From inspection of this plot, we notice that most of the studied stars seems to obey to a mildly non-canonical ML relation, with some of them lying on the canonical (no mass loss, no overshooting, no rotation) theoretical relation.

4.4. *The Hertzsprung progression*

This Hertzsprung progression (HP) phenomenon, discovered about one century ago (Hertzsprung 1926), consists of a relationship between the position of the bump along the light (and radial velocity) curves and the pulsation period, occurring in the period range from $\sim$ 6 to $\sim$ 16 d. The bump is observed on the descending branch for periods from $\sim$ 6 to $\sim$ 9 days, close to the true maximum of light/radial velocity curves for periods from $\sim$ 9 to $\sim$ 12 d, and on the rising branch for longer periods. The behaviour of pulsation amplitudes and Fourier parameters R_{21} and Φ_{21} of the bump is often considered to illustrate the phenomenon as these quantities show peculiar minima at the HP center (see Bono et al. 2000b, for details)

In Figure 7 we show an example of light (left panels) and radial pulsation velocity (right panels) curves for models of bump Cepheids varying the effective temperature and the period (see labels) at fixed mass ($M = 6.2M_{\odot}$) and luminosity ($\log L/L_{\odot} = 3.38$). We notice how the bump phase evolves from the descending to the rising branch and the curve becomes almost flat at the HP center.

As the period of the HP center is expected to depend on the stellar parameters, including the assumed ML relation and chemical composition, the comparison with observed data can provide interesting clues on the intrinsic move of observed bump Cepheids.

At the same time the interpretation of the evidence of the HP phenomenon in bump Cepheids observed by the *Gaia* satellite can be an important benchmark for pulsation models.

In Figure 8 we show the picture adapted from the ESA *Gaia* Image of the Week of May 27, 2022, showing the HP for a sample of *Gaia* Galactic Cepheids. According to this plot, the center of the progression seems to correspond to a period close to 8 days. The circle marks the center of the light curve HP, where the bump and the maximum are more or less at the same brightness level.

In a forthcoming paper (Marconi et al. in prep) we will present a very fine grid (decreased step in effective temperatures and mass) of nonlinear convective pulsation models for bump Cepheids, with metallicities ranging from 0.004 to 0.03, also increasing the helium abundance at fixed solar metallicity and different ML relations. Some preliminary results of this analysis are shown in Figure 9, where models across the HP are shown for the labelled chemical compositions and physical parameters. We notice how the period of the HP center decreases as the metallicity increases, with the $Z = 0.02$ case giving the best agreement with *Gaia* observations (see above).

5. Recent results for RR Lyrae stars

A combination of Horizontal Branch evolutionary predictions and pulsation modelling for central He burning stars is the key to understand the observed behaviours of RR Lyrae stars. As previously noted, RR Lyrae do not obey PL relations in the optical bands but are known to follow strict metal-dependent PL (PLZ) relations in the wavelength range from Cousins R to the Mid-Infrared *Spitzer* filters (see e.g. Marconi et al. 2015; Neeley et al. 2017; Marconi et al. 2022, and references therein). These PLZ relations, including a [Fe/H] term, can be inverted to derive the metallicity distributions of RR Lyrae that either share the same distance or have independently known individual distances. This approach to infer the metal abundance distribution has been applied by several authors both to Galactic Globular Clusters (see the case of ω Cen; Braga et al. 2016) and to satellites of the Milky Way (see e.g. the cases of Sculptor and Eridanus by Martínez-Vázquez et al. 2016, 2021, respectively). Metal-dependent PW relations have also been derived for RR Lyrae, with the minimum metallicity effect corresponding to some combination of optical filters, such as B,V, that is predicted to be metal-insensitive, or g,r (see Marconi et al. 2015, 2022, for details). Moreover, the predicted metallicity dependence of RR Lyrae NIR PL and PW relations is in agreement with observations in Galactic globular clusters (see Bhardwaj et al. 2023). As for the helium content, current models predict an effect on PL and PW relations in the sense that for an enhanced

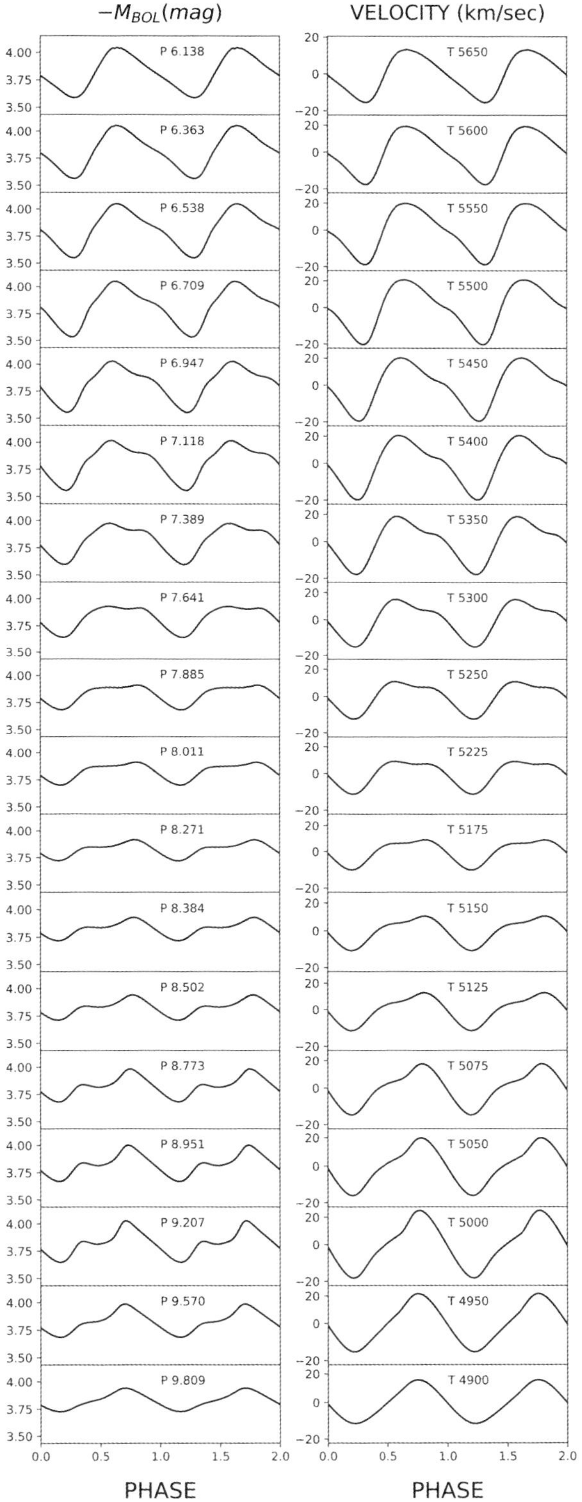

Figure 7. Light (left panels) and radial pulsation velocity (right panels) curves for models of bump Cepheids with $M = 6.2M_{\odot}$, $\log L/L_{\odot} = 3.38$. The effective temperatures and periods are labelled.

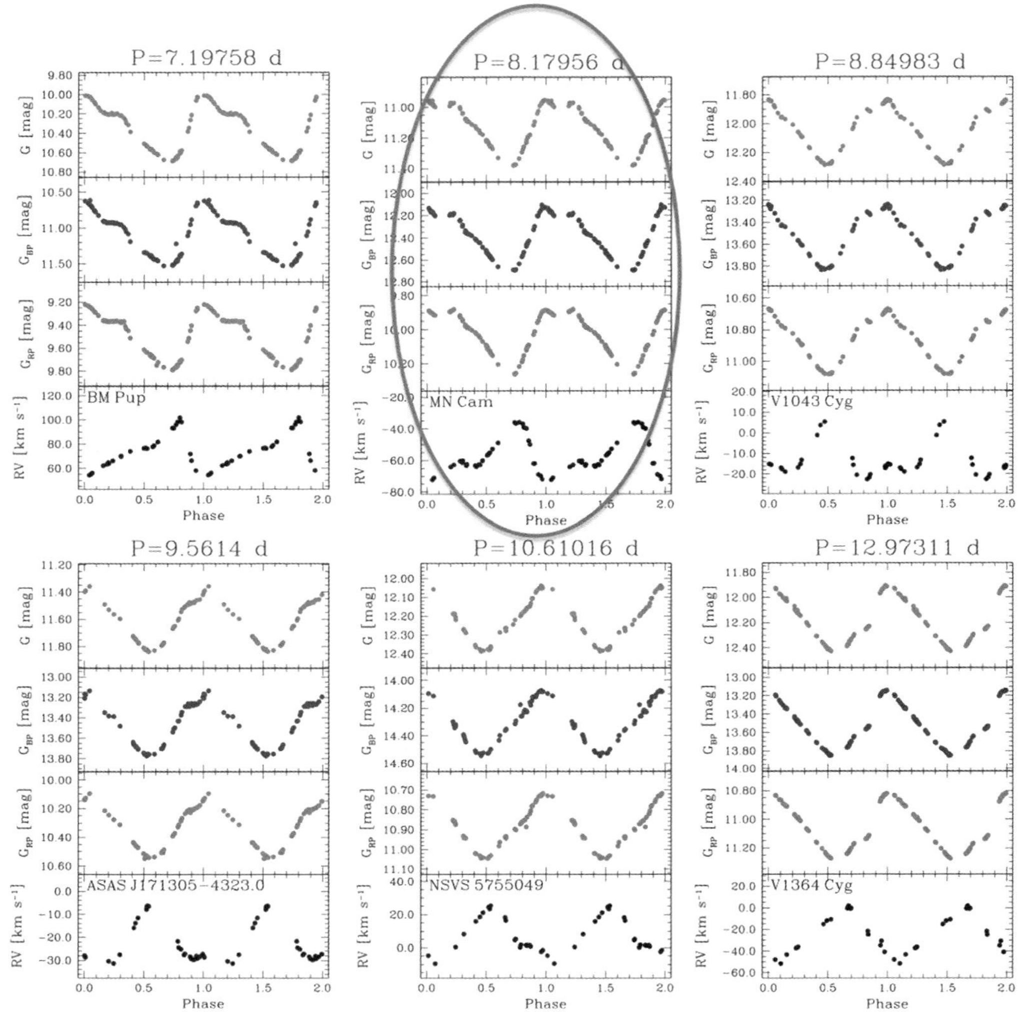

Figure 8. The HP for a subsample of *Gaia* Galactic Cepheids. The plot has been adapted from the ESA *Gaia* Image of the Week of May 27, 2022. The circle marks the center of the light curve HP.

Helium abundance longer periods and brighter magnitudes are expected on the basis of the increased luminosity (see e.g. Marconi et al. 2018, 2021, and references therein) due to the more efficient CNO-burning shell in the Horizontal Branch phase. Predicted parallaxes based on Period-Wesenheit-[Fe/H] relations were also successfully compared with EDR3 *Gaia* values (mean magnitudes from EDR3, [Fe/H] from spectroscopic literature data) for a sample of Galactic RR Lyrae (see Marconi et al. 2021, for details).

5.1. *The Bailey diagram*

Variations in the metal abundance are also expected to affect the distribution of RR Lyrae in period-amplitude planes (the so-called Bailey diagram, see Marconi et al. 2021, 2022, and references therein), in the sense that more metal-rich RR Lyrae are predicted to have shorter periods as an effect of the fainter Horizontal Branch level, and smaller amplitudes as an effect of the increased opacity and convection efficiency. This trend

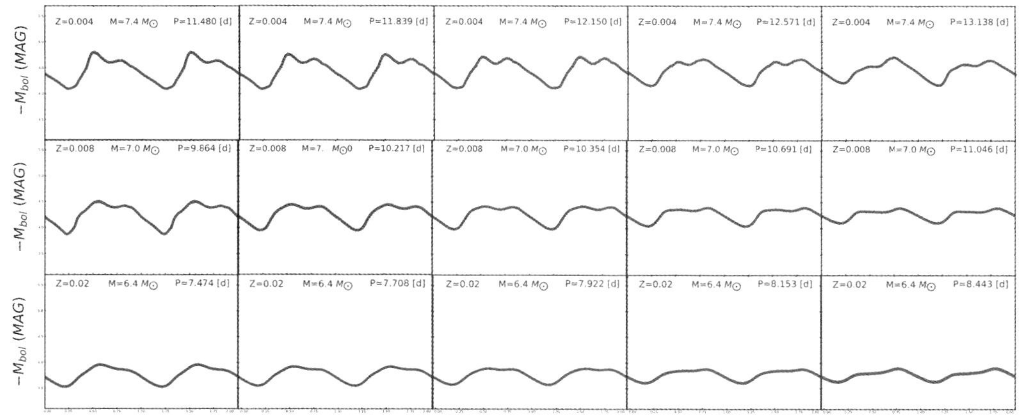

Figure 9. The HP for a subset of computed pulsation models across the HP center for the labelled chemical compositions and physical parameters.

is found to be in excellent agreement with the observational Bailey diagram based on *Gaia* data (see Clementini et al. 2022). In Figure 10 we show the quoted *Gaia* DR3 observational Bailey diagram for Galactic RR Lyrae with a metallicity color-coded scale, compared with theoretical predictions based on current nonlinear convective pulsation models for the labelled metal abundances. Solid lines are for Zero Age Horizontal Branch (ZAHB) models, dashed lines correspond to evolved RR Lyrae by 0.1 dex in log $L/L_\odot$. We notice that the same trend with metallicity is followed by theoretical results and observational data, even if the predicted amplitudes seem slightly higher than the observed ones. This effect can be due to an underestimated convective efficiency in the adopted pulsation models. Indeed increasing the mixing length parameter would reduce the pulsation amplitudes, as widely discussed in Di Criscienzo et al. (2004). However, a fine-tuning in the mixing length parameter is needed in order to simultaneously reproduce the observed instability strip width and pulsation amplitudes.

6. Conclusions and next steps

Classical pulsating stars are important distance indicators and stellar population tracers. Several theoretical efforts have been made to understand the pulsation mechanism and model the observed properties. The physical bases of stellar pulsation, of the existence of the instability strip and of PL and PLC relations are well known. Current nonlinear convective pulsation models are able to predict most of the relevant properties of classical pulsating stars, in particular Cepheids and RR Lyrae, generally in good agreement with the observations. However, model predictions depend on a number of physical and numerical assumptions that need to be constrained through the comparison with independent results. In order to improve current nonlinear convective pulsation models we need to work on three main aspects: i) To improve the input physics and the adopted atmosphere models; ii) To improve the treatment of convection, also taking into account results of 2D and 3D simulations; iii) to compare results of different pulsation codes to cross-calibrate physical and numerical assumptions; iv) to consider still open issues. Concerning the first point, accurate and updated opacity and equation of state tables or analytical formulations have to be adopted, consistently with the physical assumptions of stellar evolution models that provide the input parameters for the static envelope structures. As for the step forward in the treatment of convection, the limitations of current approaches are evident from the failure of the model fitting of observed light curves, e.g. for RR Lyrae models, close to the red edge of the instability strip (see

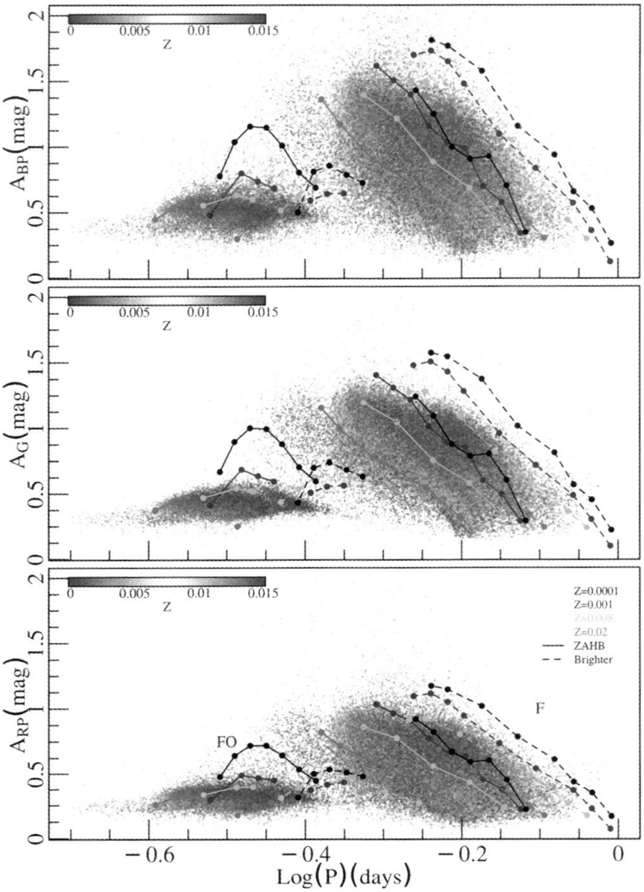

Figure 10. *Gaia* DR3 observational Bailey diagram for Galactic RR Lyrae with a metallicity color-coded scale, compared with theoretical predictions based on current nonlinear convective pulsation models for the labelled metallicities. Solid lines are for ZAHB models, dashed lines correspond to evolved RR Lyrae by 0.1 dex in $\log L/L_{\odot}$.

e.g. Marconi & Degl'Innocenti 2007), in spite of the general success of the model fitting technique. 2D and 3D simulations can provide useful inputs to model in a more realistic way the coupling between pulsation and convection in pulsating stars. Concerning the third aspect, namely the comparison among different model predictions, we think this is an important approach to better constrain the free parameters involved in the various hydrocodes and refer the interested reader to the paper by Kovács et al. (2023) who compare the results of the Budapest-Florida code with the ones obtained from the MESA RSPs module (but see also Smolec & Moskalik 2008). Finally, future models should be able to take into account open issues such as the Blazhko effect for RR Lyrae stars (see e.g. Kolenberg et al. 2011; Netzel et al. 2018) and the difficult prediction of double mode pulsators among classical Cepheids (see e.g. Smolec & Moskalik 2008).

References

Anderson, R. I., Saio, H., Ekström, S., et al. 2016, A&AS, 591, A8. doi:10.1051/0004-6361/201528031

Bhardwaj, A., Marconi, M., Rejkuba, M., et al. 2023, ApJL, 944, L51. doi:10.3847/2041-8213/acba7f

Bono, G. & Stellingwerf, R. F. 1994, ApJS, 93, 233. doi:10.1086/192054
Bono, G., Marconi, M., & Stellingwerf, R. F. 1999, ApJS, 122, 167. doi:10.1086/313207
Bono, G., Caputo, F., Cassisi, S., et al. 2000, ApJ, 543, 955. doi:10.1086/317156
Bono, G., Marconi, M., & Stellingwerf, R. F. 2000, A&A, 360, 245. doi:10.48550/arXiv.astro-ph/0006229
Bono, G., Castellani, V., & Marconi, M. 2002, ApJL, 565, L83. doi:10.1086/339420
Bono, G., Marconi, M., Cassisi, S., et al. 2005, ApJ, 621, 966. doi:10.1086/427744
Braga, V. F., Stetson, P. B., Bono, G., et al. 2016, AJ, 152, 170. doi:10.3847/0004-6256/152/6/170
Breuval, L., Riess, A. G., Kervella, P., et al. 2022 ApJ 939, 89. doi:10.3847/1538-4357/ac97e2
Caputo, F., Marconi, M., & Musella, I. 2000, A&A, 354, 610. doi:10.48550/arXiv.astro-ph/9911441
Cardelli, J. A., Clayton, G. C., & Mathis, J. S. 1989, ApJ, 345, 245. doi:10.1086/167900
Castelli, F. & Kurucz, R. L. 2003, Modelling of Stellar Atmospheres, 210, A20. doi:10.48550/arXiv.astro-ph/040508
Catelan, M., Pritzl, B. J., & Smith, H. A. 2004, ApJS, 154, 633. doi:10.1086/422916
Chiosi, C., Wood, P. R., & Capitanio, N. 1993, ApJS, 86, 541. doi:10.1086/191790
Clementini, G., Ripepi, V., Garofalo, A., et al. 2022, arXiv:2206.06278. doi:10.48550/arXiv.2206.06278
Cox, A. N. & Hodson, S. W. 1978, The HR Diagram - The 100th Anniversary of Henry Norris Russell, 80, 237
De Somma, G., Marconi, M., Molinaro, R., et al. 2020, ApJS, 247, 30. doi:10.3847/1538-4365/ab7204
De Somma, G., Marconi, M., Cassisi, S., et al. 2020, MNRAS, 496, 5039. doi:10.1093/mnras/staa1834
De Somma, G., Marconi, M., Cassisi, S., et al. 2021, MNRAS, 508, 1473. doi:10.1093/mnras/stab2611
De Somma, G., Marconi, M., Molinaro, R., et al. 2021, RR Lyrae/Cepheid 2019: Frontiers of Classical Pulsators, 529, 27
De Somma, G., Marconi, M., Molinaro, R., et al. 2022, ApJS, 262, 25. doi:10.3847/1538-4365/ac7f3b
Deupree, R. 2021, RR Lyrae/Cepheid 2019: Frontiers of Classical Pulsators, 529, 111
Di Criscienzo, M., Marconi, M., & Caputo, F. 2004, ApJ, 612, 1092. doi:10.1086/422742
Eddington, A. S. 1918, MNRAS, 79, 2
Fiorentino, G., Marconi, M., Musella, I., et al. 2007, A&A, 476, 863. doi:10.1051/0004-6361:2007758
Freedman, W. L. & Madore, B. F. 2010, ARA&A, 48, 673. doi:10.1146/annurev-astro-082708-101829
Gaia Collaboration, Clementini, G., Eyer, L., et al. 2017, A&A, 605, A79. doi:10.1051/0004-6361/201629925
Gehmeyr, M. 1992, ApJ, 399, 265. doi:10.1086/171921
Hertzsprung, E. 1926, Bulletin of the Astronomical Institutes of the Netherlands, 3, 115
Yecho, P., Kollath, Z., & Buchler, J. R. 1997, IAU Joint Discussion
Keller, S. C. & Wood, P. R. 2002, ApJ, 578, 144. doi:10.1086/342315
Keller, S. C. & Wood, P. R. 2002, ApJ, 578, 144. doi:10.1086/342315
Keller, S. C. & Wood, P. R. 2006, ApJ, 642, 834. doi:10.1086/501115
Kolenberg, K., Bryson, S., Szabó, R., et al. 2011, MNRAS, 411, 878. doi:10.1111/j.1365-2966.2010.17728.x
Kolláth, Z., Buchler, J. R., Szabó, R., et al. 2002, A&A, 385, 932. doi:10.1051/0004-6361:20020182
Kovacs, G. & Karamiqucham, B. 2021, A&A, 653, A61. doi:10.1051/0004-6361/202141100
Kovács, G. B., Nuspl, J., & Szabó, R. 2023, MNRAS, 521, 4878. doi:10.1093/mnras/stad884
Longmore, A. J., Dixon, R., Skillen, I., et al. 1990, MNRAS, 247, 684
Madore, B. F. 1982, ApJ, 253, 575. doi:10.1086/159659

Marconi, M., Bono, G., Caputo, F., et al. 2006, memSAI, 77, 67. doi:10.48550/arXiv.astro-ph/0510033

Marconi, M., Musella, I., & Fiorentino, G. 2005, ApJ, 632, 590. doi:10.1086/432790

Marconi, M. & Degl'Innocenti, S. 2007, A&A, 474, 557. doi:10.1051/0004-6361:20065840

Marconi, M., Molinaro, R., Ripepi, V., et al. 2013, MNRAS, 428, 2185. doi:10.1093/mnras/sts197

Marconi, M., Molinaro, R., Bono, G., et al. 2013, ApJL, 768, L6. doi:10.1088/2041-8205/768/1/L6

Marconi, M., Coppola, G., Bono, G., et al. 2015, ApJ, 808, 50. doi:10.1088/0004-637X/808/1/50

Marconi, M., Molinaro, R., Ripepi, V., et al. 2017, MNRAS, 466, 3206. doi:10.1093/mnras/stw3289

Marconi, M., Bono, G., Pietrinferni, A., et al. 2018, ApJL, 864, L13. doi:10.3847/2041-8213/aada17

Marconi, M., Molinaro, R., Ripepi, V., et al. 2021, MNRAS, 500, 5009. doi:10.1093/mnras/staa3558

Marconi, M., Molinaro, R., Dall'Ora, M., et al. 2022, ApJ, 934, 29. doi:10.3847/1538-4357/ac78ee

Martínez-Vázquez, C. E., Monelli, M., Cassisi, S., et al. 2021, MNRAS, 508, 1064. doi:10.1093/mnras/stab2493

Martínez-Vázquez, C. E., Stetson, P. B., Monelli, M., et al. 2016, MNRAS, 462, 4349. doi:10.1093/mnras/stw1895

Mundprecht, E., Muthsam, H. J., & Kupka, F. 2013, MNRAS, 435, 3191. doi:10.1093/mnras/stt1511

Mundprecht, E., Muthsam, H. J., & Kupka, F. 2015, MNRAS, 449, 2539. doi:10.1093/mnras/stv434

Natale, G., Marconi, M., & Bono, G. 2008, ApJL, 674, L93. doi:10.1086/526518

Neeley, J. R., Marengo, M., Bono, G., et al. 2017, ApJ, 841, 84. doi:10.3847/1538-4357/aa713d

Netzel, H., Smolec, R., Soszyński, I., et al. 2018, MNRAS, 480, 1229. doi:10.1093/mnras/sty1883

Paxton, B., Smolec, R., Schwab, J., et al. 2019, ApJS, 243, 10. doi:10.3847/1538-4365/ab2241

Ragosta, F., Marconi, M., Molinaro, R., et al. 2019, MNRAS, 490, 4975. doi:10.1093/mnras/stz2881

Riess, A. G., Casertano, S., Yuan, W., et al. 2021, ApJL, 908, L6. doi:10.3847/2041-8213/abdbaf

Riess, A. G., Breuval, L., Yuan, W., et al. 2022, ApJ, 938, 36. doi:10.3847/1538-4357/ac8f24

Ripepi, V., Catanzaro, G., Molinaro, R., et al. 2021, MNRAS, 508, 4047. doi:10.1093/mnras/stab2460

Smolec, R. & Moskalik, P. 2008, AcA, 58, 193. doi:10.48550/arXiv.0809.1979

Szabó, R., Kolláth, Z., & Buchler, J. R. 2004, A&A, 425, 627. doi:10.1051/0004-6361:20035698

Trentin, E., Ripepi, V., Catanzaro, G., et al. 2023, MNRAS, 519, 2331. doi:10.1093/mnras/stac2459

van Albada, T. S. & Baker, N. 1971, ApJ, 169, 311. doi:10.1086/151144

Verde, L., Treu, T., & Riess, A. G. 2019, Nature Astronomy, 3, 891. doi:10.1038/s41550-019-0902-0

Wood, P. R., S. Arnold, A., & Sebo, K. M. 1997, ApJL, 485, L25. doi:10.1086/310798

Zhevakin, S. A. 1963, ARA&A, 1, 367. doi:10.1146/annurev.aa.01.090163.002055

7. Discussion

Question (Anderson): Does the relatively strong temperature-dependence of the instability strip boundaries on metallicity in the Marconi/De Somma/Bono models explain why Period-Wesenheit relations (which combine one filter with one color) predict a negative sign to the metallicity effect, while each of their individual filter PL-relations have positive sign?

Answer: Exactly.

Comment (Anderson): Geneva models exhibit negative metallicity effects in each of the individual bands analyzed as well as the Period-Wesenheit relations. The dependence of the instability strip boundaries in these models is very small (Anderson et al. 2016).

Answer: The IS dependence on metallicity reflects on the PL dependence on metallicity. If you predict that the IS depends mildly on metallicity, we will have PL relations mildly dependent on metallicity. In our cases, we find quite an important metallicity effect that is in some case counterbalanced by helium variations. These have not been taken into account so far in the De Somma et al. models, but Fiorentino et al. models had already shown that if you also increase the helium content at solar or super-solar metallicity, the effect is exactly the opposite: the IS is shifted to the blue. In any case, when you correct for the color term, building Period-Wesenheit relations you are making things more accurate because you are locating the star where it should be and the fact that a negative coefficient of the metallicity term is found is in agreement with several results in the literature. The PL dependence on metallicity reflects the IS dependence; so it is important to understand what is the effect of the adopted opacity, helium-to-metal abundance ratio, there are a lot of things that affect this dependence and maybe the convective treatment should also be differential depending on the models that you are building. Certainly an important aspect to investigate, I agree with you.

Comment (Anderson): Detailed comparison with CMD with the models will help to clarify more.

At the crossroads of astrophysics and cosmology: Period–luminosity relations in the 2020s
Proceedings IAU Symposium No. 376, 2024
R. de Grijs, P. A. Whitelock & M. Catelan, eds.
doi:10.1017/S1743921323002417

Anomalous Cepheids: Updated Theoretical Period–Luminosity–Color and Period–Wesenheit Relations

Giulia De Somma[1,2,3], **Marcella Marconi**[1], **Santi Cassisi**[3,4] **and Roberto Molinaro**[1]

[1]INAF-Osservatorio astronomico di Capodimonte, Via Moiariello 16, 80131 Napoli, Italy.
email: giulia.desomma@inaf.it

[2]Istituto Nazionale di Fisica Nucleare (INFN), Sez. di Napoli, Compl. Univ. di Monte S. Angelo, Edificio G, Via Cinthia, I-80126, Napoli, Italy

[3]INAF-Osservatorio Astronomico d'Abruzzo, Via Maggini sn, 64100 Teramo, Italy

[4]Istituto Nazionale di Fisica Nucleare (INFN), Sezione di Pisa, Universitá di Pisa, Largo Pontecorvo 3, 56127 Pisa, Italy

Abstract. In the context of a project aimed to provide an updated theoretical scenario for various classes of radially pulsating stars, we present the first results obtained for anomalous Cepheids. By adopting reliable and updated evolutionary prescriptions concerning the luminosity levels for various core He-burning stellar models with masses suitable for entering the instability strip, we have computed nonlinear convective pulsation models for both fundamental and first-overtone mode anomalous Cepheids by exploring the impact of varying the metal abundance as well as the efficiency of super-adiabatic convection. These numerical simulations have allowed us to retrieve the boundaries of the instability strip and all relevant pulsation properties, namely period, amplitude, bolometric light and radial velocity curves. This theoretical scenario has been transformed into the *Gaia* photometric system to derive the first theoretical Period–Luminosity–Colour and Period–Wesenheit relations in the *Gaia* bands.

Keywords. stellar evolution, stellar oscillations, Cepheids variable stars

1. Introduction

Anomalous Cepheids (hereinafter ACs) are metal-poor, radially pulsating stars. They are called 'anomalous', because if compared, at a fixed period, with other luminous metal-poor variable stars such as Type II Cepheids, ACs appear significantly brighter. These stars have masses between $1.2M_\odot$ and $2.2M_\odot$, which means they have masses higher than the typical RR Lyrae mass ($0.6 < M/M_\odot < 0.8$) but lower than those of classical Cepheids ($3 < M/M_\odot < 15$) (Caputo et al 1998; Caputo et al. 2004; Marconi et al. 2004). Their evolutionary stage corresponds to the central helium (He)-burning phase.

Depending on their initial mass, their progenitors along the red-giant branch (RGB) develop, inside the He core, conditions of (only) partial electron degeneracy; this has the consequence that their He core mass at the beginning of the core He-burning phase is significantly lower than that characteristic of RR Lyrae stars. In addition, the He core mass strongly depends on the initial mass of the progenitor as well (see Salaris & Cassisi 2005, for a detailed discussion on this topic). Nonetheless, their progenitors will ignite helium by removing the electron degeneracy through a mild He flash at the tip of the RGB. During the central He-burning phase, stars in these mass and metallicity ($Z \leq$

0.0006) ranges are predicted to evolve from the turnover of the Zero Age Horizontal Branch (ZAHB) (see, e.g. Caputo et al 1998; Caputo et al. 2004, for details).

Owing to their mass, ACs are typically associated with intermediate-age stellar populations, an idea also supported by the evidence that they are quite abundant in Local Group dwarf galaxies characterized by an extended star formation histories (Ripepi et al. 2014; Martínez-Vázquez et al. 2021). This notwithstanding, their formation scenario is still an unsettled issue. In fact, it has also been suggested that (at least) a fraction of the observed population of ACs could be associated with mass transfer in interacting binary systems (Renzini et al. 1977; Monelli & Fiorentino 2022).

Their metallicity typically ranges between $Z = 10^{-4}$ and $Z = 6 \times 10^{-4}$ but in the last few years, observational evidence has emerged that it is possible that they could have also slightly higher metallicities (around $Z = 10^{-3}$) (Monelli & Fiorentino 2022).

ACs pulsate in both their fundamental (F) and first-overtone (FO) modes: F-mode ACs have periods from a few hours to more than two days, while FO-mode ACs have periods from about 0.3 days to slightly longer than 1 day. For a long time, knowledge of these objects was extremely limited, because very few of them had been observed. This situation first changed with the OGLE project (Soszyński et al. 2015, 2020) and then the *Gaia* mission (Gaia Collaboration et al. 2016, 2018, 2021). To provide an interpretative scenario useful to understand the huge amount of data being provided by the ongoing spectroscopic and astrometric surveys, it is crucial to update the pulsation and evolutionary scenarios for these objects.

In this context, as a first step, we computed a fine grid of nonlinear convective AC pulsation models (see, e.g. Marconi et al. 2005; De Somma et al. 2020) for two chemical compositions ($Z = 0.0001, Y = 0.245$ and $Z = 0.0003, Y = 0.245$), varying the mixing length parameter from $\alpha_{\rm ml} = 1.5$ to $\alpha_{\rm ml} = 1.7$. We investigated both the F and FO modes. Here we discuss the first results obtained from this data set only for the models with $\alpha_{\rm ml} = 1.5$. A more extensive investigation will be presented in De Somma et al. (in preparation).

2. The instability strip's boundaries

Present pulsational computations allow us to retrieve updated prescriptions about the instability strip's (IS) boundaries for all pulsation modes investigated (see, e.g. De Somma et al. 2020; De Somma et al. 2022). We present a preliminary analysis we have started of pulsation and evolutionary model predictions. The stellar model sets correspond to the updated version of the BaSTI stellar evolutionary library for the solar-scaled heavy element distribution, presented by Hidalgo et al. (2018).

Figure 1 exhibits evolutionary tracks for different progenitor masses and the ISs for $\alpha_{\rm ml} = 1.5$, for both F and FO-mode pulsators (see the legend in the figure). Each progenitor mass is associated with stellar models of different total mass but the same He-core mass as the RGB progenitor.

Examining the models shown in Figure 1, it is evident that when selecting a progenitor mass of $1.8 M_\odot$ and $Z = 0.001$, the objects that enter the IS during the He-core-burning phase have masses of $1.4 M_\odot$, $1.6 M_\odot$ and $1.8 M_\odot$ (left panels of Figure 1). Conversely, for $Z = 0.0003$ (right panels of Figure 1) and the same progenitor mass of $1.8 M_\odot$, the masses entering the IS are $1.6 M_\odot$ and $1.8 M_\odot$.

To better evaluate the probability of these objects to produce ACs, a thorough analysis of their crossing times of the IS is planned (De Somma et al., in preparation).

3. Atlas of light and radial velocity curves

From our nonlinear computation, we obtained the bolometric light curves and the radial velocity curves for each model. For example, Figures 2 and 3 show the light and radial

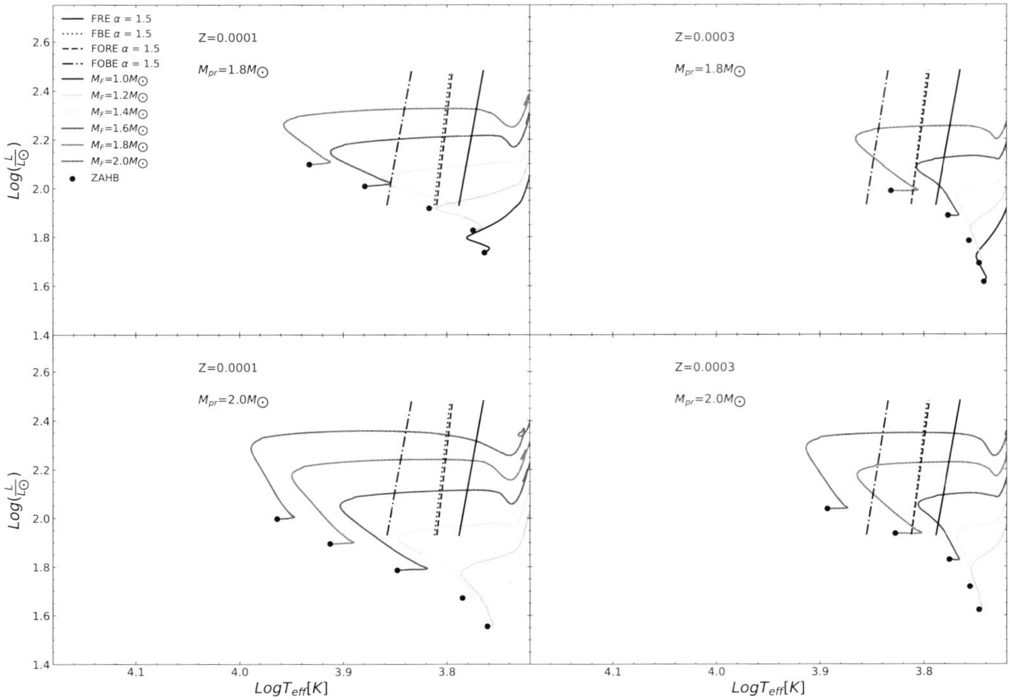

Figure 1. Location in the Hertzsprung–Russell (HR) diagram of a subsample of stellar models from the BaSTI–IAC library, for two assumptions of the initial chemical compositions. The left panels refer to $Z = 0.0001$, while the right panels correspond to $Z = 0.0003$ (see the text for more details). The predicted boundaries of the IS, the fundamental blue edge (FBE), the fundamental red edge (FRE), the FO blue edge (FOBE) and the FO red edge (FORE), for both F and FO-mode pulsators, are also shown. The black solied point indicates the ZAHB.

velocity curves obtained for $1.4M_\odot$, $Z = 0.0001$, $Y = 0.245$ and $\alpha_{\rm ml} = 1.5$, and $2.0M_\odot$, $Z = 0.0003$, $Y = 0.245$ and $\alpha_{\rm ml} = 1.5$, respectively.

4. Results in the *Gaia* filters

By employing non-overshooting ATLAS9 model atmospheres (Castelli & Kurucz 2003), we converted the bolometric light curves into the *Gaia* photometric passbands. As a result, we have generated the first theoretical catalog of *Gaia* light curves for ACs, providing a valuable resource to the scientific community.

4.0.1. The first PLC, PLCZ and PW relations

These light curves are used to derive mean $\langle G\rangle$ magnitudes and $\langle G_{\rm BP}\rangle - \langle G_{\rm RP}\rangle$ colors which, hence, are used to derive the first period–luminosity–color (PLC) and Period–Wesenheit (PW) relations in the *Gaia* filters. Table 1 provides the coefficients of the PLC relations in the *Gaia* filters ($\langle G\rangle = a + b\log P + c(\langle G_{\rm BP}\rangle - \langle G_{\rm RP}\rangle)$ for $Z = 0.0001$ (top section of the table) and $Z = 0.0003$ (bottom section of the table). To investigate the metallicity dependence in the *Gaia* filters for F and FO-mode pulsators, we derived period–luminosity–color–metallicity (PLCZ) relations (see Table 2). Based on the metallicity terms derived, we can confidently state that the metallicity dependence is almost negligible. Specifically, fixing the period and color, a 0.2 dex variation in [Fe/H] corresponds to a variation in $\langle G\rangle$ of $\sim$0.002 mag for the F mode and $\sim$0.0002 mag for the FO

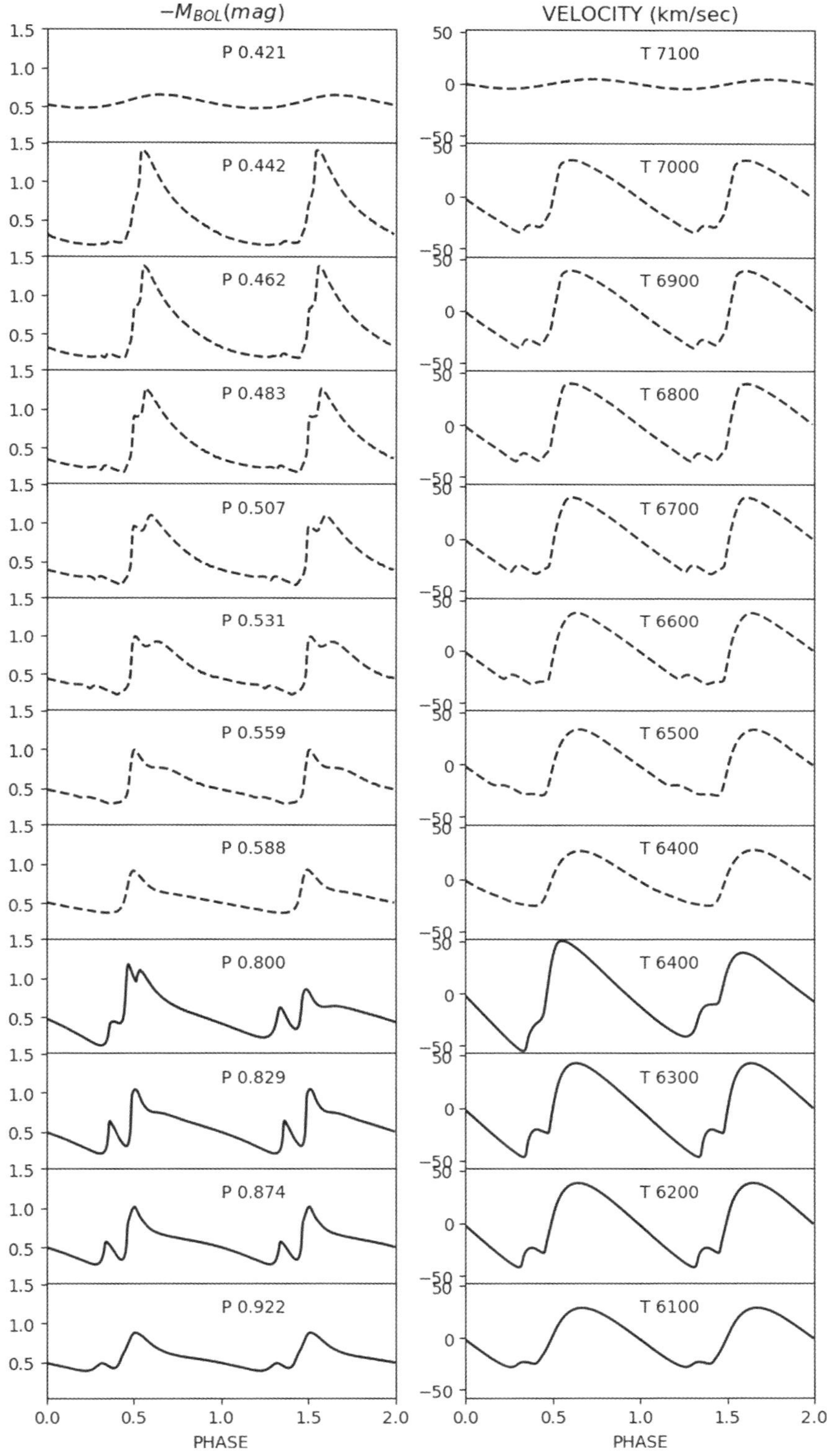

Figure 2. Bolometric light curves (left panel) and radial velocity curves (right panel) for $1.4M_{\odot}$ nonlinear F (solid line) and FO- (dashed lines) mode models derived for $Z = 0.0001, Y = 0.245$ and $\alpha_{\rm ml} = 1.5$. On each bolometric light curve, the period (in days) is labeled. On each radial velocity curve, the effective temperature (in Kelvin) is indicated.

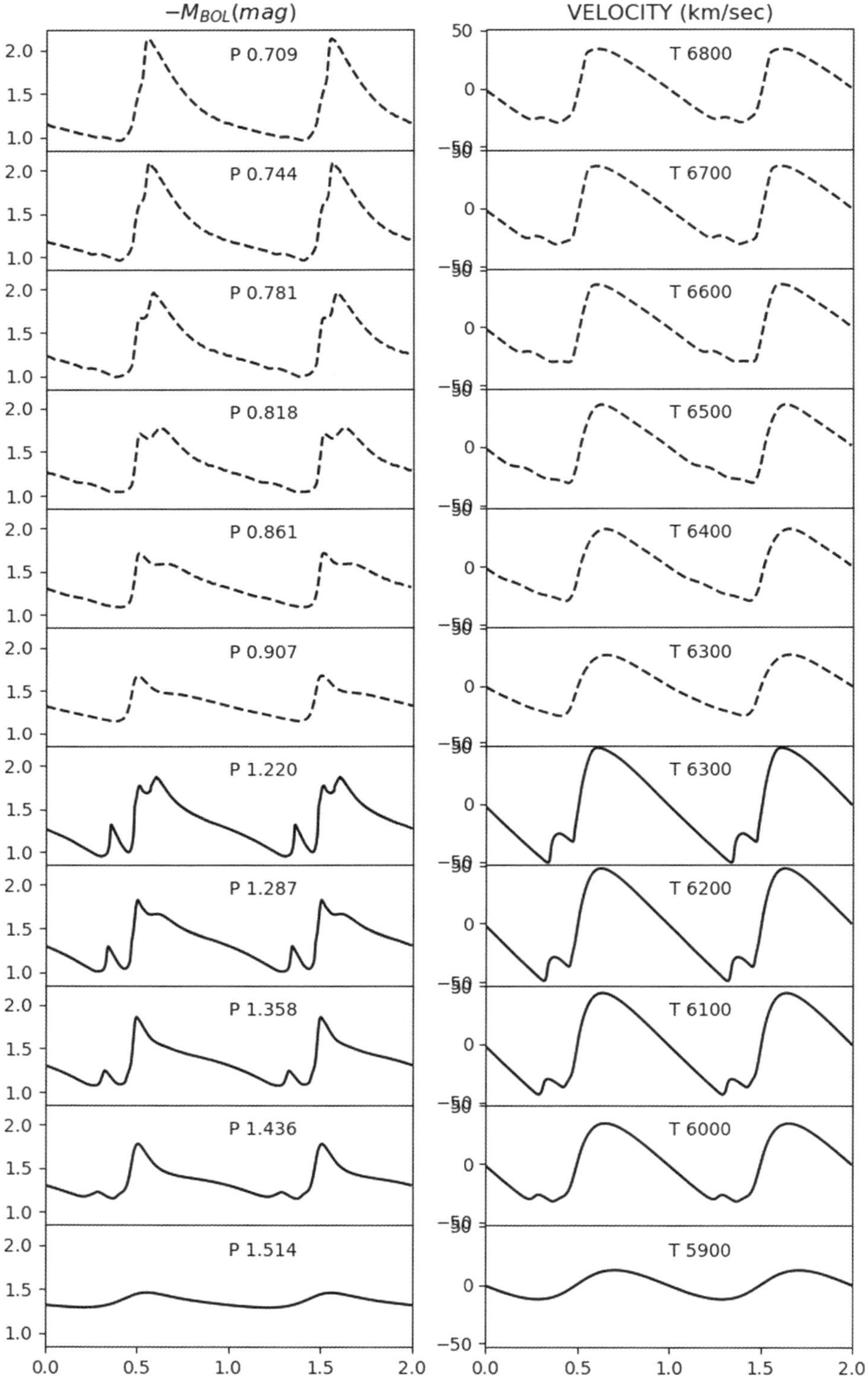

Figure 3. As Figure 2, but for $Z = 0.0003$.

Table 1. PLC coefficients in the *Gaia* EDR3 filters ($\langle G\rangle = a + b\log P + c(\langle G_{\rm BP}\rangle - \langle G_{\rm RP}\rangle)$ for F and FO ACs with $\alpha_{\rm ml} = 1.5$

mode	Z	Y	a	b	c	σ_a	σ_b	σ_c	σ	R^2
F	0.0001	0.245	−3.363	−4.742	4.511	0.053	0.028	0.095	0.010	0.999
FO	0.0001	0.245	−3.378	−4.656	3.367	0.030	0.029	0.061	0.017	0.999
F	0.0003	0.245	−3.250	−4.789	4.319	0.050	0.029	0.091	0.010	0.999
FO	0.0003	0.245	−3.357	−4.669	3.312	0.010	0.010	0.021	0.005	1.000

Table 2. PLCZ coefficients in the *Gaia* EDR3 filters ($\langle G\rangle = a + b\log P + c(\langle G_{\rm BP}\rangle - \langle G_{\rm RP}\rangle + d[{\rm Fe/H}])$ for F and FO ACs with $\alpha_{\rm ml} = 1.5$.

mode	a	b	c	d	σ_a	σ_b	σ_c	σ_d	σ	R^2
F	−3.282	−4.763	4.402	0.008	0.042	0.023	0.073	0.073	0.012	0.999
FO	−3.367	−4.663	3.342	0.001	0.021	0.016	0.034	0.034	0.013	0.999

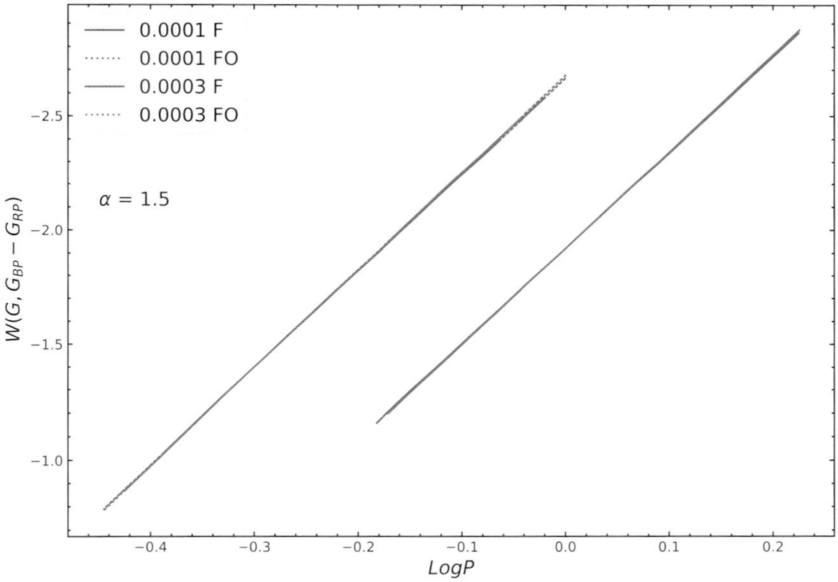

Figure 4. PW relations ($\langle W\rangle = \langle G\rangle - 1.9\langle G_{\rm BP} - G_{\rm RP}\rangle = a + b\log P$) for F and FO ACs derived for $Z = 0.0001$ (red lines) and $Z = 0.0003$ (green lines) with $\alpha_{\rm ml} = 1.5$. Solid and dashed lines represent F and FO-mode models, respectively.

mode. Figure 4 shows the PW relations ($\langle W\rangle = \langle G\rangle - 1.9\langle G_{\rm BP} - G_{\rm RP}\rangle = a + b\log P$) for F (solid lines) and FO- (dashed lines) mode pulsators, for both the chemical compositions analyzed and $\alpha_{\rm ml} = 1.5$. We observe that similar to the PLC relations, the PW relations are also almost insensitive to the specific value of the chemical compositions adopted.

In a forthcoming paper, we will extend this analysis to other photometric systems, including the Vera Rubin Observatory/Legacy Survey of Space and Time (Rubin/LSST) filters.

Acknowledgements. G.D.S. and M.M. thank the ISSI International Team project *SHoT: The Stellar Path to the H_0 Tension in the Gaia, TESS, LSST and JWST Era* (PI: G. Clementini). G.D.S. thanks the Istituto Nazionale di Fisica Nucleare (INFN), Naples section, specifically initiatives QGSKY and Moonlight2, and *Gaia* DPAC funds from the Istituto Nazionale di Astrofisica (INAF) and the Agenzia Spaziale Italiana (ASI) (PI: M.G. Lattanzi).

References

Caputo, F. 1998, Astron. Astrophys. Rev., 9, 33. doi:10.1007/s001590050014

Castelli, F. & Kurucz, R. L. 2003, Proc. IAU Symp. 210, Poster A20. doi:10.48550/arXiv.astro-ph/0405087

Caputo, F., Castellani, V., Degl'Innocenti, S., et al. 2004, Astron. Astrophys., 424, 927. doi:10.1051/0004-6361:20040307

De Somma, G., Marconi, M., Molinaro, R., et al. 2020, Astrophys. J. Suppl. Ser., 247, 30. doi:10.3847/1538-4365/ab7204

De Somma, G., Marconi, M., Molinaro, R., et al. 2022, Astrophys. J. Suppl. Ser., 262, 25. doi:10.3847/1538-4365/ac7f3b

Fiorentino, G., Limongi, M., Caputo, F., et al. 2006, Astron. Astrophys., 460, 155. doi:10.1051/0004-6361:20065349

Gaia Collaboration, Brown, A. G. A., Vallenari, A., et al. 2016, Astron. Astrophys., 595, A2. doi:10.1051/0004-6361/201629512

Gaia Collaboration, Brown, A. G. A., Vallenari, A., et al. 2018, Astron. Astrophys., 616, A1. doi:10.1051/0004-6361/201833051

Gaia Collaboration, Brown, A. G. A., Vallenari, A., et al. 2021, Astron. Astrophys., 649, A1. doi:10.1051/0004-6361/202039657

Hidalgo, S. L., Pietrinferni, A., Cassisi, S., et al. 2018, Astrophys. J., 856, 125. doi:10.3847/1538-4357/aab158

Marconi, M., Fiorentino, G., & Caputo, F. 2004, Astron. Astrophys., 417, 1101. doi:10.1051/0004-6361:20040020

Marconi, M., Musella, I., & Fiorentino, G. 2005, Astrophys. J., 632, 590. doi:10.1086/432790

Martínez-Vázquez, C. E., Monelli, M., Cassisi, S., et al. 2021, Mon. Not. R. Astron. Soc., 508, 1064. doi:10.1093/mnras/stab2493

Monelli, M. & Fiorentino, G. 2022, Universe, 8, 191. doi:10.3390/universe8030191

Renzini, A., Mengel, J. G., & Sweigart, A. V. 1977, Astron. Astrophys., 56, 369

Salaris, M. & Cassisi, S. 2005, Evolution of Stars and Stellar Populations, M. Salaris & S. Cassisi, eds, p. 400. ISBN 0-470-09220-3. Wiley-VCH.

Soszyński, I., Udalski, A., Szymański, M. K., et al. 2015, Acta Astron., 65, 233. doi:10.48550/arXiv.1508.00907

Soszyński, I., Smolec, R., Udalski, A., et al. 2020, Astrophys. J. Lett., 901, L25. doi:10.3847/2041-8213/abb817

Ripepi, V., Marconi, M., Moretti, M. I., et al. 2014, Mon. Not. R. Astron. Soc., 437, 2307. doi:10.1093/mnras/stt2047

At the crossroads of astrophysics and cosmology: Period–luminosity relations in the 2020s
Proceedings IAU Symposium No. 376, 2024
R. de Grijs, P. A. Whitelock & M. Catelan, eds.
doi:10.1017/S1743921323003538

Pulsation of chemically peculiar stars

Ernst Paunzen

Department of Theoretical Physics and Astrophysics, Masaryk University, Kotlářská 2, 611 37 Brno, Czechia. epaunzen@physics.muni.cz

Abstract. This paper reviews our current knowledge about pulsating chemically peculiar (CP) stars. CP stars are slowly rotating upper main-sequence objects, efficiently employing diffusion in their atmospheres. They can be divided into magnetic and non-magnetic objects. Magnetic activity significantly influence their pulsational characteristics. Only a handful of magnetic, classical pulsating objects are now known. The only exceptions are about 70 rapidly oscillating Ap stars, which seem to be located within a very tight astrophysical parameter space. Still, many observational and theoretical efforts are needed to understand all important physical aspects and their interrelationships. The most important steps to reach these goals are reviewed.

Keywords. chemically peculiar stars, pulsating stars, rotation, magnetic field

1. Introduction

The chemically peculiar (CP) stars of the upper main sequence (MS) have been targets of astrophysical studies since their discovery by the American astronomer Antonia Maury (1897). Most of the early research was devoted to detecting peculiar features in their spectra and characterizing their photometric behaviour. The main characteristics of classical CP stars are peculiar and often variable line strengths, the quadrature of line variability with radial velocity changes, photometric variability of the same periodicity and coincidence of extrema. Slow rotation was inferred from the sharpness of the spectral lines. Overabundances of several orders of magnitude compared to the Sun were derived for silicon, chromium, strontium, europium, and other heavy elements (Ghazaryan et al. 2018).

CP stars cover the spectral range from early B- to early F-type, i.e., stars whose stellar atmospheres are not affected by significant convection or mass loss. This led to the theory that diffusion, combined with slow rotation, is accountable for the elemental peculiarities found on their stellar surfaces (Alecian 2023). This implies that CP stars are actually Population I objects characterized by an overall solar abundance (reflecting a similar spread as in the solar neighbourhood). The incidence of CP stars among all MS objects of the same spectral types can reach up to 15%.

Babcock (1947) discovered a global dipolar magnetic field in the star 78 Virginis, which was followed by a catalogue of similar stars in which the variability of the field strength in many CP stars—including even a reversal of magnetic polarity—was found. Stibbs (1950) introduced the 'oblique rotator' concept of slowly rotating stars with non-coincidence of the magnetic and rotational axes. This model reproduces the variability and reversals of the magnetic field strength. Due to the chemical abundance concentrations at the magnetic poles, spectral and related photometric variabilities and radial velocity variations of the appearing and receding patches on the stellar surface are easily understood.

Table 1. Main characteristics of the six commonly used CP subgroups, including their designation according to Preston (1974).

Classical name	Prestonsâ€™s group	Discovery criteria	Spectral types	Temperature domain	Magnetic
λ Bootis		weak Mg II and weak metals	A0-â€“F0	7500–9000	N
Amâ€“-Fm	CP1	weak Ca II and/or Sc II; enhanced metals	A0-â€“F4	7000–10,000	N
Bpâ€“-Ap	CP2	enhanced Sr, Cr, Eu, and/or Si	B6-â€“F4	7000–16,000	Y
HgMn	CP3	enhanced Hg II and/or Mn II	A0–B6	10,500–16,000	N
Heâ€“weak	CP4	weak He I compared with colours	B2-â€“B8	14,000–20,000	Y
Heâ€“strong		enhanced He I compared with colours	B0-â€“B2	20,000–27,000	Y

Preston (1974) was the first to divide CP stars into four subclasses (CP1 to CP4). Later, these subclasses were extended by inclusion of He-strong (Pedersen & Thomsen 1977) and λ Bootis (Paunzen et al. 2002b) stars.

There are many different types of pulsating stars in the region of the Hertzsprung–Russell diagram (HRD) where CP objects are also located. This allows testing and improving of pulsational models by considering diffusion processes in a stellar magnetic field's presence or absence. Furthermore, we also have a precise knowledge of the rotational properties of CP stars, which helps to calibrate frequency splitting patterns, for example. Finally, the differences in astrophysical parameters of 'normal' and CP stars will help to understand the excitation mechanism and possible deviations of frequencies and amplitudes.

This article reviews our current knowledge of pulsating CP stars.

2. Theories explaining the CP phenomenon

Theories explaining the abundance patterns are manifold. Here, we will give a brief overview.

The observed abundance pattern of CP1 stars is defined by the diffusion of elements and the disappearance of the outer convection zone associated with helium ionization because of the gravitational settling of Helium (Théado et al. 2005). These authors predict a cut-off rotational velocity for such objects (about 100 km s^{-1}), above which meridional circulation leads to a mixing in the stellar atmosphere.

In CP2 and CP4 stars, due to additional magnetic diffusion the chemical abundance concentrations at the magnetic poles, as well as the spectral and related photometric variabilities, are also easily understood, as are radial velocity variations of the appearing and receding patches on the stellar surface (Alecian 2015).

CP3 stars are characterized by strong lines of ionized Hg and/or Mn with overabundances by up to six orders of magnitude relative to their solar abundances (Paunzen et al. 2021). Several mechanisms play significant roles in understanding these extreme peculiarities: radiatively driven diffusion, mass loss, mixing, light-induced drift, and possibly weak magnetic fields. However, there is as yet no satisfactory model to explain the abundance pattern (Adelman et al. 2003).

The λ Bootis stars are heavily depleted in refractory (iron-peak) elements, by up to 2 dex, but they exhibit solar abundances of the volatile elements C, N, O, and S (Murphy et al. 2020). To explain the peculiar chemical abundances of these stars, it has been suggested that they may originate from selective accretion of circumstellar material. One of the principal features of that hypothesis is that the observed abundance anomalies are restricted to the stellar surface (Alacoria et al. 2022).

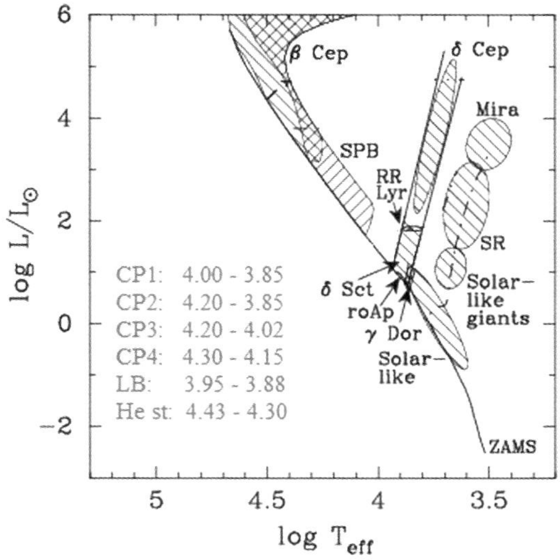

Figure 1. HRD showing the different types of pulsating variables and the temperature range where CP stars are found. CP stars populate the entire region from the zero-age main sequence to the terminal-age main sequence (luminosity classes V and IV). The $\log T_{\rm eff}$ values for the different subgroups (according to Table 1) are listed at bottom left.

Generally, CP1 and λ Bootis stars are not expected to show rotational light variability. As regards CP3 stars, Kochukhov et al. (2021) found rotational variability with amplitudes of 0.1–3 mmag in 84% of 64 objects. Using ultra-precise *Kepler* photometry, Balona et al. (2015) found that most of the Am stars they investigated show light curves indicative of rotational modulation due to star spots. With an amplitude of up to 200 ppm, observing this variability is reserved for high-precision (space) photometry. However, the interpretation of the corresponding light curves as a manifestation of surface spots is still debatable (Henriksen et al. 2023).

3. Classical pulsating variables in the CP star domain

Pulsating stars allow one to probe stellar interiors, which are not accessible to direct observations. A star's physical conditions and processes can only be observed through their influence on the pulsation periods and amplitudes. Pulsation characteristics allow one to derive basic stellar fundamental properties, such as mass, radius, and distance (Gilliland et al. 2010).

There are many different types of pulsating stars in the region of the HRD found where the CP objects are also located. Here is a summary of the subclasses.

• δ Scuti stars are multiperiodic pulsating variables located in the lower section of the Cepheid instability strip. They have periods generally between 0.02 and 0.25 days. They pulsate in low-order p modes (Breger et al. 2009).

• γ Doradus stars are pulsating intermediate-mass, late-A/early-F stars located at the cool border of the classical δ Scuti instability strip. They pulsate in high-order g modes (Guzik et al. 2000).

• hybrid δ Scuti/γ Doradus stars exhibit pulsational characteristics of both groups (Keen et al. 2015).

• β Cephei stars are short-period main-sequence variables with periods between 3 and 8 hours of spectral type O9 to B3. They pulsate in low-order p or g modes (Neilson & Ignace 2015).

• Slowly Pulsating B-type (SPB) stars are main-sequence B-type stars which exhibit multi-periodic, high-order, and low-degree g modes with periods of about one day (?)

Figure 1 shows the HRD with the different pulsating variables and the temperature range where the CP stars are found.

4. Pulsating CP stars

Analysing magnetic and non-magnetic pulsating stars helps to answer the following questions.

• Why do only a few magnetic CP stars pulsate?

• What are the main atmospheric differences between pulsating and non-pulsating CP and 'normal' stars?

• Do CP stars pulsate with the same modes, frequencies, and amplitudes as their 'normal' pulsating counterparts?

• Is it possible to select pulsating CP stars based on spectroscopic data?

• Is it possible to define an instability strip for pulsating CP stars? If so, what determines the borders?

• How does the magnetic field influence pulsational characteristics?

• Are the current pulsational models able to explain the observed frequencies and amplitudes?

• Is there a coupling between rotation and pulsation?

In the following, we will describe our current knowledge in more detail.

4.1. *CP1 stars*

Smalley et al. (2011) found about 200 CP1 stars showing low-amplitude γ Doradus and δ Scuti-type pulsations. This was very surprising because, in the narrow region of the HRD where γ Doradus stars are found, the convective zones associated with partial He and H ionization begin to merge and form a single convective envelope. The exact extension of the convective envelope is uncertain and depends on the treatment of convection (Dupret et al. 2004). The evolution of the convective core depends on the mass of the star.

The convective core grows for lower masses, leaving a discontinuity in the chemical composition. In the two cases of a shrinking and growing convective core, the determination of its exact extension is subject to significant uncertainties. Herein lies the promising diagnostic capability of gravity modes. They penetrate deep inside the star and carry information about the physical conditions close to the convective core. These waves are also sensitive to chemical gradients, influencing their periods (Dupret et al. 2007). Possible g modes in solar-type stars have their amplitudes damped by the thick envelope convection zone, and pure g modes are not predicted to be pulsationally unstable in δ Scuti stars, so this deep insight is not possible for such stars. Thus, γ Doradus stars represent a unique test bench for stellar evolution, pulsation, and convection theories.

From a large sample of 'normal' A-type and CP1 stars observed based on ground-based photometry, Smalley et al. (2017) found that δ Scuti pulsations in CP1 stars are mostly confined to a region close to the red edge of the corresponding instability strip, at temperatures cooler than 7600 K. No correlation between amplitude and metallicity was found. For unknown reasons, the incidence of pulsations in CP1 stars decreases noticeably in more peculiar objects. The amplitudes are lower than those of 'normal' A-type pulsators. These authors speculate that CP1 stars have, on average, higher radial order modes excited due to the turbulent pressure being more efficient as the depleted He is offset by an increased hydrogen number density.

4.2. *CP2 stars*

Classical CP2 and CP4 stars possess strong, large-scale organized magnetic fields of up to several kG. The observed field is always higher than a lower limit of 100 G for the longitudinal field, corresponding to a polar field of about 300 G. Sub-Gauss fields have been detected convincingly in the bright A-type star Vega and a few other A-type stars (Petit et al. 2010).

Therefore, we have to distinguish between magnetic 'normal' pulsating A- and B-type stars and classical CP2 and CP4 stars. To date, five of such magnetic pulsating δ Scuti stars have been reported: HD 432, HD 21190, HD 41641, HD 67523, and HD 188774 (Thomson-Paressant & Neiner 2021). Magnetic field detection is not always unambiguous and is still controversial for some of these objects. None of these stars show any signs of the abundance pattern typical of CP stars. However, they are essential objects to test pulsational models in the presence of an ultra-weak magnetic field.

Cunha et al. (2019) presented an analysis of photometric light curves of the Transiting Exoplanet Survey Satellite (TESS) mission, searching for pulsation characteristics among 83 CP2 stars. They found five suspected δ Scuti/γ Doradus variables, but could not confirm their true nature. Furthermore, Skarka et al. (2019) presented a pulsating CP component in an eclipsing binary system for which a detailed analysis is still pending.

In summary, these cases show how rare the combination of a stable magnetic field and stellar pulsation is.

4.3. *CP3 stars*

It is commonly accepted that many CP3 stars (up to 90%) are members of binary systems (Paunzen et al. 2021). Consequently, the tidal interaction induced by the presence of a companion star is thought to significantly impact the observed chemical peculiarities.

In general, CP3 stars fall into the region of β Cephei p-mode pulsators and the SPB g-mode variables. There are also some hybrid plusators known which are poorly understood. Kochukhov et al. (2021) investigated 64 CP3 stars using 2 minute-cadence light curves of the TESS mission. They found seven objects showing multiperidodic behaviour in the frequency domain expected for pulsation. However, the frequencies are incompatible with standard pulsational models for the corresponding astrophysical parameters. One explanation would be to take into account the possible variability of the secondary component.

The CP3 star HD 29589 was analysed by Niemczura et al. (2022), who found 20 high-order g modes and regular period spacing, which they interpreted as being due to the presence of consecutive prograde dipole modes.

Therefore, CP3 stars are excellent tracers to test the pulsational theories of β Cephei pulsators and SPB objects in the light of the CP phenomenon.

4.4. *CP4 and He-strong stars*

This subgroup mainly covers the β Cephei instability region in the HRD. What was stated for the CP2 stars is also true here. One has distinguish between magnetic 'normal' pulsators and classical CP stars.

HD 886 (γ Peg) is a β Cephei pulsator with a magnetic field of 10 G which varies with the primary pulsation frequency (Butkovskaya & Plachinda 2007). Owing to the magnetic field being frozen in the plasma, the initial hypothesis is that radial pulsations of the star result in the homothetic variation of the magnetic dipole.

The very slowly rotating (period longer than 30 years) and radially pulsating β Cephei-type star HD 46328 (ξ^1 CMa) exhibits a magnetic field of about 200 G (Erba et al. 2021).

It shows emission in Hα, which is modulated by the pulsation frequency. Together with significant variability in UV flux, this could lead to a more detailed explanation of the excitation mechanism for this stellar variability group.

No other similar pulsators have been detected to date.

4.5. λ *Bootis stars*

Paunzen et al. (2002a) found that at least 70% of all λ Bootis-type stars inside the classical instability strip pulsate, and they do so with high overtone modes (Q values smaller than 0.020 d). Only a few stars, if any, pulsate in the fundamental mode. Compared to δ Scuti stars, the cool and hot borders of the instability strip of the λ Bootis stars are shifted to the blue. Paunzen et al. (2002a) found that the period–luminosity–colour relation is, within the errors, identical to that of normal δ Scuti stars. No clear evidence for a statistically significant metallicity term was detected.

More recently, Murphy et al. (2020) analysed TESS light curves of 70 southern λ Bootis stars. They confirmed the results of Paunzen et al. (2002a) and concluded that these objects are only superficially metal-weak.

4.6. *Rapidly oscillating Ap (roAp) stars*

About 70 magnetic CP stars exhibit photometric variability in the period range from 5 to 25 minutes (high-overtone, low-degree, and non-radial pulsation modes). These stars are known as rapidly oscillating Ap (roAp) stars (Holdsworth et al. 2021). Several ground-based surveys were dedicated to searching for these objects (Paunzen et al. 2015; Joshi et al. 2016).

Observations imply a relationship exists between the excitation mechanism of pulsations in roAp stars and their heavy element distribution. Several opacity bumps are predicted to occur in the stellar atmosphere, which drive the pulsation (Théado et al. 2009). However, this cannot reproduce observations of the highest frequency modes in some stars. In these cases, Cunha et al. (2013) proposed that a mechanism linked to turbulent pressure may play a role in the excitation of these highest-frequency modes.

Murphy et al. (2020) presented a δ Scuti/roAp hybrid pulsator (KIC 11296437) with a mean magnetic field modulus of 2.8 kG. This could be a key object for understanding the different excitation modes in such stars.

Only a stringent astrophysical parameter and atmosphere regime seems to excite pulsation, which explains the rarity of these stars.

5. Conclusions and outlook

Pulsating variables which also show chemical peculiarities (mainly caused by diffusion processes), can help to understand the various driving mechanisms. Our current knowledge reveals that we know only a handful of pulsating magnetic CP stars. This suggests that a stable magnetic field prevents the excitation of pulsations. Nevertheless, modern, accurate space-based, long-term ground-based photometric data must be thoroughly analysed to find more such objects.

The situation as regards non-magnetic pulsating CP stars is much more promising. A significant number of CP1, CP3, and λ Bootis stars are known to be δ Scuti and γ Doradus pulsators. To date, no clear correlations between the degree of peculiarity and the pulsation characteristics have been found.

Still, several issues are urgently needed to study these phenomena in more detail.

- A systematic search for magnetic fields in pulsating variables;
- A more precise determination of the astrophysical parameters;

- A study of upper limits of amplitudes of non-variable CP stars;
- A study of star cluster members; and
- A study of non-variable stars.

Such comprehensive studies would help to shed more light on this very important research topic.

References

Adelman, S. J., Adelman, A. S., Pintado, O. I., 2003, A&A, 397, 267
Alacoria, J., Saffe, C., Jaque Arancibia, M., et al. 2022, A&A, 660, A98
Alecian, G., 2015, MNRAS, 454, 3143
Alecian, G., 2023, MNRAS, 519, 5913
Babcock, H. W., 1947, ApJ, 105, 105
Balona, L. A., Catanzaro, G., Abedigamba, O. P., et al. 2015, MNRAS, 448, 1378
Breger, M., Lenz, P., Pamyatnykh, A. A. 2009, MNRAS, 396, 291
Butkovskaya, V., Plachindax, S., 2007, A&A, 469, 1069
Cunha, M. S., Alentiev, D., Brandao, I. M., Perraut, K., 2013, MNRAS, 436, 1639
Cunha, M. S., Antoci, V., Holdsworth, D. L., et al. 2019, MNRAS, 487, 3523
Dupret, M.-A., Grigahcene, A., Garrido, R., et al. 2004, A&A, 414, L17
Dupret, M.-A., Miglio, A., Grigahcene, A., et al. 2007, Commun. Asteroseismol., 150, 98
Erba, C., Shultz, M. E., Petit, V., et al. 2021, MNRAS, 506, 2296
Ghazaryan, S., Alecian, G., Hakobyan, A. A., 2018, MNRAS, 480, 2953
Gilliland, R. L., Brown, T. M., Christensen-Dalsgaard, J., et al. 2010, PASP, 122, 131
Guzik, J. A., Kaye, A. B., Bradley, P. A., et al. 2000, ApJL, 542, L57
Henriksen, A. I., Antoci, V, Saio, H., et al. 2023, MNRAS, 520, 216
Holdsworth, D. L., Cunha, M. S., Kurtz, D. W., et al. 2021, MNRAS, 506, 1073
Jerzykiewicz, M., Lehmann, H., Niemczura, E., et al. 2013, MNRAS, 432, 1032
Joshi, S., Martinez, P., Chowdhury, S., et al. 2016, A&A, 590, A116
Keen, M. A., Bedding, T. R., Murphy, S. J., et al. 2015, MNRAS, 454, 1792
Kochukhov, O., Khalack, V., Kobzar, O., et al. 2021, MNRAS, 506, 5328
Maury, A., Pickering, E. C., 1897, Ann. Astron. Obs. Harvard, 28, Pt 1
Murphy, S. J.,Paunzen, E., Bedding, T. R., et al. 2020, MNRAS, 495, 1888
Murphy, S. J., Saio, H., Takada-Hidai, M., et al. 2020, MNRAS, 498, 4272
Neilson, H. R., Ignace, R. 2015 A&A, 584, A58
Niemczura, E., Walczak, P., Mikolajczyk, P., et al. 2022, MNRAS, 514, 5640
Paunzen, E., Handler, G., Weiss, W. W., et al. 2002a, A&A, 392, 515
Paunzen, E., Iliev, I. Kh., Kamp, I., et al. 2002b, MNRAS, 336, 1030
Paunzen, E., Netopil, M., Rode-Paunzen, M., et al. 2015, A&A, 575, A24
Paunzen, E., Hümmerich, S., Bernhard, K., 2021, A&A, 645, A34
Pedersen, H., Thomsen, B., 1977, A&AS, 30, 11
Petit, P., Lignieres, F., Wade, G. A., et al. 2010, A&A, 523, A41
Preston, G. W., 1974 ARA&A, 12, 257
Skarka, M., Kabath, P., Paunzen, E., et al. 2019, MNRAS, 487, 4230
Smalley, B., Kurtz, D. W., Smith, A. M. S., et al. 2011, A&A, 535, A3
Smalley, B., Antoci, V., Holdsworth, D. L., et al. 2017, MNRAS, 465, 2662
Stibbs, D. W. N., 1950, MNRAS, 110, 395
Théado, S., Vauclair, S., Cunha, M. S., 2005, A&A, 443, 627
Théado, S., Dupret, M.-A., Noels, A., et al. 2009, Commun. Asteroseismol., 158, 324
Thomson-Paressant, K., Neiner, C., 2021, MOBSTER-1 virtual conference: Stellar variability as a probe of magnetic fields in massive stars, Proceedings of the MOBSTER-1 virtual conference held 12–17 July 2020, id. 59

At the crossroads of astrophysics and cosmology: Period–luminosity relations in the 2020s
Proceedings IAU Symposium No. 376, 2024
R. de Grijs, P. A. Whitelock & M. Catelan, eds.
doi:10.1017/S1743921323002697

A study of the stellar photosphere–hydrogen ionization front interaction in δ Scuti stars

Mami Deka[1], Shashi M. Kanbur[2], Sukanta Deb[1,3] and Susmita Das[4]

[1]Department of Physics, Cotton University, Panbazar, Guwahati 781001, Assam, India.
email: mamideka8@gmail.com

[2]Department of Physics, State University of New York Oswego, Oswego, NY 13126, USA

[3]Space and Astronomy Research Center, Cotton University, Panbazar, Guwahati 781001, Assam, India

[4]Konkoly Observatory, Research Centre for Astronomy and Earth Sciences, Eötvös Loránd Research Network (ELKH), Konkoly Thege Miklós út 15-17, H-1121, Budapest, Hungary

Abstract. Pulsating variable δ Scuti stars are intermediate-mass stars with masses in the range of 1–3 $M_\odot$ and spectral types between A2 and F2. They can be found at the intersection of the Cepheid instability strip with the main sequence. They can be used as astrophysical laboratories to test theories of stellar evolution and pulsation. In this contribution, we investigate the observed period–colour and amplitude–colour (PCAC) relations at maximum/mean/minimum light of Galactic bulge and Large Magellanic Cloud δ Scuti stars for the first time and test the hydrogen ionization front (HIF)-photosphere interaction theory using the MESA-RSP code. The PCAC relations, as a function of pulsation phase, are crucial probes of the structure of the outer stellar envelope and provide insight into the physics of stellar pulsation and evolution. The observed behaviour of the δ Scuti PCAC relations is consistent with the theory of the interaction between the HIF and the stellar photosphere.

Keywords. stars, variables, delta Scuti, Galactic bulge, Magellanic Clouds

1. Introduction

The pulsating variable δ Scuti stars are intermediate-mass stars with masses in the range of 1-3 $M_\odot$ and spectral types between A2 and F2. They can be found at the intersection of the Cepheid instability strip with the main sequence. Their pulsation periods lie between 0.03 and 0.3 days. They pulsate both in single-mode and multimode. Single-mode δ Scuti stars obey a period–luminosity relation (PLR) that makes them excellent distance indicators (Nemec et al. 2017), whereas multimode δ Scuti stars are benchmark objects for asteroseismology.

The stellar photosphere is considered to be at located at an optical depth $\tau = \frac{2}{3}$ and the hydrogen ionization front (HIF) is the region of a star where the majority of hydrogen becomes ionized. The stellar photosphere and the HIF are not always co-moving during a pulsation cycle. The relative location of the HIF and the stellar photosphere is pulsation phase-dependent. The HIF interacts with the photosphere only at some particular phases where the photosphere lies at the base of HIF. This has been well-established in Classical Cepheids, RR Lyrae, BL Herculis (BL Her), and W Virginis (W Vir) stars (Simon et al. 1993; Kanbur 1995; Kanbur & Hendry 1996; Kanbur & Ngeow 2004; Bhardwaj et al. 2014; Ngeow et al. 2017; Das et al. 2018, 2020).

The correlation between the HIF and the stellar photosphere can be explained using the following equations:

(1) Saha ionization equation:

$$\frac{N_{i+1}}{N_i} = \frac{2Z_{i+1}}{n_e Z_i} \frac{2\pi m_e k T}{h^2}^{\frac{3}{2}} \exp -\frac{\chi_i}{kT}, \tag{1}$$

where N_{i+1} and N_i are the numbers of atoms in the $(i+1)^{\text{th}}$ and i^{th} ionization states, respectively; Z_{i+1} and Z_i represent the partition functions in the $(i+1)^{\text{th}}$ and i^{th} ionization states, respectively; n_e is the number of electrons; T is the temperature; m_e is the mass of the electron, χ_i represents the ionization energy, k denotes the Boltzmann constant and h is Planck constant. This equation relates stellar density and temperature.

(2) Period is dependent on the density through the period–mean density relation:

$$P \propto \sqrt{\frac{1}{\rho}}, \tag{2}$$

where P is the period and ρ represents the stellar density.

(3) Stefan–Boltzmann equation:

$$\log T_{\max} - \log T_{\min} = \frac{1}{10}(V_{\min} - V_{\max}), \tag{3}$$

where $\log T_{\max}$ and $\log T_{\min}$ are the photospheric temperature at maximum and minimum light, respectively. $V_{\min}$ and $V_{\max}$ are the minimum and maximum magnitude of the light curve, respectively. If $T_{\max}$ is independent or weakly dependent on the pulsation period, then the changes in amplitude are related to the temperature at minimum light, which will lead to a correlation between the V-band amplitude and the observed colour at minimum light. Conversely, if $T_{\min}$ is independent or weakly dependent on the period, then a correlation will exist between the V-band amplitude and the observed colour at maximum light.

The engagement between the HIF and the stellar photosphere is correlated with the temperature of the photosphere, and the temperature at which hydrogen ionizes is related to the density. As the colour is related to the temperature and the period is dependent on the stellar global parameters through the density (Eq. 2), the relative location of the HIF and the photosphere at a particular pulsation phase can explain the sloped/flat period–colour/amplitude–colour (PCAC) relation at that corresponding phase.

In the present paper, we have verified the HIF and stellar photosphere interaction theory of Simon, Kanbur & Mihalas (1993) in δ Scuti stars using their observed PCAC relation and theoretical models computed using Modules for Experiments in Stellar Astrophysics – Radial Stellar Pulsation (MESA-RSP) (Smolec & Moskalik 2008; Paxton et al. 2019).

The remaining part of this paper is organized as follows. Section 3 describes the data and methodology used. Results are discussed in Section 4. Section 4 summarizes the findings of the present work.

2. Data and Methodology

We use the optical (V, I)-band light curves of δ Scuti stars in the Galactic bulge and the Large Magellanic Cloud (LMC) from OGLE-IV (Soszyński et al. 2021) and OGLE-III (Poleski et al. 2010), respectively. The Galactic sample was cleaned from foreground and background stars following the procedure described by Pietrukowicz et al. (2015)

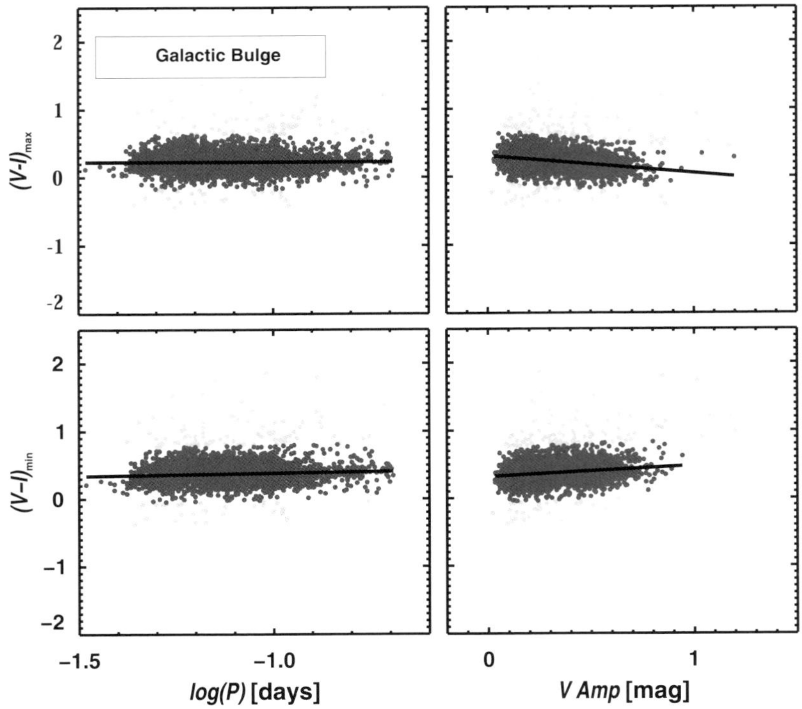

Figure 1. PCAC diagram for Galactic bulge δ Scuti stars. The left/right sides are the PC/AC diagrams at maximum (top) and minimum (bottom). The cyan points represent the removed outliers. The solid lines denote the best fits.

and Deka et al. (2022a). Outliers from the phased light curves were removed using the median absolute deviation (MAD) Leys et al. (2013); Deka et al. (2022b) criterion,

$$\frac{|m - \text{Median}(m)|}{\text{MAD}(m)} \geq 3.0,$$

where m represents the observed magnitude and MAD is the median absolute deviation.

Then, the light curves were fitted with a Fourier sine series (Deb & Singh 2009):

$$m(t) = A_0 + \sum_{i=1}^{N} A_i \sin\left[\frac{2\pi i}{P}(t_i - t_0) + \phi_i\right], \tag{4}$$

where A_0 is the mean magnitude; P and t_i represent the period of a star in days and the times of observations, respectively; t_0 is the epoch of maximum light and N is the order of the fit, which set at 4 and 3 for Galactic and LMC light curves, respectively.

The colour at maximum and minimum light is defined as:

$$(V - I)_{\text{max}} = V_{\text{max}} - I_{\text{phmax}}, \tag{5}$$

$$(V - I)_{\text{min}} = V_{\text{min}} - I_{\text{phmin}}, \tag{6}$$

where I_{phmax} and I_{phmin} correspond to the I-band magnitudes at the same phase as for V_{max} and V_{min}, respectively.

The PCAC relations for the δ Scuti stars were obtained after correcting the magnitudes and colours for extinction. We used the reddening maps of Gonzalez et al. (2012) and Haschke et al. (2011) for the Galactic bulge and LMC, respectively. The PCAC relations were considered after employing recursive 3σ outlier removal.

Table 1. Slopes and intercepts of the PCAC relation for Galactic bulge and LMC δ Scuti stars. Here, σ denotes the dispersion in the relations.

		Phase	Slope	Intercept	σ	Nature of slope
Galactic bulge	PC	Max	0.012 ± 0.017	0.239 ± 0.019	0.131	Flat
		Min	0.089 ± 0.017	0.472 ± 0.020	0.132	Flat
	AC	Max	-0.252 ± 0.013	0.305 ± 0.004	0.122	Sloped
		Min	0.166 ± 0.014	0.318 ± 0.005	0.131	Flat
LMC	PC	Max	0.318 ± 0.024	0.612 ± 0.024	0.108	Sloped
		Min	0.117 ± 0.027	0.630 ± 0.027	0.124	Sloped
	AC	Max	-0.317 ± 0.018	0.427 ± 0.008	0.100	Sloped
		Min	0.052 ± 0.021	0.488 ± 0.009	0.124	Flat

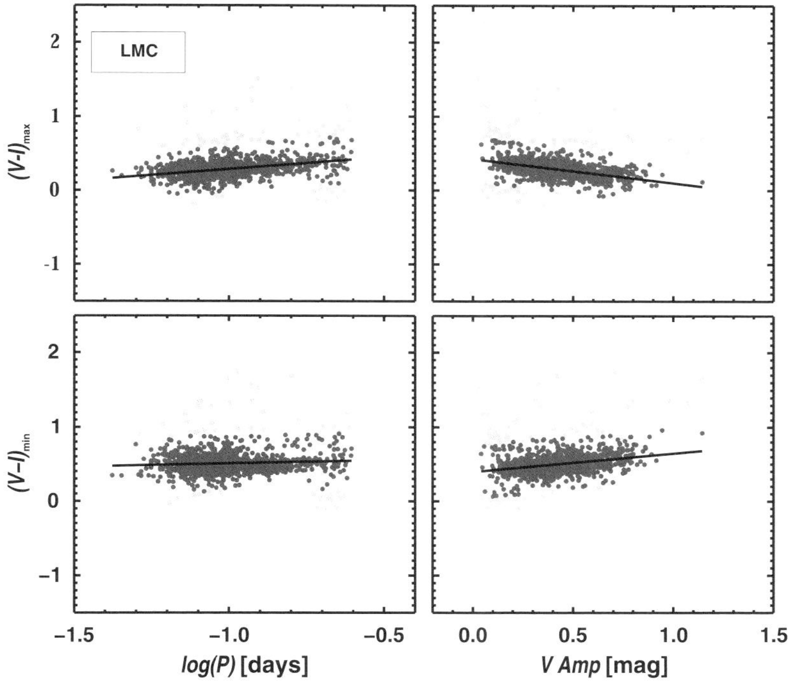

Figure 2. As Figure 1 but for the LMC.

3. Results and Discussion

3.1. *Observational results*

The PCAC relations for both the Galactic bulge and the LMC are shown in Figure 1 and Figure 2, respectively. The results are summarized in Table 1. The PC relations of the Galactic sample are shallow at both maximum and minimum light. However, for the LMC, they are sloped at maximum light and shallow at minimum light. The PC relations for the LMC and Galactic samples at maximum light show contrasting behaviour. The AC relations exhibit similar behaviour in both the Galactic and LMC samples.

To explain this contrasting behaviour of the PC relations, we have further investigated the amplitude distributions of both samples, since amplitude fluctuations are predominantly determined by temperature fluctuations. A two-component Gaussian function fits the amplitude distribution of Galactic bulge δ Scuti stars very well, as shown in the left panel of Figure 3. Separating the two Galactic distributions is beyond the scope of the present study, and we plan to carry out a more extensive analysis in a future work. Meanwhile, the LMC distribution is consistent with a single Gaussian fit, as shown in

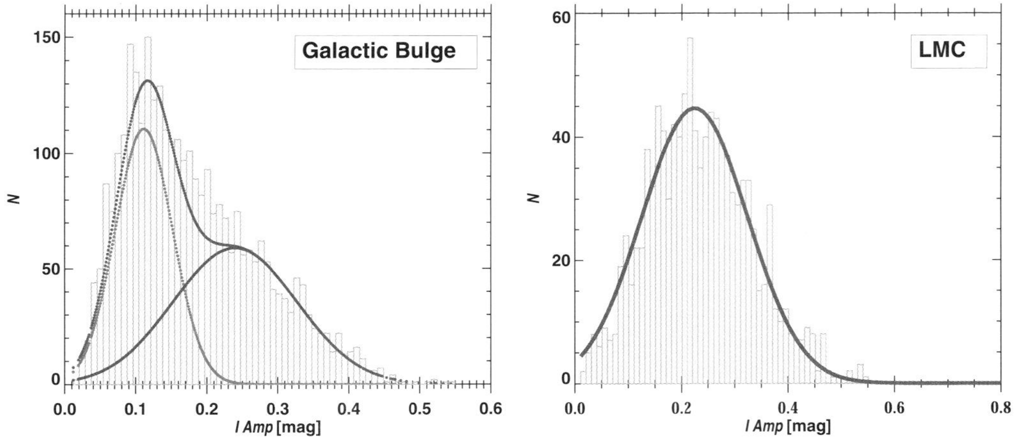

Figure 3. (left) Amplitude histograms in the Galactic I-band. The red line represents a two-component Gaussian fit to both amplitude distributions. The individual Gaussian fits are given by magenta and blue lines, respectively. (right) LMC amplitude histograms in the I band. The histogram can be explained with a single Gaussian fit (red).

the right panel of Figure 3. Hence, the presence/absence of low-amplitude stars in the Galactic/LMC samples might have led to the contrasting behaviour of the PC relations at maximum light.

3.2. *Theoretical results*

We have obtained the temperature profiles at maximum/minimum light for the two models of δ Scuti stars using MESA-RSP, version 'mesa-r15140' (Paxton et al. 2010, 2013, 2015, 2018, 2019). The input parameters taken for the two models were as follows: $Z = 0.02$, $X = 0.70$, $M = 2.0$ $M_{\odot}$, $L = 55$ $L_{\odot}$ and $T = 6950$ K (for the Galactic bulge) and $Z = 0.008$, $X = 0.736$, $M = 1.6$ $M_{\odot}$, $L = 25$ $L_{\odot}$ and $T = 6900$ K (for the LMC). The light curves obtained from both models are displayed in Figure 4. The HIF and photosphere are always found to be engaged for both compositions, as shown in Figure 5. However, they are engaged at a lower temperature at minimum light as compared to maximum light, which explains the smaller observed PC slope. For the LMC model, they are engaged at a temperature of ~ 7576 K and for the Galactic model at ~ 7269 K at maximum light. This suggests that the HIF is driven further out in the mass distribution at maximum light for the LMC model. The temperature fluctuations of Galactic δ Scuti stars are smaller than those of their LMC counterparts, leading to smaller amplitudes. Differences in the locations of the instability strip for Galactic and LMC stars might have led to the differences in their amplitudes. The flatter PC relation at maximum light for the Galactic component is due to the presence of the smaller amplitudes caused by smaller fluctuations in their photospheric temperature. Hence, the HIF stellar photosphere theory as described by Simon et al. (1993); Das et al. (2020, and references therein) is also consistent with the behaviour of δ Scuti stars.

4. Summary and Conclusions

In the present contribution, the V- and I-band light curves of δ Scuti stars from the OGLE-IV and OGLE-III databases pertaining to the Galactic bulge and the LMC have been used, respectively, to investigate the corresponding PCAC relations. The PCAC relations were obtained after employing iterative 3σ outlier clipping. The PC relations for Galactic bulge δ Scuti stars at maximum/minimum light are shallow, and the AC

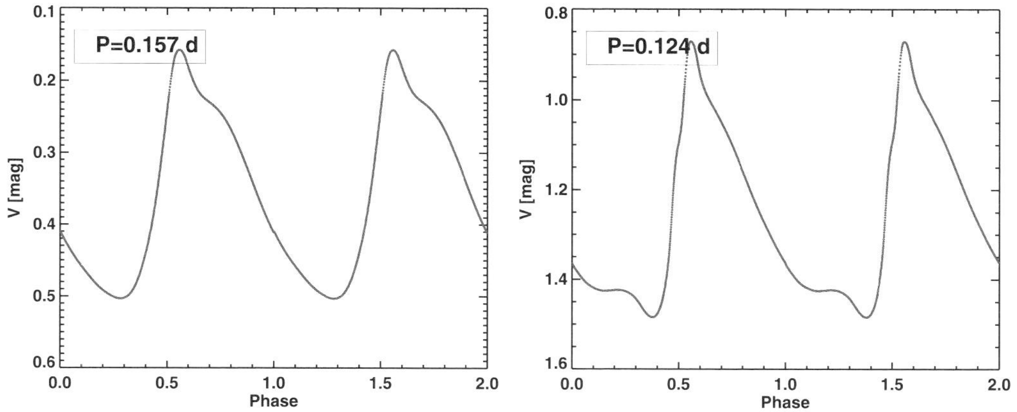

Figure 4. Theoretical light curves of δ Scuti stars obtained using MESA-RSP for (left) a Galactic composition and (right) an LMC-like composition.

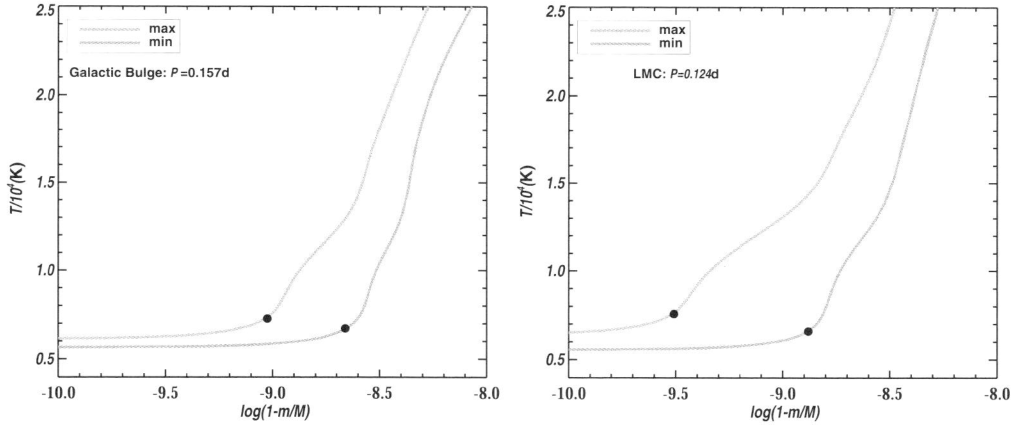

Figure 5. Temperature profile for (left) the Galactic model and (right) the LMC model.

relation is sloped at maximum/minimum light. Meanwhile, the PC relations for LMC δ Scuti stars at maximum/minimum light are sloped/flat. The behaviour of the LMC AC relations is similar to those in the bulge, but the slopes are relatively larger as compared to the Galactic ones. The amplitude distributions (Figure 3) reveal that the Galactic bulge δ Scuti sample consists of two populations (low-amplitude and high-amplitude stars), while the LMC sample consists of mostly high-amplitude stars.

Furthermore, theoretical temperature profiles corresponding to Galactic and LMC compositions were also obtained using MESA-RSP to look into the correlation between the HIF and the photosphere at maximum and minimum light. We find that the HIF and photosphere are always engaged at maximum and minimum light, for both models. The temperature at which they are engaged at minimum light is somewhat lower as compared to maximum light. This explains the flat observed PC relation at minimum light. However, the difference in the temperatures at maximum and minimum light for the Galactic model is relatively smaller as compared to that for the LMC model, which explains the contrasting behaviour of the observed PC relations to some extent. Therefore, the relative locations of the photosphere and HIF can explain the observed PCAC relations of δ Scuti stars.

Acknowledgement

MD thanks the Council of Scientific and Industrial Research (CSIR), Government of India, New Delhi, for providing a Junior Research Fellowship (JRF) through CSIR-NET through the research grant 03(1425)/18/EMR-II and she acknowledges travel support provided by SERB, Government of India vide file number ITS/2023/000385 to attend IAU Symposium 376, 'At the cross-roads of astrophysics and cosmology: Period-luminosity relations in the 2020s'. S. Deb thanks the CSIR for financial support received throught CSIR project 03(1425)/18/EMR-II. SMK acknowledges the support of SUNY Oswego and Cotton University. S. Das acknowledges the KKP-137523 'SeismoLab' Élvonal grant of the Hungarian Research, Development and Innovation Office (NKFIH). The authors acknowledge the use of the High Performance Computing facility Pegasus at IUCAA, Pune, India, and the software MESA r15140 (Paxton et al. 2010, 2013, 2015, 2018, 2019).

References

Bhardwaj A., Kanbur S. M., Singh H. P., Ngeow C.-C., 2014, MNRAS, 445, 2655
Das S., Bhardwaj A., Kanbur S. M., Singh H. P., Marconi M., 2018, MNRAS, 481, 2000
Das S., et al., 2020, MNRAS, 493, 29
Deb S., Singh H. P., 2009, A&A, 507, 1729
Deka M., Deb S., Kurbah K., 2022a, MNRAS, 514, 3984
Deka M., Kanbur S. M., Deb S., Das S., Kurbah K., Bellinger E. P., Bhardwaj A., 2022b, MNRAS, 517, 2251
Gonzalez O. A., Rejkuba M., Zoccali M., Valenti E., Minniti D., Schultheis M., Tobar R., Chen B., 2012, A&A, 543, A13
Haschke R., Grebel E. K., Duffau S., 2011, AJ, 141, 158
Kanbur S. M., 1995, A&A, 297, L91
Kanbur S. M., Hendry M. A., 1996, A&A, 305, 1
Kanbur S. M., Ngeow C.-C., 2004, MNRAS, 350, 962
Leys C., Ley C., Klein O., Bernard P., Licata L., 2013, J. Exper. Soc. Psychol., 49, 764
Nemec J. M., Balona L. A., Murphy S. J., Kinemuchi K., Jeon Y.-B., 2017, MNRAS, 466, 1290
Ngeow C.-C., Kanbur S. M., Bhardwaj A., Schrecengost Z., Singh H. P., 2017, ApJ, 834, 160
Paxton B., Bildsten L., Dotter A., Herwig F., Lesaffre P., Timmes F., 2010, ApJS, 192, 3
Paxton B., et al., 2013, ApJS, 208, 4
Paxton B., et al., 2015, ApJS, 220, 15
Paxton B., et al., 2018, ApJS, 234, 34
Paxton B., et al., 2019, ApJS, 243, 10
Pietrukowicz P., et al., 2015, ApJ, 811, 113
Poleski R., et al., 2010, Acta Astron., 60, 1
Simon N. R., Kanbur S. M., Mihalas D., 1993, ApJ, 414, 310
Smolec R., Moskalik P., 2008, Acta Astron., 58, 193
Soszyński I., et al., 2021, Acta Astron., 71, 189

At the crossroads of astrophysics and cosmology: Period–luminosity relations in the 2020s
Proceedings IAU Symposium No. 376, 2024
R. de Grijs, P. A. Whitelock & M. Catelan, eds.
doi:10.1017/S1743921323002612

A multi-wavelength analysis of BL Her stars: Models versus Observations

S. Das[1,2], L. Molnár[1,2,4], S. M. Kanbur[3,4], M. Joyce[1,2], A. Bhardwaj[5], H. P. Singh[6], M. Marconi[5], V. Ripepi[5] and R. Smolec[7]

[1]Konkoly Observatory, HUN-REN Research Centre for Astronomy and Earth Sciences, Konkoly-Thege Miklós út 15-17, H-1121, Budapest, Hungary. email: susmitadas130@gmail.com

[2]CSFK, MTA Centre of Excellence, Budapest, Konkoly Thege Miklós út 15-17., H-1121, Hungary

[3]Department of Physics, State University of New York Oswego, Oswego, NY 13126, USA

[4]ELTE Eötvös Loránd University, Institute of Physics and Astronomy, 1117, Pázmány Péter sétány 1/A, Budapest, Hungary

[5]INAF-Osservatorio Astronomico di Capodimonte, Salita Moiariello 16, 80131, Naples, Italy

[6]Department of Physics & Astrophysics, University of Delhi, Delhi 110007, India

[7]Nicolaus Copernicus Astronomical Center, Polish Academy of Sciences, Bartycka 18, PL-00-716 Warsaw, Poland

Abstract. We present new theoretical period–luminosity (PL) and period–radius (PR) relations at multiple wavelengths (Johnson–Cousins–Glass and *Gaia* passbands) for a fine grid of BL Herculis models computed using MESA-RSP. The non-linear models were computed for periods typical of BL Her stars, i.e. $1 \leq P(\text{days}) \leq 4$, covering a wide range of input parameters: metallicity (-2.0 dex $\leq$ [Fe/H] $\leq$ 0.0 dex), stellar mass (0.5–0.8 $M_\odot$), luminosity (50–300 $L_\odot$) and effective temperature (full extent of the instability strip; in steps of 50K). We investigate the impact of four sets of convection parameters on multi-wavelength properties. Most empirical relations match well with theoretical relations from the BL Her models computed using the four sets of convection parameters. No significant metallicity effects are seen in the PR relations. Another important result from our grid of BL Her models is that it supports combining PL relations of RR Lyrae and Type II Cepheids together as an alternative to classical Cepheids for the extragalactic distance scale calibration.

Keywords. stars: oscillations (including pulsations), stars: Population II, stars: variables: Cepheids, stars: low-mass

1. Introduction

Classical pulsators, in particular RR Lyrae stars, classical and Type II Cepheids are extremely important astrophysical objects; they are commonly used to estimate extragalactic distances, thanks to their well-defined period–luminosity (PL) relations, especially at longer wavelengths (for a review, see Bhardwaj 2020). However, although the effect of metallicity on the PL relations of RR Lyrae stars and classical Cepheids is minimised at near-infrared wavelengths, the precise calibration of PL-metallicity (PLZ) relations is one of the most important current topics of research in stellar variability studies (for example, see Ripepi et al. 2021, 2022; Breuval et al. 2022; De Somma et al.

2022). Type II Cepheids, on the other hand, are shown to exhibit little to no metallicity dependence on the PL relations (Matsunaga et al. 2011; Groenewegen and Jurkovic 2017; Das et al. 2021). Type II Cepheids are located in the classical instability strip of the Hertzsprung–Russell diagram. They have luminosities brighter than RR Lyrae but fainter than classical Cepheids. This makes them useful tracers, especially in regions with faint RR Lyrae and/or scarce classical Cepheids (for a review, see Wallerstein 2002).

Type II Cepheids are further subclassified on the basis of their pulsational periods (Soszyński et al. 2018) into BL Her stars ($1 \lesssim P(\text{days}) \lesssim 4$), W Vir stars ($4 \lesssim P(\text{days}) \lesssim 20$) and RV Tau stars ($P \gtrsim 20$ days). An additional class includes the peculiar W Vir (pW Vir) stars, which are brighter and bluer than the normal W Vir stars (Soszyński et al. 2008). In this study, we analyse only the shortest-period Type II Cepheids, i.e., BL Her stars, from both theoretical and empirical points of view. This is because MESA-RSP (Paxton et al. 2019) can reliably model BL Her stars but not the longer-period Type II Cepheids (W Vir and RV Tau stars) owing to their highly non-adiabatic nature. As alluded to already, both empirical (Matsunaga et al. 2006, 2009, 2011; Groenewegen and Jurkovic 2017) and theoretical (Di Criscienzo et al. 2007; Das et al. 2021) studies provide evidence of largely negligible effects of metallicity on the PL relations of BL Her stars and Type II Cepheids, in general. However, recent results from Wielgórski et al. (2022) indicate a significant effect of metallicity on the PL relations of field Type II Cepheids at near-infrared wavelengths, albeit with the caveat that their sample size was rather small, containing only 23 Type II Cepheids. If confirmed with more observational data using high-resolution spectroscopy and multiband photometry, this discrepancy between cluster and field Type II Cepheids could potentially turn into an open problem.

In this study, we use the state-of-the-art radial stellar pulsation code MESA-RSP to compute a fine grid of BL Her models using the four sets of convection parameters, derive new theoretical PL, period–Wesenheit (PW) and period–radius (PR) relations at multiple wavelengths (Johnson–Cousins–Glass $UBVRIJHKLL'M$ and *Gaia* $GG_{\rm BP}G_{\rm RP}$ bands) and thereby study the effects of metallicity and convection parameters on these relations. We also compare our theoretical results with empirical results of BL Her stars in the LMC from *Gaia* Data Release 3 (DR3; Ripepi et al. 2019) and from Groenewegen and Jurkovic (2017) using OGLE data.

2. Data and Methodology

The computation of BL Her models using MESA-RSP and the subsequent final selection of the models used for this analysis have been described in Das et al. (2021). In brief, the BL Her models were computed using MESA r11701, which uses the Kuhfuss (1986) theory of turbulent convection and the method of stellar pulsation as prescribed by Smolec and Moskalik (2008). The four sets of convection parameters are from Table 4 of Paxton et al. (2019), each with increasing complexities: set A corresponds to the simplest convection model, set B has radiative cooling added, set C included turbulent pressure and turbulent flux and set D has all of these effects added simultaneously.

We begin with the linear computations covering a wide range of input parameters: metallicity (-2.0 dex $\leq$ [Fe/H] ≤ 0.0 dex), stellar mass (0.5–0.8 $M_\odot$), luminosity (50–300 $L_\odot$) and effective temperature (full extent of the instability strip; in steps of 50K) using the four sets of convection parameters, resulting in a combination of 20,412 models per convection set over the entire range of input parameters. The linear stability analysis yields the linear periods and the growth rates of the models in different radial pulsation modes. This helps us estimate the red and blue edges of the instability strip as traced by the positive growth rates of the fundamental-mode BL Her pulsators (see Figure 1; Das et al. 2021). We proceed with the non-linear computations of the models with linear

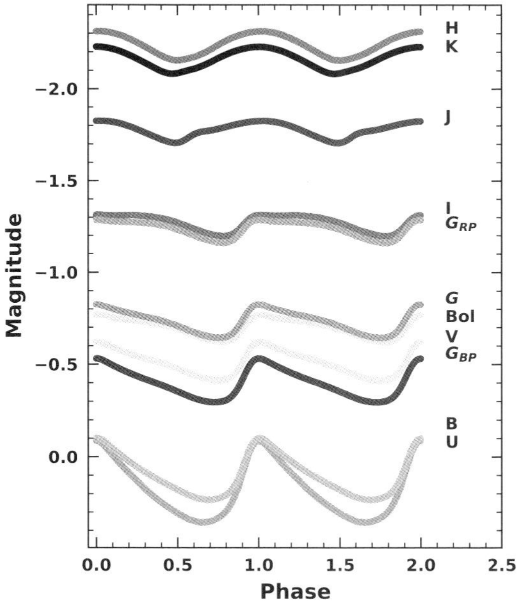

Figure 1. Example of multi-wavelength light curves for the BL Her model with input parameters, $Z = 0.00135$, $X = 0.74806$, $M = 0.6\ M_\odot$, $L = 150\ L_\odot$ and $T = 5500$ K, based on convective parameter set A.

periods between 0.8 and 4.2 days. The non-linear integrations are computed over 4000 pulsation cycles each and only those models are accepted for the final analysis that have non-linear periods between 1 and 4 days and satisfy the condition of full-amplitude stable pulsations (the amplitude of radius variation δR, the pulsation period P, and the fractional growth of the kinetic energy per pulsation period Γ do not vary by more than 0.01 over the last 100 cycles of the 4000-cycle integrations: see Figure 2; Das et al. 2021). Models that did not converge within 4000 cycles were not included in this study. After these conditions are met, we finally have 3266 BL Her models computed using set A, 2260 models using set B, 2632 models using set C and 2122 models using set D.

The bolometric luminosities generated as output from the non-linear computations of MESA-RSP are converted into absolute bolometric magnitudes and also into absolute magnitudes M_λ in a given passband λ, using either pre-computed (for transformation into Johnson–Cousins–Glass bands using Lejeune et al. 1998) or user-provided (for transformation into *Gaia* passbands using the Packaged Model Grids from MESA Isochrones & Stellar Tracks, MIST†) bolometric correction tables. The bolometric correction tables are defined as a function of the stellar photosphere in terms of $T_{\rm eff}$ (K), $\log g({\rm cm\,s^{-2}})$, and metallicity, [M/H]. An example of the light curves in both the bolometric magnitudes and in the different passbands for a BL Her model computed using convection parameter set A is presented in Figure 1.

The multi-wavelength theoretical light curves are fitted with the Fourier sine series (for an example, see Das et al. 2018) of the form,

$$m(x) = m_0 + \sum_{k=1}^{N} A_k \sin(2\pi k x + \phi_k), \qquad (1)$$

where x is the pulsation phase, m_0 the mean magnitude and N the order of the fit ($N = 20$).

† `https://waps.cfa.harvard.edu/MIST/index.html`

3. Period–luminosity relations

The mean magnitudes (M_λ) from the Fourier-fitted light curves of the BL Her models are used to obtain multi-wavelength PL relations of the mathematical form,

$$M_\lambda = a\log(P) + b \tag{2}$$

in a given band, λ. The multi-wavelength PL relations of the BL Her models computed using the four sets of convection parameters are summarised in Table 1 and compared using the standard t-test. In brief, we define a T statistic for the comparison of two linear regression slopes, $\hat{W}$, with sample sizes, n and m, as follows:

$$T = \frac{\hat{W}_n - \hat{W}_m}{\sqrt{\mathrm{Var}(\hat{W}_n) + \mathrm{Var}(\hat{W}_m)}}, \tag{3}$$

where $\mathrm{Var}(\hat{W})$ is the variance of the slope. The null hypothesis of equivalent slopes is rejected if $T > t_{\alpha/2,\nu}$ or the probability of the observed value of the T statistic is $p < 0.05$ where $t_{\alpha/2,\nu}$ is the critical value under the two-tailed t-distribution for the 95% confidence limit (α=0.05) and degrees of freedom, $\nu = n + m - 4$.

From Table 1 we find that the BL Her models computed using sets B and D (with radiative cooling) exhibit statistically similar PL slopes at most wavelengths. In addition, although there exist differences in the PL slopes of the BL Her models computed using different convection parameters, the empirical PL slopes of BL Her stars in the LMC seem to match well with the theoretical PL slopes at most wavelengths. However, for a few particular cases (HK_sGG_{BP}), the empirical PL slopes of BL Her stars in the LMC seem to match better with the theoretical PL slopes obtained using models with radiative cooling. Note here that the *Gaia* apparent magnitudes used to obtain the empirical PL relations have not been corrected for extinction; however, to account for uncertainties arising from extinction, we also obtain empirical PW relations using Wesenheit magnitudes as defined by Ripepi et al. (2019) for *Gaia* DR3. We find the empirical PW slope of the BL Her stars in the LMC using *Gaia* DR3 for calibration to be statistically similar to the theoretical PW slopes computed using the four sets of convection parameters.

The variation in the theoretical PL and PW relations obtained from the BL Her models as a function of metallicity is presented in Figure 2. We find strong effects of metallicity on the theoretical PL relations in the U and B bands, possibly because of increased sensitivity of bolometric corrections to metallicity at wavelengths shorter than the V band (Gray 2005; Kudritzki et al. 2008). In the other bands, the theoretical PL relations exhibit little to no metallicity dependence; this is consistent with empirical PL relations from Matsunaga et al. (2006, 2009, 2011) and Groenewegen and Jurkovic (2017). An interesting result to note is the small but significant effect of metallicity on the theoretical PW relations using the *Gaia* passbands, although there seems to be no effect of metallicity on the individual *Gaia* passbands. We will probe into the possible reasons of this metallicity dependence in a future paper (Das et al., in prep.).

In addition, we also compare the PL relations obtained for the BL Her models with those from the recent grid of RR Lyrae models from Marconi et al. (2015) in the $RIJHK_s$ bands. From the lowest panel of Table 1, we find the PL slopes of the RR Lyrae models to be statistically similar in the $RIJK_s$ bands to the BL Her models computed without radiative cooling (sets A and C). This equivalence of the PL relations from RR Lyrae and BL Her models is in agreement with empirical evidence from Majaess (2010), Bhardwaj et al. (2017) and Braga et al. (2020) and with previous theoretical analysis (for example, see Di Criscienzo et al. 2007; Marconi and Di Criscienzo 2007) and supports the claim that it could be useful to obtain common PL relations from RR Lyrae

Table 1. Comparison of the slopes of the PL and PW relations for BL Her stars of the mathematical form $M_\lambda = a\log(P) + b$. N is the total number of stars, T represents the observed value of the t-statistic, and $p(t)$ gives the probability of acceptance of the null hypothesis (equal slopes). The bold-faced entries indicate that the null hypothesis of the equivalent PL slopes can be rejected.

								$(T,\ p(t))$ w.r.t.			
Band	**Source**	a	b	σ	N	**Reference**‡	**Theoretical/ Empirical**	**Set A**	**Set B**	**Set C**	**Set D**
Bolometric	$Z_{\rm all}$ (Set A)	−1.799±0.028	−0.181±0.01	0.253	3266	TW	Theoretical	…	…	…	…
Bolometric	$Z_{\rm all}$ (Set B)	−1.581±0.03	−0.18±0.01	0.231	2260	TW	Theoretical	**(5.245,0.0)**	…	…	…
Bolometric	$Z_{\rm all}$ (Set C)	−1.693±0.031	−0.094±0.011	0.256	2632	TW	Theoretical	**(2.532,0.006)**	**(2.609,0.005)**	…	…
Bolometric	$Z_{\rm all}$ (Set D)	−1.559±0.032	−0.103±0.011	0.246	2122	TW	Theoretical	**(5.625,0.0)**	(0.512,0.304)	**(3.051,0.001)**	…
Bolometric	LMC	−1.749±0.200	0.141±0.051	0.274	57(4)	G17	Empirical	(0.248,0.402)	(0.831,0.203)	(0.277,0.391)	(0.938,0.174)
U	$Z_{\rm all}$ (Set A)	−0.841±0.044	0.185±0.015	0.391	3266	TW	Theoretical	…	…	…	…
U	$Z_{\rm all}$ (Set B)	−0.596±0.046	0.215±0.015	0.353	2260	TW	Theoretical	**(3.846,0.0)**	…	…	…
U	$Z_{\rm all}$ (Set C)	−0.422±0.051	0.298±0.018	0.428	2632	TW	Theoretical	**(6.219,0.0)**	**(2.512,0.006)**	…	…
U	$Z_{\rm all}$ (Set D)	−0.369±0.053	0.309±0.018	0.409	2122	TW	Theoretical	**(6.851,0.0)**	**(3.213,0.001)**	(0.722,0.235)	…
B	$Z_{\rm all}$ (Set A)	−1.166±0.04	0.187±0.014	0.351	3266	TW	Theoretical	…	…	…	…
B	$Z_{\rm all}$ (Set B)	−0.896±0.041	0.209±0.013	0.311	2260	TW	Theoretical	**(4.764,0.0)**	…	…	…
B	$Z_{\rm all}$ (Set C)	−0.942±0.043	0.33±0.015	0.359	2632	TW	Theoretical	**(3.843,0.0)**	(0.791,0.214)	…	…
B	$Z_{\rm all}$ (Set D)	−0.805±0.044	0.324±0.015	0.339	2122	TW	Theoretical	**(6.1,0.0)**	(1.508,0.066)	**(2.235,0.013)**	…
V	$Z_{\rm all}$ (Set A)	−1.616±0.032	−0.14±0.011	0.284	3266	TW	Theoretical	…	…	…	…
V	$Z_{\rm all}$ (Set B)	−1.374±0.034	−0.134±0.011	0.256	2260	TW	Theoretical	**(5.221,0.0)**	…	…	…
V	$Z_{\rm all}$ (Set C)	−1.487±0.034	−0.031±0.012	0.288	2632	TW	Theoretical	**(2.761,0.003)**	**(2.343,0.01)**	…	…
V	$Z_{\rm all}$ (Set D)	−1.337±0.036	−0.043±0.012	0.275	2122	TW	Theoretical	**(5.829,0.0)**	(0.757,0.225)	**(3.023,0.001)**	…
I	$Z_{\rm all}$ (Set A)	−2.043±0.025	−0.592±0.008	0.219	3266	TW	Theoretical	…	…	…	…
I	$Z_{\rm all}$ (Set B)	−1.848±0.027	−0.599±0.009	0.203	2260	TW	Theoretical	**(5.386,0.0)**	…	…	…
I	$Z_{\rm all}$ (Set C)	−1.932±0.027	-0.534±0.009	0.226	2632	TW	Theoretical	**(3.05,0.001)**	**(2.223,0.013)**	…	…
I	$Z_{\rm all}$ (Set D)	−1.81±0.028	−0.545±0.01	0.218	2122	TW	Theoretical	**(6.225,0.0)**	(0.976,0.165)	**(3.129,0.001)**	…
J	$Z_{\rm all}$ (Set A)	−2.303±0.021	−0.914±0.007	0.186	3266	TW	Theoretical	…	…	…	…
J	$Z_{\rm all}$ (Set B)	−2.131±0.023	−0.928±0.008	0.177	2260	TW	Theoretical	**(5.505,0.0)**	…	…	…
J	$Z_{\rm all}$ (Set C)	−2.239±0.023	−0.877±0.008	0.19	2632	TW	Theoretical	**(2.067,0.019)**	**(3.332,0.0)**	…	…
J	$Z_{\rm all}$ (Set D)	−2.122±0.024	−0.89±0.008	0.187	2122	TW	Theoretical	**(5.617,0.0)**	(0.243,0.404)	**(3.497,0.0)**	…
J	LMC	−2.164±0.240	17.131±0.038 (@0.3)*	0.25	55	M09	Empirical	(0.577,0.282)	(0.137,0.446)	(0.311,0.378)	(0.174,0.431)
J	LMC	−2.294±0.153	15.375±0.113 (@1.0)†	0.202	55	B17	Empirical	(0.058,0.477)	(1.054,0.146)	(0.355,0.361)	(1.111,0.133)
H	$Z_{\rm all}$ (Set A)	−2.57±0.018	−1.17±0.006	0.157	3266	TW	Theoretical	…	…	…	…
H	$Z_{\rm all}$ (Set B)	−2.429±0.02	−1.192±0.007	0.154	2260	TW	Theoretical	**(5.236,0.0)**	…	…	…
H	$Z_{\rm all}$ (Set C)	−2.529±0.019	−1.162±0.007	0.16	2632	TW	Theoretical	(1.568,0.058)	**(3.587,0.0)**	…	…
H	$Z_{\rm all}$ (Set D)	−2.432±0.021	−1.175±0.007	0.162	2122	TW	Theoretical	**(5.048,0.0)**	(0.081,0.468)	**(3.439,0.0)**	…
H	LMC	−2.259±0.248	16.857±0.039 (@0.3)*	0.26	54	M09	Empirical	(1.251,0.106)	(0.683,0.247)	(1.086,0.139)	(0.695,0.244)
H	LMC	−2.088±0.214	15.218±0.163 (@1.0)†	0.296	52	B17	Empirical	**(2.244,0.012)**	(1.587,0.056)	**(2.053,0.02)**	(1.6,0.055)
K	$Z_{\rm all}$ (Set A)	−2.528±0.018	−1.124±0.006	0.16	3266	TW	Theoretical	…	…	…	…
K	$Z_{\rm all}$ (Set B)	−2.383±0.021	−1.144±0.007	0.157	2260	TW	Theoretical	**(5.308,0.0)**	…	…	…
K	$Z_{\rm all}$ (Set C)	−2.483±0.02	−1.112±0.007	0.164	2632	TW	Theoretical	**(1.7,0.045)**	**(3.526,0.0)**	…	…
K	$Z_{\rm all}$ (Set D)	−2.383±0.021	−1.125±0.007	0.165	2122	TW	Theoretical	**(5.194,0.0)**	(0.008,0.497)	**(3.453,0.0)**	…
$K_{\rm s}$	LMC	−1.992±0.278	16.733±0.040 (@0.3)*	0.26	47	M09	Empirical	**(1.924,0.027)**	(1.402,0.081)	**(1.762,0.039)**	(1.402,0.081)
$K_{\rm s}$	LMC	−2.083±0.154	15.162±0.114 (@1.0)†	0.262	47	B17	Empirical	**(2.87,0.002)**	**(1.93,0.027)**	**(2.576,0.005)**	**(1.93,0.027)**

Table 1. (Continued)

								(T, p(t)) w.r.t.			
Band	**Source**	a	b	σ	N	**Reference**‡	**Theoretical/ Empirical**	**Set A**	**Set B**	**Set C**	**Set D**
G	$Z_{\rm all}$ (Set A)	−1.76±0.029	−0.263±0.01	0.262	3266	TW	Theoretical	...	...	...	...
G	$Z_{\rm all}$ (Set B)	−1.531±0.031	−0.261±0.01	0.237	2260	TW	Theoretical	**(5.352,0.0)**	...	...	...
G	$Z_{\rm all}$ (Set C)	−1.63±0.032	−0.171±0.011	0.267	2632	TW	Theoretical	**(2.996,0.003)**	**(2.231,0.026)**	...	...
G	$Z_{\rm all}$ (Set D)	−1.49±0.033	−0.183±0.011	0.255	2122	TW	Theoretical	**(6.095,0.0)**	(0.895,0.371)	**(3.049,0.002)**	...
G	LMC	−1.248±0.149	18.549±0.043	0.159	61	*Gaia* DR3	Empirical	**(3.373,0.001)**	(1.86,0.063)	**(2.507,0.012)**	(1.586,0.113)
$G_{\rm BP}$	$Z_{\rm all}$ (Set A)	−1.543±0.033	−0.077±0.011	0.293	3266	TW	Theoretical	...	...	...	...
$G_{\rm BP}$	$Z_{\rm all}$ (Set B)	−1.296±0.034	−0.068±0.011	0.262	2260	TW	Theoretical	**(5.193,0.0)**	...	...	...
$G_{\rm BP}$	$Z_{\rm all}$ (Set C)	−1.382±0.036	0.034±0.012	0.299	2632	TW	Theoretical	**(3.33,0.001)**	(1.731,0.084)	...	...
$G_{\rm BP}$	$Z_{\rm all}$ (Set D)	−1.238±0.037	0.024±0.013	0.284	2122	TW	Theoretical	**(6.178,0.0)**	(1.151,0.25)	**(2.804,0.005)**	...
$G_{\rm BP}$	LMC	−0.759±0.254	18.628±0.074	0.247	55	*Gaia* DR3	Empirical	**(3.061,0.002)**	**(2.095,0.036)**	**(2.428,0.015)**	(1.866,0.062)
$G_{\rm RP}$	$Z_{\rm all}$ (Set A)	−2.015±0.025	−0.588±0.009	0.225	3266	TW	Theoretical	...	...	...	...
$G_{\rm RP}$	$Z_{\rm all}$ (Set B)	−1.81±0.027	−0.594±0.009	0.209	2260	TW	Theoretical	**(5.503,0.0)**	...	...	...
$G_{\rm RP}$	$Z_{\rm all}$ (Set C)	−1.909±0.028	−0.523±0.01	0.231	2632	TW	Theoretical	**(2.819,0.005)**	**(2.565,0.01)**	...	...
$G_{\rm RP}$	$Z_{\rm all}$ (Set D)	−1.779±0.029	−0.536±0.01	0.223	2122	TW	Theoretical	**(6.144,0.0)**	(0.783,0.434)	**(3.271,0.001)**	...
$G_{\rm RP}$	LMC	−1.782±0.216	18.041±0.065	0.219	56	*Gaia* DR3	Empirical	(1.072,0.284)	(0.129,0.898)	(0.583,0.56)	(0.014,0.989)
$W(G, G_{\rm BP}-G_{\rm RP})$	$Z_{\rm all}$ (Set A)	−2.656±0.018	−1.234±0.006	0.159	3266	TW	Theoretical	...	...	...	...
$W(G, G_{\rm BP}-G_{\rm RP})$	$Z_{\rm all}$ (Set B)	−2.507±0.021	−1.261±0.007	0.159	2260	TW	Theoretical	**(5.432,0.0)**	...	...	...
$W(G, G_{\rm BP}-G_{\rm RP})$	$Z_{\rm all}$ (Set C)	−2.633±0.019	−1.231±0.007	0.163	2632	TW	Theoretical	(0.887,0.375)	**(4.412,0.0)**	...	...
$W(G, G_{\rm BP}-G_{\rm RP})$	$Z_{\rm all}$ (Set D)	−2.517±0.021	−1.247±0.007	0.165	2122	TW	Theoretical	**(4.98,0.0)**	(0.341,0.733)	**(3.994,0.0)**	...
$W(G, G_{\rm BP}-G_{\rm RP})$	LMC	−2.362±0.205	17.312±0.061	0.206	58	*Gaia* DR3	Empirical	(1.429,0.153)	(0.704,0.482)	(1.316,0.188)	(0.752,0.452)
RR Lyrae PL relations compared with BL Her PL relations											
R	$Z_{\rm all}$	−1.756±0.077	−0.114±0.014	0.196	226	M15	Theoretical	(1.19,0.117)	(1.485,0.069)	(0.208,0.418)	**(1.882,0.03)**
I	$Z_{\rm all}$	−1.973±0.068	−0.415±0.013	0.175	226	M15	Theoretical	(0.966,0.167)	**(1.709,0.044)**	(0.561,0.287)	**(2.217,0.013)**
J	$Z_{\rm all}$	−2.245±0.06	−0.778±0.011	0.155	226	M15	Theoretical	(0.902,0.184)	**(1.769,0.039)**	(0.098,0.461)	$textbf(1.898,0.029)
H	$Z_{\rm all}$	−2.206±0.118	−1.043±0.022	0.302	226	M15	Theoretical	**(3.056,0.001)**	**(1.867,0.031)**	**(2.708,0.003)**	**(1.889,0.03)**
K	$Z_{\rm all}$	−2.514±0.057	−1.11±0.011	0.147	226	M15	Theoretical	(0.24,0.405)	**(2.149,0.016)**	(0.507,0.306)	**(2.149,0.016)**

‡ TW = This work; M09 = Matsunaga et al. (2009); M15 = Marconi et al. (2015); B17 = Bhardwaj et al. (2017); G17 = Groenewegen and Jurkovic (2017)

* Zero point at $\log(P) = 0.3$

† Zero point at $\log(P) = 1.0$

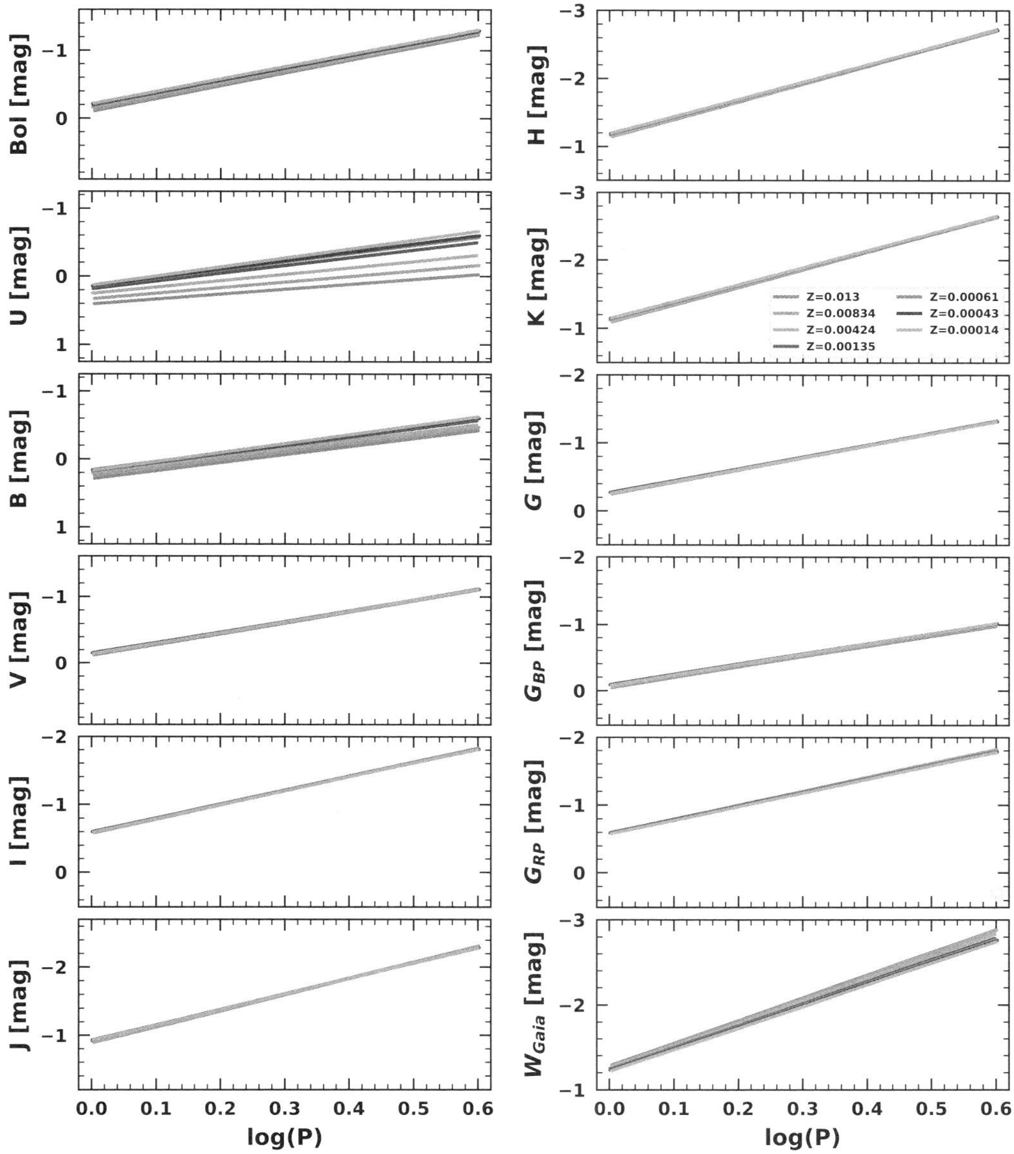

Figure 2. PL relations of the BL Her models with different chemical compositions at different wavelengths for the convective parameter set A. The y-axis scale is same (2.5 mag) in each panel to facilitate a relative comparison. The other convection parameter sets (B, C and D) show similar PL relations as a function of metallicity and wavelength.

and Type II Cepheids as an alternative to classical Cepheids for the calibration of the extragalactic distance scale.

4. Period–radius relations

We derive theoretical PR relations for the BL Her models of the mathematical form,

$$\log(R/R_\odot) = \alpha \log(P) + \beta, \tag{4}$$

where R is the mean radius of the BL Her model obtained by averaging the radius over a pulsation cycle.

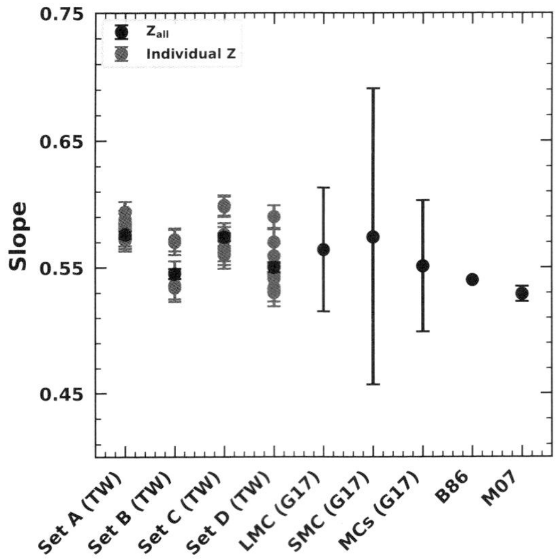

Figure 3. Comparison of the slopes of the PR relations for the BL Her stars obtained from this work (TW), using four different convective parameter sets, with those obtained from the literature. The red dots refer to the slopes of the PR relations obtained for the different chemical compositions individually, while the blue dots refer to the results obtained from considering the entire range of chemical compositions ($Z = 0.00014$ to $Z = 0.013$) combined. G17, B86 and M07 refer to Groenewegen and Jurkovic (2017), Burki and Meylan (1986) and Marconi and Di Criscienzo (2007), respectively.

A comparison of the slopes of the PR relations for the BL Her models computed using the four different convective parameter sets with those obtained from previous results from Burki and Meylan (1986), Marconi and Di Criscienzo (2007) and Groenewegen and Jurkovic (2017) is presented in Figure 3. We find that the BL Her models computed without radiative cooling (sets A and C) exhibit similar PR slopes while those computed using radiative cooling (sets B and D) have statistically similar PR slopes. The theoretical PR slopes of the BL Her models computed using the four sets of convection parameters are in broad agreement with the PR slopes from earlier empirical and theoretical results.

We also derive theoretical PRZ relations for the BL Her models of the form,

$$\log(R/\mathrm{R}_\odot) = \alpha + \beta \log(P) + \gamma[\mathrm{Fe/H}] \tag{5}$$

to test for the effects of metallicity on the PR relations. We find the following relations for set A, the simplest convection parameter set:

$$\begin{aligned}\log(R/\mathrm{R}_\odot) =& (0.879 \pm 0.001) + (0.581 \pm 0.003)\log(P) \\ & - (0.006 \pm 0.001)[\mathrm{Fe/H}]\ (N = 3266; \sigma = 0.029),\end{aligned} \tag{6}$$

and for set D, the most complex convection parameter set:

$$\begin{aligned}\log(R/\mathrm{R}_\odot) =& (0.886 \pm 0.002) + (0.554 \pm 0.004)\log(P) \\ & - (0.006 \pm 0.001)[\mathrm{Fe/H}]\ (N = 2122; \sigma = 0.029).\end{aligned} \tag{7}$$

The metallicity coefficient terms in Equations 6 and 7 indicate a weak dependence of the PR relations on metallicity, in agreement with previous empirical results from Burki and Meylan (1986) and Groenewegen and Jurkovic (2017).

5. Summary and Conclusions

We computed a very fine grid of BL Her models, the shortest period Type II Cepheids using MESA-RSP covering a wide range of input parameters: metallicity (-2.0 dex $\leq$ [Fe/H] ≤ 0.0 dex), stellar mass (0.5–0.8 $M_{\odot}$), luminosity (50–300 $L_{\odot}$) and effective temperature (full extent of the instability strip; in steps of 50K) with non-linear pulsational periods typical of BL Her stars, i.e., $1 \leq P(\text{days}) \leq 4$. The bolometric luminosities that result as an output of the non-linear computations were converted into absolute bolometric magnitudes and transformed into absolute magnitudes in the respective passbands (Johnson–Cousins–Glass $UBVRIJHKLL'M$ and *Gaia* bands $GG_{\rm BP}G_{\rm RP}$) using pre-computed or user-provided bolometric correction tables. The mean magnitudes obtained from Fourier fitting the light curves were then used to derive theoretical PL and PW relations.

As a function of different sets of convection parameters, the BL Her models computed using sets B and D (with radiative cooling) were found to exhibit statistically similar PL slopes at most wavelengths. While there exist differences among the theoretical PL relations of the BL Her models computed using the different sets of convection parameters, the empirical PL relations of BL Her stars in the LMC are in broad agreement with their theoretical counterparts, with some preference for models computed using radiative cooling especially in the $HK_{\rm s}GG_{\rm BP}$ bands. The empirical PW slope of the BL Her stars in the LMC using *Gaia* DR3 is statistically similar to the theoretical PW slopes computed using the four sets of convection parameters. As a function of metallicity, there exists a strong effect of metallicity on the theoretical PL relations in the U and B bands, with weak to negligible effects of metallicity at longer wavelengths, which is in agreement with earlier empirical results. We will probe the cause of the small but significant metallicity effect on the theoretical PW relations using the *Gaia* passbands in a future paper.

We also found a weak dependence of the PR relations on metallicity. In addition, the PL slopes of BL Her models computed without radiative cooling (sets A and C) using MESA-RSP are statistically similar to those for RR Lyrae models from Marconi et al. (2015) in the $RIJK_{\rm s}$ bands, thereby supporting the claim that a common PL relation using RR Lyrae and Type II Cepheids could be used as an alternative to classical Cepheids for extragalactic distance scale calibrations. We also find the results from the MESA-RSP-computed grid of BL Her models to be in good agreement with previous theoretical predictions from Di Criscienzo et al. (2007); Marconi and Di Criscienzo (2007) for BL Her stars.

6. Acknowledgements

S. D. and L. M. acknowledge the KKP-137523 'SeismoLab' Élvonal grant of the Hungarian Research, Development and Innovation Office (NKFIH). M. J. gratefully acknowledges funding from MATISSE: *Measuring Ages Through Isochrones, Seismology, and Stellar Evolution*, awarded through the European Commission's Widening Fellowship. This project has received funding from the European Union's Horizon 2020 research and innovation program.

References

Bhardwaj, A., Macri, L. M., Rejkuba, M., et al. 2017, AJ, 153, 154
Bhardwaj, A. 2020, JAA, 41, 23
Braga, V. F., Bono, G., Fiorentino, G., et al. 2020, A&A, 644, A95
Breuval, L., Riess, A. G., Kervella, P., Anderson, R. I., & Romaniello, M. 2022, ApJ, 939, 89
Burki, G., & Meylan, G. 1986, A&A, 159, 261
Das, S., Bhardwaj, A., Kanbur, S. M., Singh, H. P., & Marconi, M. 2018, MNRAS, 481, 2000
Das, S., Kanbur, S. M., Smolec, R., et al. 2021, MNRAS, 501, 875

De Somma, G., Marconi, M., Molinaro, R., et al. 2022, ApJS, 262, 25
Di Criscienzo, M., Caputo, F., Marconi, M., & Cassisi, S. 2007, A&A, 471, 893
Gray, D. F. 2005, The Observation and Analysis of Stellar Photospheres
Groenewegen, M. A. T., & Jurkovic, M. I. 2017, A&A, 604, A29
Kudritzki, R.-P., Urbaneja, M. A., Bresolin, F., et al. 2008, ApJ, 681, 269
Kuhfuss, R. 1986, A&A, 160, 116
Lejeune, T., Cuisinier, F., & Buser, R. 1998, A&AS, 130, 65
Majaess, D. J. 2010, JAVSO, 38, 100
Marconi, M., & Di Criscienzo, M. 2007, A&A, 467, 223
Marconi, M., Coppola, G., Bono, G., et al. 2015, ApJ, 808, 50
Matsunaga, N., Fukushi, H., Nakada, Y., et al. 2006, MNRAS, 370, 1979
Matsunaga, N., Feast, M. W., & Menzies, J. W. 2009, MNRAS, 397, 933
Matsunaga, N., Feast, M. W., & Soszyński, I. 2011, MNRAS, 413, 223
Paxton, B., Smolec, R., Schwab, J., et al. 2019, ApJS, 243, 10
Ripepi, V., Molinaro, R., Musella, I., et al. 2019, A&A, 625, A14
Ripepi, V., Catanzaro, G., Molinaro, R., et al. 2021, MNRAS, 508, 4047
Ripepi, V., Catanzaro, G., Clementini, G., et al. 2022, A&A, 659, A167
Smolec, R., & Moskalik, P. 2008, AcA, 58, 193
Soszyński, I., Udalski, A., Szymański, M. K., et al. 2008, AcA, 58, 293
Soszyński, I., Udalski, A., Szymański, M. K., et al. 2018, AcA, 68, 89
Wallerstein, G. 2002, PASP, 114, 689
Wielgórski, P., Pietrzyński, G., Pilecki, B., et al. 2022, ApJ, 927, 89

At the crossroads of astrophysics and cosmology: Period–luminosity relations in the 2020s
Proceedings IAU Symposium No. 376, 2024
R. de Grijs, P. A. Whitelock & M. Catelan, eds.
doi:10.1017/S1743921323003411

Impact of the *Gaia* ESA mission on the primary Period–Luminosity Calibrators in the Milky Way: Cepheids and RR Lyrae

Gisella Clementini[1,2]

[1]Istituto Nazionale di Astrofisica-Osservatorio di Astrofisica e Scienza dello Spazio, Via Piero Gobetti 93/3, Bologna, 40129, Italy; gisella.clementini@inaf.it

[2]*Gaia* Data Processing and Analysis Consortium (DPAC), Coordination Unit 7 (CU7; Variability Processing)

Abstract. We discuss the impact of *Gaia*, the cornerstone mission of the European Space Agency (ESA), on the calibration of the period–luminosity and luminosity–metallicity relations of Cepheids and RR Lyrae stars, with specific reference to data published as part of the most recent *Gaia* releases: Early Data Release 3 (EDR3), on 19 December 2020, and Data Release 3 (DR3) on 13 June 2022. We provide future perspectives for the *Gaia* mission, including extensions approved by ESA and a tentative schedule of the data releases that will take place in the next few years. We briefly present plans for cross-Coordination Unit processing of *Gaia* data of Cepheids and RR Lyrae stars for DR4 and conclude by outlining the expected improvement in astrometry at the end of the extended *Gaia* mission, which will help to further strengthen the calibration of the first rung of the cosmic distance ladder.

Keywords. Stars: oscillations – Stars: variables: RR Lyrae – Stars: variables: Cepheids – Methods: data analysis

---

1. Introduction

Cepheids and RR Lyrae stars are the most important stellar standard candles at the basis of the cosmological distance ladder through their period–luminosity(–metallicity) and luminosity–metallicity relations. They are also excellent tracers of the Milky Way's (MW) structure and of its formation via accretion and mergers.

The *Gaia* mission is producing the largest, most homogeneous and parameter-rich catalogues of Cepheids and RR Lyrae stars ever published, hence causing a step change in many astrophysical contexts. This is thanks to *Gaia*'s capability to scan the whole sky (at an average of 14 times per year) and collect astrometry nearly simultaneously to multiband photometry of all sources brighter than $G \sim 21$ mag and spectroscopy of sources brighter than $G \sim 16.5$ mag.

The *Gaia* focal plane, with physical dimensions of 1.0×0.4 m^2, is the largest ever developed for a space mission. It hosts an array of 106 CCDs (Gaia Collaboration et al. 2016; their Figure 4), featuring almost 1,000 million pixels in total, which are devoted to the following main scientific functions: (i) astrometry measurements and white-light photometry in the G band (330–1050 nm) on the CCDs of the astrometric field; (ii) blue (330–680 nm) and red (640–1050 nm) spectro-photometry employing low-resolution ($R \sim$ 20–90) prism spectra obtained by the blue and red photometers; and (iii) intermediate-resolution ($R = 11,500$) slitless spectroscopy in the calcium triplet region (847–874 nm) with the radial-velocity spectrometer (RVS; Cropper et al. 2018).

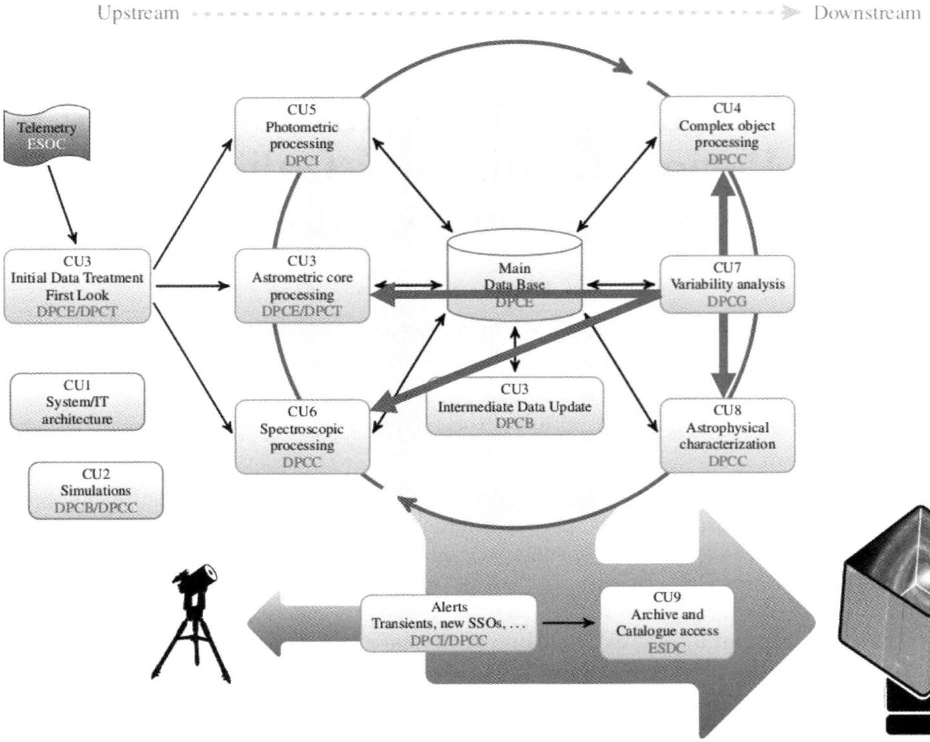

Figure 1. Schematic overview of the *Gaia* Data Processing and Analysis Consortium (DPAC) showing its organisation into different Coordination Units (CUs). Blue arrows show the inter-CUs exchange (cross-CUs processing) planned for DR4 to improve the processing of the *Gaia* data for Cepheids and RR Lyrae stars (see Section 4.1). (Adapted from `https://www.cosmos.esa.int/web/gaia/dpac`).

The *Gaia* Data Processing and Analysis Consortium (DPAC), an international team composed of 400–450 European scientists and software developers, is in charge of developing the data processing algorithms and software to turn the raw telemetry from *Gaia* into the final scientific data products to be released to the scientific community. DPAC activities involve cyclic, iteratively improved processing of all data collected by *Gaia* which is carried out within dedicated Coordination Units (CUs). Figure 1, shows a schematic view of the *Gaia* DPAC and its organisation into CUs. Three upstream 'core' CUs perform the astrometric (CU3), photometric (CU5) and spectroscopic (CU6) processing. An additional three downstream CUs process complex sources such as binaries or galaxies (CU4), perform variability analysis (CU7) and carry out astrophysical characterisation (CU8) of the sources.

Three releases of *Gaia* data products have occurred since the start of the mission in 2013. *Gaia* data for Cepheids and RR Lyrae stars were published already as part of the first data release (DR1) in 2016 (Clementini et al. 2016), showcasing the extraordinary quality of *Gaia*'s light curves and the astrometric quality already achieved by employing the Tycho–*Gaia* astrometric solution (TGAS; Lindegren et al. 2016) for an unprecedented number of these standard candles defining the astronomical distance ladder (Gaia Collaboration et al. 2017). The third and most recent *Gaia* release was split into two separate instalments, both covering 34 months of *Gaia* observations obtained from 25 July 2014 to 28 May 2017. *Gaia* Early Data Release 3 (EDR3; Gaia Collaboration et al. 2021) on 3 December 2020 published positions on the sky, parallaxes and proper motions for about 1.468 billion sources with G magnitudes in the range from ~3 to ~21 mag, G photometry for about 1.806 billion sources and integrated $G_{\rm BP}$ and $G_{\rm RP}$ magnitudes for about 1.5 billion. Data Release 3 (DR3; Gaia Collaboration et al. 2023), on 13 June 2022, complemented EDR3 with new products (for a full list, see `https://www.cosmos.esa.int/web/gaia/dr3`), among which in particular were included results from the variability analysis, together with the underlying epoch photometry, for a total sample of 10.5 million sources, including around 15,000 Cepheids and 271,000 RR Lyrae stars.

2. Variability processing

Identification and characterisation of the variable sources observed by *Gaia* is performed within DPAC Coordination Unit 7 (CU7) based on analysis of the time-series data of all sources observed by the satellite. For DR3 the integrated G, $G_{\rm BP}$ and $G_{\rm RP}$ time-series photometry (Riello et al. 2021) of more than 1.8 billion sources was fed into the CU7 pipelines, along with their EDR3 astrometry (positions on the sky, parallaxes and proper motions). For a small sample of Cepheids and RR Lyrae stars comprising 799 Cepheids and 1196 RR Lyrae stars, the CU7 pipelines also analysed the epoch radial velocities obtained with the RVS (Sartoretti et al. 2022).

Statistical and Machine Learning methods are used to identify, characterise and classify the candidate variable sources into different types (Eyer et al. 2017, 2023; Rimoldini et al. 2023). Candidate variables of the different types are then ingested into the Specific Object Study (SOS) pipelines, which are specialised to validate the classification into types and derive specific attributes for the variable sources, which are confirmed to actually belong to the classes they have been assigned to. A summary of the variability processing steps and the variability content of *Gaia* DR3 is presented by Eyer et al. (2023) to which the interested reader is referred for details.

2.1. *Cepheids and RR Lyrae stars in Gaia DR3*

Variable stars classified as Cepheid and RR Lyrae candidates are processed by the SOS Cep&RRL pipeline (Clementini et al. 2016, 2019, 2023; Ripepi et al. 2023), which was specifically devised to test the classification into types and derive characteristic parameters of variable sources confirmed as Cepheids and RR Lyrae stars. Attributes computed by the SOS Cep&RRL pipeline include: pulsation period and secondary periodicities in case of multimode pulsation modes; identification of RR Lyrae/Cepheid types and pulsation modes; G, $G_{\rm BP}$, $G_{\rm RP}$ and RV amplitudes of the light and radial velocity (RV) curves computed as peak-to-peak amplitudes of the non-linear Fourier modelling of the light and RV variations, respectively; G, $G_{\rm BP}$ and $G_{\rm RP}$ intensity-averaged mean magnitudes; mean RV; parameters of the Fourier decomposition of the light and RV curves; metal abundance of fundamental-mode classical Cepheids (DCEP_Fs) with periods shorter than 6.3 days inferred from the R_{21} and R_{31} Fourier parameters of the G-band light curves and the calibration of Klagyivik et al. (2013); metal abundance of RR Lyrae stars inferred from the ϕ_{31} Fourier parameter of the G-band light curves and the calibration of Nemec et al. (2013); and G absorption of fundamental-mode RR Lyrae stars (RRab) inferred from the pulsation period, the amplitude of the G-band light curve and the $(G - G_{\rm RP})$ colour, based on the relation of Piersimoni et al. (2002) and the procedure described by Clementini et al. (2019).

We refer the reader to Clementini et al. (2023) and Ripepi et al. (2023) for detailed descriptions of the SOS Cep&RRL processing of RR Lyrae stars and Cepheids released as part of DR3. Attributes computed by the SOS Cep&RRL pipeline for *bona fide* Cepheids and RR Lyrae can be retrieved from the `vari_rrlyrae` and `vari_cepheid` tables of the

Table 1. Number of SOS Cepheids released in DR3 divided by type, pulsation mode and sky region[1].

Cepheids (type/mode)	LMC	SMC	M31	M33	All-Sky	Total
DCEP_F	2357	2487	309	173	2008	7334
DCEP_FO	1931	1803	10	12	1101	4857
DCEP_MULTI	58	110	-	-	195	363
DCEP (Total)	4346	4400	319	185	3304	**12,554**
ACEP_F	69	87	-	-	150	306
ACEP_FO	32	80	-	-	132	244
ACEP (Total)	101	167	-	-	282	**550**
T2CEP BL Her	66	16	-	-	579	661
T2CEP W Vir	120	20	-	-	795	935
T2CEP RV Tau	30	13	2	-	261	306
T2CEP (Total)	216	49	2	-	1635	**1902**
Cepheids (Grand total)	**4663**	**4616**	**321**	**185**	**5221**	**15,006**

Notes: [1] All numbers in this table refer to the clean DR3 SOS Cepheid sample (see Ripepi et al. 2023).

Gaia DR3 archive. The epoch photometry and the RV time-series data can be downloaded by querying the *Gaia* data release archive (`http://archives.esac.esa.int/gaia/`).

2.1.1. Gaia DR3 catalogue of SOS Cepheids

The *Gaia* DR3 catalogue of Cepheids validated and characterised by the SOS Cep&RRL pipeline lists 15,006 *bona fide* Cepheids (hereafter SOS Cepheids). Among these, 474 are new Cepheids discovered by *Gaia* and an additional 327 are variable stars with a different classification in the literature, which have been re-classified as Cepheids based on our analysis of the *Gaia* data (Ripepi et al. 2023). They are subdivided into classical Cepheids (DCEPs), anomalous Cepheids (ACEPs) and Type 2 Cepheids (T2CEPs). According to their pulsation modes, DCEPs and ACEPs are further subdivided into fundamental-mode (DCEP_F, ACEP_F), first-overtone (DCEP_FO, ACEP_FO) and multi-mode (only for DCEPs, DCEP_ MULTI) when more than one mode is excited, whilst T2CEPs are subdivided into three distinct types: BL Herculis (BL Her), W Virginis (W Vir) and RV Tauri (RV Tau) based on the Cepheid pulsation period and their evolutionary status.

Table 1 summarises the number of SOS Cepheids, divided by type, pulsation mode and position on the sky, according to the following five regions defined by Ripepi et al. (2023): Large Magellanic Cloud (LMC), Small Magellanic Cloud (SMC), Andromeda galaxy (M31), Triangulum galaxy (M33) and All-Sky, which mainly contains MW Cepheids.

General statistics and information pertaining to the number of SOS Cepheids with astrophysical parameters computed by the SOS Cep&RRL pipeline or by other CUs are provided in Table 2.

2.1.2. Gaia DR3 catalogue of SOS RR Lyrae stars

The *Gaia* DR3 catalogue of RR Lyrae stars validated and characterised by the SOS Cep&RRL pipeline contains 270,905 entries (hereafter SOS RR Lyrae), of which 174,947 are fundamental-mode pulsators (RRab), 93,952 are first-overtone pulsators (RRc) and 2006 are RR Lyrae stars that pulsate in both modes simultaneously. Around 70,000 of the SOS RR Lyrae were not known previously and are new discoveries by *Gaia*. Figure 30 of Clementini et al. (2023) shows the distribution on the sky of the SOS RR Lyrae released

Table 2. Information about the DR3 SOS Cepheids.

Characteristic	**N1**	**N2**
Total sample	15,006	–
Known	14,205	–
Reclassified	327	–
New discoveries	474	–
Range in mean magnitude	$3.4 < G < 20.9$ mag	–
MW DCEPs with $\sigma_\pi/\pi \leq 0.1$ and RUWE$\leq$1.4[1]	~1060	–
$[Fe/H]_{SOS}$[2]	5123	–
Epoch RVs	>799	–
Mean RVs (CU6)	>2000	–
$E(B-V)$ from GSP_phot (CU8)	~1700	278[3]
[M/H] from GSP_spec (CU8)	~1000	400[4]

Notes: [1] RUWE is a parameter that measures the quality of the parallax estimate. Reliable parallaxes have RUWE values < 1.4 (see Section 14.1.2 of *Gaia* Data Release 2 Documentation release 1.2; `https://gea.esac.esa.int/archive/documentation/GDR2/`); [2] only for DCEP_Fs with $P < 6.3$ days and less reliable for values more metal-poor than [Fe/H] ~ 0.3 dex (for details, see Ripepi et al. 2023); [3] with $\Delta E(B-V) = \pm 0.2$–0.25 mag from literature values; [4] with Δ[M/H] $= \pm 0.2$-0.3 dex from literature values.

Table 3. Number of SOS RR Lyrae stars released in DR3 by sky region or host system[1].

LMC	**SMC**	**95 GCs**	**7 dSphs & 16 UFDs**	**All-Sky**	**Total**
31,379	4788	1676	1114	231,948	**270,905**

Note: [1] All numbers in the table refer to the clean DR3 SOS RR Lyrae sample (see Clementini et al. 2023).

in DR3. They populate all Galactic components, from the halo to the bulge to the disc of the MW. The new RR Lyrae stars are mainly located in the MW disc and bulge, where crowding is very high. Hence, their light curves may be more noisy and have their amplitudes reduced owing to blending with other sources. Concentrations of sources seen in the DR3 SOS RR Lyrae sky distribution correspond to the variables in the Magellanic Clouds, the Sagittarius dwarf spheroidal galaxy (dSph), six other dSphs (among which Sculptor, Ursa Minor and Carina), 16 ultra-faint dwarf (UFD) MW satellites and 95 globular clusters (GCs). Table 3 provides the number of SOS RR Lyrae stars released in DR3 split by sky region and/or host system. General statistics and information about the number of SOS RR Lyrae with astrophysical parameters computed by the SOS Cep&RRL pipeline or by other CUs is provided in Table 4.

3. *Gaia* DR3 Cepheids and RR Lyrae stars as standard candles of the cosmic distance ladder

The dramatic improvements in the Cepheid and RR Lyrae astrometry as *Gaia* data releases proceed (see Figure 2) and the large number of Cepheids and RR Lyrae stars released as part of DR3 have allowed to derive an improved absolute geometric calibration using parallaxes of the fundamental relations that make Cepheids and RR Lyrae stars primary standard candles of the cosmic distance ladder.

More than one thousand of the SOS DCEPs in the MW have DR3 parallaxes better than 10% and RUWE$\leq$1.4 (see Table 2). They were used to calibrate, with unprecedented accuracy, the DCEP period–luminosity–metallicity (PLZ) and period–Wesenheit–metallicity (PWZ) relations (see, e.g., Breuval et al. 2021, 2022; Ripepi et al. 2021, 2022a). This represents a crucial step forward in constraining the first rung of the cosmic distance ladder and reduce the uncertainty in the estimate of the Hubble constant, H_0, obtained from distance ladder measurements (see, e.g., Riess et al. 2021, 2022a,b).

Table 4. Information about the DR3 SOS RR Lyrae stars.

Characteristic	Number
Total sample	270,905
Known	200,294
New discoveries	70,611
Fundamental mode (RRab)	174,947
First overtone (RRc)	93,952
Double mode (RRd)	2006
Range in mean magnitude	$7.64 < G < 21.14$
RR Lyrae with $\sigma_\pi/\pi \leq 0.1$ and RUWE$\leq$1.4	$\sim$300
$[Fe/H]^{[1]}_{SOS}$	135,559
Epoch RVs	1096
Mean RVs (CU6)	5096
$A(G)^{[2]}_{(SOS)}$	142,660
$A(G)$ from GSP_phot (CU8)	84,500
[M/H] from GSP_spec (CU8)	193[3]
[M/H] from GSP_phot (CU8)	87,273

Notes: [1] Metallicity errors published in the `vari_rrlyrae` table are slightly overestimated; [2] $A(G)$ estimates are available only for RRab stars; [3] Selected using the flags_gspspec flag to filter out sources with GSP_spec abundances of poor quality.

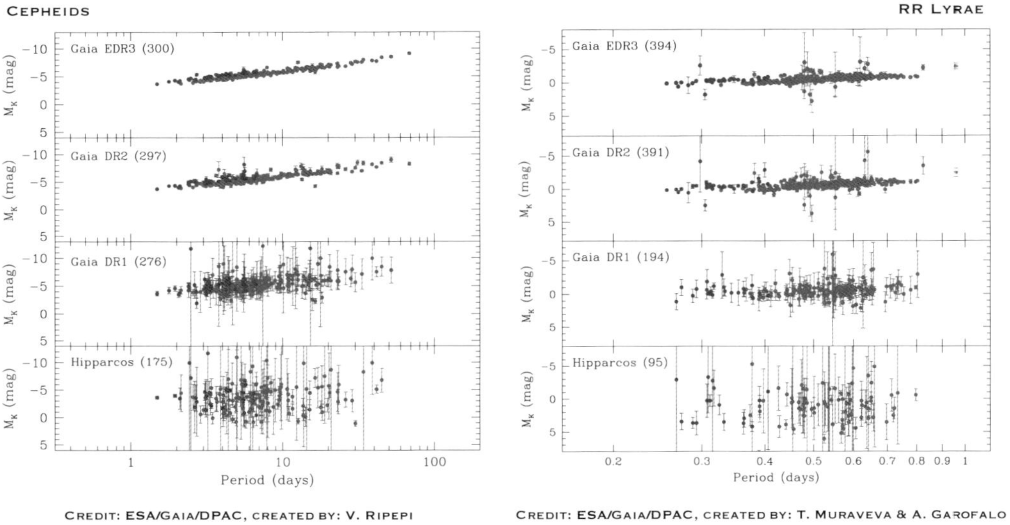

Figure 2. Near-infrared period–luminosity relations showing the improvement of the astrometry for MW Cepheids (left panels) and RR Lyrae stars (right panels) from *Hipparcos* (*Gaia*'s predecessor; bottom panels) to *Gaia* DR1, DR2 and EDR3. Red and blue points correspond to fundamental-mode and first-overtone pulsators, respectively. The reduction in the K absolute magnitude error and the increasing number of sources with reliable parallaxes are impressive. (Figures produced to showcase the improvement of *Gaia*'s parallaxes on the occasion of the EDR3 release. They update the figures originally published by Gaia Collaboration et al. 2017; `https://www.cosmos.esa.int/web/gaia/edr3-vs-dr2`)

With typical ages greater than 9–10 billion years, RR Lyrae stars are standard candles for systems mainly composed of old stellar populations. RR Lyrae stars obey luminosity-metallicity (LZ) relations in optical bands such as the Johnson V and the *Gaia* G band, PLZ relations in the near-infrared (NIR) and PWZ relations in optical and NIR bands. Around 300 SOS RR Lyrae stars in the MW have parallaxes better than 10% and RUWE$\leq$1.4 (see Table 4). They were used to recalibrate the RR Lyrae optical and NIR relations (Muraveva et al. 2018; Bhardwaj et al. 2021; Garofalo et al. 2022; Li et al. 2023).

Table 5. Estimates of the parallax zero-point offset published after *Gaia* EDR3.

ZPO value (μas)	Source/Method	Reference
-14 ± 6	Classical Cepheids	Riess et al. (2021)
-15 ± 3[1]	Asteroseismology of red giants	Zinn (2021)
-10 ± 7	RR Lyrae stars	Gilligan et al. (2021)
-39[2]	Comparison with *HST* parallaxes	Groenewegen (2021)
$-7 \pm 3/-32 \pm 4$[3]	RR Lyrae stars	Bhardwaj et al. (2021)
-28[4]	RR Lyrae stars	Garofalo et al. (2022)
-22 ± 4[5]	Classical Cepheids	Molinaro et al. (2023)

Notes: [1] For $G \geq 10.8$ mag; [2] Mean value of individual ZPOs; [3] Without or with Lindegren et al. (2021) corrections, respectively; [4] Mean value without Lindegren et al. (2021) corrections; [5] With Lindegren et al. (2021) corrections.

In spite of the impressive improvement in quality of *Gaia*'s parallaxes and the increasing number of Cepheids and RR Lyrae stars with reliable parallax values, a number of issues still affect the calibration of the fundamental relations followed by these standard candles. Some of these issues are discussed in the following sections.

3.1. *Gaia parallax zero point offset*

Gaia's parallaxes are known to be underestimated due to unexpectedly large variations of the 'basic angle' between the viewing directions of the two *Gaia* telescopes (Lindegren et al. 2021). The exact value of the *Gaia* parallax zero-point offset (ZPO) has been found to depend on the magnitude, colour and position on the sky of the sources.

Lindegren et al. (2018) estimated a median ZPO of the *Gaia* DR2 parallaxes of about -29 μas with respect to a sample of bright quasars. The DR2 average ZPO was found to be ~ -33 μas for Cepheids and ~ -56 μas for RR Lyrae stars (Arenou et al. 2018; 2018). However, a larger ZPO of -46 ± 13 μas was estimated for Cepheids by Riess et al. (2018), and a smaller value of -42 ± 13 μas was determined by Layden et al. (2019) for RR Lyrae stars. In *Gaia* EDR3, the parallax ZPO was reduced to ~ -17 μas, with systematic variations on the order of $\sim$10 μas (Lindegren et al. 2021). These latter authors also developed a code to compute individual parallax zero-point offsets for sources with five and six-parameter solutions, separately. The code is availabe as a Jupyter notebook, at `https://gitlab.com/icc-ub/public/gaiadr3$_$zeropoint`.

Different estimates of the ZPO value have also been published after the release of the *Gaia* EDR3 parallaxes. Some of these estimates are summarised in Table 5, including in particular the ZPO values inferred from Cepheids and RR Lyrae stars. Suggestions have also been made that the Lindegren et al. (2021) corrections might overestimate the ZPO of the EDR3 parallaxes (see, e.g., Riess et al. 2021; Molinaro et al. 2023).

The ZPO remains one of the most important issues affecting the *Gaia* parallaxes that limits our precision in measuring the Hubble constant with the distance ladder method. While the existence of the ZPO limits *Gaia*'s ability to determine absolute parallaxes, improving the characterisation and modelling of the ZPO so as to mitigate systematic effects such as the dependence on magnitude, colour and position on sky should be possible.

3.2. *Metallicity dependence of the Cepheid PL/PW relations*

A significant source of uncertainty that limits the use of Cepheids as standard candles of the astronomical distance ladder is our poor knowledge of the dependence on metal abundance of their PL/PW relations. Rather discordant estimates are found in the literature, with recent values ranging from -0.221 ± 0.051 mag dex^{-1} (Breuval et al. 2021) up to -0.456 ± 0.099 mag dex^{-1} (Ripepi et al. 2021) for the Cepheid PL relation in the K_s band. The contribution to the H_0 error budget of the uncertainty in the behaviour of the Cepheid PL/PW relations as a function of chemical composition is estimated between 0.5% (out of 1.2%; Riess et al. 2022) for the LMC and 0.15% for other anchors of the cosmic distance ladder, such as NGC 4258 or M31.

Observing programmes are in progress (see, e.g., Ripepi et al. 2021) to directly measure the metallicity dependence of the DCEP PL/PW relations using only Cepheids in the MW with accurate metallicities from abundance analysis of high-resolution spectra and spanning a large metallicity range. The metallicities derived from abundance analysis of the *Gaia* RVS spectra of Cepheids may also contribute to study the metallicity dependence of the DCEP PL/PW relations (see Section 3.3).

3.3. *Astrophysical parameters of Cepheids and RR Lyrae stars in Gaia DR3*

Astrophysical parameters such as metal abundance and reddening are needed, in addition to the parallax, to compute the period–luminosity and luminosity–metallicity relations. The astrophysical parameters of Cepheids and RR Lyrae stars published in the *Gaia* DR3 archive were derived independently by CU7 as part of the variability processing, and by CU8 as part of the astrophysical characterisation of the sources, using different pipelines and input data from *Gaia*. Specifically, photometric metallicities and G absorption values of the Cepheids and RR Lyrae stars in the `vari_cepheid` and `vari_rrlyrae` tables of the *Gaia* DR3 archive were determined by the CU7 SOS Cep&RRL from the characteristic parameters of the light variation along the source pulsation cycle (pulsation period, amplitude, colour and parameters of the Fourier decomposition of the light curve). Astrophysical parameters of Cepheids and RR Lyrae stars published in the `astrophysical_parameters` tables were determined instead by the GSP_spec and GSP_phot modules of the CU8 Apsis pipeline (Creevey et al. 2023, and references therein) from analysis of the stacked RVS and BP/RP prism spectra of the sources, in practice 'absorbing' the variation owing to the pulsation as an increased measurement uncertainty.

The DR3 `vari_cepheid` table contains photometric [Fe/H] abundances for 5123 fundamental-mode DCEPs with periods shorter than 6.3 days. They were inferred by the SOS Cep&RRL pipeline from the R_{21} and R_{31} Fourier parameters of the G-band light curves based on the Klagyivik et al. (2013) calibration. The comparison with metal abundances from high-resolution spectra that are available in the literature for 185 DCEPs with photometric metallicities from CU7 shows the latter to be about 0.08 dex higher than the literature values, but still reliable, at least in the metallicity range $-0.3 <$ [Fe/H] $< +0.4$ dex (Ripepi et al. 2023). Spectroscopic [M/H] abundances estimated with the GSP_spec pipeline are available for about 1000 DR3 Cepheids. A comparison with the abundances from high-resolution spectra in the literature shows that for around 400, the two estimates agree within ± 0.20–0.35 dex. While the agreement among Cepheids abundances from high-resolution spectroscopy and metallicities from the either the SOS Cep&RRL or the GSP_spec pipelines are already quite good, further improvement of the latter may result, in particular, from analysis based on the RVS per-transit spectra.

Photometric metal abundances are published in the DR3 `vari_rrlyrae` table for 133,559 RR Lyrae stars. They were computed by the SOS Cep&RRL pipeline from the ϕ_{31} Fourier parameter of the G-band light curves and the stellar pulsation period (P), adopting Nemec et al. (2013)'s calibration of the [Fe/H]–P–ϕ_{31} relation. There have been claims that the Nemec et al. (2013) calibration produces significantly biased metal abundances (e.g., Li et al. 2023). However, a comparison of the $[\mathrm{Fe/H}]_{\mathrm{SOS}}$ values with RR Lyrae abundances from high-resolution spectroscopic studies (Clementini et al. 1995; Lambert et al. 1996; Pancino et al. 2015; Crestani et al. 2021) shows reasonably good agreement within the errors, except that the $[Fe/H]_{SOS}$ estimates are about 0.2 dex higher around $[Fe/H] \sim -1.5$ dex. Similarly, the GSP_spec [M/H] estimates obtained for 193 RR Lyrae stars from the RVS spectra (Recio-Blanco et al. 2023) are also in satisfactory agreement with the $[Fe/H]_{SOS}$ values, within the errors (see Clementini et al. 2023; their Figure 23). Finally, about 87,000 DR3 RR Lyrae stars have [M/H] abundances estimated from the BP/RP low-resolution spectra using the GSP_phot pipeline (Andrae et al. 2023). However, the GSP_phot metallicities of RR Lyrae stars are affected by very large uncertainties and systematic effects that prevent a meaninful comparison with the $[Fe/H]_{SOS}$ abundances.

$E(B-V)$ reddening estimates obtained by the GSP_phot pipeline from the BP/RP prism spectra are available for about 1700 DR3 Cepheids. The GSP_phot values are in reasonably good agreement with the literature values: for a subsample of 278 Cepheids the difference between the two estimates is on the order of 0.2–0.25 mag.

Interstellar G-band absorption values were obtained by the SOS Cep&RRL pipeline for 142,660 fundamental-mode DR3 RR Lyrae stars. They were compared with the G-band absorption values computed by the GSP_phot pipeline, which are available for about 84,500 DR3 RRab stars. This comparison showed that the GSP_Phot DR3 absorption estimates of RR Lyrae stars are often affected by the temperature–absorption degeneracy and by misidentification of RR Lyrae stars for much hotter sources (see Clementini et al. 2023; their Section 5.2.1).

4. Future perspectives

The inconsistent results obtained for Cepheids and RR Lyrae stars by different DPAC pipelines clearly showed the need for more coordinated joint processing of these most important calibrators of the cosmic distance ladder by the different CUs. In light of *Gaia* DR4, a significant joint effort of the DPAC CUs is thus being devoted to better characterise the parallax ZPO and derive consistent astrophysical parameters (temperature, metallicity, gravity and absorption values) for Cepheids and RR Lyrae stars.

4.1. *Cross-CUs processing of Cepheids and RR Lyrae stars for Gaia DR4*

Gaia's Cepheids and RR Lyrae stars are identified and characterised by CU7's processing. However, this information is not subsequently taken into account by other DPAC analysis pipelines. This is largely because of how the processing of the *Gaia* data naturally develops, with the CUs processing the astrometry, photometry and spectroscopy starting their processing and freezing their results much earlier than CU7, and CU4 and CU8 running almost in parallel to the variability processing. In addition, the rather tight schedule of the *Gaia* data releases so far has made such cross-processing practically impossible.

The lack of internal coherence introduces systematic errors and uncontrolled uncertainties in the DPAC products. The magnitude and colour variations of Cepheids and RR Lyrae stars implies variations in effective temperature of up to more than one thousand degrees and variations in gravity values that can be as large as 1 dex during a pulsation cycle. So far, the analysis of the RVS and BP/RP prism spectra performed within CU8 has neglected the per-transit variations of the atmospheric parameters of variable sources. This likely explains the inconsistent results described in the previous sections.

Cross-processing of the RVS data for Cepheids and RR Lyrae stars was already successfully implemented between CU6 and CU7 for *Gaia* DR3, leading to the first release of RVS RV curves for a small sample of these variable stars (see Section 2). This collaborative effort will continue for DR4, extending also to other CUs. A DPAC Task Force has been created, including members from CU8, CU7, CU6, CU4 and CU3, to discuss and coordinate the cross-CUs processing of a restricted, selected sample of Cepheids and RR Lyrae stars for which an optimised per-transit treatment can be performed to increase the precision and reduce systematic errors in the *Gaia* data products (astrometric measurements, chemical abundances, epoch and mean RVs, absorption values, etc.) published for these sources in future releases, starting from *Gaia* DR4. This cross-CU processing is schematically illustrated in Figure 1 by the blue arrows connecting the CUs involved.

Processing of the astrometry that will be released as part of DR4 is well advanced. A number of iteratively improved astrometric solutions have already been realised. Tests are underway using the PL(Z), PW(Z) and LZ relations to select the best solution for Cepheids and RR Lyrae stars. In parallel, astrophysical parameters derived by the SOS Cep&RRL pipeline for DR3 Cepheids and RR Lyrae stars, complemented by literature values were fed back to the pipelines processing the RVS spectra to improve the accuracy of the RV determinations. Tests of abundance analysis of the per-transit RVS spectra are also in progress.

4.2. *Gaia's next data releases*

The *Gaia* mission is producing unprecedented astrometry (positions, parallaxes and proper motions), multiband photometry reaching 1 mmag precision, radial velocities and astrophysical parameters for increasing numbers of Cepheids and RR Lyrae stars, with a tremendous impact on the calibration of these standard candles of the cosmic distance ladder. Precision and accuracy of the data products for Cepheids and RR Lyrae stars are expected to further improve thanks to the longer time baseline and the improved processing of the data sets that will be published in future releases.

Two releases of *Gaia* data products will take place within the next few years, a Focused Product Release (FPR) on 10 October 2023 and Data Release 4 (DR4) at the end of 2025. *Gaia* FPR will contain three data products that may be of relevance for calibrations of the distance scale:

1. Astrometry and photometry from engineering images taken in selected regions of high source density (the GC ω Cen, for this FPR);
2. RV epoch data from the RVS for a subsample of the DR3 long-period variables (LPVs);
3. Diffuse interstellar bands (DIBs) from stacked RVS spectra.

Gaia DR4, at the end of 2025, will publish data products from observations collected in the 66 months from July 2014 to January 2020. This will consist of:

1. Full astrometric, photometric and RV catalogues;
2. All available variable-star and non-single-star solutions;
3. Source classifications plus multiple astrophysical parameters (derived from BP/RP, RVS and astrometry) for stars, unresolved binaries, galaxies and quasars;
4. All epoch and transit data for all sources.

The lifetime of the *Gaia* mission is dictated by exhaustion of its cold-gas propellant, which is fundamental to maintain its precision pointing. This will occur in the second quarter of 2025. A first two-year extension of *Gaia*'s operations was already approved by ESA's Science Programme Committee (SPC), extending the collection of data from *Gaia* until the end of 2022. A further extension until mid-2025 was approved in the spring of 2023, thus turning *Gaia* a mission with continued observations over a 10 years total period. This will result in a 30% improvement in position and parallax measurements and a 70% improvement in proper motions. The ESA SPC also extended the Multilateral Agreement concerning *Gaia* data processing. *Gaia*'s post-operations phase will be completed by the end of 2030, and the fifth and last *Gaia* data release, DR5, currently scheduled not before the end of 2030, is anticipated to contain all data collected by the mission over its 10-year lifetime.

4.3. *Synergies with future surveys*

Large spectroscopic surveys of Cepheids and RR Lyrae stars are planned with the new generation spectrographs WEAVE, 4MOST and MOONS. They will use *Gaia* sources as input catalogues and will provide, within the next few years, abundances and radial velocities for *Gaia*'s Cepheids and RR Lyrae stars well beyond the magnitude limit that can be reached with the RVS. This will allow us to further improve the calibration of the PL and PW relations by better constraining their dependence on metal abundance.

A strong synergy will also exist with the Legacy Survey of Space and Time (LSST) at the Vera Rubin Observatory, which in 2024 will start collecting data reaching ~5 mag fainter than *Gaia* in photometry and ~3 mag in astrometry (Ivezić et al. 2014; their Figure 1).

4.4. *Gaia Near-infrared*

Gaia Near Infra-Red (GaiaNIR) is one of the proposals that have been selected as a potential future large- or medium-class mission involving international cooperation, among those received in response to the 2016 Call for New Science Ideas in ESA's Science Programme. It is a *Gaia*-type mission covering wavelengths from the optical to the NIR, with a wavelength cutoff in the K band. Proposed for launch around 2045, GaiaNIR would extend the astrometric achievements of *Gaia* to astronomical sources that are only visible in the NIR, allowing to probe deeper through the Galactic dust in the MW disc, the spiral arms and the bulge region, at the same time maintaining the accuracy of the *Gaia* optical reference frame and improving the stellar parallax and proper motion accuracy by revisiting the astronomical sources about 20 years after *Gaia*. Efforts are now underway to identify suitable detectors and to prepare for writing mission proposals.

Such a new all-sky NIR astrometric mission will expand and improve on the science cases of *Gaia*, including the calibration of the Cepheids ad RR Lyrae relations that are known to be much tighter in the NIR.

Acknowledgments

The author warmly thanks colleagues of the SOS Cep&RRL team (V. Ripepi, A. Garofalo, R. Molinaro, S. Leccia and T. Muraveva), as well as the whole *Gaia* CU7 team of the Geneva Data Processing Centre (DPCG). This work presents results from the ESA space mission *Gaia*, processed by the *Gaia* Data Processing and Analysis Consortium (DPAC). Funding for the DPAC has been provided by national institutions participating in the *Gaia* Multilateral Agreement. In particular, the Italian participation in DPAC has been supported by Istituto Nazionale di Astrofisica (INAF) and the Agenzia Spaziale Italiana (ASI) through grants I/037/08/0, I/058/10/0, 2014-025-R.0, 2014-025-R.1.2015 and 2018-24-HH.0 to INAF (PI M. G. Lattanzi).

References

Andrae, R., Fouesneau, M., Sordo, R., et al. 2023, A&A, 674, A27
Arenou, F., Luri, X., Babusiaux, C., et al. 2018, A&A, 616, A17
Bhardwaj, A., Rejkuba, M., de Grijs, R., et al. 2021, ApJ, 909, 200
Breuval, L., Kervella, P., Wielgórski, P., et al. 2021, ApJ, 913, 38
Breuval, L., Riess, A. G., Kervella, P., et al. 2022, ApJ, 939, 89
Clementini, G., Carretta, E., Gratton, R., et al. 1995, AJ, 110, 2319
Clementini, G., Ripepi, V., Garofalo, A., et al. 2023, A&A, 674, A18

Clementini, G., Ripepi, V., Leccia, S., et al. 2016, A&A, 595, A133
Clementini, G., Ripepi, V., Molinaro, R., et al. 2019, A&A, 622, A60
Creevey, O. L., Sordo, R., Pailler, F., et al. 2023, A&A, 674, A26
Crestani, J., Braga, V. F., Fabrizio, M. et al., 2021, ApJ, 914, 10
Cropper, M., Katz, D., Sartoretti, P., et al. 2018, A&A, 616, A5
Eyer, L., Mowlavi, N., Evans, D. W., et al. 2017, arXiv:1702.03295
Eyer, L., Audard, M., Holl, B., et al. 2023, A&A, 674, A13
Gaia Collaboration, Brown, A. G. A., Vallenari, A., et al. 2021, A&A, 649, A1
Gaia Collaboration, Clementini, G., Eyer, L., et al. 2017, A&A, 605, A79
Gaia Collaboration, Prusti, T., de Bruijne, J. H. J., et al. 2016, A&A, 595, A1
Gaia Collaboration, Vallenari, A., Brown, A. G. A., et al. 2023, A&A, 674, A1
Garofalo, A., Delgado, H. E., Sarro, L. M., et al. 2022, MNRAS, 513, 788
Gilligan, C. K., Chaboyer, B., Marengo, M., et al. 2021, MNRAS, 503, 4719
Ivezić, Ž., Kahn, S. M., & Eliason, P. 2014, EAS Publ. Ser., 67–68, 211
Klagyivik, P., Szabados, L., Szing, A., et al. 2013, MNRAS, 434, 2418
Lambert, D. L., Heath, J. E., Lemke, M., et al. 1996, ApJS, 103, 183
Layden, A. C., Tiede, G. P., Chaboyer, B., et al. 2019, AJ, 158, 105
Groenewegen, M. A. T. 2021, A&A, 654, A2
Li, X.-Y., Huang, Y., Liu, G.-C., et al. 2023, ApJ, 944, 88
Lindegren, L., Bastian, U., Biermann, M., et al. 2021, A&A, 649, A4
Lindegren, L., Hernández, J., Bombrun, A., et al. 2018, A&A, 616, A2
Lindegren, L., Lammers, U., Bastian, U., et al. 2016, A&A, 595, A4
Molinaro, R., Ripepi, V., Marconi, M., et al. 2023, MNRAS, 520, 4154
Muraveva, T., Delgado, H. E., Clementini, G., et al. 2018, MNRAS, 481, 1195
Nemec, J. M., Cohen, J. G., Ripepi, V., et al. 2013, ApJ, 773, 181
Pancino, E., Britavskiy, N., Romano, D., et al. 2015, MNRAS, 447, 2404
Piersimoni, A. M., Bono, G., & Ripepi, V. 2002, AJ, 124, 1528
Recio-Blanco, A., de Laverny, P., Palicio, P. A., et al. 2023, A&A, 674, A29
Riello, M., De Angeli, F., Evans, D. W., et al. 2021, A&A, 649, A3
Riess, A. G., Breuval, L., Yuan, W., et al. 2022a, ApJ, 938, 36
Riess, A. G., Casertano, S., Yuan, W., et al. 2018, ApJ, 861, 126
Riess, A. G., Casertano, S., Yuan, W., et al. 2021, ApJL, 908, L6
Riess, A. G., Yuan, W., Macri, L. M., et al. 2022, ApJL, 934, L7
Rimoldini, L., Holl, B., Gavras, P., et al. 2023, A&A, 674, A14
Ripepi, V., Catanzaro, G., Clementini, G., et al. 2022, A&A, 659, A167
Ripepi, V., Catanzaro, G., Molinaro, R., et al. 2021, MNRAS, 508, 4047
Ripepi, V., Clementini, G., Molinaro, R., et al. 2023, A&A, 674, A17
Sartoretti, P., Blomme, R., David, M., et al. 2022, *Gaia* DR3 documentation, European Space Agency; *Gaia* Data Processing and Analysis Consortium.
Zinn, J. C. 2021, AJ, 161, 214

Discussion

Question (Kiss): Can we go back please to the Galactic Cepheid PL relation? Maybe this is a very naive question, but I noticed an excess of stars at periods 2 to 3 days in several talks, in other words at luminosities around −4 to −5 mag the PL relation is thicker. Why is that so? Do we understand this excess in the luminosity function of Cepheids? Is this the luminosity where the tracks turn, and is that why there are many stars there?

Answer: Im afraid we need the theoreticians to provide an answer here.

Question (Kiss): OK, then I ask the theoreticians in the audience: do we understand this feature?

Answer (Ripepi): If I understand the question well, it's simply a matter of the timescale in relation to the tracks. It is the period of time that the stars spend in the instability strip. So, you have more stars at shorter periods, simply because they are slower, and there are more [of them], also.

Answer (Clementini): Yes, because one of the improvements of *Gaia* is exactly the possibility of seeing the short-period Cepheids, so basically you are increasing this region, probably. Now, this feature shows up more clearly, but it should be an evolution effect.

Question (Riess): Very nice talk. What are the prospects for improving the parallax offset now for DR4?

Answer: Well, there have been calculations already; we are on a good track. As I said before, there will be specific work done at least for Cepheids and RR Lyrae in that context, particularly for bright Cepheids. So, we'll definitely do a more tailored processing of those sources, so there is hope. Already the first results showed that the errors are going down, but it is too early to say. There have been a number of processes already, but those are not final.

Question (Riess): But parallax offsets, are they related to this reprocessing?

Answer: Well, if you try to control the systematics, we still hope that we can improve it a little bit. The first tests show that there is still room for improvement. But how much, it is difficult to say.

Question (Skowron): As regards the forthcoming release of the time-series spectroscopy and radial velocities for long-period variables, is it going to be for all long-period variables or just for a subsample?

Answer: I don't know exactly the number. It will be for a subsample of the long-period variables (LPVs) that were already released in DR3. So, it's not something new, but it's a subsample of the LPVs for which time-series radial velocity data will be published. I don't know exactly the number. Laurent [Eyer], do you remember the number?

Answer (Eyer): Ten thousand.

Answer (Clementini): The problem is that the RVS only goes to sources brighter than about 14 mag in G_{RVS}. So, it's the brightest subsample. This is a small sample of bright LPVs for which there is enough time-series RV data.

Question (Baruch): [I'm] very interested in the precision of the parallax measurements. Lindegren, in an earlier paper, DR3, said that for Cepheids that were close to quasars they were getting a couple of μas in precision. And so, the precision is really important, especially from my point of view, where I think there is absorption going on anyway. Do you think DR4 will produce a much better version?

Answer: Well, much better may be a big promise. Better, I'd say. I'd stay on the safe side. Because the processing is going on these days; we don't really have the final results yet.

Question (Baruch): How confident was Lindegren in saying that these were a couple of μas for Cepheids near quasars?

Answer: This is a difficult question, because there are so many effects in play. A couple of μas is really at the limit, I would say.

At the crossroads of astrophysics and cosmology: Period–luminosity relations in the 2020s
Proceedings IAU Symposium No. 376, 2024
R. de Grijs, P. A. Whitelock & M. Catelan, eds.
doi:10.1017/S1743921323002995

Primary Period–Luminosity-Relation Calibrators in the Milky Way: Cepheids and RR Lyrae

Physical basis, Calibration, and Applications

M. A. T. Groenewegen

Koninlijke Sterrenwacht van België, Ringlaan 3, 1180 Brussels, Belgium
martin.groenewegen@oma.be

Abstract. In this invited review I discuss the calibration and applications of the period–luminosity relation of classical Cepheid and RR Lyrae stars. After a brief introduction, starting with results from *Hipparcos* and discussing some post-*Hipparcos* era developments, I focus on recent results using *Gaia* Data Release 3 data. I present an overview of the most recent period–luminosity relations, a discussion and some new results on Cepheids in open clusters. I also discuss the effect of reddening and that the use of Wesenheit indices is actually an oversimplification to dealing with the problem of reddening.

Keywords. Cepheids, RR Lyrae, Distance sale, Galactic structure

---

1. Introduction

Period–luminosity (PL) relations are key tools to determine distances to classical variables like RR Lyrae (RRL) and classical Cepheids (but also to Type II Cepheids (T2Cs), Mira and δ Scuti variables), and therefore, tracers of Galactic structure. The calibration of these PL relations is an important step, and even a crucial one in, arguably, the most important application of PL relations, i.e., anchoring the local distance scale, in particular for classical Cepheids (DCEPs).

In this review I invert the order of my talk in Budapest and start by discussing the calibration of PL relations, focussing on recent results obtained with *Gaia* Data Release 3 (DR3) data (Gaia Collaboration et al. 2021, 2023b), and I then discuss applications to Galactic structure. A similar review was given at IAU Symposium 330 (Groenewegen 2018a) when *Gaia* DR1 had just become available (Gaia Collaboration et al. 2016b,a).

1.1. *The Hipparcos and pre-Gaia era*

Let us turn back the clock a quarter of a century. *Hipparcos* provided parallax and proper motion (PM) data for a predefined sample of about 110,000 objects (ESA 1997). Feast and Catchpole (1997) derived the PL relation for DCEPs in the V band†. *Hipparcos* data were available for 220 DCEPs, although most weight was in a subset of 26 stars, which itself was dominated by one Cepheid, α UMi (Polaris). Although revolutionary at the time, the parallax data were not good enough to determine the slope of the PL relation (it was fixed to the value derived for the LMC), let alone to explore any metallicity dependence (which was not considered).

† $M_V = -2.81 \log P + (-1.43 \pm 0.10)$

They were also the first to work in 'parallax space', avoiding any issues with taking the logarithm of a distance in case of negative parallaxes. The (photometric) parallax is defined as

$$\pi = 10^{0.2(m_0 - F - 5)}, \tag{1}$$

where π is the parallax, m_0 the dereddened magnitude, and F any (non-)linear function of period, metallicity, or any other quantity. This equation can be solved using maximum-likelihood techniques or χ^2 minimisation for the parameters of the function F, taking into account the errors in the parallax and the observables.

Shortly after Feast and Catchpole (1997), Madore and Freedman (1998) used $BVIJHK$ data of 8–19 Cepheids (depending on the combination of available magnitudes) to determine reddening values and the distance to the LMC†. Ten years later, van Leeuwen et al. (2007) used the same sample of 220 DCEPs to derive a PL relation in the K band‡.

For RR Lyrae stars, even fewer (accurate) *Hipparcos* parallaxes were available and Gratton (1998) used 3 RR Lyrae, 9 red horizontal branch (HB), and 10 blue HB stars to derive a relation between V-band magnitude and metallicity§.

A step forward in parallax accuracy came about when it was realised that the *Hubble Space Telescope* (*HST*) could be used to determine parallaxes. A paper on the namesake RR Lyrae was published by Benedict et al. (2002), followed by papers on nine DCEPs (Benedict et al. 2007), five RRL, and two T2Cs (Benedict et al. 2011), and the Cepheid η Aql (Benedict et al. 2022). Over one hundred objects have had their parallaxes determined using the *HST*/Fine Guidance Sensor (FGS); see the review by Benedict et al. (2017). Then, Riess et al. (2014) developed a spatial scanning technique using the Wide-Field Camera 3 (WFC3) on board the *HST* to derive the parallax of SY Aur, and of seven other DCEPs (Riess et al. 2018a).

Table 1 lists these DCEPs and indicates the progress moving from the revised *Hipparcos* parallaxes to the *HST*-based parallaxes to *Gaia* (Gaia Collaboration et al. 2016b), in particular DR3 (GDR3; Gaia Collaboration et al. 2021). Table 2 shows the same for the RRL and T2Cs. The last column lists the goodness-of-fit parameter (GoF) which indicates the quality of the astrometric solution (in a way similar to the 'renormalised unit weight error', RUWE). The expectation value for the GoF is zero, with an error of unity.

Several things are noteworthy: (1) The improvement in accuracy from *HST*/FGS to GDR3 is about a factor of 10 for the RRL and T2Cs, but not really apparent for the DCEPs observed with the FGS. (2) For the DCEPs observed with *HST*/WFC3 and for the RRL/T2C, the GDR3 parallax errors are comparable, and the GoF values indicate good astrometric solutions (except for κ Pav). (3) For most of the brighter DCEPs observed with *HST*/FGS the GoF values are poor, and those with GoF >8 would typically be excluded from any fitting. (4) The *HST* parallaxes are, on average, larger than the GDR3 ones. This is directly related to the parallax zero-point offset (PZPO; see below) in the *Gaia* data, and this allowed Groenewegen (2018b) to derive a global offset of -49 ± 18 μas for DCEP in GDR2, based on nine DCEPs with GoF < 8—including RS Pup and its geometric distance from Kervella et al. (2014)—and non-*Gaia* parallax errors comparable to the GDR2 parallax errors.

The advent of the revised *Hipparcos* parallaxes and *HST*/FGS measurements led to new PL relations that were probably the standard ones until the arrival of *Gaia*.

† $E(B-V)$ ranging from 0.14 ± 0.08 mag to 0.17 ± 0.07 mag and distance moduli ranging from 18.44 ± 0.35 mag to 18.57 ± 0.11 mag.

‡ $M_K = -3.258 \log P + (-2.40 \pm 0.05)$.

§ $M_V = (0.22 \pm 0.22)([\mathrm{Fe/H}] + 1.5) + (0.66 \pm 0.11)$

Table 1. Cepheid parallaxes (in mas) from *Hipparcos* to *Gaia*.

Name	$\pi \pm \sigma_\pi$ (rev. *Hipparcos*)	$\pi \pm \sigma_\pi$ (*HST*)	$\pi \pm \sigma_\pi$ (GDR3)	GoF
η Aql	2.36 ± 1.04	3.71 ± 0.07	3.67 ± 0.19	29
β Dor	3.24 ± 0.36	3.14 ± 0.16	2.93 ± 0.14	47
δ Cep	3.77 ± 0.17	3.66 ± 0.15	3.56 ± 0.15	31
FF Aql	2.11 ± 0.33	2.81 ± 0.18	1.91 ± 0.07	1.5
l Car	2.09 ± 0.29	2.01 ± 0.20	1.98 ± 0.11	24
RT Aur	-1.10 ± 1.41	2.40 ± 0.19	1.81 ± 0.12	66
T Vul	2.71 ± 0.43	1.90 ± 0.23	1.69 ± 0.06	5.3
Y Sgr	2.64 ± 0.45	2.13 ± 0.29	1.97 ± 0.06	13
X Sgr	3.31 ± 0.26	3.00 ± 0.18	2.81 ± 0.14	5.1
ζ Gem	2.37 ± 0.30	2.78 ± 0.18	3.07 ± 0.22	22
W Sgr	3.75 ± 1.12	2.28 ± 0.20	2.37 ± 0.18	36
SY Aur	-1.84 ± 1.72	0.428 ± 0.054	0.427 ± 0.020	2.5
SS CMa	0.40 ± 1.78	0.389 ± 0.029	0.287 ± 0.013	4.0
XY Car	-1.02 ± 0.88	0.438 ± 0.048	0.378 ± 0.014	2.1
VX Per	0.87 ± 1.52	0.420 ± 0.076	0.364 ± 0.017	3.5
VY Car	0.36 ± 1.42	0.586 ± 0.045	0.554 ± 0.017	-2.4
WZ Sgr	3.50 ± 1.22	0.512 ± 0.039	0.574 ± 0.028	-1.1
S Vul		0.322 ± 0.040	0.205 ± 0.020	1.1
X Pup	1.97 ± 1.26	0.277 ± 0.048	0.376 ± 0.020	1.2

Notes. Col. 2: Revised *Hipparcos* parallaxes (van Leeuwen 2007, 2008); Col. 3: *HST* parallaxes – Benedict et al. (2022) for η Aql, Benedict et al. (2007) for β Dor/W Sgr, Riess et al. (2014) for SY Aur, and Riess et al. (2018a) for SS CMa/X Pup; Col. 4: GDR3 parallaxes, with the goodness-of-fit (GoF) parameter in Col. 5.

Table 2. RRL and T2C parallaxes (in mas) from *Hipparcos* to *Gaia*.

Name	Type	$\pi \pm \sigma_\pi$ (rev. *Hipparcos*)	$\pi \pm \sigma_\pi$ (*HST*)	$\pi \pm \sigma_\pi$ (GDR3)	GoF
RR Lyr	RRab	3.46 ± 0.64	3.77 ± 0.13	3.985 ± 0.027	0.9
SU Dra	RRab	0.20 ± 1.13	1.42 ± 0.16	1.332 ± 0.014	0.4
UV Oct	RRab	2.44 ± 0.81	1.71 ± 0.10	1.838 ± 0.012	0.0
XZ Cyg	RRab	2.29 ± 0.84	1.67 ± 0.17	1.586 ± 0.015	4.1
RZ Cep	RRc	0.59 ± 1.48	2.54 ± 0.19	2.401 ± 0.012	-0.3
VY Pyx	BL Her	5.01 ± 0.44	6.44 ± 0.23	3.950 ± 0.019	-3.6
κ Pav	W Vir	6.52 ± 0.77	5.57 ± 0.28	5.245 ± 0.122	38

Col. 2: Pulsation Type. Col. 3: Revised *Hipparcos* parallaxes (van Leeuwen 2007, 2008). Col. 4: *HST* parallaxes from Benedict et al. (2011). Col. 5: GDR3 parallaxes, with the GoF parameter in Col. 6.

Fouqué et al. (2007) presented PL relations in seven bands, from B to K, and in two Wesenheit colours using 59 DCEPs with parallaxes based on the revised *Hipparcos* catalogue, *HST*/FGS, Baade–Wesselink techniques, and distances to open clusters (OCs) containing DCEPs.

Feast et al. (2008) used the revised *Hipparcos* parallaxes for the 142 RRL in the *Hipparcos* catalogue and the *HST*/FGS parallax of RR Lyrae (Benedict et al. 2002) to derive relations in the V and K band†. Benedict et al. (2011) presented PL relations based on the five RRL they observed with the FGS‡.

† $M_V = 0.214([\mathrm{Fe/H}] + 1.38) + (0.54 \pm 0.10)$ and $M_K = -2.41(\log P + 0.252) + (-0.63 \pm 0.10)$, for fixed slopes.

‡ $M_V = 0.214([\mathrm{Fe/H}] + 1.5) + (0.45 \pm 0.05)$ and, $M_K = -2.38(\log P + 0.28) + (-0.57 \pm 0.03)$ [solution 2 in their table 9], for fixed slopes.

2. The *Gaia* DR3 era

Numerous studies have been published related to PL relations of variable stars since the first *Gaia* data release. One recurrent issue has been the fact that point sources at, essentially, infinity (like quasi-stellar objects; QSOs) do not have, on average, a *Gaia* parallax of zero; there is a PZPO. For the QSOs this global offset was PZPO $= -29$ μas in DR2 (Lindegren et al. 2018) and -17 μas in DR3 (Lindegren et al. 2021a), in the sense $\pi_{\rm true} = \pi_{Gaia} - {\rm PZPO}$.

After the publication of GDR2, numerous studies appeared that determined the PZPO using other classes of stars, typically at brighter magnitudes than the QSOs. Examples using GDR2 data for DCEPs are Riess et al. (2018b) and Groenewegen (2018b), who found a more negative value (-46 ± 13 μas and -49 ± 18 μas, respectively), and for RRL Muraveva et al. (2018) and Layden et al. (2019) (~ -56 mas and -42 ± 13 mas, respectively). Many other classes of objects have been used to study the PZPO, like red clump stars, (detached) eclipsing binaries, red giant asteroseismic parallaxes, stars with parallaxes from Very Long Baseline Interferometry, W UMa-type binaries, and so on, and this continued when G(E)DR3 was released (Huang et al. 2021b; Stassun and Torres 2021; Ren et al. 2021; Vasiliev and Baumgardt 2021; Zinn 2021; Flynn et al. 2022; Maíz Apellániz 2022; Wang et al. 2022; Khan et al. 2023).

Lindegren et al. (2021a, hereafter L21) provided a python script to the community which returned the PZPO (without an error bar) as a function of input parameters, namely ecliptic latitude (β), G-band magnitude, the `astrometric_params_solved` parameter, and either the effective wavenumber of the source used in the astrometric solution ($\nu_{\rm eff}$, `nu_eff_used_in_astrometry` for the five-parameter solution `astrometric_params_solved` = 31) or the astrometrically estimated pseudo-colour of the source (`pseudocolour`) for the six-parameter solution (`astrometric_params_solved` = 95). The module is defined in the range $6 < G < 21$ mag, $1.24 < \nu_{\rm eff} < 1.72 \mu{\rm m}^{-1}$, corresponding to about $0.15 < (Bp - Rp) < 3.0$ mag where G, Bp, and Rp are the magnitudes in the *Gaia* G, BP, and RP band, respectively. Maíz Apellániz (2022, hereafter MA22) provided an IDL script to determine the PZPO (without an error bar) using the same mathematical framework as in L21, but with updated coefficients.

Groenewegen (2021), hereafter G21, took a different approach, in particular regarding the spatial component of the PZPO. While in L21 and MA22 the spatial component is parameterised using a second-order polynomial fit, G21 determined the spatial component (for a fiducial magnitude of $G = 20$ mag) based on the HEALPix formalism (Górski et al. 2005)† The magnitude-dependent component is a combination of linear functions and based on wide binary systems, as in L21. Groenewegen (2021) found no real colour dependence of the PZPO.

Claims have been made that the L21 correction overcorrects the parallax offset (there are also claims to the contrary; see below). In some papers, one therefore considers a correction to the PZPO recipes, like

$$\pi_{\rm true} = \pi_{Gaia} - {\rm PZPO} + \Delta\pi \tag{2}$$

(note that in some papers the sign of the last term is reversed).

Some papers do not use the (corrected) parallax but rather the Bayesian estimate of the distance from Bailer-Jones et al. (2021). This latter work includes the L21 correction and uses a prior on the distance. This prior is pre-calculated based on a mock catalogue of GDR3 data for HEALPix level 5. This prior may not be optimal for specific populations,

† The HEALPix formalism is a convenient way to divide the sky into equal-area pixels. At HEALPix level 0 there are 12 pixels, and this increases by a factor of four for every next higher level. The HEALPix formalism is used by the *Gaia* team and is encoded in the `source_id` as pixel number = `source_id`$/(2^{35} \times 4^{(12-{\rm level})})$ for a given HEALPix level.

Table 3. Recent PL relations for DCEPs.

α^1	β	γ^1	Band	GAL	N	σ	Remarks[2]
			Ripepi et al. (2023)				
-2.793 ± 0.015	17.333 ± 0.010	0.0	G	LMC	2477	0.169	FU
-2.830 ± 0.033	17.757 ± 0.026	0.0	G	SMC	843	0.251	$P > 2.95$ d
-3.317 ± 0.007	15.998 ± 0.005	0.0	W_G	LMC	2477	0.075	
-3.382 ± 0.021	16.592 ± 0.026	0.0	W_G	SMC	839	0.156	$P > 2.95$ d
		Ripepi et al. (2022); DR3, L21 + Riess et al. (2021a) correction, $1.1\sigma_\pi$					
-3.294 ± 0.040	-6.042 ± 0.013	0.0	W_G	MW	372	0.017	FU/ABL
-3.178 ± 0.048	-5.971 ± 0.017	-0.661 ± 0.077	W_G	MW	372	0.016	
-3.176 ± 0.044	-5.988 ± 0.018	-0.520 ± 0.090	W_G	MW	435	0.015	F+1O 'best'
	Breuval et al. (2022); DR3, BJ21 + correction (L21 + Riess et al. (2021a) correction)						
$-2.715F$	-4.153 ± 0.060		V	MW	183	0.22	
$-2.715F$	-4.129 ± 0.046	-0.311 ± 0.082	V	MCs/MW	2083		
$-2.816F$	-4.468 ± 0.064		G	MW	446	0.49	
$-2.816F$	-4.384 ± 0.047	-0.462 ± 0.089	G	MCs/MW	2083		
$-3.222F$	-5.899 ± 0.033		K	MW	65	0.17	
$-3.222F$	-5.827 ± 0.034	-0.321 ± 0.068	K	MCs/MW	2017		
$-3.324F$	-5.965 ± 0.032		[3.6]	MW	21	0.20	
$-3.324F$	-5.912 ± 0.031	-0.292 ± 0.057	[3.6]	MCs/MW	110		
$-3.338F$	-6.006 ± 0.021		W_G	MW	596	0.32	
$-3.338F$	-5.959 ± 0.025	-0.384 ± 0.051	W_G	MCs/MW	2473		
$-3.291F$	-5.882 ± 0.038		W_{VI}	MW	157	0.18	
$-3.291F$	-5.851 ± 0.035	-0.201 ± 0.071	W_{VI}	MCs/MW	2051		
$-3.323F$	-6.185 ± 0.038		W_{JK}	MW	63	0.18	
$-3.323F$	-6.134 ± 0.042	-0.322 ± 0.079	W_{JK}	MCs/MW	2014		
$-3.305F$	-5.955 ± 0.024		W_H	MW	60	0.16	
$-3.305F$	-5.931 ± 0.027	-0.280 ± 0.078	W_H	LMC/MW	130		
		Owens et al. (2022); DR3, BJ21 + offset					
$-3.265F$	-5.693 ± 0.023		K	MW	37	0.142	+18 F 5–60 d
$-3.284F$	-5.793 ± 0.022		[3.6]	MW	37	0.087	+18 F
-3.401 ± 0.091	-5.986 ± 0.023		W_{JK}	MW	37	0.140	+18 F
-3.339 ± 0.099	-6.009 ± 0.025		W_{VI}	MW	37	0.151	+18 F
		Molinaro et al. (2023); DR3, L21 + fitted offset					
-3.167 ± 0.053	-5.887 ± 0.035	-0.43 ± 0.12	K	MW	443	0.017	-24.8 ± 4.6
-3.333 ± 0.057	-6.178 ± 0.038	-0.40 ± 0.12	W_{JK}	MW	443	0.016	-21.9 ± 5.0
-3.191 ± 0.042	-6.018 ± 0.028	-0.32 ± 0.10	$W_H(T)$	MW	430	0.016	-23.6 ± 4.1
		Riess et al. (2022); DR3, L21 + fitted offset, MW cluster Cepheids					
$-3.299F$	-5.902 ± 0.026	$-0.217F$	W_H	MW	17	0.06	-3 ± 4, G+S
$-3.299F$	-5.890 ± 0.018	$-0.217F$	W_H	MW	17	0.06	$0F$
-3.36 ± 0.07	-5.893 ± 0.018	$-0.217F$	W_H	MW	17	0.06	$0F$
	Cruz Reyes and Anderson (2023); DR3, L21 + fitted offset, MW cluster+field Cepheids						
$-3.299F$	-5.914 ± 0.017	$-0.217F$	W_H	MW	82		-13 ± 5
-3.383 ± 0.052	-5.930 ± 0.020	$-0.217F$	W_H	MW	82		-17 ± 5
		Riess et al. (2021a); DR3, L21 + fitted offset, MW field Cepheids					
-3.28 ± 0.06	-5.915 ± 0.030	-0.20 ± 0.13	W_H	MW	75		-14 ± 6
$-3.26F$	-5.915 ± 0.022	$-0.17F$	W_H	MW	75		-14 ± 6
$-3.26F$	-5.865 ± 0.013	$-0.17F$	W_H	MW	75		$0F$

like the Cepheids that mainly reside in the Galactic disc, or for RR Lyrae that are mainly found in the halo.

2.1. *Recent PL relations.*

Table 3 lists recent PL relations for DCEPs. Always consult the original reference, since typically additional PL relations are also given, e.g. in other photometric bands, fundamental (FU) mode versus first overtone (FO) versus FU+FO pulsators. After each reference some keywords are listed as to how the distances or parallaxes were obtained. In most cases, this is based on GDR3 data and application of the L21 correction to individual objects. Sometimes an additional constant offset is fitted, or one used distances from BJ21. Always consult the original reference for details on the methodology.

Table 3. Recent PL relations for DCEPs (Continued) and T2C.

α^1	β	γ^1	Band	GAL	N	σ	Remarks[2]
			Lin et al. (2022); DR3, L21, MW cluster Cepheids				
-2.94 ± 0.12	-5.87 ± 0.11		W_G	MW OC	46	0.25	
			Wang et al. (2018); distances from JHK PL relations of 31 DCEP in OCs (Chen et al. 2017)				
-3.248 ± 0.018	-5.781 ± 0.017		$W1$	MW	255	0.081	FU
-3.266 ± 0.027	-5.811 ± 0.027		$W2$	MW	190	0.107	FU
-3.298 ± 0.024	-5.765 ± 0.024		[3.6]	MW	85	0.066	FU
-3.245 ± 0.024	-5.747 ± 0.027		[4.5]	MW	99	0.073	FU
			Lemasle et al. (2022)				
-2.436 ± 0.013	-5.632 ± 0.019		W_{W1W2}	LMC	2326	0.149	FU
			T2C[3]				
			Ripepi et al. (2023)				
-1.905 ± 0.036	18.731 ± 0.033		G	LMC	205	0.275	
-2.140 ± 0.100	19.105 ± 0.098		G	SMC	42	0.372	
-2.577 ± 0.019	17.516 ± 0.033		W_G	LMC	197	0.138	
-2.505 ± 0.054	17.843 ± 0.052		W_G	SMC	42	0.190	
			Ngeow et al. (2022b); distances GCC from Baumgardt and Vasiliev (2021)				
-2.41 ± 0.03	-1.09 ± 0.03		K	GCC	48	0.10	
-2.46 ± 0.03	-1.28 ± 0.03		W_{JK}	GCC	46	0.08	
			Wielgórski et al. (2022); DR3 + L21 (also for G21) / LMC				
-2.387 ± 0.030	-1.013 ± 0.011		K	LMC	62	0.087	BLH+WVIR
-2.544 ± 0.029	-1.166 ± 0.011		W_{JK}	LMC	61	0.087	BLH+WVIR
-2.225 ± 0.089	-1.169 ± 0.024		K	MW	14	0.065	BLH+WVIR
-2.399 ± 0.095	-1.355 ± 0.022		W_{JK}	MW	15	0.061	BLH+WVIR

Notes. [1] F stands for a parameter that was fixed. [2] Any parallax offsets quoted in Col. 8 are in μas. [3] For T2C zero points are at $P = 1$ day.

PL relations are parameterised in the literature typically as $M = \alpha(\log P - p1) + \beta + \gamma([\mathrm{Fe/H}] - p2)$, where $p1$ and $p2$ are constants. Zero points β in the table have been rescaled to a period of 10 days and [Fe/H] = 0 dex where appropriate. Wesenheit functions are defined as follows: $W_G = G - 1.90(Bp - Rp)$; $W_{VI} = I - \epsilon(V - I)$, with $\epsilon = 1.387$ (Breuval et al. 2022) or 1.61 (Owens et al. 2022); $W_{JK} = K - \epsilon(J - K)$, with $\epsilon = 0.735$ (Breuval et al. 2022), 0.71 (Owens et al. 2022), 0.69 (Molinaro et al. 2023; Wielgórski et al. 2022), or 0.618 (Ngeow et al. 2022b); $W_{W1W2} = W2 - 2.0(W1 - W2)$; and $W_H = \mathrm{F160W} - 0.386(\mathrm{F555W} - \mathrm{F814W})$. $W_H(T)$ means that VIH-band photometry was transformed to the *HST* system. In particular for W_{VI} and W_{JK}, different functional forms are adopted in the literature that make a direct comparison of PL relations quasi-impossible. The different coefficients relate to differences in the adopted reddening laws (see Section 2.3).

One of the lingering open questions regarding DCEPs was the metallicity dependence of the PL relation. Significant progress has been made over the past years. In their derivation of the metallicity term Breuval et al. (2022) adopt a single metallicity for the LMC DCEPs (-0.409 dex from Romaniello et al. 2022, with an uncertainty of 0.020 dex) and SMC DCEPs (-0.75 ± 0.05 dex from Gieren et al. 2018), but also for the Milky Way (MW) a single metallicity was adopted, typically ranging from ($+0.087 \pm 0.056$) dex to ($+0.099 \pm 0.089$) dex from the V to K bands and the Wesenheit indices, depending on the exact sample per filter†. The approximation of a single metallicity for the MW sample does not make use of the actual range in metallicities that MW DCEPs cover.

Ripepi et al. (2022) and Molinaro et al. (2023) use exclusively MW DCEPs to constrain the metallicity term. The full range in [Fe/H] is from -0.83 to $+0.55$ dex (median = $+0.05$ dex, dispersion = 0.07 dex, from the 499 stars used by Ripepi et al. 2022). These two

† For comparison, Gieren et al. (2018) adopted [Fe/H] = -0.34 ± 0.06 dex for the LMC and [Fe/H] = $+0.00 \pm 0.05$ dex for the MW.

Table 4. Recent PL relations for RRL.

α	β	γ	Band	GAL	N	σ	Remarks[2]
		Bhardwaj et al. (2023); DR3, L21					
-2.37 ± 0.02	-0.80 ± 0.03	0.18 ± 0.01	K	MW/GC	1077	0.05	RRabcd
-2.73 ± 0.02	-1.06 ± 0.03	0.16 ± 0.02	W_{JK}	MW/GC	1096	0.06	
		Zgirski et al. (2023); DR3, BJ21					
-3.03 ± 0.25	-1.06 ± 0.03	0.083 ± 0.025	K	MW	28	0.073	RRabc
-3.26 ± 0.23	-0.80 ± 0.03	0.083 ± 0.024	W_{JK}	MW	28	0.070	
		Looijmans et al. (2023); DR3, fit global offset					
$+0.393 \pm 0.049$	$+0.76 \pm 0.05$	0.273 ± 0.022	V	MW/GC	200		-14 ± 9
-2.45 ± 0.11	-0.86 ± 0.04	0.171 ± 0.018	K	MW/GC	200		-15 ± 5
-2.48 ± 0.07	-0.93 ± 0.02	0.121 ± 0.010	W_G[1]	MW/GC	200		-8 ± 3
		Mullen et al. (2023); DR3, L21, error inflation factor					
-2.44 ± 0.10	-0.841 ± 0.008	0.144 ± 0.014	$W1$	MW/GC	1052	0.02	
-2.54 ± 0.10	-0.857 ± 0.009	0.151 ± 0.014	$W2$	MW/GC	397	0.02	
		Li et al. (2023); DR3, L21 + Bayesian estimate					
	$+1.106 \pm 0.021$	0.350 ± 0.016	G	MW	205	0.12	RRab
-2.465 ± 0.084	-0.792 ± 0.043	0.161 ± 0.011	K	MW	159	0.14	
-2.452 ± 0.080	-0.834 ± 0.031	0.179 ± 0.011	$W1$	MW	164	0.09	
		Garofalo et al. (2022); DR3, fit global offset					
	$+1.13 \pm 0.03$	0.33 ± 0.02	V	MW	270	0.02	-33 ± 2
-2.49 ± 0.21	-0.88 ± 0.09	0.14 ± 0.03	W_G	MW	190	0.09	-33 ± 3
		Gilligan et al. (2021); DR3, L21 + fit global offset, error inflation					
-2.78 ± 0.12	-1.018 ± 0.020	0.115 ± 0.016	$W1$	MW	150		10 ± 7
		Ngeow et al. (2022a); GCC (DR3)					
-1.43 ± 0.14	$+0.243 \pm 0.055$	0.144 ± 0.018	i	GCC	321	0.15	RR0 (= FU)
-2.60 ± 0.13	$+0.093 \pm 0.051$	0.193 ± 0.017	W_{ri}[3]	GCC	325	0.15	
		Neeley et al. (2019); DR2, PZPO = −30 mas					
-2.40 ± 0.27	-0.87 ± 0.02	0.18 ± 0.03	[3.6]	MW	55	0.18	
-2.45 ± 0.28	-0.89 ± 0.02	0.18 ± 0.03	[4.5]	MW	55	0.18	
-2.60 ± 0.25	-1.01 ± 0.02	0.13 ± 0.04	W_{VI}[4]	MW	55	0.18	

Notes. [1] They used $W_G = G - 1.85(Bp - Rp)$. [2] Any parallax offsets quoted in Col. 8 are in μas. [3] $W_{ri} = r - 4.051(r - i)$. [4] $W_{VI} = I - 1.467(V - I)$.

papers find steeper slopes than Breuval et al. (2022) but with larger error bars, and, formally, the slopes agree at the 2σ level. One should also keep in mind that, in general, the adopted or derived PZPO correlate with the coefficients α, β, and γ (see, e.g., table 7 of Groenewegen 2018b or table 7 of Breuval et al. 2020 based on GDR2. Parallaxes are intrinsically more accurate in GDR3, but this correlation should still hold to some extent also for GDR3 data).

Grosso modo, PL relations derived by different authors appear to be in reasonably good agreement. A noticeable exception are the differences between Breuval et al. (2022) and Owens et al. (2022), in particular the additional PZPO that has a different sign; but see footnote 8 in Breuval et al. (2022) for an explanation that is related to the metallicity term.

Table 4 lists recent PL relations for RRL. Zero points β have been rescaled to a period of one day and [Fe/H] = 0 dex where appropriate. Coefficients of 0.69 and 1.90 were adopted for the W_{JK} and W_G relations, unless noted otherwise. From a procedural point of view, the work of Mullen et al. (2023) is interesting as it takes into account that the errors in the astrometric solution are underestimated, and they use Fabricius et al. (2021) to correct their results (see also Ripepi et al. 2022, who apply a correction to their Cepheid sample). Li et al. (2023) adopt a Bayesian estimate to estimate the distance from the parallax, although not from the work of Bailer-Jones et al. (2021); instead, they use a prior that is more appropriate for RRL, namely a power law in Galactocentric distance (Huang et al. 2021a).

2.2. *Cepheids in Open Clusters*

Since the release of GDR2 there has been a revolution in the census and study of OCs; see Hunt and Reffert (2023) and references therein. Not surprisingly, there has been renewed interest specifically for DCEPs in OCs as well, extending pre-*Gaia* work by, e.g., Anderson et al. (2013). Recent papers on the subject are by Medina et al. (2021), Zhou and Chen (2021), Lin et al. (2022), Riess et al. (2022), Hao et al. (2022), and Cruz Reyes and Anderson (2023). In addition some new results are presented in this work.

Table 5 contains the derived parallaxes and errors of the clusters containing DCEPs, where the recent work by Cruz Reyes and Anderson (2023) has been taken as reference for the list of clusters. Riess et al. (2022) considered clusters for which *HST* photometry is available. The difference between Riess et al. (2022) and Cruz Reyes and Anderson (2023) is in the membership list. Riess et al. (2022) use cluster members derived from GDR2 data Cantat-Gaudin et al. (2018), although parallaxes from GDR3 were used in the distance determination, while Cruz Reyes and Anderson (2023) independently determined cluster members from GDR3 data. There are also small differences in selecting the final sample, e.g., Cruz Reyes and Anderson (2023) only consider members in the range $12.5 < G < 17$ mag and $0.23 < Bp - Rp < 2.75$ mag, where the L21 correction results in the smallest residuals. Indeed, both works include the L21 correction. The differences in parallaxes are small (Col. 3 − Col. 2 = 5.7 ± 2.0 μas for 15 clusters). Column 4 gives the parallaxes as quoted by Hao et al. (2022). They do not include a PZPO.

Column 5 lists the results obtained by me using the membership list of Cruz Reyes and Anderson (2023), and they do not include a PZPO either. No selection based on magnitude or colour was made; however a selection based on the GoF as well as 4σ clipping on parallaxes and proper motions (PMs) in Right Ascension and Declination† relative to the median values and using (1.486× median-absolute-deviation) as an estimate for σ was performed. These two estimates agree very well with (Col. 5 − Col. 4) $= 0.8 \pm 2.5$ μas for 20 clusters.

A few remaining remarks on the results in Table 5: First, the difference between Cols. 2/3 and Cols. 4/5 is the inclusion of the L21 correction. The average PZPO (Col. 5 − Col. 3) = -35.2 ± 2.0 μas for 30 clusters. Second, Cepheids can be assigned to *different* clusters by different authors. This is the meaning of the pair of curly brackets at the end of the table. Membership is typically based on position, parallax and PM in contemporary searches. Typically, there is not enough accurate (*Gaia*) radial velocity (RV) data available to be used as an additional criterion. Third, one notices that there is a lower limit to the error in the mean cluster parallax of 6–8 μas in Riess et al. (2022) and Cruz Reyes and Anderson (2023) and 10–12 μas in the present work.

The total parallax uncertainty adds in quadrature the statistical uncertainty determined as the error on the median‡, the uncertainty of where the Cepheid is located in the cluster§, and the systematic contribution due to the angular covariance of parallaxes at small scales. The first term is typically small, as this goes as $1/\sqrt{N}$. When replacing the parallax of the Cepheid with the parallax of the cluster, there is an uncertainty of where the Cepheid is located with respect to the centre of the cluster. To estimate this effect, the radius on the sky containing 68% of cluster members is converted into a distance. This effect is also small in most cases. The total error is dominated by the last term. Parallaxes are correlated on all spatial scales, but the effect is largest on small scales. Riess et al. (2022) and Cruz Reyes and Anderson (2023) largely follow Maíz Apellániz et al. (2021). There are other studies on the matter (e.g., Lindegren et al. 2021b), and I followed

† Cruz Reyes and Anderson (2023) use 6σ clipping.

‡ Cruz Reyes and Anderson (2023) use the weighted mean.

§ This effect is not considered in either Riess et al. (2022) or Cruz Reyes and Anderson (2023).

Table 5. Clusters containing Cepheids.

Cluster	$\pi \pm \sigma_\pi$ (1)	$\pi \pm \sigma_\pi$ (2)	$\pi \pm \sigma_\pi$ (3)	$\pi \pm \sigma_\pi$ (4)	N	R (′)
Berkeley 58	336 ± 8	336 ± 7	290 ± 20	302 ± 11	85	4.0
NGC 129	559 ± 7	557 ± 7	530 ± 30	529 ± 10	290	6.5
FSR 0951	594 ± 7	610 ± 7	570 ± 30	566 ± 11	128	6.8
vandenBergh 1	579 ± 10	585 ± 10	550 ± 40	543 ± 13	50	1.9
Ruprecht 79	274 ± 8	281 ± 7	240 ± 20	244 ± 11	99	2.8
NGC 5662	1322 ± 7	1336 ± 6	1300 ± 30	1292 ± 12	225	18.1
Lynga 6	408 ± 9	421 ± 8	380 ± 40	383 ± 11	157	5.2
NGC 6067	496 ± 7	513 ± 7	470 ± 30	473 ± 11	671	4.5
NGC 6087	1057 ± 7	1073 ± 7	1020 ± 30	1025 ± 11	154	7.7
IC 4725	1540 ± 7	1554 ± 6	1520 ± 50	1514 ± 12	325	11.6
NGC 6649	508 ± 8	514 ± 7	470 ± 50	474 ± 11	301	3.1
UBC 129	886 ± 7	880 ± 7	850 ± 20	852 ± 11	116	11.7
UBC 130	428 ± 8	425 ± 9	400 ± 20	396 ± 11	36	2.4
NGC 7790	331 ± 8	322 ± 7	290 ± 30	293 ± 11	114	3.0
Czernik 41		407 ± 8	370 ± 40	373 ± 11	103	2.7
FSR 0384		520 ± 8		492 ± 12	56	3.3
NGC 103		317 ± 7	280 ± 20	284 ± 11	104	3.1
NGC 6664		504 ± 7	460 ± 40	469 ± 11	286	6.3
UBC 106		443 ± 7	410 ± 30	407 ± 10	328	6.8
UBC 290		639 ± 6	610 ± 20	604 ± 11	204	9.9
UBC 375		562 ± 7		536 ± 12	142	8.3
UBC 533		878 ± 9		834 ± 12	52	6.0
CWNU 0175		732 ± 9		709 ± 12	31	7.0
Cl XPup		363 ± 6		330 ± 10	122	16.2
Cl IQNor		544 ± 9		504 ± 12	42	7.7
Cl STTau		953 ± 8		921 ± 11	66	9.9
Cl SXVel		497 ± 7		459 ± 10	73	12.6
Cl V378Cen		518 ± 8		485 ± 11	95	6.3
{ UBC 231	356 ± 8	345 ± 8		312 ± 11	56	7.9
{ OC -0717			370 ± 10	370 ± 10	64	24.4
{ Ruprecht 93		482 ± 7		448 ± 11	169	4.7
{ NGC 3496	439 ± 8			408 ± 10	464	7.6

Notes. Parallaxes in μas. Col. 2: Riess et al. (2022); Col. 3: Cruz Reyes and Anderson (2023); Col. 4: Hao et al. (2022) (the original work quotes parallaxes in mas to two decimal places); Col. 5: This work. Columns 2 and 3 include the L21 PZPO correction; Cols. 4–5 include no PZPO. Col. 6: Number of stars entering the distance calculation in the present work. Col. 7: Size of the cluster in arcminutes defined as the radius that includes 68% of its members. Names in Col. 1 starting with "Cl" are new clusters proposed by Cruz Reyes and Anderson (2023).

Vasiliev and Baumgardt (2021, their Eq. 1). A better understanding of the behaviour of the spatial covariance and smaller values are expected in further *Gaia* releases.

Studies have typically used the L21 correction when considering the PZPO. Table 6 lists results using other recipes. Columns 2 and 6 have been copied from Table 5. They reflect the results from Cruz Reyes and Anderson (2023) (which includes L21) and this work (without PZPO). Columns 3–5 are from this work using the recipe from L21, MA22, and G21 for HEALPix level 2.

The differences between Columns 2 and 3 are minor (Col. 3 − Col. 2 = -0.3 ± 2.0 μas for 30 clusters). They are solely related to minor differences in the selection of the final sample (differences in selection of magnitude and colour, differences in the clipping in parallaxes and PMs, etc.), since both sets use the L21 correction. The differences between using MA22 or L21 is also small (Col. 4 − Col. 3 = -1.3 ± 2.0 μas). The differences between using G21 and L21 is larger: (Col. 5 − Col. 3) = $+1.9 \pm 2.0$ μas. Using no PZPO gives (Col. 6 − Col. 3) = -35.0 ± 2.0 μas.

Figure 1 provides another view of these differences (similar to figure 2 in Cruz Reyes and Anderson 2023). It shows the difference between individual parallaxes and the mean cluster parallax for all cluster members summed over all clusters as a function of G magnitude. Plotted are the models with no correction (top left), L21 (top right),

Table 6. Comparing different PZPO models in clusters.

Cluster	$\pi \pm \sigma_\pi$ (1)	$\pi \pm \sigma_\pi$ (2)	$\pi \pm \sigma_\pi$ (3)	$\pi \pm \sigma_\pi$ (4)	$\pi \pm \sigma_\pi$ (5)
Berkeley 58	336 ± 7	330 ± 11	329 ± 11	338 ± 11	302 ± 11
NGC 129	557 ± 7	558 ± 10	556 ± 10	564 ± 10	529 ± 10
FSR 0951	610 ± 7	607 ± 11	607 ± 11	588 ± 11	566 ± 11
vandenBergh 1	585 ± 10	585 ± 12	582 ± 12	576 ± 13	543 ± 13
Ruprecht 79	281 ± 7	275 ± 11	274 ± 11	256 ± 11	244 ± 11
NGC 5662	1336 ± 6	1328 ± 12	1329 ± 12	1358 ± 12	1292 ± 12
Lynga 6	421 ± 8	420 ± 11	417 ± 11	400 ± 11	383 ± 11
NGC 6067	513 ± 7	514 ± 11	513 ± 11	495 ± 11	473 ± 11
NGC 6087	1073 ± 7	1065 ± 11	1066 ± 11	1065 ± 11	1025 ± 11
IC 4725	1554 ± 6	1552 ± 12	1551 ± 12	1544 ± 12	1514 ± 12
NGC 6649	514 ± 7	514 ± 11	511 ± 11	487 ± 11	474 ± 11
UBC 129	880 ± 7	882 ± 11	880 ± 11	886 ± 11	852 ± 11
UBC 130	425 ± 9	426 ± 11	423 ± 12	431 ± 12	396 ± 11
NGC 7790	322 ± 7	321 ± 11	320 ± 11	341 ± 11	293 ± 11
Czernik 41	407 ± 8	407 ± 11	404 ± 11	406 ± 11	373 ± 11
FSR 0384	520 ± 8	520 ± 12	519 ± 12	541 ± 11	492 ± 12
NGC 103	317 ± 7	312 ± 11	312 ± 11	320 ± 11	284 ± 11
NGC 6664	504 ± 7	507 ± 11	504 ± 11	477 ± 11	469 ± 11
UBC 106	443 ± 7	445 ± 10	443 ± 10	511 ± 10	407 ± 10
UBC 290	639 ± 6	641 ± 11	641 ± 11	626 ± 11	604 ± 11
UBC 375	562 ± 7	566 ± 12	563 ± 11	575 ± 12	536 ± 12
UBC 533	878 ± 9	871 ± 12	870 ± 12	896 ± 12	834 ± 12
CWNU 0175	732 ± 9	736 ± 12	734 ± 12	727 ± 11	709 ± 12
Cl XPup	363 ± 6	368 ± 10	366 ± 10	381 ± 10	330 ± 10
Cl IQNor	544 ± 9	542 ± 12	541 ± 12	571 ± 12	504 ± 12
Cl STTau	953 ± 8	962 ± 12	960 ± 12	935 ± 12	921 ± 11
Cl SXVel	497 ± 7	494 ± 10	495 ± 10	478 ± 10	459 ± 10
Cl V378Cen	518 ± 8	520 ± 11	519 ± 11	538 ± 11	485 ± 11
UBC 231	345 ± 8	350 ± 12	349 ± 12	371 ± 12	312 ± 11
Ruprecht 93	482 ± 7	484 ± 11	484 ± 11	473 ± 11	448 ± 11

Notes. Parallaxes in μas. Col. 2: Cruz Reyes and Anderson (2023); Cols. 3–6: This work, with the PZPOs from L21 (Col. 3), MA22 (Col. 4), G21 for HEALPix level 2 (Col. 5), no PZPO (Col. 6).

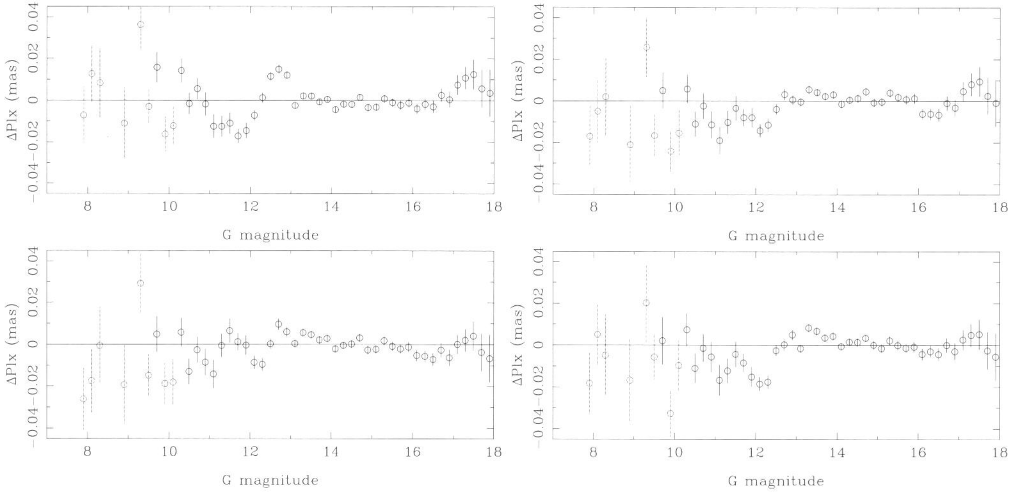

Figure 1. Residuals between individual and cluster parallaxes summed over all clusters and plotted against G magnitude. Bins plotted in red contain less than 0.1% of all stars. PZPO models are: no correction (top left), L21 (top right), MA22 (bottom left), G21 (bottom right).

Table 7. Root-mean-square residuals between different PZPO models as a function of magnitude.

G (mag)	no PZPO (μas)	L21 (μas)	MA22 (μas)	G21 (μas)
All	9.78	9.37	9.45	9.41
$G \leq 13$ mag	13.37	12.83	13.10	13.06
$13 < G < 17$ mag	2.35	3.48	3.60	3.39
$17 \leq G \leq 18$ mag	8.62	6.06	3.97	4.33

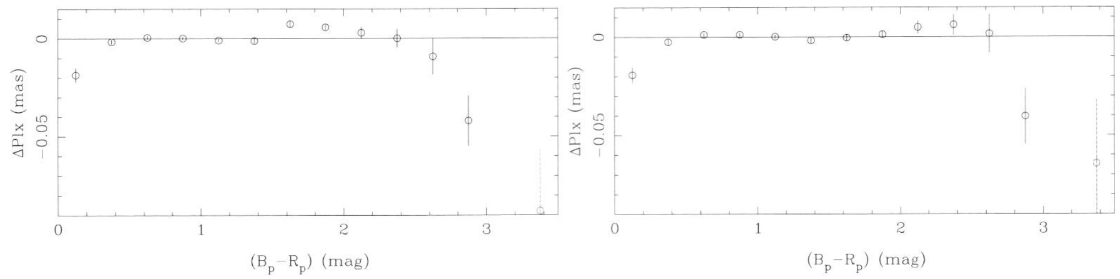

Figure 2. Residuals between individual and cluster parallaxes summed over all clusters and plotted against $(Bp - Rp)$ colour for no PZPO correction (left) and the L21 correction (right). Bins plotted in red contain less than 0.1% of all stars.

MA22 (bottom left), and G21 (bottom right). The model without PZPO shows the characteristic behaviour as a function of magnitude; see figure 20 in L21 (or figure 7 in G21 or figure 8 in MA22).

All three correction schemes reduce the scatter, in particular the trends between about 11.5 and 13th magnitude, but none does so in a perfect manner. Some residual trends and offsets remain. Table 7 lists the r.m.s. values in different magnitude bins and overall. It is difficult to draw firm conclusions, although the 11.5–13 magnitude range remains problematic even after applying any of the PZPO corrections, at least for the sample studied here.

Figure 2 shows the same plotted against $(Bp - Rp)$ colour but only for the no-PZPO and the L21 correction as the other two corrections are very similar. Although the L21 correction includes a dependence on colour, the difference with no correction seems small. The offsets for the bluest and reddest colours appear real (see the discussion in Cruz Reyes and Anderson 2023)). L21 recommend not to use their correction outside the range corresponding to 0.15–3.0 mag in $(Bp - Rp)$.

2.3. *Reddening and the Wesenheit function*

Dereddening the observed magnitudes for the effect of dust extinction in the MW foreground or internal to a galaxy represents a crucial step in obtaining absolute calibrated magnitudes and PL relations. The Wesenheit function (Madore 1982) was introduced to eliminate the effects of reddening, but the use of different coefficients in the literature mentioned in Section 2.1 reminds us of the fact that any given coefficient is only valid for one particular reddening law. For a Wesenheit function $W = a - R(b - c)$, R is given by $R = A_a/(A_b - A_c)$.

I consider here the reddening laws of Cardelli et al. (1989) with the update in the optical published by O'Donnell (1994, hereafter COD), Fitzpatrick (1999, hereafter F99)†, and Wang and Chen (2023, hereafter W23). The reddening laws are continuous functions of wavelength and, therefore, in order to calculate the coefficient of a Wesenheit function one has to adopt certain a wavelength for a certain filter. This characteristic wavelength

† Using the routine `https://idlastro.gsfc.nasa.gov/ftp/pro/astro/fm_unred.pro`

Table 8. Effective wavelengths (in nm) for in several filters under different conditions.

Filter	Model 1	2	3	4	5	6	7
G	702	769	658	701	700	669	736
Bp	563	578	538	561	559	549	577
Rp	794	826	785	795	796	781	811
V	556	560	551	556	555	553	560
I_c	790	795	788	790	790	787	793
V_o	541	543	537	541	540	538	544
I_o	798	803	796	799	799	795	802
J	1237	1240	1231	1236	1236	1234	1238
H	1643	1648	1641	1643	1643	1641	1645
K	2153	2154	2153	2163	2153	2152	2154
F555W	545	548	540	544	544	541	549
F814W	814	828	810	814	814	807	823
F160W	1532	1541	1527	1532	1532	1531	1534

Standard is $\log g = 2.0$ and $M = 5$ $M_\odot$ for all MARCS models. Model 1: (default) $T_{\rm eff} = 5000$ K, $[Z] = +0.0$, $E(B-V) = 0.5$ mag, Model 2: $T_{\rm eff} = 3600$ K, $[Z] = +0.0$, $E(B-V) = 0.5$, Model 3: $T_{\rm eff} = 7000$ K, $[Z] = +0.0$, $E(B-V) = 0.5$, Model 4: $T_{\rm eff} = 5000$ K, $[Z] = -0.5$, $E(B-V) = 0.5$, Model 5: $T_{\rm eff} = 5000$ K, $[Z] = -1.0$, $E(B-V) = 0.5$, Model 6: $T_{\rm eff} = 5000$ K, $[Z] = +0.0$, $E(B-V) = 0.1$. Model 7: $T_{\rm eff} = 5000$ K, $[Z] = +0.0$, $E(B-V) = 1.0$.

depends on the underlying spectrum of the star, but also on the reddening itself. I have calculated $\lambda_{\rm eff}$ defined as

$$\lambda_{\rm eff} = \frac{\int \lambda R(\lambda) M(\lambda) 10^{-0.4 A_\lambda}\, d\lambda}{\int R(\lambda) M(\lambda) 10^{-0.4 A_\lambda}\, d\lambda}, \tag{3}$$

where $R(\lambda)$ is the filter response curve, $M(\lambda)$ the flux of a MARCS model (Gustafsson et al. 2008), and A_λ the reddening for the following filters: *Gaia* G, Bp, Rp, Johnson V, Cousins I (I_c), OGLE-IV (V_o, I_o), 2MASS JHK, and *HST* F555W, F814W, and F160W.

I took 5000 K as a representative effective temperature of DCEPs, and 3600 K and 7000 K as extremes at the longest and shortest periods, respectively. The impact of choosing a metallicity and reddening was also investigated. The results are listed in Table 8. As expected, increased reddening and cooler stars lead to longer effective wavelengths. The effect of metallicity is negligible (at most 2 nm in the Bp band). Overall, the effect is most noticeable in the broadband *Gaia* G filter and in the optical in general, and it essentially vanishes in the infrared (IR).

Table 9 lists the relative reddenings for the three different reddening laws (and the indicated values for R_V) for models 1/2/3 that cover the extreme range in changes in effective wavelength. The bottom part gives the relevant coefficient for different Wesenheit indices. The conclusion is that the coefficients vary considerably depending on the reddening law. In the case of the Wesenheit index based on the *Gaia* bands there is even a strong dependence on the temperature for the Cepheids (and probably also on the reddening) for a given reddening law.

What is the effect in practice, and how does one deal with Wesenheit coefficients that differ per star? One can still take a fiducial relation, say a Wesenheit relation and consider it to be just another period–luminosity–colour relation where you need to deredden explicitly, and calibrate against that. Table 10 gives an example for CR Car which is a 9.7 day DCEP. The table lists the observed magnitudes (the reddening coefficients for the OGLE filters were used for V and I), the dereddened magnitudes, and the unreddened and dereddened magnitudes for the 'period–luminosity–colour' relations, $W_G = G - 1.90(Bp - Rp)$, $W_{VI} = I - 1.387(V - I)$, $W_{JK} = K - 0.735(J - K)$, and $W_H = \mathrm{F160W} - 0.386(\mathrm{F555W} - \mathrm{F814W})$.

Table 9. Relative reddening values for different reddening laws for Models 1/2/3.

Filter	$\lambda_{\rm eff}$/Model	A_λ/A_V F99 $R_V = 3.3$	A_λ/A_V COD $R_V = 3.1$	A_λ/A_V W23 $R_V = 3.4$
G	7020/7690/6580	0.697/0.601/0.769	0.751/0.655/0.810	0.703/0.610/0.776
Bp	5630/5780/5380	0.950/0.919/1.008	0.969/0.937/1.029	0.968/0.935/1.027
Rp	7940/8260/7850	0.570/0.532/0.581	0.617/0.569/0.631	0.579/0.543/0.590
$V_{\rm o}$	5410/5430/5370	1.001/0.996/1.011	1.021/1.016/1.032	1.020/1.015/1.030
$I_{\rm o}$	7980/8030/7960	0.565/0.559/0.567	0.611/0.603/0.614	0.575/0.569/0.577
J	12370/12400/12310	0.261/0.260/0.263	0.287/0.286/0.289	0.249/0.247/0.252
H	16430/16480/16410	0.165/0.164/0.166	0.182/0.181/0.182	0.138/0.138/0.139
$K_{\rm s}$	21530/21540/21530	0.112/0.112/0.112	0.118/0.117/0.118	0.079/0.079/0.079
F555W	5450/5480/5400	1.011/0.984/1.003	1.011/1.004/1.024	1.011/1.003/1.022
F814W	8140/8280/8100	0.587/0.530/0.551	0.587/0.566/0.593	0.557/0.541/0.561
F160W	15320/15410/15270	0.203/0.182/0.185	0.203/0.201/0.204	0.160/0.158/0.161
			R	
W_G		1.831/1.557/1.798	2.136/1.781/2.034	1.807/1.558/1.775
W_{VI}		1.296/1.279/1.279	1.490/1.463/1.471	1.291/1.275/1.276
W_{JK}		0.751/0.756/0.740	0.694/0.698/0.685	0.465/0.468/0.459
W_H		0.414/0.402/0.409	0.479/0.461/0.474	0.352/0.342/0.349

Table 10. Observed and dereddened magnitudes for CR Car and different reddening laws.

Filter	Observed	Dereddened magnitudes F99/ Model 1 $R_V = 3.3$ $E(B\text{-}V) = 0.5$	COD/ Model 1 $R_V = 3.1$ $E(B\text{-}V) = 0.5$	W23/ Model 3 $R = 3.4$ $E(B\text{-}V) = 0.1$
G	11.063	9.909	9.875	10.655
Bp	11.880	10.311	10.347	11.341
Rp	10.137	9.199	9.161	9.827
V	11.511	9.859	9.928	10.986
I	10.007	9.074	9.040	9.704
J	8.846	8.415	8.392	8.714
H	8.188	7.916	7.900	8.115
K	8.058	7.873	7.871	8.017
F555W	11.750	10.080	10.150	11.213
F814W	9.973	9.003	9.044	9.678
F160W	8.384	8.049	8.063	8.299
W_G	7.751	7.795	7.621	7.780
W_{VI}	7.921	7.984	7.808	7.926
W_{JK}	7.479	7.475	7.489	7.504
W_H	7.698	7.633	7.636	7.707

In this particular case, assuming that the adopted Wesenheit relation takes care of reddening and is universal and thus using the observed magnitudes, compared to actually dereddening the photometry assuming different reddening values introduces differences that can be as large as almost 0.1 mag, especially in the optical. In the IR the effect is smaller but may still be 0.05 mag. The exact effects will depend on a star-to-star basis and the choice of the fiducial period–luminosity–colour relations.

The effect of different reddening laws will also potentially affect other standard candles, e.g., the tip of the red-giant branch method that requires I (and a $(V-I)$ colour), or the JAGB method that requires J (and a $(J-K)$ colour). For example, the often used reddening map of the Magellanic Clouds of Skowron et al. (2021) is given in terms of $E(V-I)$ and needs to be transformed to the reddening in other bands†

† The value to transform $E(V-I)$ to $E(B-V)$ given by Skowron et al. (2021) is only valid for one particular reddening law.

3. Galactic structure

Once accurate and precise PL relations are available, distances can be computed (and those will be more accurate than distances derived from the currently available *Gaia* data, particularly at larger distances) and an obvious application is in the context of Galactic structure. Many other tracers than DCEPs and RR Lyrae have been used to study Galactic structure (e.g., OB stars, OCs) but those will not be covered here. More extended reviews on this topic were given by Dorota Skowron and Vasiliy Belokurov at this conference.

3.1. *Distance to the Galactic Centre*

An obvious application is to use variable stars to determine the structure of the Galactic Centre (GC) region and in particular the distance to the GC. Majaess et al. (2018) used 4194 RRab stars from the VISTA Variables in the Vía Láctea survey (VVV; Minniti et al. 2010), in the region $350 \lesssim l \lesssim 10^\circ$, $-10 \lesssim b \lesssim$ -2.7°, $2.7 \lesssim b \lesssim 5^\circ$, an M_K (based on an LMC distance modulus of 18.43 mag and no metallicity term) and a $(J-K)_0$ versus $\log P$ relation to derive $d_{\rm GC} = 8.30 \pm 0.36$ kpc (for a subsample with $|\, b\, |> 4^\circ$), uncorrected for the cone effect.

Braga et al. (2018) presented IR data for 894 T2C in the VVV survey (with periods from OGLE) and derived an $M_K - \log P$ relation that also includes terms depending on Galactic longitude and latitude. They obtained $d_{\rm GC} = 8.46 \pm 0.14$ kpc (combining statistical and systematic errors). Griv et al. (2021) used 715 high-latitude ($|\, b\, |> 1^\circ$) and centrally symmetrically concentrated T2Cs identified in the VVV survey to derive $d_{\rm GC} = 8.35 \pm 0.10$ kpc, concluding that the T2C distribution is an ellipsoid with axial ratios of about 1:0.7:0.6.

These distances agree with that derived from the stars orbiting Sgr A*, $d_{\rm GC} = 8277 \pm 34$ pc (adding random and systematic errors; see GRAVITY Collaboration et al. 2022, 2021, 2019).

3.2. *RR Lyrae*

Pietrukowicz et al. (2020) used the OGLE survey to discriminate between MW and background RRL to study samples of 53,000 Galactic bulge and disc and 951 halo RRab stars (plus 2288 Sgr dSph, 26,900 LMC and 4600 SMC RRL). They used photometric metallicities to derive characteristic metallicites for the halo/bulge/disc of [Fe/H] = -1.2, -1.0, and -0.6 dex, respectively. They found that 1/3 of the RRL within the bulge area actually belong to the halo.

Iorio and Belokurov (2021) used 5D *Gaia* DR2 data to study the kinematical properties of halo and disc RRL to find a dominant, non-rotating, halo-like population and a much smaller rotating disc-like population.

Li et al. (2023) used a sample of 2700 RRL with precise metallicities to calibrate a P–ϕ_{31}–R_{21}–[Fe/H] relation, and 205 RRab and 31 RRc to calibrate an M_G–[Fe/H] relation using parallaxes from GDR3. This was then applied to 115,410 RRab and 20,463 RRc stars in GDR3 and used to derive the distances to the LMC and SMC, the mean metallicities and metallicity gradients in the Magellanic Clouds, and the values were compared with independent metallicity and distance determinations of RRL in globular clusters.

3.3. *Classical Cepheids*

Classical Cepheids, being quite young objects, can be used to study the properties of the spiral arms in the MW, the presence and properties of the warp, when velocities are available, properties of the Galactic rotation curve, and the metallicity gradient in the Galactic disc. An updated list of Galactic Cepheids is given by Pietrukowicz et al. (2021)†.

Minniti et al. (2021) use 50 classical Cepheids from the VVV to study the spiral structure in the poorly explored far side of the disc (quadrants I and IV).

Ablimit et al. (2020) use classical Cepheids from several photometric surveys (OGLE, ASAS-SN, *Gaia*, *WISE*, and the Zwicky Transient Facility) and calculate distances based on the *WISE* PL relations from Wang et al. (2018); 3483 objects with distances from the Galactic plane of less than 4 kpc are kept. The warp is clearly visible (as already shown in earlier work). Using PMs from GDR2 and RVs from GDR2 and LAMOST DR6 applied to a cleaned sample of 968 DCEPs they derive a rotation curve of the form $\Theta(R) = (232.5 \pm 0.8) + (-1.3 \pm 0.1)(R - \mathrm{R}_0)$ km s^{-1} (for $\mathrm{R}_0 = 8.122$ kpc). Kinematics studies have also been performed by Bobylev and Bajkova (2023) (also by Bobylev et al. 2021) but using fewer (2000) stars and PL relations that ultimately are tied GDR2. Similar studies, not only using DCEPs but also OB stars, OCs, and RGB stars, are Gaia Collaboration et al. (2023a) and Drimmel et al. (2023).

Lemasle et al. (2022) studied the warp and spiral arms using unWISE (Meisner et al. 2021) photometry for 3260 MW classical Cepheids and a Wesenheit index using *WISE* filters, $W_{W1W2} = W2 - 2.0(W1 - W2)$, calibrated on a sample of LMC DCEPs and the canonical LMC distance. This was applied to the MW sample without any correction for metallicity. They then fitted a model of the form (Skowron et al. 2019) $z(r, \Theta) = \begin{cases} z_0 & R_{\mathrm{GC}} < r_0 \\ z_0 + (R_{\mathrm{GC}} - r_0)^2 \cdot [z_1 \sin(\Theta - \Theta_1) + z_2 \sin(2\,(\Theta - \Theta_2))] & R_{\mathrm{GC}} \geq r_0, \end{cases}$ where z is the vertical distance from the Galactic plane, R_{GC} is the distance from the GC, and Θ the Galactocentric azimuth. Warping starts at $R_{\mathrm{GC}} \sim 10$–12 kpc. Earlier studies have been carried out by Chen et al. (2019) and Skowron et al. (2019), while another very recent study is Dehnen et al. (2023), who also conclude that there is no warping inside about 11 kpc.

Another classical application is to study abundance patterns throughout the MW. Trentin et al. (2023) use 637 DCEPs and a M_{W_G} relation to derive [Fe/H] = $(-0.060 \pm 0.002)R_{\mathrm{GC}} + (0.573 \pm 0.017)$ dex (for $R_{\mathrm{GC}} > 4$ kpc), with a possible break at 9.25 kpc.

Hocdé et al. (2023) took the interesting approach to derive [Fe/H] from the Cepheids' light curve Fourier decomposition (a standard method for RRL). For short-period Cepheids ($2.5 < P < 6.3$ days) they suggest to use a relation based on the amplitudes A_1 and A_2, and for long-period Cepheids ($12 < P < 40$ days) a relation using A_1, ϕ_{21}, and R_{41}. The metallicity gradient they derive is in agreement with the literature. Lemasle et al. (2022) show the radial gradient as well and refer to the companion paper by da Silva et al. (2022) who give [Fe/H] = $(-0.055 \pm 0.003)R_{\mathrm{GC}} + (+0.43 \pm 0.03)$. Kovtyukh et al. (2022) also present the metallicity gradient, but they do not cite a linear coefficient and their figure 2 seems to suggest a slope that steepens in the inner disc.

4. Conclusions

Compared to a few years ago, the metallicity dependence of the Cepheid PL relations is now better determined. What remains to be seen if the dependence derived using the SMC, LMC, and using a single average metallicity for the MW, and that derived from MW DCEPs using the individually derived metallicities can be brought into better agreement.

† 3666 as per the version of September 2022; see `https://www.astrouw.edu.pl/ogle/ogle4/OCVS/allGalCep.listID`

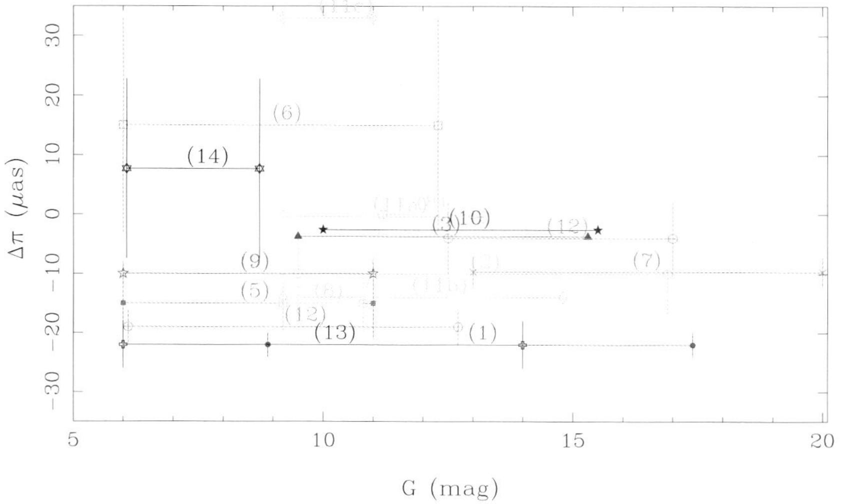

Figure 3. Countercorrection $\Delta\pi$ plotted against magnitude for (1) Bhardwaj et al. (2021), RRL; (2) Gilligan et al. (2021), RRL; (3) Huang et al. (2021b), red clump (5p solution); (4) Ren et al. (2021), WUMa EBs (5p solution); (5) Riess et al. (2021a), DCEP LMC; (6) Stassun and Torres (2021), EBs; (7) Vasiliev and Baumgardt (2021), globular clusters; (8) Zinn (2021), asteroseismology; (9) Flynn et al. (2022), open/globular clusters; $Bp - Rp > 1$ (10) Wang et al. (2022), giants; (11) Khan et al. (2023) red giants: asteroseismology, for the (a) *Kepler*, (b) *K2*, and (c) *TESS*-SCVZ fields ('E20', APOGEE DR17 values); (12) Cruz Reyes and Anderson (2023) Clusters ($G = 12.5 - 17$ mag, $0.8 < Bp - Rp < 2.75$ mag) and MW DCEPs; (13) Molinaro et al. (2023) MW DCEPS; (14) Groenewegen (2023) dynamical parallax of binary systems with spectroscopic and astrometric orbits.

Ongoing and future spectroscopic surveys in both the optical and the near-IR (NIR; ground-based as well as future *Gaia* releases) will provide many more accurate metallicity determinations (as well as radial velocities) that will help in addressing this question, as well as be useful to study abundance gradients (radial and vertical) in more detail and for kinematics studies. Reddening remains an important component, and blindly using Wesenheit indices may lead to systematic effects.

Two topics have not been addressed here. The first is the role of binarity, leading to observed magnitudes that are those of the Cepheid plus a companion. In terms of calibration it becomes a differential effect. This was recently tackled using population synthesis (Karczmarek et al. 2022, 2023) with the conclusion that there is essentially no effect on the derived slope of the PL relation, and a small effect on the zeropoint, depending on filter and binary fraction. However, further investigations are required.

The second point is the role of any excess emission in the NIR and mid-IR. This is known to exist in Galactic DCEPs, e.g., from direct interferometric observations in the optical or NIR (e.g., Kervella et al. 2006; Mérand et al. 2006; Gallenne et al. 2012; Nardetto et al. 2016), modelling with the SPIPS code (e.g., Breitfelder et al. 2016; Trahin 2019; Trahin et al. 2021), or modelling the SEDs of Galactic DCEPs (Gallenne et al. 2013; Groenewegen 2020). Groenewegen (2020) found evidence of an IR excess in 5% of the sample. Groenewegen & Lub (in prep.) studied the spectral energy distributions of LMC and SMC DCEPs and found evidence of an IR excess in 1/142 LMC and 0/77 SMC Cepheids studied. However, there is less mid- and far-IR data available for the Magellanic Clouds, and so it is harder to detect significant excess emission than for MW DCEPs. At face value, a lower percentage of stars with IR excess is consistent with the fact that there is less condensible material available in the LMC and SMC. However, the interpretation of the IR excess in terms of dust emission has also been questioned, and Hocdé et al. (2020) proposed an alternative origin, namely free–free emission from a thin hydrogen shell. In that case, the abundance argument would not play an important role.

A final comment relates to the 'countercorrections' (Eq. 2) that have been derived in the recent literature, i.e., the additional correction one has to apply after applying the L21 correction to bring the parallaxes of one's favourite sample of objects in agreement with an absolute calibration relation for that specific class. Figure 3 collects recent determinations related to GDR3. Over a large range in G magnitude the derived countercorrections do not agree. Khan et al. (2023) also showed that $\Delta\pi$ depends on sky position.

Uncorrected parallaxes have offsets of -17 μas (Lindegren et al. 2021a) at the faint magnitudes of the QSOs, or ~ -24 to -28 μas (Riess et al. 2021b; Huang et al. 2021b; Ren et al. 2021; Wang et al. 2022) at brighter magnitudes. This implies that some countercorrections almost completely counter the L21 correction itself, which is hard to understand given its physical basis.

Acknowledgements

I thank Mauricio Cruz Reyes for kindly making the membership lists of OCs containing Cepheids available before publication.

This work has made use of data from the European Space Agency (ESA) mission *Gaia* (`https://www.cosmos.esa.int/gaia`), processed by the *Gaia* Data Processing and Analysis Consortium (DPAC, `https://www.cosmos.esa.int/web/gaia/dpac/consortium`). Funding for the DPAC has been provided by national institutions, in particular the institutions participating in the *Gaia* Multilateral Agreement.

References

Ablimit, I., Zhao, G., Flynn, C., & Bird, S. A. 2020, *ApJL*, 895, L12.

Anderson, R. I., Eyer, L., & Mowlavi, N. 2013, *MNRAS*, 434, 2238–2261.

Bailer-Jones, C. A. L., Rybizki, J., Fouesneau, M., Demleitner, M., & Andrae, R. 2021, *AJ*, 161, 147.

Baumgardt, H. & Vasiliev, E. 2021, *MNRAS*, 505, 5957–5977.

Benedict, G. F., Barnes, T. G., Evans, N. R., Cochran, W. D., Anderson, R. I., McArthur, B. E., & Harrison, T. E. 2022, *AJ*, 163, 282.

Benedict, G. F., McArthur, B. E., Feast, M. W., Barnes, T. G., Harrison, T. E., Bean, J. L., Menzies, J. W., Chaboyer, B., Fossati, L., Nesvacil, N., Smith, H. A., Kolenberg, K., Laney, C. D., Kochukhov, O., Nelan, E. P., Shulyak, D. V., Taylor, D., & Freedman, W. L. 2011, *AJ*, 142, 187.

Benedict, G. F., McArthur, B. E., Feast, M. W., Barnes, T. G., Harrison, T. E., Patterson, R. J., Menzies, J. W., Bean, J. L., & Freedman, W. L. 2007, *AJ*, 133, 1810–1827.

Benedict, G. F., McArthur, B. E., Fredrick, L. W., Harrison, T. E., Lee, J., Slesnick, C. L., Rhee, J., Patterson, R. J., Nelan, E., Jefferys, W. H., van Altena, W., Shelus, P. J., Franz, O. G., Wasserman, L. H., Hemenway, P. D., Duncombe, R. L., Story, D., Whipple, A. L., & Bradley, A. J. 2002, *AJ*, 123, 473–484.

Benedict, G. F., McArthur, B. E., Nelan, E. P., & Harrison, T. E. 2017, *PASP*, 129, 012001.

Bhardwaj, A., Marconi, M., Rejkuba, M., de Grijs, R., Singh, H. P., Braga, V. F., Kanbur, S., Ngeow, C.-C., Ripepi, V., Bono, G., De Somma, G., & Dall'Ora, M. 2023, *ApJL*, 944, L51.

Bhardwaj, A., Rejkuba, M., de Grijs, R., Yang, S.-C., Herczeg, G. J., Marconi, M., Singh, H. P., Kanbur, S., & Ngeow, C.-C. 2021, *ApJ*, 909, 200.

Bobylev, V. V. & Bajkova, A. T. 2023, *RAA*, 23, 045001.

Bobylev, V. V., Bajkova, A. T., Rastorguev, A. S., & Zabolotskikh, M. V. 2021, *MNRAS*, 502, 4377.

Braga, V. F., Bhardwaj, A., Contreras Ramos, R., Minniti, D., Bono, G., de Grijs, R., Minniti, J. H., & Rejkuba, M. 2018, *A&A*, 619, A51.

Breitfelder, J., Mérand, A., Kervella, P., Gallenne, A., Szabados, L., Anderson, R. I., & Le Bouquin, J.-B. 2016, *A&A*, 587, A117.

Breuval, L., Kervella, P., Anderson, R. I., Riess, A. G., Arenou, F., Trahin, B., Mérand, A., Gallenne, A., Gieren, W., Storm, J., Bono, G., Pietrzyński, G., Nardetto, N., Javanmardi, B., & Hocdé, V. 2020, *A&A*, 643, A115.

Breuval, L., Riess, A. G., Kervella, P., Anderson, R. I., & Romaniello, M. 2022, *ApJ*, 939, 89.

Cantat-Gaudin, T., Jordi, C., Vallenari, A., Bragaglia, A., Balaguer-Núñez, L., Soubiran, C., Bossini, D., Moitinho, A., Castro-Ginard, A., Krone-Martins, A., Casamiquela, L., Sordo, R., & Carrera, R. 2018, *A&A*, 618, A93.

Cardelli, J. A., Clayton, G. C., & Mathis, J. S. 1989, *ApJ*, 345, 245.

Chen, X., de Grijs, R., & Deng, L. 2017, *MNRAS*, 464, 1119.

Chen, X., Wang, S., Deng, L., de Grijs, R., Liu, C., & Tian, H. 2019, *Nat. Astron.*, 3, 320.

Cruz Reyes, M. & Anderson, R. I. 2023, *A&A*, 672, A85.

da Silva, R., Crestani, J., Bono, G., Braga, V. F., D'Orazi, V., Lemasle, B., Bergemann, M., Dall'Ora, M., Fiorentino, G., François, P., Groenewegen, M. A. T., Inno, L., Kovtyukh, V., Kudritzki, R. P., Matsunaga, N., Monelli, M., Pietrinferni, A., Porcelli, L., Storm, J., Tantalo, M., & Thévénin, F. 2022, *A&A*, 661, A104.

Dehnen, W., Semczuk, M., & Schönrich, R. 2023, *MNRAS*, 523, 1556–1564.

Drimmel, R., Khanna, S., D'Onghia, E., Tepper-García, T., Bland-Hawthorn, J., Chemin, L., Ripepi, V., Romero-Gómez, M., Ramos, P., Poggio, E., Andrae, R., Blomme, R., Cantat-Gaudin, T., Castro-Ginard, A., Clementini, G., Figueras, F., Fouesneau, M., Frémat, Y., Lobel, A., Marshall, D., & Muraveva, T. 2023, *A&A*, 670, A10.

ESA 1997, *VizieR Online Data Catalog*, 1239.

Fabricius, C., Luri, X., Arenou, F., Babusiaux, C., Helmi, A., Muraveva, T., Reylé, C., Spoto, F., Vallenari, A., Antoja, T., Balbinot, E., Barache, C., Bauchet, N., Bragaglia, A., Busonero, D., Cantat-Gaudin, T., Carrasco, J. M., Diakité, S., Fabrizio, M., Figueras, F., Garcia-Gutierrez, A., Garofalo, A., Jordi, C., Kervella, P., & et al. 2021, *A&A*, 649, A5.

Feast, M. W. & Catchpole, R. M. 1997, *MNRAS*, 286, L1.

Feast, M. W., Laney, C. D., Kinman, T. D., van Leeuwen, F., & Whitelock, P. A. 2008, *MNRAS*, 386, 2115.

Fitzpatrick, E. L. 1999, *PASP*, 111, 63.

Flynn, C., Sekhri, R., Venville, T., Dixon, M., Duffy, A., Mould, J., & Taylor, E. N. 2022, *MNRAS*, 509, 4276–4284.

Fouqué, P., Arriagada, P., Storm, J., Barnes, T. G., Nardetto, N., Mérand, A., Kervella, P., Gieren, W., Bersier, D., Benedict, G. F., & McArthur, B. E. 2007, *A&A*, 476, 73.

Gaia Collaboration, Brown, A. G. A., Vallenari, A., Prusti, T., de Bruijne, J. H. J., Babusiaux, C., Biermann, M., Creevey, O. L., Evans, D. W., Eyer, L., Hutton, A., Jansen, F., Jordi, C., Klioner, S. A., Lammers, U., Lindegren, L., Luri, X., Mignard, F., Panem, C., Pourbaix, D., Randich, S., & et al. 2021, *A&A*, 649, A1.

Gaia Collaboration, Brown, A. G. A., Vallenari, A., Prusti, T., de Bruijne, J. H. J., Mignard, F., Drimmel, R., Babusiaux, C., Bailer-Jones, C. A. L., Bastian, U., & et al. 2016,a *A&A*, 595a, A2.

Gaia Collaboration, Drimmel, R., Romero-Gomez, M., Chemin, L., Ramos, P., Poggio, E., Ripepi, V., Andrae, R., Blomme, R., Cantat-Gaudin, T., Castro-Ginard, A., Clementini, G., Figueras, F., Fouesneau, M., Fremat, Y., Jardine, K., Khanna, S., Lobel, A., Marshall, D. J., Muraveva, T., Brown, A. G. A., Vallenari, A., Prusti, T., de Bruijne, J. H. J., Arenou, F., Babusiaux, C., Biermann, M., & et al. 2023,a *A&A*, 674a, A37.

Gaia Collaboration, Prusti, T., de Bruijne, J. H. J., Brown, A. G. A., Vallenari, A., Babusiaux, C., Bailer-Jones, C. A. L., Bastian, U., Biermann, M., Evans, D. W., & et al. 2016,b *A&A*, 595b, A1.

Gaia Collaboration, Vallenari, A., Brown, A.G.A., Prusti, T., & et al. 2023,b *A&A*, 674b, A1.

Gallenne, A., Kervella, P., & Mérand, A. 2012, *A&A*, 538, A24.

Gallenne, A., Mérand, A., Kervella, P., Chesneau, O., Breitfelder, J., & Gieren, W. 2013, *A&A*, 558, A140.

Garofalo, A., Delgado, H. E., Sarro, L. M., Clementini, G., Muraveva, T., Marconi, M., & Ripepi, V. 2022, *MNRAS*, 513, 788.

Gieren, W., Storm, J., Konorski, P., Górski, M., Pilecki, B., Thompson, I., Pietrzyński, G., Graczyk, D., Barnes, T. G., Fouqué, P., Nardetto, N., Gallenne, A., Karczmarek, P., Suchomska, K., Wielgórski, P., Taormina, M., & Zgirski, B. 2018, *A&A*, 620, A99.

Gilligan, C. K., Chaboyer, B., Marengo, M., Mullen, J. P., Bono, G., Braga, V. F., Crestani, J., Dall'Ora, M., Fiorentino, G., Monelli, M., Neeley, J. R., Fabrizio, M., Martínez-Vázquez, C. E., Thévenin, F., & Sneden, C. 2021, *MNRAS*, 503, 4719–4733.

Górski, K. M., Hivon, E., Banday, A. J., Wandelt, B. D., Hansen, F. K., Reinecke, M., & Bartelmann, M. 2005, *ApJ*, 622, 759.

Gratton, R. G. 1998, *MNRAS*, 296, 739.

GRAVITY Collaboration, Abuter, R., Aimar, N., Amorim, A., Ball, J., Bauböck, M., Berger, J. P., Bonnet, H., Bourdarot, G., Brandner, W., Cardoso, V., Clénet, Y., Dallilar, Y., Davies, R., de Zeeuw, P. T., Dexter, J., Drescher, A., Eisenhauer, F., Förster Schreiber, N. M., Foschi, A., Garcia, P., Gao, F., Gendron, E., Genzel, R., Gillessen, S., Habibi, M., Haubois, X., Heißel, G., Henning, T., Hippler, S., Horrobin, M., Jochum, L., Jocou, L., Kaufer, A., Kervella, P., Lacour, S., Lapeyrère, V., Le Bouquin, J. B., Léna, P., Lutz, D., Ott, T., Paumard, T., Perraut, K., Perrin, G., Pfuhl, O., Rabien, S., Shangguan, J., Shimizu, T., Scheithauer, S., Stadler, J., Stephens, A. W., Straub, O., Straubmeier, C., Sturm, E., Tacconi, L. J., Tristram, K. R. W., Vincent, F., von Fellenberg, S., Widmann, F., Wieprecht, E., Wiezorrek, E., Woillez, J., Yazici, S., & Young, A. 2022, *A&A*, 657, L12.

GRAVITY Collaboration, Abuter, R., Amorim, A., Bauböck, M., Berger, J. P., Bonnet, H., Brandner, W., Clénet, Y., Coudé Du Foresto, V., de Zeeuw, P. T., Dexter, J., Duvert, G., Eckart, A., Eisenhauer, F., Förster Schreiber, N. M., Garcia, P., Gao, F., Gendron, E., Genzel, R., Gerhard, O., Gillessen, S., Habibi, M., Haubois, X., Henning, T., Hippler, S., Horrobin, M., Jiménez-Rosales, A., Jocou, L., Kervella, P., & et al. 2019, *A&A*, 625, L10.

GRAVITY Collaboration, Abuter, R., Amorim, A., Bauböck, M., Berger, J. P., Bonnet, H., Brandner, W., Clénet, Y., Davies, R., de Zeeuw, P. T., Dexter, J., Dallilar, Y., Drescher, A., Eckart, A., Eisenhauer, F., Förster Schreiber, N. M., Garcia, P., Gao, F., Gendron, E., Genzel, R., Gillessen, S., Habibi, M., Haubois, X., Heißel, G., Henning, T., Hippler, S., Horrobin, M., Jiménez-Rosales, A., Jochum, L., Jocou, L., & et al. 2021, *A&A*, 647, A59.

Griv, E., Gedalin, M., Pietrukowicz, P., Majaess, D., & Jiang, I.-G. 2021, *MNRAS*, 502, 4194.

Groenewegen, M. A. T. In Recio-Blanco, A., de Laverny, P., Brown, A. G. A., & Prusti, T., editors, *Astrometry and Astrophysics in the Gaia Sky* 2018,a, volume 330, 287.

Groenewegen, M. A. T. 2018,b *A&A*, 619b, A8.

Groenewegen, M. A. T. 2020, *A&A*, 635, A33.

Groenewegen, M. A. T. 2021, *A&A*, 654, A20.

Groenewegen, M. A. T. 2023, *A&A*, 669, A4.

Gustafsson, B., Edvardsson, B., Eriksson, K., Jørgensen, U. G., Nordlund, Å., & Plez, B. 2008, *A&A*, 486, 951.

Hao, C. J., Xu, Y., Wu, Z. Y., Lin, Z. H., Bian, S. B., Li, Y. J., & Liu, D. J. 2022, *A&A*, 668, A13.

Hocdé, V., Nardetto, N., Lagadec, E., Niccolini, G., Domiciano de Souza, A., Mérand, A., Kervella, P., Gallenne, A., Marengo, M., Trahin, B., Gieren, W., Pietrzyński, G., Borgniet, S., Breuval, L., & Javanmardi, B. 2020, *A&A*, 633, A47.

Hocdé, V., Smolec, R., Moskalik, P., Ziółkowska, O., & Singh Rathour, R. 2023, *A&A*, 671, A157.

Huang, Y., Li, Q., Zhang, H., Li, X., Sun, W., Chang, J., Dong, X., & Liu, X. 2021,a *ApJL*, 907a, L42.

Huang, Y., Yuan, H., Beers, T. C., & Zhang, H. 2021,b *ApJL*, 910b, L5.

Hunt, E. L. & Reffert, S. 2023, *A&A*, 673, A114.

Iorio, G. & Belokurov, V. 2021, *MNRAS*, 502, 5686–5710.

Karczmarek, P., Hajdu, G., Pietrzyński, G., Gieren, W., Narloch, W., Smolec, R., Wiktorowicz, G., & Belczynski, K. 2023, *arXiv e-prints*,, arXiv:2303.15664.

Karczmarek, P., Smolec, R., Hajdu, G., Pietrzyński, G., Gieren, W., Narloch, W., Wiktorowicz, G., & Belczynski, K. 2022, *ApJ*, 930, 65.

Kervella, P., Bond, H. E., Cracraft, M., Szabados, L., Breitfelder, J., Mérand, A., Sparks, W. B., Gallenne, A., Bersier, D., Fouqué, P., & Anderson, R. I. 2014, *A&A*, 572, A7.

Kervella, P., Mérand, A., Perrin, G., & Coudé du Foresto, V. 2006, *A&A*, 448, 623.

Khan, S., Miglio, A., Willett, E., Mosser, B., Elsworth, Y. P., Anderson, R. I., Girardi, L., Belkacem, K., Brown, A. G. A., Cantat-Gaudin, T., Casagrande, L., Clementini, G., & Vallenari, A. 2023, *arXiv e-prints*,, arXiv:2304.07158.

Kovtyukh, V., Lemasle, B., Bono, G., Usenko, I. A., da Silva, R., Kniazev, A., Grebel, E. K., Andronov, I. L., Shakun, L., & Chinarova, L. 2022, *MNRAS*, 510, 1894.

Layden, A. C., Tiede, G. P., Chaboyer, B., Bunner, C., & Smitka, M. T. 2019, *AJ*, 158, 105.

Lemasle, B., Lala, H. N., Kovtyukh, V., Hanke, M., Prudil, Z., Bono, G., Braga, V. F., da Silva, R., Fabrizio, M., Fiorentino, G., François, P., Grebel, E. K., & Kniazev, A. 2022, *A&A*, 668, A40.

Li, X.-Y., Huang, Y., Liu, G.-C., Beers, T. C., & Zhang, H.-W. 2023, *ApJ*, 944, 88.

Lin, Z., Xu, Y., Hao, C., Liu, D., Li, Y., & Bian, S. 2022, *ApJ*, 938, 33.

Lindegren, L., Bastian, U., Biermann, M., Bombrun, A., de Torres, A., Gerlach, E., Geyer, R., Hernández, J., Hilger, T., Hobbs, D., Klioner, S. A., Lammers, U., McMillan, P. J., Ramos-Lerate, M., Steidelmüller, H., Stephenson, C. A., & van Leeuwen, F. 2021,a *A&A*, 649a, A4.

Lindegren, L., Hernández, J., Bombrun, A., Klioner, S., Bastian, U., Ramos-Lerate, M., de Torres, A., Steidelmüller, H., Stephenson, C., Hobbs, D., Lammers, U., Biermann, M., Geyer, R., Hilger, T., Michalik, D., Stampa, U., McMillan, P. J., Castañeda, J., Clotet, M., Comoretto, G., Davidson, M., Fabricius, C., Gracia, G., Hambly, N. C., Hutton, A., Mora, A., Portell, J., van Leeuwen, F., & et al. 2018, *A&A*, 616, A2.

Lindegren, L., Klioner, S. A., Hernández, J., Bombrun, A., Ramos-Lerate, M., Steidelmüller, H., Bastian, U., Biermann, M., de Torres, A., Gerlach, E., Geyer, R., Hilger, T., Hobbs, D., Lammers, U., McMillan, P. J., Stephenson, C. A., Castañeda, J., Davidson, M., Fabricius, C., Gracia-Abril, G., Portell, J., Rowell, N., Teyssier, D., Torra, F., Bartolomé, S., Clotet, M., Garralda, N., González-Vidal, J. J., Torra, J., & et al. 2021,b *A&A*, 649b, A2.

Looijmans, K., Lub, J., & Brown, A. G. A. 2023, *arXiv e-prints*,, arXiv:2303.02211.

Madore, B. F. 1982, *ApJ*, 253, 575.

Madore, B. F. & Freedman, W. L. 1998, *ApJ*, 492, 110.

Maíz Apellániz, J. 2022, *A&A*, 657, A130.

Maíz Apellániz, J., Pantaleoni González, M., & Barbá, R. H. 2021,.

Majaess, D., Dékány, I., Hajdu, G., Minniti, D., Turner, D., & Gieren, W. 2018, *Ap&SS*, 363, 127.

Medina, G. E., Lemasle, B., & Grebel, E. K. 2021, *MNRAS*, 505, 1342.

Meisner, A. M., Lang, D., Schlafly, E. F., & Schlegel, D. J. 2021, *RNAAS*, 5, 168.

Mérand, A., Kervella, P., Coudé du Foresto, V., Perrin, G., Ridgway, S. T., Aufdenberg, J. P., ten Brummelaar, T. A., McAlister, H. A., Sturmann, L., Sturmann, J., Turner, N. H., & Berger, D. H. 2006, *A&A*, 453, 155.

Minniti, D., Lucas, P. W., Emerson, J. P., Saito, R. K., Hempel, M., Pietrukowicz, P., Ahumada, A. V., Alonso, M. V., Alonso-Garcia, J., Arias, J. I., Bandyopadhyay, R. M., Barbá, R. H., Barbuy, B., Bedin, L. R., Bica, E., Borissova, J., Bronfman, L., Carraro, G., Catelan, M., Clariá, J. J., Cross, N., de Grijs, R., Dékány, I., Drew, J. E., Fariña, C., Feinstein, C., Fernández Lajús, E., Gamen, R. C., Geisler, D., Gieren, W., Goldman, B., Gonzalez, O. A., Gunthardt, G., Gurovich, S., Hambly, N. C., Irwin, M. J., Ivanov, V. D., Jordán, A., Kerins, E., Kinemuchi, K., Kurtev, R., López-Corredoira, M., Maccarone, T., Masetti, N., Merlo, D., Messineo, M., Mirabel, I. F., Monaco, L., Morelli, L., Padilla, N., Palma, T., Parisi, M. C., Pignata, G., Rejkuba, M., Roman-Lopes, A., Sale, S. E., Schreiber, M. R., Schröder, A. C., Smith, M., , Jr., L. S., Soto, M., Tamura, M., Tappert, C., Thompson, M. A., Toledo, I., Zoccali, M., & Pietrzynski, G. 2010, *NewA*, 15, 433.

Minniti, J. H., Zoccali, M., Rojas-Arriagada, A., Minniti, D., Sbordone, L., Contreras Ramos, R., Braga, V. F., Catelan, M., Duffau, S., Gieren, W., Marconi, M., & Valcarce, A. A. R. 2021, *A&A*, 654, A138.

Molinaro, R., Ripepi, V., Marconi, M., Romaniello, M., Catanzaro, G., Cusano, F., De Somma, G., Musella, I., Storm, J., & Trentin, E. 2023, *MNRAS*, 520, 4154.

Mullen, J. P., Marengo, M., Martínez-Vázquez, C. E., Chaboyer, B., Bono, G., Braga, V. F., Dall'Ora, M., D'Orazi, V., Fabrizio, M., Monelli, M., & Thévenin, F. 2023, *ApJ*, 945, 83.

Muraveva, T., Delgado, H. E., Clementini, G., Sarro, L. M., & Garofalo, A. 2018, *MNRAS*, 481, 1195.

Nardetto, N., Mérand, A., Mourard, D., Storm, J., Gieren, W., Fouqué, P., Gallenne, A., Graczyk, D., Kervella, P., Neilson, H., Pietrzynski, G., Pilecki, B., Breitfelder, J., Berio, P., Challouf, M., Clausse, J. M., Ligi, R., Mathias, P., Meilland, A., Perraut, K., Poretti, E., Rainer, M., Spang, A., Stee, P., Tallon-Bosc, I., & ten Brummelaar, T. 2016, *A&A*, 593, A45.

Neeley, J. R., Marengo, M., Freedman, W. L., Madore, B. F., Beaton, R. L., Hatt, D., Hoyt, T., Monson, A. J., Rich, J. A., Sarajedini, A., Seibert, M., & Scowcroft, V. 2019, *MNRAS*, 490, 4254–4270.

Ngeow, C.-C., Bhardwaj, A., Dekany, R., Duev, D. A., Graham, M. J., Groom, S. L., Mahabal, A. A., Masci, F. J., Medford, M. S., & Riddle, R. 2022,a *AJ*, 163a, 239.

Ngeow, C.-C., Bhardwaj, A., Henderson, J.-Y., Graham, M. J., Laher, R. R., Medford, M. S., Purdum, J., & Rusholme, B. 2022,b *AJ*, 164b, 154.

O'Donnell, J. E. 1994, *ApJ*, 422, 158.

Owens, K. A., Freedman, W. L., Madore, B. F., & Lee, A. J. 2022, *ApJ*, 927, 8.

Pietrukowicz, P., Soszyński, I., & Udalski, A. 2021, *AcA*, 71, 205.

Pietrukowicz, P., Udalski, A., Soszyński, I., Skowron, D. M., Wrona, M., Szymański, M. K., Poleski, R., Ulaczyk, K., Kozłowski, S., Skowron, J., Mróz, P., Rybicki, K., Iwanek, P., & Gromadzki, M. 2020, *AcA*, 70, 121.

Ren, F., Chen, X., Zhang, H., de Grijs, R., Deng, L., & Huang, Y. 2021, *ApJL*, 911, L20.

Riess, A. G., Breuval, L., Yuan, W., Casertano, S., Macri, L. M., Bowers, J. B., Scolnic, D., Cantat-Gaudin, T., Anderson, R. I., & Cruz Reyes, M. 2022, *ApJ*, 938, 36.

Riess, A. G., Casertano, S., Anderson, J., MacKenty, J., & Filippenko, A. V. 2014, *ApJ*, 785, 161.

Riess, A. G., Casertano, S., Yuan, W., Bowers, J. B., Macri, L., Zinn, J. C., & Scolnic, D. 2021,a *ApJL*, 908a, L6.

Riess, A. G., Casertano, S., Yuan, W., Bowers, J. B., Macri, L., Zinn, J. C., & Scolnic, D. 2021,b *ApJL*, 908b, L6.

Riess, A. G., Casertano, S., Yuan, W., Macri, L., Anderson, J., MacKenty, J. W., Bowers, J. B., Clubb, K. I., Filippenko, A. V., Jones, D. O., & Tucker, B. E. 2018,a *ApJ*, 855a, 136.

Riess, A. G., Casertano, S., Yuan, W., Macri, L., Bucciarelli, B., Lattanzi, M. G., MacKenty, J. W., Bowers, J. B., Zheng, W., Filippenko, A. V., Huang, C., & Anderson, R. I. 2018,b *ApJ*, 861b, 126.

Ripepi, V., Catanzaro, G., Clementini, G., De Somma, G., Drimmel, R., Leccia, S., Marconi, M., Molinaro, R., Musella, I., & Poggio, E. 2022, *A&A*, 659, A167.

Ripepi, V., Clementini, G., Molinaro, R., Leccia, S., Plachy, E., Molnár, L., Rimoldini, L., Musella, I., Marconi, M., Garofalo, A., Audard, M., Holl, B., Evans, D. W., Jevardat de Fombelle, G., Lecoeur-Taibi, I., Marchal, O., Mowlavi, N., Muraveva, T., Nienartowicz, K., Sartoretti, P., Szabados, L., & Eyer, L. 2023, *A&A*, 674, A17.

Romaniello, M., Riess, A., Mancino, S., Anderson, R. I., Freudling, W., Kudritzki, R.-P., Macrì, L., Mucciarelli, A., & Yuan, W. 2022, *A&A*, 658, A29.

Skowron, D. M., Skowron, J., Mróz, P., Udalski, A., Pietrukowicz, P., Soszyński, I., Szymański M. K., Poleski, R., Kozłowski, S., Ulaczyk, K., Rybicki, K., Iwanek, P., Wrona, M., & Gromadzki, M. 2019, *AcA*, 69, 305.

Skowron, D. M., Skowron, J., Udalski, A., Szymański, M. K., Soszyński, I., Wyrzykowski, Ł., Ulaczyk, K., Poleski, R., Kozłowski, S., Pietrukowicz, P., Mróz, P., Rybicki, K., Iwanek, P., Wrona, M., & Gromadzki, M. 2021, *ApJS*, 252, 23.

Stassun, K. G. & Torres, G. 2021, *ApJL*, 907, L33.

Trahin, B. 2019,. *Étalonnage de l'échelle des distances dans l'ère Gaia: Les étoiles pulsantes RR Lyrae et Céphéides.* PhD thesis, L'Université PSL, l'Observatoire de Paris.

Trahin, B., Breuval, L., Kervella, P., Mérand, A., Nardetto, N., Gallenne, A., Hocdé, V., & Gieren, W. 2021, *A&A*, 656, A102.

Trentin, E., Ripepi, V., Catanzaro, G., Storm, J., Marconi, M., De Somma, G., Testa, V., & Musella, I. 2023, *MNRAS*, 519, 2331.

van Leeuwen, F. 2007, *A&A*, 474, 653.

van Leeuwen, F. 2008, *VizieR Online Data Catalog*, 1311.

van Leeuwen, F., Feast, M. W., Whitelock, P. A., & Laney, C. D. 2007, *MNRAS*, 379, 723.

Vasiliev, E. & Baumgardt, H. 2021, *MNRAS*, 505, 5978–6002.

Wang, C., Yuan, H., & Huang, Y. 2022, *AJ*, 163, 149.

Wang, S. & Chen, X. 2023, *ApJ*, 946, 43.

Wang, S., Chen, X., de Grijs, R., & Deng, L. 2018, *ApJ*, 852, 78.

Wielgórski, P., Pietrzyński, G., Pilecki, B., Gieren, W., Zgirski, B., Górski, M., Hajdu, G., Narloch, W., Karczmarek, P., Smolec, R., Kervella, P., Storm, J., Gallenne, A., Breuval, L., Lewis, M., Kałuszyński, M., Graczyk, D., Pych, W., Suchomska, K., Taormina, M., Rojas Garcia, G., Kotek, A., Chini, R., Pozo Nũnez, F., Noroozi, S., Sobrino Figaredo, C., Haas, M., Hodapp, K., Mikołajczyk, P., Kotysz, K., Moździerski, D., & Kołaczek-Szymański, P. 2022, *ApJ*, 927, 89.

Zgirski, B., Pietrzyński, G., Górski, M., Gieren, W., Wielgórski, P., Karczmarek, P., Hajdu, G., Lewis, M., Chini, R., Graczyk, D., Kałuszyński, M., Narloch, W., Pilecki, B., Rojas García, G., Suchomska, K., & Taormina, M. 2023, *arXiv e-prints*,, arXiv:2305.09414.

Zhou, X. & Chen, X. 2021, *MNRAS*, 504, 4768–4784.

Zinn, J. C. 2021, *AJ*, 161, 214.

Discussion

Question (Anderson): In the comparison of the cluster parallaxes you based your conclusions on Mauricio Cruz Reyes' membership list to determine the parallax of the cluster. What conclusion did you draw with respect to the zero-point offset relative to what he did? Are these things consistent or what do you see in there?

Answer: I didn't actually look at that in detail. I also, of course, applied the Lindegren correction, so I could make the direct comparison. I had no time to investigate that and I wanted to do the other comparison using no offset, just to make the message clear what happens. [The comparison of the average cluster parallaxes using other offsets is now included in these proceedings.]

Question (Riess): You're saying the countercorrections are worrisomely large but as it pertains to Cepheids, which you're looking at, the Cepheids are much brighter than where *Gaia* has done their calibration, so in that brightness range is it so surprising?

Answer: No, but if you have countercorrections of about -20 μas at 18 mag, it's almost like saying quasars are not at infinity. At the bright end, you see a large dispersion, as you know. We have to acknowledge that there is this 10 μas residual offset, no matter how you look at things. Countercorrections of ± 15 μas, nobody can rule those out at the moment, to be fair.

At the crossroads of astrophysics and cosmology: Period–luminosity relations in the 2020s
Proceedings IAU Symposium No. 376, 2024
R. de Grijs, P. A. Whitelock & M. Catelan, eds.
doi:10.1017/S1743921323004258

Cepheids with giant companions: A new, abundant source of Cepheid astrophysics

Bogumił Pilecki

Centrum Astronomiczne im. Mikołaja Kopernika, PAN, Bartycka 18, 00-716 Warsaw, Poland.
email: pilecki@camk.edu.pl

Abstract. We present a progress report of our project aiming to increase the number of known Cepheids in double-lined binary (SB2) systems from six to 100 or more. This will allow us, among other goals, to accurately measure masses for a large sample of Cepheids. Currently, only six accurate Cepheid masses are available, which hinders our understanding of their physical properties and renders the Cepheid mass–luminosity relation poorly constrained. At the same time, Cepheids are widely used for essential measurements (e.g., extragalactic distances, the Hubble constant). To examine Cepheid period–luminosity relations, we selected as binary candidates Cepheids that are too bright for their periods. To date, we have confirmed 56 SB2 systems, including the detection of significant orbital motions of the components for 32. We identified systems with orbital periods up to five times shorter than the shortest reported period to date, as well as systems with mass ratios significantly different from unity (suggesting past merger events). Both features are essential to understand how multiplicity affects the formation and destruction of Cepheid progenitors and what effect this has on global Cepheid properties. We also present eight new systems composed of two Cepheids (only one such system was known before). Among confirmed SB2 Cepheids, there are also several wide-orbit systems. In the future, these may facilitate independent accurate geometric distance measurements to the Large and Small Magellanic Clouds.

Keywords. Cepheids, binary systems, physical properties, P-L relation

1. Introduction

Classical Cepheids are perhaps the most important objects in astrophysics, given that they are crucial to gain improved physical insights in various fields, including as regards stellar oscillations and the evolution of intermediate-mass and massive stars. They wield enormous influence on modern cosmology. Since the discovery of the relationship between their pulsation period and their luminosity more than a century ago (the Leavitt Law; Leavitt and Pickering 1912), Cepheids have been used extensively to measure distances both within and outside of our Galaxy. The recent local Hubble constant determination accurate to 1.8% (Riess et al. 2022), which shows a significant discrepancy with the value inferred from *Planck* data, depends sensitively on this period–luminosity (P–L) relation.

Although theoretical studies of Cepheids are quite advanced (e.g., Bono et al. 1999, 2005; Valle et al. 2009), our empirical knowledge is still limited. Theory predicts masses of Cepheids in the range of 3–11 $M_\odot$, but their measured masses clump between 3.6 $M_\odot$ and 5 $M_\odot$, with only one higher but uncertain value of 6 $M_\odot$ (Pilecki et al. 2013, 2018; Evans et al. 2018; Gallenne et al. 2019a). This makes the Cepheid mass–luminosity relation very poorly constrained (Anderson et al. 2016). Nevertheless, it is crucial for our theoretical understanding of the P–L relation and basic stellar physics regarding, e.g., convection, mass loss, and rotation. Moreover, the blue loops predicted by current evolution theory are too short to explain the existence of low-mass, short-period Cepheids (see, e.g., Espinoza-Arancibia and Pilecki, these proceedings).

Masses covering a wider range would be of the utmost importance to resolve these issues, but we in practice can measure them only for Cepheids in spectroscopic double-lined binaries (SB2), for which lines of companions are easily detectable. Unfortunately, such systems are very rare. Most binary Cepheids, and all found in the Milky Way, have an early-type main-sequence companion (exhibiting few and broad lines) which is typically 2–5 mag fainter in the V band, making it extremely hard to determine its velocity and thus the mass of the Cepheid.

One good source for more binary Cepheids would be to look for eclipses, but this solution has three serious limitations. First, such systems are not numerous, because a very specific orbital orientiation is needed for eclipses to occur. Second, according to current statistics, only about half of eclipsing systems involving Cepheids are double-lined, and third, the best places to look for them (the Galactic disk, Magellanic Clouds) have already been examined, and we do not expect many more such binaries to be found there any time soon. Therefore, we decided to look for another possible source that had not yet been considered and remained unexplored.

In the first paper of the series (Pilecki et al. 2021) we described such a source, highlighting its potential, and we presented the first results for the Large Magellanic Cloud (LMC). Here, we present the current stage of the project, providing updates based on new observations and expansion of the sample to the Small Magellanic Cloud (SMC) and the Milky Way (MW).

2. The project

The basic idea of the project was to look for Cepheids in binary systems, for which lines of both components are present in their spectra. To meet these conditions, one has to find Cepheids accompanied by stars of similar luminosity and preferentially of late spectral types, i.e., subgiants or later stages of stellar evolution. To identify such candidates, we considered three observable features caused by such companions. Compared with single Cepheids (or Cepheids with significantly fainter companions), for Cepheids with giant companions we expect:

- their total observed brightness to increase significantly;
- their photometric pulsation amplitude to decrease; and
- their color to be either similar or redder.

Looking at the P–L diagram for LMC Cepheids (Figure 1), we noticed that all previously confirmed eclipsing giant–giant SB2 systems including Cepheids (Pilecki et al. 2018) lie significantly above the corresponding P–L relation, being at least 50% (0.44 mag) brighter than a typical Cepheid for its period. At the same time, their amplitudes are about half the typical ones, and their colors are typical or redder. From the number of known eclipsing binary Cepheids we also estimated how many similar binary Cepheids may exist but have not yet been detected in photometric studies due to a lack of eclipses. Assuming random inclinations, we expected on the order of 50 SB2 systems composed of a Cepheid and a giant companion exhibiting a range of orbital periods from 1 to $\sim$5 years, i.e., similar to the periods for eclipsing Cepheids.

Indeed, when we investigated the P–L diagram for all Cepheids from the OGLE-3 catalog (Soszyński et al. 2008), we identified 41 additional Cepheids that lie 0.44 mag (4.7σ) above the P–L relation, but for which eclipses were not detected (Figure 1). Virtually all Cepheids selected this way have low amplitudes and colors that are on average redder than typical by 0.12 mag in the $(V - I)$ index. As stated above, these three features together clearly indicate that the overbright Cepheids have luminous, late-type giant companions. This, in turn, made them perfect candidates for SB2 systems

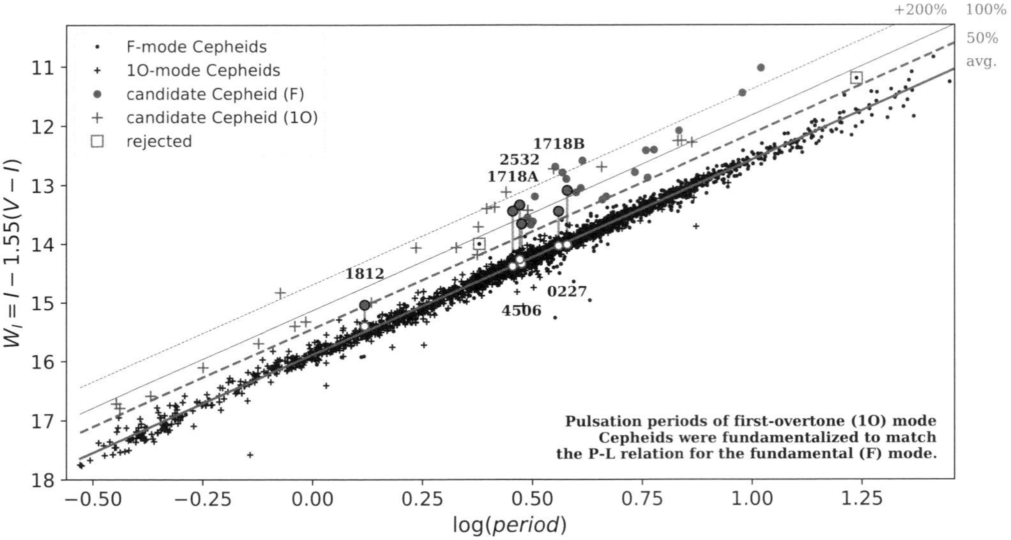

Figure 1. P–L relation based on the reddening-free Wesenheit index. Filled red circles show known eclipsing binaries involving Cepheids, while empty circles mark their brightness without any companion light. Our candidates for binary Cepheids are ≥50% brighter than an average Cepheid for a given period. Note that their distribution parallel to the P–L relation advocates strongly against blending by random stars.

composed of giants, for which lines of both components could be easily detected and their radial velocities (RVs) measured.

The Cepheids in our sample are distributed along the full P–L relation, with periods from 0.26 to 10.5 days. Among them, 20 pulsate in the fundamental (F) mode and 21 in the first-overtone (1O) mode. Three of the latter are actually double-mode Cepheids pulsating in the second overtone (2O) as well. In Figure 1 the periods of 1O Cepheids (P_{1O}) were fundamentalized to match the periods of F-mode Cepheids (P_F) using the prescription of Pilecki et al. (2021).

Meanwhile, the project has been expanded by inclusion of new candidates in the LMC from the OGLE-4 catalog (Soszyński et al. 2017) and a dozen candidates in the SMC (Soszyński et al. 2010). In the samples selected as described above we found all seven known double Cepheids (i.e., objects for which pulsation of two Cepheids was detected at the same coordinates), which were all selected as candidates. This statistical result suggests that they share the same properties as other P–L-overbright Cepheids. This allows us to add to our SB2 candidate list double Cepheids from the MW, where the P–L relation method of detecting SB2 Cepheids cannot be applied directly (individual distances to the objects are needed to do so).

3. Data

Spectroscopic monitoring of our first SB2 candidates started in October 2020, with new objects added subsequently. Observations were performed using three of the world's best instruments mounted on telescopes located at observatories in Chile. The brighest targets ($V \leq 15.7$ mag) were observed with the High Accuracy Radial velocity Planet Searcher (HARPS) instrument mounted on the 3.6 meter telescope at the La Silla observatory. Most of the acquired spectra were obtained with the Magellan Inamori Kyocera Echelle (MIKE) spectrograph mounted on the 6.5 m Magellan Clay telescope at the Las Campanas observatory. A significant fraction of data were also acquired with the

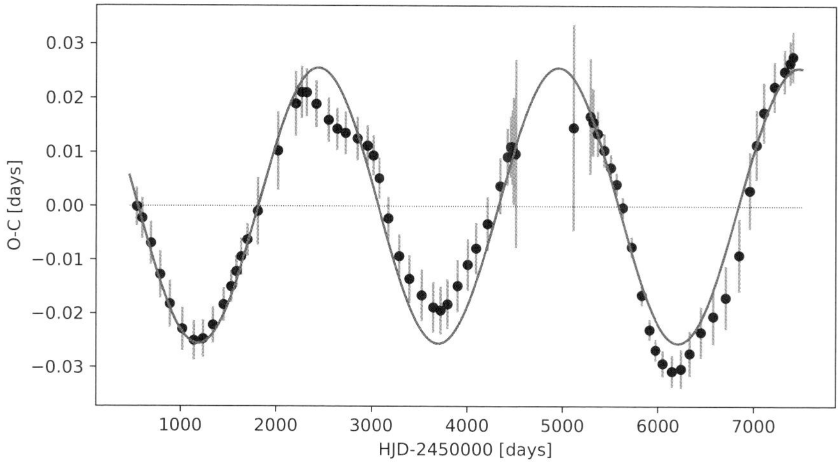

Figure 2. O–C diagram for one of the SB2 Cepheid candidates. A clear LTTE modulation due to binary motion is present. The red line is the best fit to the Cepheid's orbit.

European Southern Observatory's UV–Visual Echelle Spectrograph (UVES) on the 8.2 m Very Large Telescope (VLT) at Paranal Observatory. HARPS and MIKE were used in visitor mode, so that the observations were confined to the specific nights of the corresponding runs. UVES observations were obtained in service mode and could thus cover the orbital cycles more uniformly, while several-day-long HARPS runs were useful for covering pulsation phases of bright, long-period Cepheids.

4. Results

In a preliminary study, we analyzed photometric data from the OGLE catalogs for our initial sample of 41 Cepheids. The study of their O–C diagrams led to detection of the light travel-time effect (LTTE) due to binary motion for four candidates. One example, with an orbital period of about 2500 days, is shown in Figure 2.

4.1. *Double-lined binary Cepheids*

During the pilot program, we found 16 of 18 (~90%) of our brightest candidate Cepheids to be components of SB2 systems, showing that the majority of these overbright pulsators are normal Cepheids with luminous giant companions. Additional observations of fainter candidates led to the discovery of a Cepheid in a binary system with an orbital period of only 59 days, five times shorter than any measured before for a firmly confirmed binary Cepheid. For a more detailed description of the system, please refer to Pilecki et al. (2022). Updated RV curves for this object are shown in Figure 3.

The observations have continued and, to date, 56 candidate Cepheids in the LMC, SMC, and MW have been confirmed as members of SB2 systems. For 32 of them, the anticorrelated orbital motions of the Cepheid and its companion were detected, which represents the ultimate proof that they are gravitationally bound. For 23, preliminary orbital solutions could be obtained, although in general the data coverage is not yet satisfactory due to long orbital periods and the complexity of the variabilities involved. For a few cases we managed to cover the orbital cycles well, and the solutions are not expected to change significantly (see Figure 4). Moreover, eight of the confirmed systems are composed of two Cepheids (see the next subsection).

For systems with preliminary orbital solutions we calculated the mass ratios of the components and the minimum Cepheid masses, $M_{\rm Cep}\sin^3(i)$. The minimum masses obtained

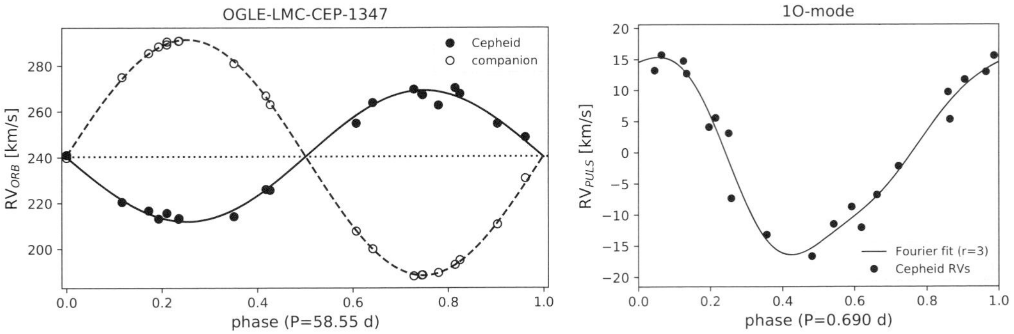

Figure 3. Updated RV curves for OGLE-LMC-CEP-1347. The scatter in the Cepheid RVs is due to unaccounted-for 2O-mode pulsations.

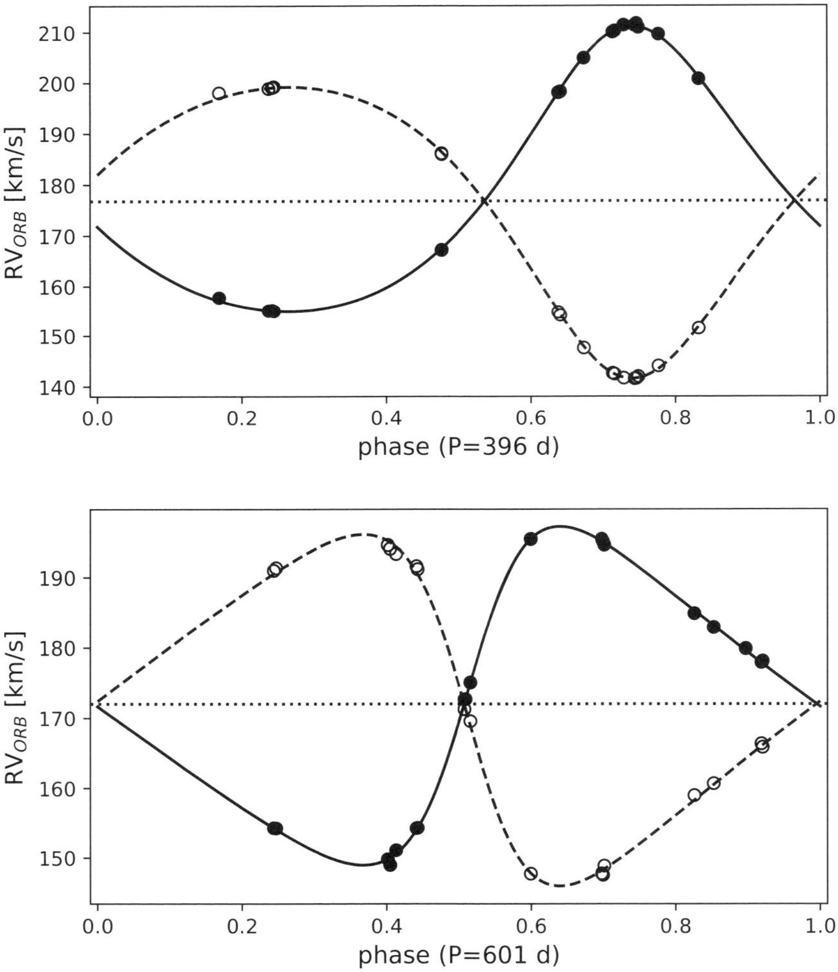

Figure 4. Preliminary orbits for two example double-lined binary Cepheids with best-covered orbital cycles. More data are still needed to cover the pulsation curves and improve the precision of the derived orbital parameters but no major change is expected to the solution.

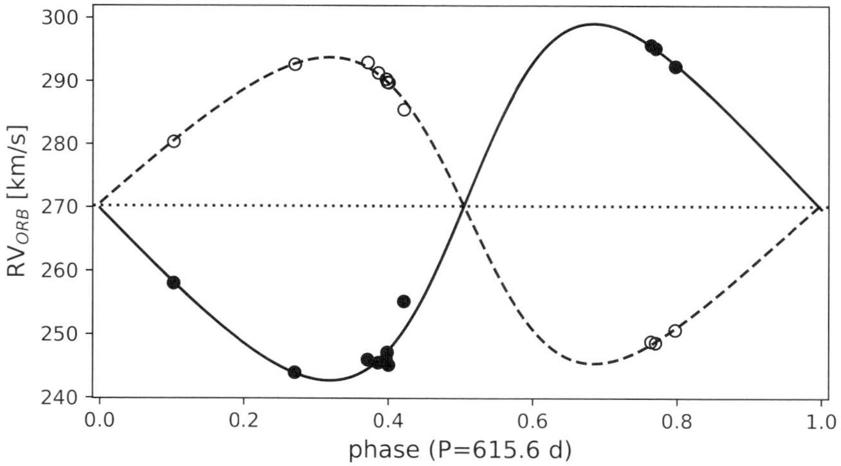

Figure 5. Example of a binary Cepheid with a preliminary mass ratio that is significantly different from unity.

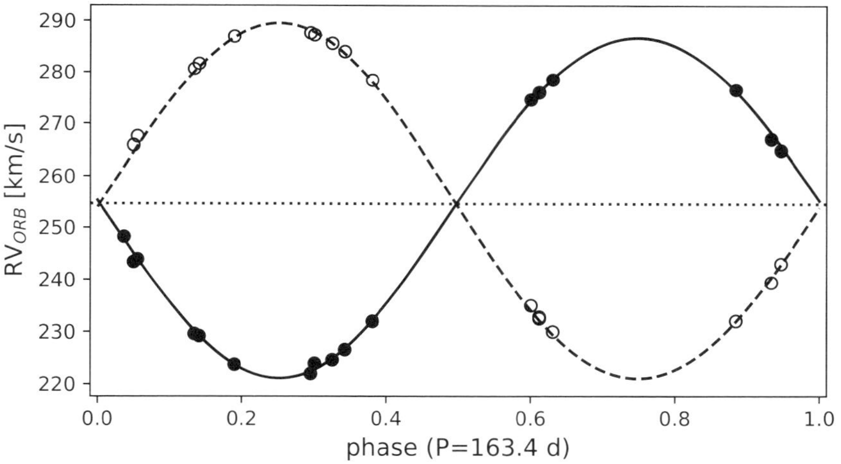

Figure 6. Another binary Cepheid with an orbital period below the 200-day limit.

for several systems are close to the masses derived for eclipsing Cepheids, showing that the inclinations of these systems are high and that they are good targets for accurate and precise mass measurements. Mass ratios obtained for some of the systems are different from unity. As $q = M_2/M_1 \sim 1$ is expected for components at similar evolutionary stages and the same age, which means that they could have passed through binary interactions in the past, either by means of a merger event or via mass transfer. One example of such a system is shown in Figure 5.

Among the binary Cepheids with orbital solutions, a new system was found with an orbital period below 200 days, which is the lower limit obtained by Neilson et al. (2015) for binary Cepheids that passed through their evolution on the RGB. As systems with periods below this limit would be destroyed at that evolutionary stage, their components are probably younger, and the Cepheids are likely on their first crossing through the instability strip (in the Hertzsprung gap). Our newly discovered system has a period of only 163 days, which is the second shortest period ever measured for a binary Cepheid (Figure 6).

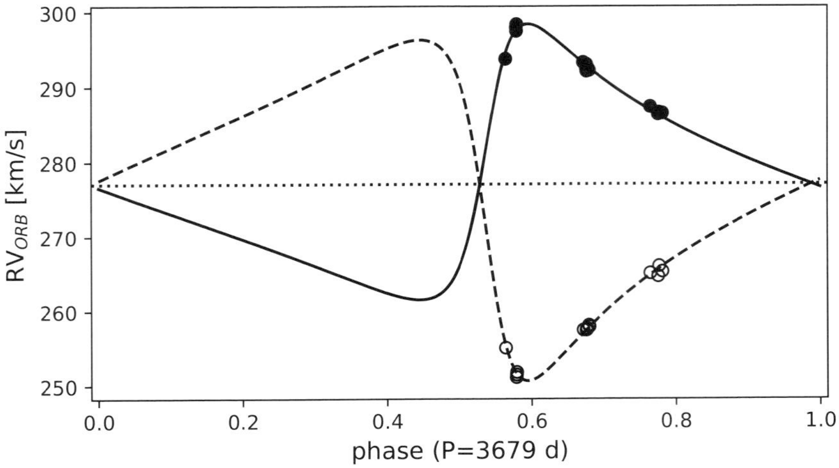

Figure 7. Example of a wide-orbit binary Cepheid with a long orbital period. The current solution suggests that its period may reach 10 years or even longer.

The longest orbital period measured so far for an extragalactic binary system is $P = 1550$ days ($\sim$4 yr; Gieren et al. 2015). Meanwhile, preliminary solutions for some of our new systems suggest orbital periods longer than 4 years, even up to about 10 years or more (see Figure 7). These binary systems, for which we can obtain spectroscopic orbits, are currently the best candidates for direct geometric distance measurements using interferometric methods, offering the best possible accuracy. And although at the moment we do not expect high precision for extragalactic measurements, with upcoming improvements to instrumentation, such systems may eventually become our best and most direct tools for distance measurements to nearby galaxies, surpassing previously used methods (e.g., Pietrzyński et al. 2019).

4.2. *Binary double Cepheids*

An interesting subsample of double-lined binary Cepheids form those binary systems composed of two Cepheid components. As part of our project we observed nine candidates for such systems. Binary double (BIND) Cepheids are unique laboratories for pulsation and evolution studies. However, only one system of this kind was known before. To date, we have confirmed orbital motions for eight additional systems of a similar nature. Preliminary orbits for two are shown in Figure 8. Altough the orbital solutions may change once a full orbital cycle has been covered, the anti-correlated orbital motion of both components is already clear, proving binarity of these double Cepheids.

Interestingly, some preliminary solutions suggest mass ratios significantly different from unity, while for others the pulsation periods are very different (after fundamentalization of the 1O periods, for comparison). Both features may indicate past binary interaction events, which would affect the physical properties of the same-age components. Another interesting feature of the new BIND Cepheids is their long orbital periods. For most, periods longer than five years are expected according to the current models (based on only partially covered cycles). Although long periods make the observations more challenging, in the future these systems will also offer an opportunity to measure direct distances to their host galaxies, as described in the previous section.

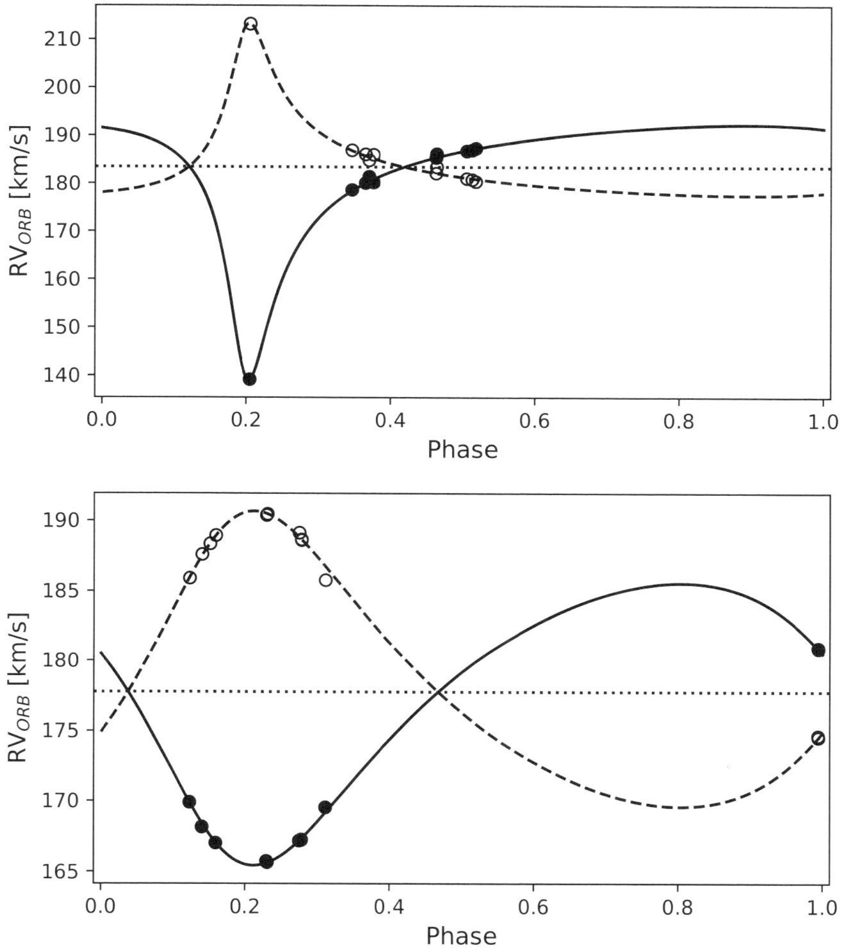

Figure 8. Preliminary orbits for two example binary double Cepheids show a clear change of the components' orbital velocities. However, as the orbital cycle has not yet been fully covered, the final, exact solutions (including the orbital periods) may change significantly.

5. Conclusions

Analysis of spectroscopic data we have collected so far clearly shows that the Cepheids selected as described in Section 2 are indeed components of double-lined binary systems. This has important implications for the interpretation of period–luminosity relations and for our general knowledge of Cepheids and their evolution. We know now that overbright Cepheids, which are often rejected as P–L relation outliers, are in the great majority Cepheids with red, luminous giant companions. These conclusions can also likely be extended to other pulsating stars with well-defined P–L relations, e.g., Type II Cepheids or RR Lyrae stars.

With 56 confirmed candidates, we have already ten-fold increased the number of Cepheids in spectroscopic double-lined binaries. This includes over 90% of all candidates in the LMC. In Figure 1 one can see that many of them have periods longer than those of known SB2 Cepheids, reaching up to 10.5 days. Extrapolation of the number of confirmed cases to the SMC suggests that about 50 new SMC Cepheids in SB2 systems can be expected once all SMC objects have been studied. This would increase the total number of known SB2 Cepheids by a further 80% (i.e., 20-fold compared with the state before we commenced the project).

Eventually, our study will yield firm mass estimates for a large sample of Cepheids, including long-period, high-mass Cepheids for which the lack of data is the most severe and for those from the uncertain low-mass end. Similar evolutionary stages for both components imply mass ratios close to unity. Any significant deviation from unity may indicate a probable past merger event. This is the only way merged Cepheids can be unambiguously detected (as compared with, e.g., dubious chemical composition peculiarities) and characterized. From the simulations of Sana et al. (2012) we can estimate that up to 30% of Cepheids may be merger products. In our study we will be able to estimate this fraction observationally and compare it with simulations.

Geometric methods provide the most accurate and direct distances to astronomical objects. One such method is based on the determination of both astrometric (angular) and spectroscopic (absolute) orbits of binary systems, which combined give directly the distance (see, e.g., Gallenne et al. 2019b). Unfortunately, at increasingly large distances, it becomes commensurately harder to measure orbital angular sizes, as one needs systems of increasingly higher luminosity and larger component separations. Using traditional methods, such systems are extremely hard to identify in other galaxies, as one needs decades of monitoring to find just a few. However, our method of detection of binary Cepheids is independent of the orbital size, and several systems with very wide orbits have already been identified. They are currently our best candidates for direct geometric distance determinations to the LMC and SMC, which can eventually lead to the ultimate calibration of the first rung of the cosmic distance ladder.

Acknowledgements. The research leading to these results received funding from the Polish National Science Center (grant SONATA BIS 2020/38/E/ST9/00486). This work is based on observations collected at the Las Campanas Observatory and the European Southern Observatory, ESO (ESO programs: 106.21GB, 108.229Z, 108.22BS, 110.2434, and 110.2436). We thank the Carnegie Institution and the CNTAC for the allocation of observing time for this project. This research has made use of NASA's Astrophysics Data System Service.

References

Anderson, R. I., Saio, H., Ekström, S., et al. 2016, A&A, 591, A8.
Bono, G., Caputo, F., Castellani, V., & Marconi, M. 1999, ApJ, 512, 711–723.
Bono, G., Marconi, M., Cassisi, S., et al. 2005, ApJ, 621, 966–977.
Evans, N. R., Proffitt, C., Carpenter, K. G., et al. 2018, ApJ, 866, 30.
Gallenne, A., Kervella, P., Borgniet, S., et al. 2019a, A&A, 622a, A164.
Gallenne, A., Pietrzyński, G., Graczyk, D., et al. 2019b, A&A, 632b, A31.
Gieren, W., Pilecki, B., Pietrzyński, G., *et al.* 2015, ApJ, 815, 28.
Leavitt, H. S. & Pickering, E. C. 1912, Harvard College Obs. Circ., 173, 1–3.
Neilson, H. R., Schneider, F. R. N., Izzard, R. G., et al. 2015, A&A, 574, A2.
Pietrzyński, G., Graczyk, D., Gallenne, A., et al. 2019, Nature, 567(7747), 200–203.
Pilecki, B., Gieren, W., Pietrzyński, G., et al. 2018, ApJ, 862, 43.
Pilecki, B., Graczyk, D., Pietrzyński, G., et al. 2013, MNRAS, 436, 953–967.
Pilecki, B., Pietrzyński, G., Anderson, R. I., et al. 2021, ApJ, 910, 118.
Pilecki, B., Thompson, I. B., Espinoza-Arancibia, F., *et al.* 2022, ApJL, 940, L48.
Riess, A. G., Yuan, W., Macri, L. M., et al. 2022, ApJL, 934, L7.
Sana, H., de Mink, S. E., de Koter, A., et al. 2012, Science, 337, 444.
Soszyński, I., Poleski, R., Udalski, A., et al. 2008, AcA, 58, 163–185.
Soszyński, I., Poleski, R., Udalski, A., et al. 2010, AcA, 60(1), 17–39.
Soszyński, I., Udalski, A., Szymański, M. K., et al. 2017, AcA, 67(2), 103–113.
Valle, G., Marconi, M., Degl'Innocenti, S., & Prada Moroni, P. G. 2009, A&A, 507, 1541–1554.

At the crossroads of astrophysics and cosmology: Period–luminosity relations in the 2020s
Proceedings IAU Symposium No. 376, 2024
R. de Grijs, P. A. Whitelock & M. Catelan, eds.
doi:10.1017/S1743921323003022

Implication of the period-magnitude relation for massive AGB stars and its astronomical applications

Akiharu Nakagawa[1], Tomoharu Kurayama[2], Hiroshi Sudou[3] and Gabor Orosz[4]

[1]Graduate School of Science and Engineering, Kagoshima University, 1-21-35 Korimoto, Kagoshima, Kagoshima 890-0065, Japan. email: nakagawa@sci.kagoshima-u.ac.jp

[2]Teikyo University of Science, 2-2-1 Senju-Sakuragi, Adachi-ku, Tokyo 120-0045, Japan

[3]Department of General Engineering, National Institute of Technology, Sendai College, Natori Campus, 48 Nodayama, Medeshima-Shiote, Natori-shi, Miyagi 981-1239, Japan

[4]Joint Institute for VLBI ERIC (JIVE), Oude Hoogeveensedijk 4, 7991 PD, Dwingeloo, Netherlands

Abstract. We present astrometric very long baseline interferometry (VLBI) studies of AGB stars. To understand the properties and evolution of AGB stars, distances are an important parameter. The distribution and kinematics of their circumstellar matter are also revealed with the VLBI method. We used the VERA array to observe 22 GHz H_2O masers in various subclasses of AGB stars. Parallaxes of the three OH/IR stars NSV17351, OH39.7+1.5, IRC−30363, and the Mira-type variable star AW Tau were newly obtained. We present the circumstellar distribution and kinematics of H_2O masers around NSV17351. The absolute magnitudes in mid-infrared bands of OH/IR stars with very long pulsation periods were investigated and a period-magnitude relation in the WISE W3 band, $M_{\mathrm{W3}} = (-7.21 \pm 1.18) \log P + (9.25 \pm 3.09)$, was found for the Galactic AGB stars. The VLBI is still a powerful tool for parallax measurements of the Galactic AGB stars surrounded by thick dust shells.

Keywords. VLBI, Astrometry, Masers, AGB stars, OH/IR stars

1. Introduction

1.1. *Evolution of AGB stars*

Asymptotic giant branch (AGB) stars are known to be the final stage of the evolution of stars with an initial mass of 0.8-10 $M_\odot$ (e.g., Karakas & Lattanzio 2014). Considering the shape of the initial mass function (IMF), a significant portion of stars will spend a period of their lifetime as AGB stars. They are surrounded by thick circumstellar dust shells and frequently present stellar pulsations. Since AGB stars return various elements into interstellar space by stellar winds, they are important objects that contribute to the chemical composition of the universe and galaxy (e.g., Höfner & Olofsson 2018). AGB stars exhibit a wide range of pulsation periods, with the shortest periods being around 100 days and occasionally reaching 3000 days (e.g., Habing 1996). Hence, they are often referred to as long-period variable stars (LPVs). A more detailed picture of the evolutionary process on the AGB phase reveals several stages depending on their growth.

During the early AGB phase, stars have relatively thin dust layers, making them observable in both optical and infrared bands. The Mira-type variables, visible in both optical and infrared bands, are a well-known class of AGB stars. As they progress, strong absorption caused by thick circumstellar dust shells makes them faint and difficult to be detected in optical bands (e.g., Habing 1996; Kamizuka *et al.* 2020). On the other hand, they become brighter in the infrared bands as a result of re-radiation from the outer dust layer. At this stage, many objects also present OH maser emission from their outermost layer, therefore they are referred to as OH/IR stars. The OH/IR stars, a sub-class of AGB stars, are considered to be in the late stage of the AGB phases before progressing to become planetary nebulae (te Lintel Hekkert *et al.* 1991). Additionally, there are other sub-classes that exist in the late stages of the AGB phase. Massive stars with initial masses ranging from 7 to 10 $M_\odot$ are known to go through a phase referred to as the super AGB phase, characterized by high luminosity and long pulsation periods (Karambelkar *et al.* 2019). OH/IR stars with intermediate mass and large mass loss rates ($>10^{-4}\,M_\odot$/yr) are classified as extreme-OH/IR stars (Justtanont *et al.* 2015).

Mira-type variables, surrounded by thin dust shells, represent the early stage of the AGB phase and are often associated with SiO and H_2O masers. Following the Mira phase, H_2O molecules are transported to the outer side of the circumstellar envelope and are photodissociated, resulting in the production of OH maser emissions (e.g., Goldman *et al.* 2017). Due to an excess of infrared emission from the circumstellar dust shell and the presence of OH maser emission, they will be recognized as OH/IR stars (e.g., Nyman *et al.* 1998). As they progress towards the post-AGB phase, the amplitude of their stellar pulsation gradually decreases, eventually leading to a phase known as non-variable OH/IR stars (e.g., Habing 1996; Kamizuka *et al.* 2020). Subsequently, they move into the post-AGB phase. To understand the sequential evolution from early to late stages in the AGB phases, it is crucial to conduct studies on AGB stars with various properties. Target sources of our very long baseline interferometry (VLBI) studies cover a variety of masses and pulsation periods. This also means that the target sources that we are studying are representative of a wide range of ages.

1.2. *Masses of AGB stars and pulsation periods*

There is a correlation between pulsation periods and masses in AGB stars. AGB stars with longer pulsation periods are generally considered to be more massive than those with shorter pulsation periods. For instance, Feast (2009) suggests that AGB stars with pulsation periods of 1000 days have masses ranging from 3 $M_\odot$ to 4 $M_\odot$. The pulsation periods and mean densities of pulsating stars are known to be coupled (e.g., Cox 1980), and Takeuti *et al.* (2013) also reported a correlation between masses and pulsation periods of AGB stars. Therefore, to study AGB stars with various masses, observations of AGB stars exhibiting a wide range of pulsation periods are necessary.

In our previous VLBI observations conducted with the VLBI Exploration of Radio Astrometry (VERA) array from 2003 to 2017, we have carried out many observations towards dozens of AGB stars. Most of the target sources are classified as Mira-type variables and SR variables, with pulsation periods typically shorter than 400 days. Using H_2O masers at 22 GHz, we have investigated the structures and kinematics of circumstellar matter around Mira and SR variables. The study of the SR variable S Crt by Nakagawa *et al.* (2008) was the initial outcome of our VLBI observation program, revealing the parallax and anisotropic circumstellar distribution of H_2O masers, as well as detailed investigations of other Mira-type variables. Subsequent studies on the Mira-type variables, such as SY Scl by Nyu *et al.* (2011), R UMa by Nakagawa *et al.* (2018), and BX Cam by Matsuno *et al.* (2020), have also been produced from our VLBI observations. However, to date, there have been a limited number of VLBI studies focusing on OH/IR stars. Notable examples are the research on U Her by Vlemmings & van Langevelde

(2007) and WX Psc and OH138.0+7.2 by Orosz *et al.* (2017), which represent valuable contributions to the OH/IR star studies using the VLBI method.

To obtain a diverse sample of AGB stars covering various evolutionary phases and a range of masses in our observations, it is crucial to include a wide range of pulsation periods. Since 2017, we have expanded our VLBI observations to include OH/IR stars with longer pulsation periods. The primary objective of our long-term VLBI study is to obtain astrometric and physical properties of OH/IR stars, including distance determination by parallax measurements, proper motion analysis, internal maser motion, luminosity estimation, mass loss rate analysis, and more. We think that comparing these properties between OH/IR stars and Mira-type variables can provide insights into their evolutionary relationship. In this proceedings paper, we present the current status and results from our astrometric VLBI observations. In particular, we will discuss the latest research results on the OH/IR star NSV17351 and present preliminary parallax values for other OH/IR stars.

1.3. *Finding a new period-magnitude relation of Galactic AGB stars*

It is known that there is a relation between K-band apparent magnitudes (m_{K}) and the logarithm of pulsation periods ($\log P$) of Mira-type variables in the Large Magellanic Cloud (LMC) (e.g., Wood *et al.* 1999; Ita *et al.* 2004). By utilizing the known distance to the LMC, this relation can be converted into a relation between absolute magnitude in the K band (M_{K}) and $\log P$, thus it can be used as a distance estimator for variable stars. However, considering the difference of metallicity between the LMC and our Galaxy, it is crucial to establish a period-magnitude relation using sources in our galaxy. To convert apparent magnitudes to absolute magnitudes, parallax distances of the Galactic variable stars are required. Over the past decade, we have conducted astrometric VLBI observations for Mira and SR variables, and reported their period-magnitude relation as $M_{\mathrm{K}} = -3.52 \log P + (1.09 \pm 0.14)$ (Nakagawa *et al.* 2016). As mentioned in the previous subsection, our previous observations had a limited period coverage of shorter than approximately 400 days.

Compared to typical Mira-type variables, OH/IR stars tend to exhibit longer pulsation periods, sometimes exceeding 1000 days. Since the OH/IR stars are surrounded by thick circumstellar dust shells, there is a large amount of extinction in optical bands. Sometimes the extinction effects extend into the near-infrared region, including the K band. We think this effect leads to a scattering of estimated K-band absolute magnitudes (Nakagawa *et al.* 2018). Conversely, the extinction diminishes at longer wavelengths, and the re-radiation effect from the dust shell becomes more dominant. To mitigate the impact of circumstellar extinction, we aim to validate the existence of the period-magnitude relation in the mid-infrared region. We plan to utilize data from the Wide-field Infrared Survey Explorer (WISE)†, especially in the W3 band ($\lambda = 12\,\mu$m). If a period-magnitude relation is confirmed in the mid-infrared bands, it can serve as a new distance estimator for sources with pulsation periods longer than those typically observed in the Mira-type variables. With our new target sources, given in Section 2.4, we will explore this relation in the longer pulsation period range.

1.4. *Astrometry of OH/IR stars to study the Galactic dynamics*

In recent studies of spiral arms in disk galaxies, there has been a long-standing question about how spiral arms are formed and maintained. The quasi-stationary density wave theory (e.g., Lin, & Shu 1964) and the dynamic spiral theory (e.g., Sellwood & Carlberg

† `http://wise.ssl.berkeley.edu/index.html`

Table 1. Models and observations for study of the Galactic dynamics.

Time scale	Phenomena and model	Target source	VLBI Obs.
$\sim 10^6$ yr	Spiral arm	SFRs, giants	Well studied
$\sim 10^8$ yr	Bifurcating/merging arm	Heavy OH/IR stars	***Few cases***
$\sim 10^9$ yr	Thick disk stars	Miras	Well studied

1984; Baba 2015) are two major theories under discussion. The studies of the Galactic spiral arms based on three-dimensional N-body simulations support a picture of non-steady spiral arms (Baba *et al.* 2013). The spiral arms do not show rigid rotating patterns but rather differentially rotating dynamic patterns. In the dynamic spiral theory, the amplitudes, pitch angles, and pattern speeds of the spiral arms are not constant, but change within a time span of 1-2 rotation periods at each radius (Baba 2015). Characteristic behavior predicted from recent studies is the bifurcating or merging of the Galactic spiral arms on a time scale of $\sim 10^8$ yr (e.g., Baba 2015). In the Milky Way galaxy, rotation periods at the position of the Sun also correspond to the time scale of $\sim 10^8$ years.

The OH/IR stars with the longer periods are assumed to have larger masses, i.e. variable stars with $P \simeq 1000$ days have initial masses of 3 to 4 $M_\odot$ (Feast 2009). Assuming this mass range and the relation between the main sequence lifetime $\tau_{\rm MS}$ and mass given in Sparke & Gallagher (2000), we obtained $\tau_{\rm MS}$ of 1.6×10^8 to 3.5×10^8 years. A recent study by Nikzat *et al.* (2022) also supports this estimation. Now, we find that the age of OH/IR stars with very long pulsation period ($P \gtrsim 1000$ days) is similar to the characteristic time scale of $\sim 10^8$ yr in the dynamic spiral theory. This consideration also implies that the age of OH/IR stars with pulsation periods longer than $\sim$1000 days is $\sim 10^8$ years, which is two orders of magnitude larger than the typical age of high-mass star-forming regions (SFRs) associated with spiral arms.

In the last 20 years, VLBI astrometry has measured more than two hundred parallaxes of SFRs (e.g., Burns *et al.* 2016; Motogi *et al.* 2016; Reid *et al.* 2019; VERA Collaboration *et al.* 2020) and evolved stars (e.g., Kamezaki *et al.* 2016a; Nakagawa *et al.* 2016, 2018, 2019; Sudou *et al.* 2019). However, the ages of almost all VLBI targets fall mainly into two time scales, on the order of 10^6 years for the SFRs and 10^9 years for the Mira-type variable stars. To fully understand the mechanism of spiral arm formation, observations of sources of different ages are now needed, and the very long-period OH/IR stars with estimated ages of $\sim 10^8$ years can be good probes to fill the time scale gap between 10^6 years and 10^9 years (Table 1). Astrometric VLBI is a promising tool to determine the three-dimensional positions and kinematics of the OH/IR stars. The OH/IR stars that we have selected, which have pulsation periods $P \gtrsim 1000$ days, can contribute to this study.

1.5. *Why are VLBI observations important?*

Annual parallaxes can be used to derive distances of celestial sources without making any assumptions about their chemical and/or physical properties. Recently, *Gaia* Data Release 3† (DR3; Gaia Collaboration *et al.* 2023) has provided a huge amount of astrometric measurements. Most of the VLBI parallax measurements made so far have been towards SFRs. They are very close to the Galactic plane and are deeply obscured by dense dust and molecular clouds. As a result, the optical emission from the stars is intercepted by their surroundings and interstellar matter. For this reason, it is relatively difficult to find counterparts to SFRs in *Gaia* catalogs. Compared to the SFRs, it is easier to identify AGB stars in *Gaia* catalogs because there are many AGB stars distributed at high Galactic latitude. Because they are distributed far from the Galactic plane, they

† `https://www.cosmos.esa.int/web/gaia/data-release-3`

are not as heavily obscured by interstellar dust or molecular clouds. Mira-type variables, which are thought to be in the early stages of AGB phases, are bright in both optical and infrared bands. Parallaxes for a large number of Mira-type variables are available in *Gaia* catalogs. Since the AGB stars have both *Gaia* and VLBI parallax values, they are suitable sources for verifying *Gaia* and the VLBI parallax measurements.

After the release of *Gaia* Data Release 2‡ (DR2), Chiavassa *et al.* (2018) conducted a study using three-dimensional radiative hydrodynamics simulations of convection to explore the impact of convection-related surface structures in AGB stars on their photometric variability. They extracted parallax errors in DR2 for SR variables in the solar neighbourhood and compared them with synthetic predictions of photocenter displacements. As a result, they reported that the position of the photocenter has a temporal excursion between 0.077 – 0.198 au (5 to 11% of the corresponding stellar radius), depending on the simulation considered. Since the distances of the sources in our VLBI studies are in the order of a hundred pc to a few kpc, the angular size of the excursion in Chiavassa *et al.* (2018) can be expected to be 0.1 – 1 mas. In addition, the time variation of the surface brightness degrades the accuracy of parallax measurements on optical images. Therefore, in DR2, and even in its updated version DR3, *Gaia*'s parallax measurements of AGB stars can be expected to suffer from this effect.

If the central star is surrounded by thick dust layers in the late stage of the AGB phase, the source becomes very faint in the optical bands and cannot be observed with Gaia. For example, an OH/IR star, OH127.8+0.0, is known to have a thick circumstellar dust shell, a high mass loss rate (Kemper *et al.* 2002), and a long pulsation period of 1994 days (VSX†; Watson *et al.* 2006). But we cannot find it in DR3 catalog. OH127.8+0.0 is very bright in the infrared, but faint in the optical bands due to the strong influence of circumstellar extinction by the dust layer. The VLBI method is still a very effective and promising tool for making parallax measurements of this kind of stars.

2. Observation

2.1. *Single dish monitoring of H_2O and SiO masers*

Using the 20 m aperture telescope at the VERA Iriki station, we have been observing H_2O and SiO maser emissions to obtain their spectra and time variability. Since the pulsation periods of a large number of dusty OH/IR stars are not found in the literature or in databases, we have to determine the pulsation period ourselves from single-dish observations.

The integration time of our single-dish observations is 10 to 40 minutes, to reduce the noise level (antenna temperature in K) in each observation to less than 0.05 K. The time interval of single-dish observations is approximately one month. The conversion factor from antenna temperature to flux density is 19.6 $\mathrm{Jy\,K^{-1}}$. The acquired signal with a bandwidth of 32 MHz is split into 1024 spectral channels with a frequency resolution of 31.25 kHz, which corresponds to a velocity resolution of 0.42 $\mathrm{km\,s^{-1}}$ at 22 GHz and 0.21 $\mathrm{km\,s^{-1}}$ at 43 GHz. We used a signal-to-noise ratio (S/N) of 3 to 5 as a detection criterion in our single-dish observations.

To measure the overall activity of the circumstellar masers, we use the integrated intensities I in unit of $\mathrm{K\,km\,s^{-1}}$ as an integration of the total maser components over a detected velocity range. We use the value I to estimate the pulsation period. Antenna temperatures have relative uncertainties of 5-20%, and we have uniformly applied uncertainties of 10% to all the integrated intensities. A simple sinusoidal function I_{model}, defined as

$$I_{\mathrm{model}} = \Delta I \sin\left[\frac{2\pi(t+\theta)}{P}\right] + I_0, \quad (1)$$

‡ https://www.cosmos.esa.int/web/gaia/dr2
† https://www.aavso.org/vsx/

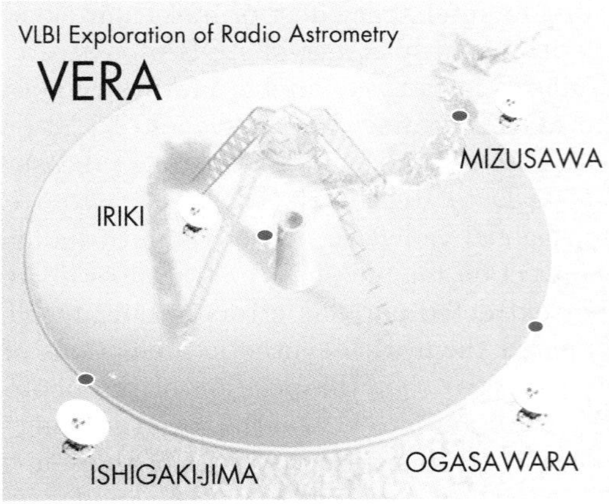

Figure 1. Locations of the 4 antennas of the VERA array, which consists of four 20-metre aperture antennas spread across Japan, with the longest baseline length of 2270 km between Ishigaki-jima and Mizusawa.

is used to estimate the pulsation period. ΔI is the amplitude of the variation, t is the time, θ is a zero-phase time lag, P is the period of the variation, and I_0 is the average.

2.2. *VLBI observations*

For parallax measurements and mapping of circumstellar masers, we carry out continuous VLBI observations. VERA, operated by the National Astronomical Observatory of Japan (NAOJ), has been used to observe 22 GHz H_2O and 43 GHz SiO maser emission from Mira-type variables and OH/IR stars. The VERA array consists of four 20-metre aperture radio telescopes at Mizusawa, Iriki, Ogasawara, and Ishigaki-jima (Figure 1). The array, including the longest baseline length of 2270 km between Ishigaki-jima and Mizusawa, gives a typical synthesised beam size of $\sim$1.2 mas and $\sim$0.6 mas at 22 GHz and 43 GHz, respectively. Each antenna of the VERA array is equipped with a dual beam system (e.g., Kawaguchi *et al.* 2000), which allows us to simultaneously observe a target maser source and an extragalactic continuum source at a separation angle between 0.3° and 2.2°. The extragalactic sources are used as a position reference. Using the dual beam system, we can calibrate short-term tropospheric fluctuations using the phase referencing technique (Honma *et al.* 2008). The relative position of the target maser spots with respect to the position reference source can then be determined with an accuracy of better than 0.1 mas. By tracking the celestial motions of the maser spots, the annual parallax of the target is derived.

The signals of left-handed circular polarization from the target and position reference source are acquired with a total data acquisition rate of 1 giga-bit per second (Gbps). This gives a total bandwidth of 256 MHz. The data were recorded on the hard disks of the "OCTADISK" system (Oyama *et al.* 2016). This total bandwidth is divided into 16 intermediate frequency (IF) channels. Each IF channel has a width of 16 MHz. One IF channel (16 MHz) was allocated to the maser source and the remaining 15 IF channels (16 MHz $\times$ 16 = 240 MHz) were allocated to the position reference sources. Correlation processing was performed using the Mizusawa software correlator at the Mizusawa VLBI observatory, NAOJ. In the final output of the correlator, the 16 MHz bandwidth data of the H_2O or SiO masers were divided into 512 channels with a frequency resolution of 31.25 kHz. This corresponds to a velocity resolution of 0.42 km s^{-1} at 22 GHz and 0.21 km s^{-1} at 43 GHz.

For reduction of the VLBI data we use the Astronomical Image Processing System† (AIPS; Fomalont 1981) developed by the National Radio Astronomy Observatory (NRAO). A detailed description of the phase referencing analysis is given in Nakagawa *et al.* (2023a).

2.3. *Phase referencing with the VERA dual beam system*

We will give a simplified explanation of the phase referencing method using the dual beam system equipped with the VERA array. Letting the phases obtained by a VLBI observation towards the target maser source A and the reference continuum source B be $\phi^{\rm A}_{\rm obs}$ and $\phi^{\rm B}_{\rm obs}$, respectively, we can express the phases as follows:

$$\phi^{\rm A}_{\rm obs} = \phi^{\rm A}_{\rm pos} + \phi^{\rm A}_{\rm struc} + \phi^{\rm A}_{\rm atm} + \phi^{\rm A}_{\rm inst} + \phi^{\rm A}_{\rm clock} \tag{2}$$

and

$$\phi^{\rm B}_{\rm obs} = \phi^{\rm B}_{\rm pos} + \phi^{\rm B}_{\rm struc} + \phi^{\rm B}_{\rm atm} + \phi^{\rm B}_{\rm inst} + \phi^{\rm B}_{\rm clock}, \tag{3}$$

where the terms on the right-hand side are the residual phase due to the sky plane position of the sources ($\phi^{\rm A}_{\rm pos}$), the residual phase due to the source structure ($\phi_{\rm struc}$), the residual phase due to the unpredictable atmospheric path length ($\phi_{\rm atm}$), the residual phase due to the difference between the two signal paths ($\phi_{\rm inst}$), and the residual phase due to the clock offset ($\phi_{\rm clock}$) (see, e.g., Thompson *et al.* 2001). Since the two sources A and B have a small separation angle on the sky plane, $\phi^{\rm A}_{\rm atm}$ and $\phi^{\rm B}_{\rm atm}$ are considered to be equal. The clock offsets $\phi^{\rm A}_{\rm clock}$ and $\phi^{\rm B}_{\rm clock}$ can be eliminated in the ordinary reduction process. However, since the signal paths for sources A and B are independent in the VERA receiver system, $\phi^{\rm A}_{\rm inst}$ and $\phi^{\rm AB}_{\rm inst}$ are not the same,

$$\phi^{\rm A}_{\rm inst} \neq \phi^{\rm B}_{\rm inst}. \tag{4}$$

By taking the difference between the phases from sources A and B, we will obtain

$$\phi^{\rm A}_{\rm obs} - \phi^{\rm B}_{\rm obs} = \phi^{\rm A}_{\rm pos} - \phi^{\rm B}_{\rm pos} + \phi^{\rm A}_{\rm inst} - \phi^{\rm B}_{\rm inst}. \tag{5}$$

Here we use the calibration information obtained from the dual beam system. Artificial noise sources are installed on the surface of the VERA antenna and the same noise is sent to the two receivers along with the signal from the celestial source (Honma *et al.* 2008). By correlating two signals from A and B, the phase error caused by the difference in path length $\phi^{\rm A}_{\rm inst} - \phi^{\rm B}_{\rm inst}$ can be obtained. The phase $\phi^{\rm B}_{\rm pos}$ is assumed to be zero for the source with point-like structure. Even if source B has some structure, we can estimate and eliminate the $\phi^{\rm B}_{\rm pos}$ term by the self-calibration reduction process. Finally, we obtain

$$\phi^{\rm A}_{\rm obs} - \phi^{\rm B}_{\rm obs} = \phi^{\rm A}_{\rm pos}. \tag{6}$$

This gives the phase information that reflects the celestial position of the target maser source A. By performing VLBI observations for 1.5 year to 2 years with an interval of about one month, we can track the position of the target maser source A to derive its parallax and proper motions.

2.4. *Target sources*

From 2003 to 2017, we have carried out many VLBI observations of dozens of AGB stars using the VERA array. The main targets of the previous studies were Mira-type variables and SR variables, which in most cases have pulsation periods shorter than 400 days (Nakagawa *et al.* 2018). From 2017, we started VLBI observations of OH/IR stars or Mira-type variables, which have longer pulsation periods than those in the previous

† `http://www.aips.nrao.edu/index.shtml`

Table 2. DR3 parallaxes of our VLBI target sources.

Source name (2017–2022)†	$\Pi_{\rm DR3}$ [mas]	Err. [%]	Period [days]	Source name (2023–)†	$\Pi_{\rm DR3}$ [mas]	Err. [%]	Period [days]
NSV17351	0.088±0.147	166	1122	V697 Her	1.029±0.129	13	497
OH 127.8−0.0	n/a	n/a	1994	NSV 23099	0.209±0.102	49	431
NSV 25875	n/a	n/a	1535	OH 358.667−0.044	0.207±0.142	69	300
RAFGL 5201	−0.131±0.253	−194	600	OH 358.23+0.11	−0.061±0.190	−313	704
OH 83.4−0.9	0.836±0.556	66	1428	OH 0.66−0.07	n/a	n/a	n/a
OH 141.7+3.5	n/a	n/a	1750	IRAS 18039−1903	n/a	n/a	n/a
CU Cep	0.231±0.057	25	700	OH 9.097−0.392	0.261±0.232	89	634
RAFGL 2445	−1.548±0.369	−24	n/a	RAFGL 1686	1.053±0.359	34	500
OH 39.7+1.5	n/a	n/a	1260	IRAS 18176−1848	2.404±0.618	26	n/a
OH 26.5+0.6	n/a	n/a	1589	OH 44.8−2.3	0.918±0.631	69	n/a
OH 42.3−0.1	n/a	n/a	1650				
IRC−30363	0.241±0.130	54	720				
IRC+10322	0.553±0.183	33	570				
IRC+10451	0.818±0.196	24	730				
OH 26.2−0.6	n/a	n/a	1330				
OH 51.8−0.1	n/a	n/a	1270				
OH 358.16+0.49	n/a	n/a	1507				
V1018 Sco	n/a	n/a	n/a				

† Duration of our VLBI observations using VERA.

studies. Among them, the OH/IR stars are shown in particular in Table 2. The first four columns show the sources for which observations have been made between the years 2017 and 2022. Although successful observations of all sources have been difficult, data acquisition for some stars has been completed and data reduction is currently underway. In this proceedings paper, we will present some parallax measurements as preliminary results. In early 2023 we proposed VLBI observations of new sources, which are presented in the last four columns of Table 2. The parallaxes from DR3 ($\Pi_{\rm DR3}$) and their relative errors are also shown. For those sources for which no parallax values were found in DR3, we have indicated "n/a" in their respective "$\Pi_{\rm DR3}$" and "Err." columns. For some sources (RAFGL 5201, RAFGL 2445, and OH 358.23+0.11), the parallaxes have negative values. Several sources have errors greater than 100 % (NSV17351, RAFGL 5201, and OH 358.23+0.11). From this table, we can interpret that it is very difficult for *Gaia* to determine accurate parallaxes of OH/IR stars. Therefore, the VLBI method is still important for parallax measurements of these dust-obscured OH/IR stars. Pulsation periods are also given in columns 4 and 8. We have selected sources that cover a wide range in pulsation periods. Sources at low Galactic latitudes with longer pulsation periods can be expected to be young and more massive than typical Mira-type variables.

3. Results and discussion

3.1. *Parallaxes from VLBI and* Gaia

We will now compare parallax measurements of AGB stars from VLBI and *Gaia*. In Table 3 we have presented 44 Galactic LPVs whose parallaxes have been determined from astrometric VLBI, in order of their right ascension (RA). In the third, fourth, and fifth columns, the parallax values determined from astrometric VLBI, DR2 and DR3 are presented as $\Pi_{\rm VLBI}$, $\Pi_{\rm DR2}$, and $\Pi_{\rm DR3}$, respectively. Although the number of digits of the original parallaxes and their formal errors in DR2/DR3 is large, we have presented the values in digits of 0.001 mas. The species of the maser molecules observed in each VLBI observation are given in the sixth column. In the last column, the references of their VLBI parallaxes are given using abbreviations, whose meaning is explained in the table footnote. Regarding R Aqr, since the VLBI parallax is published in two independent

Table 3. Parallaxes from the VLBI and DR2, DR3.

Source	Var. type	Π_{VLBI} [mas]	Π_{DR2} [mas]	Π_{DR3} [mas]	Maser	Ref.† Π_{VLBI}
SY Scl	Mira	0.75±0.03	0.675±0.227	0.525±0.122	H_2O	nyu11
WX Psc	OH/IR	5.3 [b]	n/a	n/a	OH	oro17
S Per	SRc	0.413±0.017	0.222±0.121	−0.503±0.081	H_2O	asa10
OH138.0+7.2	OH/IR	0.52±0.09	n/a	n/a	OH	oro17
V637 Per	Mira	0.94±0.02	1.846±0.152	0.845±0.097	H_2O	ver20
BX Eri	Mira	2.116±0.105	2.477±0.110	2.349±0.063	H_2O	ver20
T Lep	Mira	3.06±0.04	2.958±0.189	3.086±0.103	H_2O	nak14
BW Cam	Mira	0.749±0.189	1.187±0.214	0.956±0.105	H_2O	ver20
RW Lep	Mira	1.62±0.16	2.355±0.134	2.539±0.075	H_2O	kam14
BX Cam	Mira	1.73±0.03	4.134±0.255	1.764±0.101	H_2O	mat20
U Lyn	Mira	1.27±0.06	0.580±0.224	1.014±0.083	H_2O	kam16a
NSV17351	Mira	0.247±0.035	0.353±0.228	0.088±0.147	H_20	nak23a
VY CMa	SRc	0.88±0.08	−5.917±0.825	0.419±0.408	H_2O	cho08
OZ Gem	Mira	0.806±0.039	−0.961±0.456	0.458±0.325	H_2O	ura20
OH231.8+4.2	OH/IR	0.61±0.03	0.096±0.182	0.030±0.160	H_2O	nak23b
HU Pup	Mira	0.308±0.042	0.182±0.057	0.294±0.030	H_2O	ver20
R Cnc	Mira	3.84±0.29	4.435±0.549	3.938±0.179	H_2O	ver20
X Hya	Mira	2.07±0.05	1.891±0.276	2.531±0.111	H_2O	ver20
R UMa	Mira	1.97±0.05	2.045±0.202	1.747±0.086	H_2O	nak16
W Leo	Mira	1.03±0.02	n/a	0.878±0.108	H_2O	ver20
HS UMa	Mira	2.816±0.095	3.215±0.144	3.202±0.101	H_2O	ver20
S Crt	SRb	2.33±0.13	2.646±0.146	0.061±0.097	H_2O	nak08
T UMa	Mira	0.96±0.15	0.748±0.105	0.989±0.065	H_2O	nak18
U CVn	Mira	0.911±0.031	0.921±0.167	0.563±0.077	H_2O	ver20
RT Vir	SRb	4.417±0.134	2.050±0.291	4.137±0.227	H_2O	zha17
R Hya	Mira	7.93±0.18	4.468±0.894	6.736±0.464	H_2O	ver20
W Hya	Mira	10.18±2.36	6.091±0.816	n/a	OH	vle03
RX Boo	SRb	7.31±0.50	7.829±0.300	6.424±0.231	H_2O	kam12
FV Boo	Mira	0.97±0.06	0.573±0.181	1.014±0.091	H_2O	kam16b
Y Lib	Mira	0.855±0.050	n/a	0.832±0.083	H_2O	chi19
S CrB	Mira	2.39±0.17	2.322±0.285	2.596±0.114	OH	vle07
S Ser	Mira	1.25±0.04	−0.512±0.317	0.768±0.129	H_2O	ver20
U Her	Mira	3.76±0.27	1.749±0.149	2.357±0.077	OH	vle07
VX Sgr	SRc	0.64±0.04	0.787±0.229	0.050±0.187	H_2O	xu18
RR Aql	Mira	1.58±0.40	3.146±0.298	1.953±0.113	OH	vle07
SY Aql	Mira	1.10±0.07	3.433±0.206	1.067±0.091	H_2O	ver20
NML Cyg	SRc	0.62±0.047	1.526±0.568	0.528±0.348	H_2O	zha12
UX Cyg	Mira	0.54±0.06	0.176±0.167	0.701±0.094	H_2O	kur05
SV Peg	SRb	3.00±0.06	1.124±0.283	2.586±0.170	H_2O	sud19
IRAS22480+6002	SRc	0.400±0.025	0.479±0.078	0.363±0.029	H_2O	ima12
R Peg	Mira	2.76±0.28	2.830±0.254	2.629±0.117	H_2O	ver20
R Aqr	Mira	4.7±0.8	3.122±0.278	2.593±0.333	SiO	kam10
R Aqr	Mira	4.59±0.24	3.122±0.278	2.593±0.333	SiO	min14
PZ Cas	SRc	0.356±0.026	0.420±0.081	0.355±0.041	H_2O	kus13
R Cas	Mira	5.67±1.95	5.342±0.245	5.742±0.201	OH	vle03

† Reference of VLBI parallax : (nyu11) Nyu *et al.* (2011), (oro17) Orosz *et al.* (2017), (asa10) Asaki *et al.* (2010), (ver20) VERA Collaboration *et al.* (2020), (nak14) Nakagawa *et al.* (2014), (kam14) Kamezaki *et al.* (2014), (mat20) Matsuno *et al.* (2020), (kam16a) Kamezaki *et al.* (2016a), (nak23a) Nakagawa *et al.* (2023a), (cho08) Choi *et al.* (2008), (ura20) Urago *et al.* (2020), (nak23b) This work, preliminary, (nak16) Nakagawa *et al.* (2016), (nak08) Nakagawa *et al.* (2008), (nak18) Nakagawa *et al.* (2018), (zha17) Zhang *et al.* (2017), (vle03) Vlemmings *et al.* (2003), (kam12) Kamezaki *et al.* (2012), (kam16b) Kamezaki *et al.* (2016b), (chi19) Chibueze *et al.* (2019), (vle07) Vlemmings & van Langevelde (2007), (xu18) Xu *et al.* (2018), (zha12) Zhang *et al.* (2012), (kur05) Kurayama *et al.* (2005), (sud19) Sudou *et al.* (2019), (ima12) Imai *et al.* (2012), (kam10) Kamohara *et al.* (2010), (min14) Min *et al.* (2014), (kus13) Kusuno *et al.* (2013).

studies by Kamohara *et al.* (2010) and Min *et al.* (2014), it has two lines in the table. For the two sources WX Psc and OH138.0+7.2 we were not able to find the corresponding parallaxes in either DR2 or DR3. For W Leo and Y Lib we could not find their parallaxes in DR2, but found them in DR3. It is assumed that the data quality has improved from

DR2 to DR3. In the case of W Hya, the annual parallax was listed in DR2, but we could not find it in DR3. The parallax of S Ser shows a negative value in DR3.

In Figure 2 we present the parallaxes of AGB stars determined from VLBI ($\Pi_{\rm VLBI}$) and DR2/DR3 ($\Pi_{\rm Gaia\,DR2}$, $\Pi_{\rm Gaia\,DR3}$). A total of 41 sources are included in the figure. The horizontal and vertical axes represent the parallax values on logarithmic scales. Open circles represent the comparison between the parallaxes from the VLBI and DR2. Filled circles represent the comparison between the VLBI and DR3. A dotted line shows a relation of the form $\Pi_{\rm VLBI} = \Pi_{\rm Gaia\,DR2/DR3}$. In Figure 2, source names are added near the comparison data between the VLBI and DR3 (filled circles). In the case of W Hya, we do not have a DR3 parallax, so the name is found near the open circle. The dispersion of the filled circles from the $\Pi_{\rm VLBI} = \Pi_{\rm Gaia\,DR2/DR3}$ relation is significantly smaller than that of the open circles. This suggests that, for many sources, DR3 parallax measurements are closer to those carried out with the VLBI than are those obtained with DR2. However, for some sources there are still large discrepancies between the VLBI and DR3 parallaxes. For three sources, NSV17351, OH231.8+4.2, and VX Sgr, DR3 measurements are smaller than the VLBI measurements. The types of the three LPVs are OH/IR stars (NSV17351 and OH231.8+4.2) and red supergiant (VX Sgr). Although we are curious about many other OH/IR stars in Table 2, they could not be shown in Figure 2 because *Gaia* and/or VLBI parallaxes of many dust-obscured OH/IR stars are not available.

3.2. *Parallax and error*

We will consider the parallaxes and their errors. In Figure 3 we have presented the parallaxes and their formal errors obtained from the VLBI (filled circles), DR2 (open triangles), and DR3 (grey squares). From this figure we can see a positive correlation between the parallaxes and the errors for all data sets from the VLBI and DR2/DR3. We fitted the distribution to a linear function using a least-squares analysis and obtained three lines as follows,

$$\log \sigma_{\Pi_{\rm VLBI}} = (0.96 \pm 0.13) \log \Pi_{\rm VLBI} - 1.20 \pm 0.06\,, \tag{7}$$

$$\log \sigma_{\Pi_{\rm Gaia\,DR2}} = (0.29 \pm 0.07) \log \Pi_{\rm Gaia\,DR2} - 0.72 \pm 0.04\,, \tag{8}$$

$$\log \sigma_{\Pi_{\rm Gaia\,DR3}} = (0.53 \pm 0.09) \log \Pi_{\rm Gaia\,DR3} - 1.07 \pm 0.03\,. \tag{9}$$

It should be noted that five data points from DR3, presented with open squares, were excluded from this consideration because they were distributed outside the main group of data. Equations (7) to (9) show the relationships between parallax and errors for the VLBI (solid line), DR2 (one-dotted chain line), and DR3 (dotted line), respectively. Equations (7) and (8) intersect at $\log \Pi = 0.716$, corresponding to a distance of 192 pc. Equations (7) and (9) intersect at $\log \Pi = 0.302$, corresponding to a distance is 499 pc. Using the error ratio σ_Π/Π as an indicator of the effectiveness of parallax measurements, VLBI measurements can derive better distance estimates than DR3 for the LPVs further than ~ 499 pc. We can understand that the VLBI and *Gaia* are complementary for distance measurements of dusty AGB stars, and the distance of $\sim$500 pc would be a boundary of validity between the VLBI and *Gaia* for parallax measurement of AGB stars.

3.3. *Results for individual sources*

3.3.1. Pulsation period of an OH/IR star NSV17351

NSV17351 is classified as an OH/IR star (Le Squeren *et al.* 1979). From long-term monitoring of our single-dish program, we find that this source can be a variable star. Although we searched for its pulsation period in online databases and the literature, we

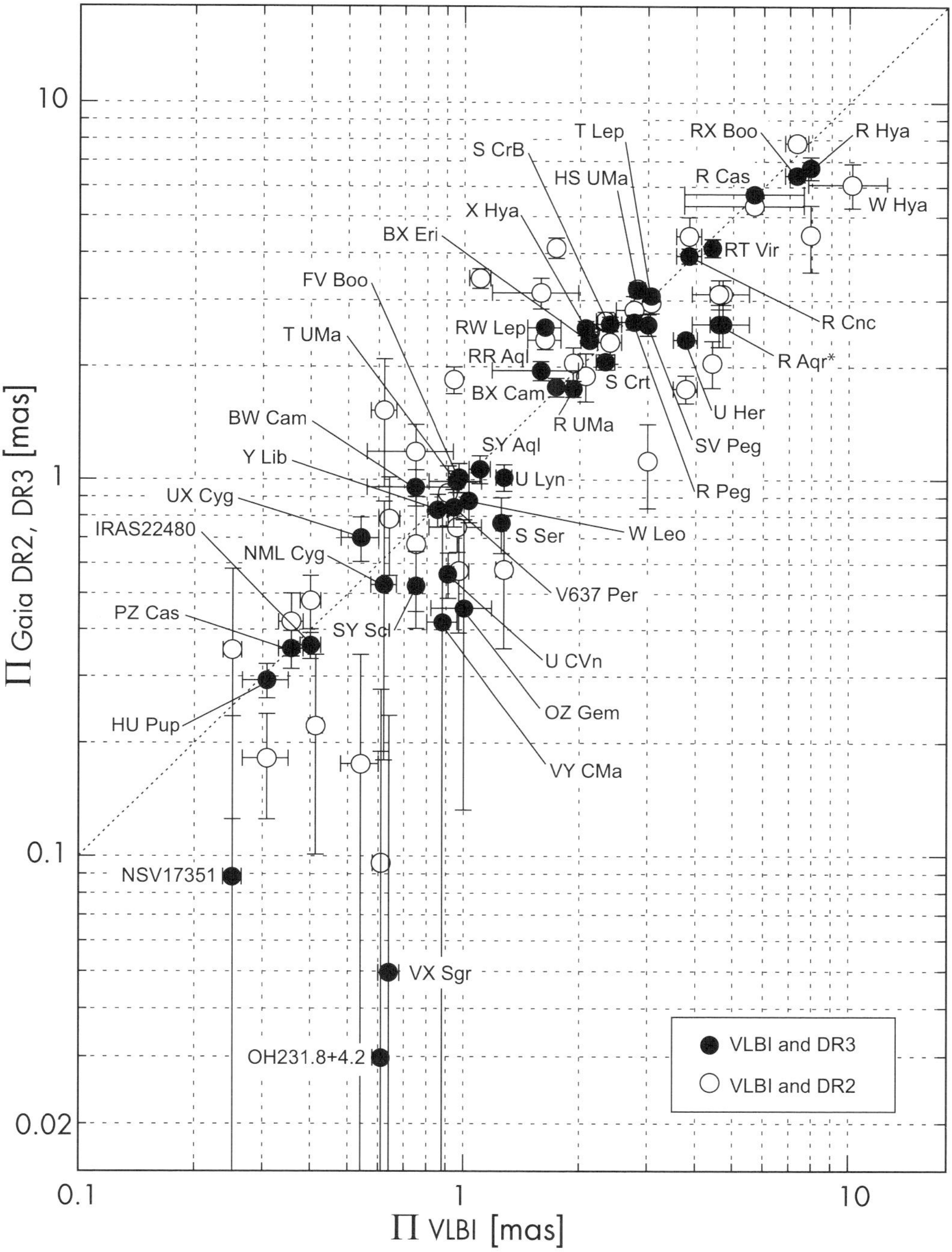

Figure 2. Annual parallaxes of 41 LPVs determined from the VLBI (horizontal axis) and DR2/DR3 (vertical axis) on a logarithmic scale. Open circles correspond to the comparison between the VLBI and DR2, whereas filled circles correspond to the comparison between the VLBI and DR3. Error bars are also shown on both axes.

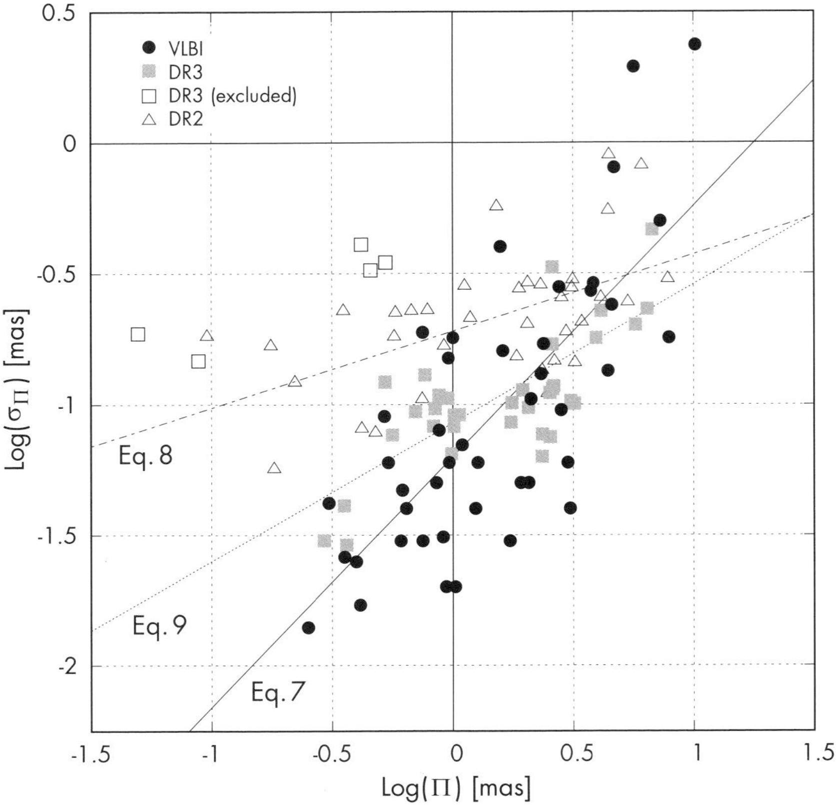

Figure 3. Parallaxes and their formal errors obtained with the VLBI (filled circles), DR2 (triangles), and DR3 (squares). Three lines represent the relationships between parallax and errors for the VLBI (solid line; Eq. 7), DR2 (one-dotted chain line; Eq. 8), and DR3 (dotted line; Eq. 9). The points of intersection are $\log \Pi = 0.716$ for equations (7) and (8), and $\log \Pi = 0.302$ for equations (7) and (9).

could not find it. We determined the pulsation period of NSV17351 from our single-dish monitoring of the H_2O maser at 22 GHz. From our least-squares analysis assuming a simple sinusoidal function as presented in Section 2.1, the pulsation period of NSV17351 was solved to be 1122±24 days (Nakagawa *et al.* 2023a). The fit of the model is shown with a solid curve in the left panel of Figure 4. As there is no prior information on this period in the literature or online databases, this is the first time we have successfully measured its periodicity. Since NSV17351 has a very long periodicity, we think that it is a candidate for an extreme OH/IR star.

In the right panel of Figure 4, we superimposed the H_2O maser spectrum obtained on 22 April 2018 (solid line) and a 1612 MHz OH maser spectrum obtained in February 1978 (dotted line). We can see that the cut-off velocity in the blue-shifted component shows exactly the same value ($38\,\mathrm{km\,s^{-1}}$ to $40\,\mathrm{km\,s^{-1}}$). Since it is thought that OH molecules are supplied by photodissociation of H_2O molecules carried to the outer part of the circumstellar envelope, we understand that the H_2O molecules have been carried to the outermost region and the H_2O gas has accelerated to the terminal velocity.

3.3.2. Parallax of NSV17351

To estimate the annual parallax, we track the positions of 22 GHz H_2O maser spots obtained from multiple VLBI images. In Figure 5 we show examples of maser spot images

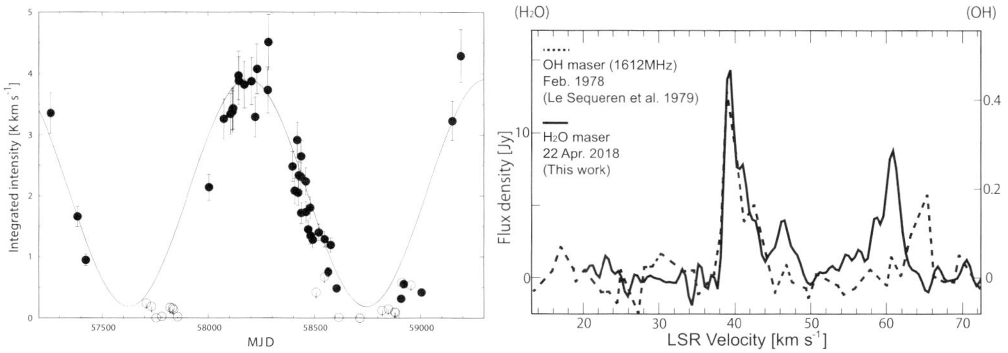

Figure 4. (Left): Time variation of the integrated H_2O maser intensities of NSV17351. Filled circles represent successful detections. In the case of non-detections, open circles with downward arrows represent the upper limits of detection. The solid line is the model indicating a pulsation period of 1122±24 days. (Right): Superpositions of the H_2O maser (solid line) and OH maser (dotted line) of NSV17351 obtained in 2018 and 1978, respectively. The cut-off velocity of the blue-shifted side appears to be exactly the same in both spectra.

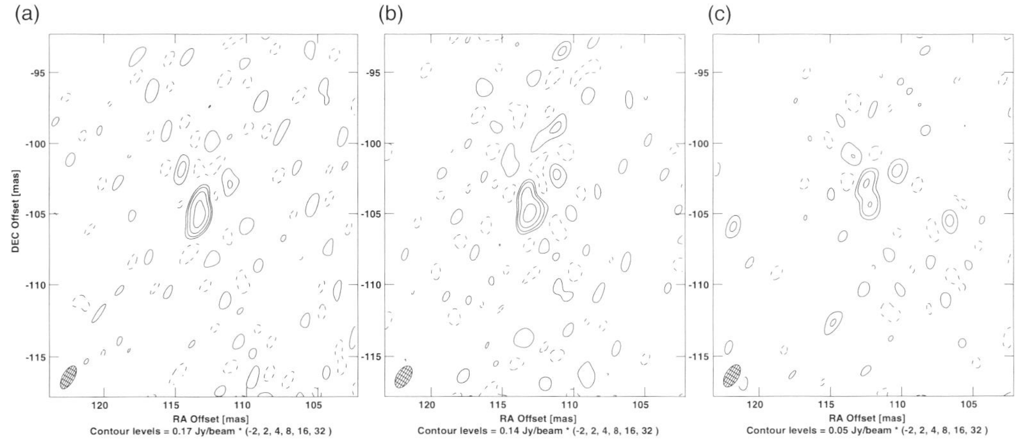

Figure 5. VLBI images of maser spots of NSV17351 at a $V_{\rm LSR}$ of 39.15 km s^{-1} detected on (a) 16 April 2018, (b) 1 November 2018 and (c) 12 March 2019 (Nakagawa *et al.* 2023a). The synthesised beams are presented in the lower left of each map.

in the same velocity channel. Since the shape of the spot changes gradually with time, we carefully examined the maser structure, its time variation and continuity. In this velocity channel, we concluded that the southern components in the maps of Figure 5 (b) and (c) are identical to the peak in the map of Figure 5 (a). Using the 2018-2019 VERA observations of the H_2O maser at 22 GHz, we derived a parallax of 0.247±0.035 mas for NSV17351, corresponding to a distance of 4.05±0.59 kpc. Figure 6 shows the position offsets after removal of proper motions and the fitted parallax along the RA (top) axis and DEC (bottom) axes. The observed data are indicated as filled circles, with their grey scales representing the local standard of rest (LSR) velocities of each maser spot. Error bars are 0.05 mas and 0.09 mas in RA and DEC, respectively. The solid curves are the best-fit models of the parallax. A systemic proper motion of $(\mu_\alpha \cos\delta, \mu_\delta) = (-1.19 \pm 0.11,\ 1.30 \pm 0.19)$ mas yr^{-1} was also obtained.

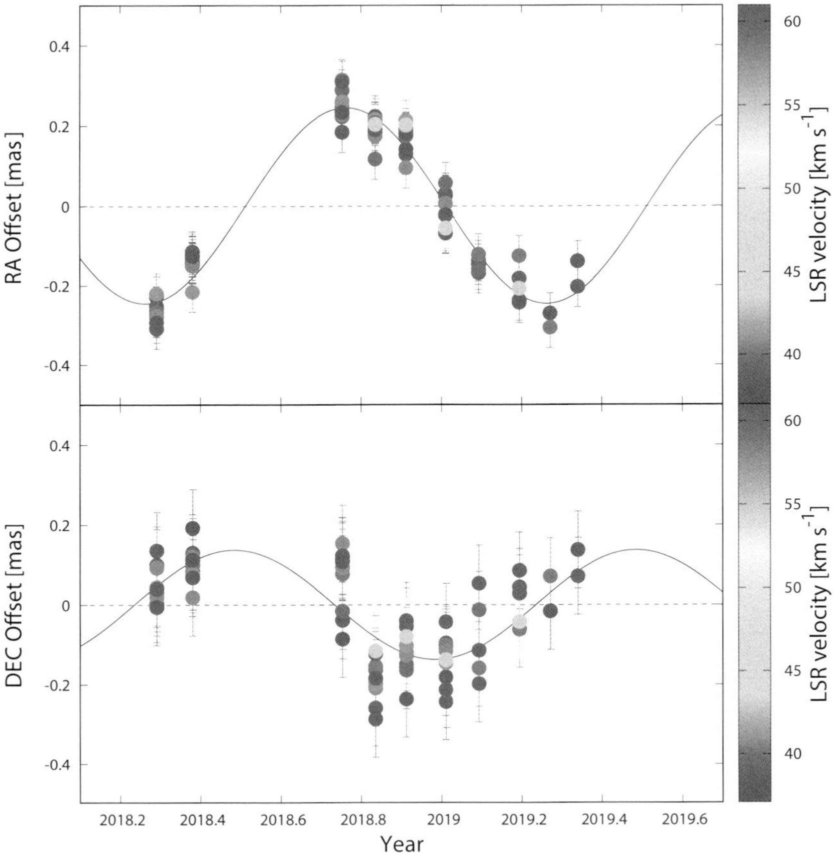

Figure 6. The annual parallax of NSV17351 along RA (top) and DEC (bottom) (Nakagawa *et al.* 2023a). Results from ten observations are shown with filled circles. The grey scale indicates the LSR velocity $V_{\rm LSR}$ of each maser spot. The solid curves are the best-fit models obtained from our fit, showing the 0.247 mas parallax.

3.3.3. H_2O maser distribution of NSV17351

The circumstellar distribution and motions of H_2O masers in an 80 $\times$ 120 au region around the OH/IR star NSV17351 are shown in the left panel of Figure 7. Since the angular proper motions of each maser spot obtained from our VLBI observations are with respect to a position reference source, we have to convert them to motions on the frame fixed to NSV17351. To perform this conversion, we take the average motion of the whole maser spots and then subtract it from the original proper motions of each maser spot. The detailed procedure has been presented in Nakagawa *et al.* (2014). The estimated stellar position of NSV17351 is indicated by a cross symbol whose length represents the position error. We can see that the maser spots at different radial velocities are moving outwards from the expected position of the central star. As an average, we have derived a three-dimensional outward expansion velocity of the H_2O masers of 15.7$\pm$3.3 km s^{-1}.

We can see that the bluest maser spots overlap the estimated position of the central star. In the case of OH masers in OH/IR stars, it is known that the most blue- and red-shifted maser spots are seen at the same position of the central star along the plane of the sky. For example, Orosz *et al.* (2017) revealed that the blue- and red-shifted OH masers coincide with the same position where the central star is assumed to exist. Thus, as suggested by the left panel of Figure 7, we can interpret that the most blue-shifted

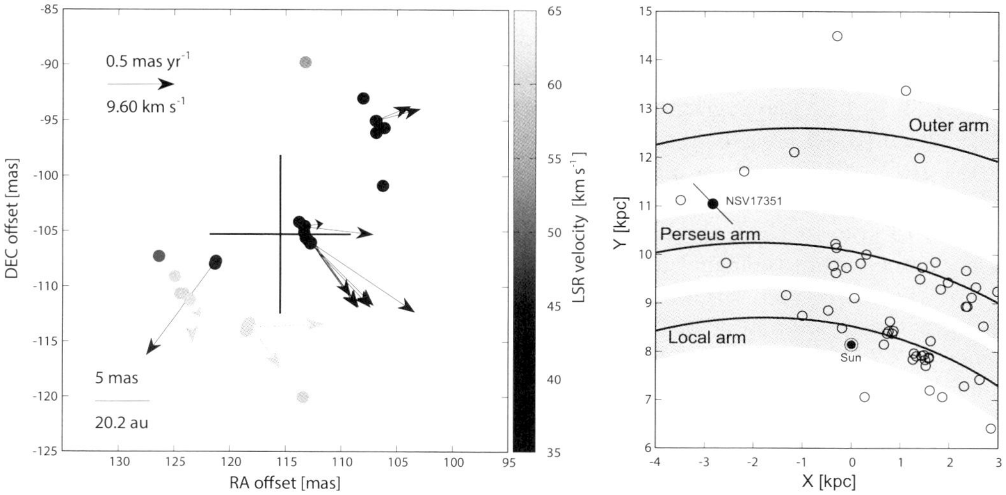

Figure 7. (Left): sky plane distribution and expansion motions of the H_2O maser spots of NSV17351 (Nakagawa *et al.* 2023a). Filled circles indicate maser spots and arrows indicate their internal motions. A cross symbol indicates the estimated position of the central star. (Right): Position of NSV17351 on the face-on view of the Milky Way. The Galactic center is at (0, 0) kpc and the Sun is indicated by the symbol ($\odot$) at (0, 8.15) kpc. The filled circle with an error bar indicates the position of NSV17351. Open circles indicate maser sources with Galactocentric distances > 7 kpc. Solid lines and grey regions indicate the centers of three spiral arms and their widths, respectively, reproduced from a study by Reid *et al.* (2019).

maser spots are superimposed on the position of the central star of NSV17351 along the line of sight. They can possibly be explained by the emission being excited along the line of sight to the central star.

3.3.4. Position of NSV17351 in our galaxy

From the systemic proper motion of NSV17351 derived from our VLBI observations, we can estimate a motion of the source in Galactocentric coordinates. In Nakagawa *et al.* (2023a), we derived a three-dimensional position of NSV17351 as $(X, Y, Z) = (-2.83\pm0.12,\ 11.05\pm0.12,\ -0.09\pm0.01)$ kpc, where the origin of the coordinate system is the Galactic center. We confirm that the Z value of NSV17351, $Z = -0.09 \pm 0.01$ kpc, is within the Z range of SFRs (i.e., $-0.12 < Z < 0.11$ kpc). In the right panel of Figure 7 we have shown the position of NSV17351 using a filled circle. NSV17351 is located a little bit outside of the Perseus arm. The location of NSV17351 can be understood by considering its age. Feast (2009) reported that Mira-type variables with a period of 1000 days have an initial mass of 3 to 4 $M_\odot$. Assuming this mass, we can estimate the age of NSV17351 to be 1.6×10^8 to 3.5×10^8 years. The age of NSV17351 can be two orders of magnitude greater than the typical age of high-mass SFRs associated with spiral arms, and we can say that we are observing a state in which NSV17351 is leaving the arm in which it was born, but is not yet sufficiently dynamically relaxed. With more samples of OH/IR stars representing very long pulsation periods, we can provide observational results for studies of the Galactic spiral arms (Nakagawa *et al.* 2023a).

3.3.5. OH39.7+1.5, IRC−30363 (OH/IR star), and AW Tau (Mira)

In addition to NSV17351, we have observed H_2O masers in two OH/IR stars, OH39.7+1.5 and IRC−30363, and a Mira-type variable, AW Tau, at 22 GHz.

For OH39.7+1.5, using two maser spots with radial velocities of 34.6 and 8.6 $\mathrm{km\,s^{-1}}$, a preliminary parallax of 0.54±0.03 mas was obtained. This corresponds to a distance of 1.85±0.10 kpc. This parallax measurement is the only one available, as DR3 has no data for OH39.7+1.5. The averaged proper motion of the two maser spots is $(\mu_\alpha \cos\delta, \mu_\delta) = (-0.22 \pm 0.13, -1.53 \pm 0.13)$ $\mathrm{mas\,yr^{-1}}$.

For IRC−30363, using a maser spot with a radial velocity of 9.72 $\mathrm{km\,s^{-1}}$, a parallax of 0.562±0.201 mas was obtained from our VLBI observations. This gives a distance of 1.78±0.73 kpc. In DR3, the parallax of IRC−30363 is reported to be 0.241±0.130 mas. They agree within their margins of error.

For the Mira-type variable AW Tau, a parallax of 0.449±0.032 mas was obtained using a maser spot at a radial velocity of −9.54 $\mathrm{km\,s^{-1}}$. This corresponds to a distance of 2.23±0.16 kpc. In DR3, the parallax of AW Tau is reported to be 0.434±0.113 mas. For this source, the two measurements from the VLBI and DR3 are in very good agreement.

The pulsation periods of OH39.7+1.5, IRC−30363 (OH/IR star), and AW Tau are 1260, 720, and 672 days, respectively. All sources show longer pulsation periods than typical Mira-type variables, e.g., lack of Mira-type variables with periods of about 500 days or longer are reported in Habing (1996). We are currently working on a more detailed analysis of the sources with very long pulsation periods.

3.4. *Finding a new period-magnitude relation for OH/IR stars in mid-infrared*

Based on the parallax measurements presented in Section 3.3.5, we estimated the absolute K-band magnitudes ($M_{\rm K}$) of three AGB stars, AW Tau, IRC−30363, and OH39.7+1.5. If we compare the three $M_{\rm K}$ values with a period-$M_{\rm K}$ diagram of the Galactic LPVs, we find that the $M_{\rm K}$ value of OH39.7+1.5 is far below the expected one. About 6 years ago, we investigated the same problem. Using published literature results, we compiled ∼20 OH/IR stars with very long periods of more than 1000 days. The distances of some of the sources were determined using the "phase-lag method" (e.g., Engels *et al.* 2015; Etoka *et al.* 2018). For the other sources for which no distance estimate is available, Nakagawa *et al.* (2018) derived kinematic distances from their radial velocities. Using these distances, the K-band apparent magnitudes were converted to absolute magnitudes ($M_{\rm K}$) and presented on a period-$M_{\rm K}$ diagram in Figure 1(b) of Nakagawa *et al.* (2018). As a result, the distribution of $M_{\rm K}$ values of the OH/IR stars shows large scatter. We could not find a clear relationship between periods and $M_{\rm K}$ values. The OH/IR stars are thought to be surrounded by thick layers of circumstellar dust. The activity of the central star also affects the outer layers. In addition, the spatial structure of the dust layers surrounding the central star is unstable and anisotropic. We think that this is one reason for the large scatter seen in the K-band magnitudes. In the case of the recent result obtained for the OH/IR star OH39.7+1.5, if we assume a thick dust layer or a high mass loss rate, we can possibly explain this darkening by circumstellar absorption.

At longer wavelengths, re-radiation from the dust shell is known to be dominant. To minimise the circumstellar extinction, we have estimated the absolute magnitudes $M_{\rm W3}$ in the mid-infrared using the W3 band data from the Wide-field Infrared Survey Explorer (WISE). The central wavelength of the WISE W3 band is 12 μm (Wright *et al.* 2010). In Table 4 we have compiled 36 Galactic LPVs whose parallaxes have been determined by VLBI observations. Some red giants and SR variables are also included in the table. Sources with pulsation periods shorter than 200 days have been excluded. Using the parallaxes from VLBI observations $\Pi_{\rm VLBI}$ in the third column, we derived the absolute magnitude in the W3 band ($M_{\rm W3}$) and present it in the seventh column.

Table 4. WISE W3 band absolute magnitude.

Source	Var. type	$\Pi_{\rm VLBI}$ [mas]	$\sigma\Pi_{\rm VLBI}$ [mas]	Period (P) [d]	$\log P$	M_{W3} [mag]	$\sigma M_{\rm W3}$ [mag]
SY Scl	Mira	0.75	0.03	411	2.61	−10.46	0.16
OH127.8+0.0	OH/IR	0.22†	0.08	1591	3.20	−14.32	1.23
S Per	SRc	0.413	0.017	822	2.92	−13.92	0.50
OH138.0+7.2	OH/IR	0.52†	0.09	1410	3.15	−11.76	0.63
R Tau	Mira	2.04	0.05	321	2.51	−9.09	0.40
T Lep	Mira	3.06	0.04	372	2.57	−8.94	0.29
BW Cam	Mira	0.749	0.189	628	2.80	−12.46	0.93
AW Tau	Mira	0.45	0.03	672	2.83	−10.71	0.21
RAFGL 5201	OH/IR	0.61†	0.04	600	2.78	−11.26	0.47
AP Lyn	Mira	2.01	0.04	433	2.64	−10.46	0.51
U Lyn	Mira	1.27	0.06	434	2.64	−9.90	0.45
NSV17351	OHIR	0.247	0.035	1100	3.04	−13.31	0.45
OZ Gem	Mira	0.806	0.039	598	2.78	−11.27	0.36
OH231.8+4.2	OH/IR	0.61†	0.03	548	2.74	−11.16	0.44
R Cnc	Mira	3.84	0.29	357	2.55	−8.81	0.48
X Hya	Mira	2.07	0.05	300	2.48	−9.04	0.32
R UMa	Mira	1.97	0.05	302	2.48	−9.21	0.36
W Leo	Mira	1.03	0.02	392	2.59	−9.99	0.28
T UMa	Mira	0.96	0.15	257	2.41	−9.02	0.49
U CVn	Mira	0.911	0.031	346	2.54	−9.97	0.19
R Hya	Mira	7.93	0.18	380	2.58	−8.40	0.07
FV Boo	Mira	0.97	0.06	313	2.50	−10.12	0.21
Y Lib	Mira	0.855	0.050	277	2.44	−9.44	0.19
S CrB	Mira	2.39	0.17	360	2.56	−9.68	0.47
S Ser	Mira	1.25	0.04	372	2.57	−9.50	0.26
U Her	Mira	3.76	0.27	406	2.61	−9.14	0.52
IRC−30363	OH/IR	0.56†	0.2	720	2.86	−12.35	1.21
OH39.7+1.5	OH/IR	0.55†	0.03	1259	3.10	−12.90	0.40
RAFGL 2445	OH/IR	0.64†	0.01	626	2.80	−12.11	0.36
RR Aql	Mira	1.58	0.4	396	2.60	−10.94	0.93
SY Aql	Mira	1.10	0.07	356	2.55	−9.64	0.24
UX Cyg	Mira	0.54	0.06	565	2.75	−12.67	0.50
NSV 25875	OH/IR	0.38†	0.13	1535	3.19	−14.62	1.13
R Peg	Mira	2.76	0.28	378	2.58	−8.94	0.45
R Aqr	Mira	4.7	0.8	390	2.59	−9.51	0.53
	Mira	4.59	0.24	390	2.59	−9.56	0.16
PZ Cas	SRc	0.356	0.026	925	2.97	−14.10	0.50
R Cas	Mira	5.67	1.95	430	2.63	−9.33	1.12

† The parallax values are preliminary results from our analysis and have not yet been published.

In this table, $\sigma M_{\rm W3}$ is the error of $M_{\rm W3}$ estimated by taking into account the parallax error ($\sigma\Pi_{\rm VLBI}$) and the WISE measurement error. The pulsation period P and its logarithm $\log P$ are also given. The $M_{\rm W3}$ values of the latest VLBI observation sources AW Tau, IRC−30363, and OH39.7+1.5 were found to be -10.71 ± 0.21, -12.35 ± 1.21, and -12.90 ± 0.40, respectively.

In Figure 8, we present the pulsation periods and the W3-band absolute magnitudes ($M_{\rm W3}$) of the sources in Table 4. The Mira-type and SR variables are presented with open circles. The OH/IR stars are represented by filled circles. First, all 37 sources were used to derive the period-$M_{\rm W3}$ relation. The derived relation is

$$M_{\rm W3} = (-7.21 \pm 1.18) \log P + (9.25 \pm 3.09), \tag{10}$$

which is indicated by a solid line in the figure. We then derived another relationship using only the 9 OH/IR stars, represented by filled circles, and obtained a relation of the form

$$M_{\rm W3} = (-4.20 \pm 1.33) \log P + (0.01 \pm 3.89), \tag{11}$$

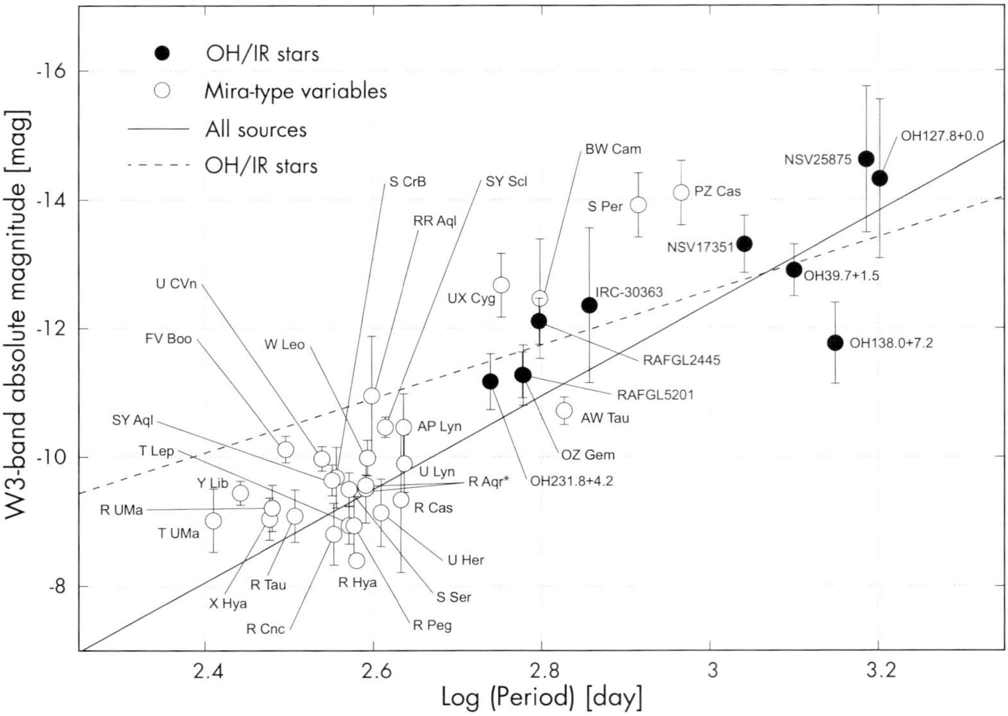

Figure 8. Absolute magnitudes of the WISE W3 band estimated from VLBI parallaxes. Filled circles and open circles represent OH/IR stars and Mira-type variables respectively. Two lines represent period-M_{W3} relations in the mid-infrared. The solid line and dotted lines represent the relations of $M_{W3} = (-7.21 \pm 1.18) \log P + (9.25 \pm 3.09)$ and $M_{W3} = (-4.20 \pm 1.33) \log P + (0.01 \pm 3.89)$, respectively.

which is shown with a dotted line. At this stage, it is difficult to say that there is an obvious period-M_{W3} relation. However, if a clear relationship is confirmed, it can be used as a new distance estimator for the AGB sources along the Galactic plane, or for the sources deeply obscured by circumstellar dust. It is known that there is a deep silicate absorption feature at $\lambda \simeq 8$ to 18 μm in the spectral energy distribution of AGB stars. So, ideally, it would be necessary to use bolometric absolute magnitudes $M_{\rm bol}$ to avoid the effects of absorption. We have tried to do this, but the variance of the $M_{\rm bol}$ values is greater than that in the W3 band, and so far our attempt has not been successful. For a more detailed investigation of the period-M_{W3} relation, we are continuing astrometric VLBI observations of OH/IR stars to increase the number of sources covering a wide period range.

4. Summary

We have been observing Mira-type variables and OH/IR stars with the astrometric VLBI method. The obtained distances and stellar parameters help us to understand the evolutionary relationship of the subclasses in the AGB phases. The VERA array was used for all VLBI observations in our study. The masers used are the H_2O one at 22 GHz and the SiO one at 43 GHz. The phase referencing technique was used to accurately measure the distances of stars ranging from a few hundred pc to several kpc. Due to the properties of the circumstellar matter and its time variability, the parallax measurements of stars surrounded by thick dust shells in optical bands is sometimes very difficult. This can sometimes make parallax measurements with *Gaia* difficult. The VLBI and *Gaia* are complementary for distance measurements of AGB stars.

The results for NSV17351, AW Tau, IRC−30363, and OH39.7+1.5 are presented. The pulsation period of NSV17351 was determined from the time variability of the H_2O maser. Parallax, circumstellar distribution of the masers and kinematics of NSV17351 were also presented. Absolute magnitudes in the near- and mid-infrared bands of OH/IR stars with very long pulsation periods were studied. Using the WISE W3 band data, a period-magnitude relation in the WISE W3 band $M_{W3} = (-7.21 \pm 1.18) \log P + (9.25 \pm 3.09)$ was found for the Galactic AGB stars. For further understanding, we need more detailed measurements of the circumstellar masers of AGB stars of different types, pulsation periods, and masses.

Discussion

Question (Whitelock): These VLBI measurements are very important for deriving distances for the stars that *Gaia* can't reach. Most of their luminosity will be in the dust and the amplitudes, even at long wavelengths, are very large, several magnitudes. You can only observe the O-rich AGB stars with H_2O and SiO Masers. Metal-weak stars are C-rich and will not have these lines, so I wonder if you can do anything with the CO line from the C-rich stars.

Answer: Thank you for your comment, Patricia. One thing that I wanted to share at this conference was the importance of VLBI measurements of the parallaxes of dusty AGB stars, and I think we did that to some extent. The CO lines are usually observed in dusty AGB stars. The existence of CO masers has also been reported by Vlemmings *et al.* (2021). However, at present I think that astrometric VLBI observations of the CO maser are difficult due to the accuracy of the absolute position and the sensitivity. So I do not expect to be able to do the same studies with the CO masers any time soon. Regarding the CO line, if we can somehow select OH/IR stars with very long pulsation periods, they could be massive and young, and they are expected to show CO line emission. Since observing the CO line can give a better estimate of radial velocities than maser emissions, we can estimate their distances using the kinematic distance method and use them for the PL relation or to study the Galactic dynamics.

Question (Jiang): It's very nice to see that the maser observation can be more accurate than *Gaia* for the nearby dust-obscured stars. I just wonder how the other astrometric parameters, proper motions, measured using maser compare to *Gaia*?

Answer: Thank you for your comment, Biwei. VLBI measures the sky plane motion of each maser spot, which is distributed around the central AGB star. The maser spots show outward motion with respect to the central star. Thus, the proper motions of each maser spot obtained from VLBI observations are the sum of (1) the systemic proper motion of the central star and (2) the outward motion of the maser with respect to the central star. We average all the proper motions of the detected maser spots to estimate the systemic proper motion of the central star, which *Gaia* can measure directly. The difference in proper motions between *Gaia* and VLBI gives us information about the outward motion of the maser spots with respect to the central star. In Nakagawa *et al.* (2014), we presented the procedure to derive the systemic proper motion of the star from our VLBI measurements.

References

Asaki, Y., Deguchi, S., Imai, H., *et al.* 2010, *ApJ*, 721, 267
Baba, J., Saitoh, T. R., & Wada, K. 2013, *ApJ*, 763, 46
Baba, J. 2015, *MNRAS*, 454, 2954
Burns, R. A., Handa, T., Nagayama, T., *et al.* 2016, *MNRAS*, 460, 283

Chiavassa, A., Freytag, B., & Schultheis, M. 2018, *A&A*, 617, L1
Chibueze, J. O., Omodaka, T., Urago, R., *et al.* 2019, *PASJ*, 71, 92
Choi, Y. K., Hirota, T., Honma, M., *et al.* 2008, *PASJ*, 60, 1007
Cox, J. P. 1980, *Theory of Stellar Pulsation* (Princeton: Princeton University Press)
Engels, D., Etoka, S., Gérard, E., *et al.* 2015, in: F. Kerschbaum, R. F. Wing, & J. Hron (eds.), *Why Galaxies Care about AGB Stars III: A Closer Look in Space and Time*, ASP Conf. Ser., 497, p. 473
Etoka, S., Engels, D., Gérard, E., *et al.* 2018, in: A. Tarchi, M. J. Reid, & P. Castangia (eds.), *Astrophysical Masers: Unlocking the Mysteries of the Universe*, Proc. IAU Symp. No. 336 (Cambridge: Cambridge University Press), p. 381
Feast, M. W. 2009, in: T. Ueta, N. Matsunaga, & Y. Ita. (eds.), *AGB Stars and Related Phenomena* (Tokyo: NAOJ), p. 48
Fomalont, E. 1981, *NRAO Newsletter*, 3, 3
Gaia Collaboration, Vallenari, A., Brown, A. G. A., *et al.* 2023, *A&A*, 674, A1
Goldman, S. R., van Loon, J. T., Zijlstra, A. A., *et al.* 2017, *MNRAS*, 465, 403
Habing, H. J. 1996, *A&AR*, 7, 97
Höfner, S. & Olofsson, H. 2018, *A&AR*, 26, 1
Honma, M., Kijima, M., Suda, H., *et al.* 2008, *PASJ*, 60, 935
Imai, H., Sakai, N., Nakanishi, H., *et al.* 2012, *PASJ*, 64, 142
Ita, Y., Tanabé, T., Matsunaga, N., *et al.* 2004, *MNRAS*, 347, 720
Justtanont, K., Barlow, M. J., Blommaert, J., *et al.* 2015, *A&A*, 578, A115
Kamezaki, T., Nakagawa, A., Omodaka, T., *et al.* 2012, *PASJ*, 64, 7
Kamezaki, T., Kurayama, T., Nakagawa, A., *et al.* 2014, *PASJ*, 66, 107
Kamezaki, T., Nakagawa, A., Omodaka, T., *et al.* 2016, *PASJ*, 68, 71
Kamezaki, T., Nakagawa, A., Omodaka, T., *et al.* 2016, *PASJ*, 68, 75
Kamizuka, T., Nakada, Y., Yanagisawa, K., *et al.* 2020, *ApJ*, 897, 42
Kamohara, R., Bujarrabal, V., Honma, M., *et al.* 2010, *A&A*, 510, A69
Karakas, A. I. & Lattanzio, J. C. 2014, *PASA*, 31, e030
Karambelkar, V. R., Adams, S. M., Whitelock, P. A., *et al.* 2019, *ApJ*, 877, 110
Kawaguchi, N., Sasao, T., & Manabe, S. 2000, *Proc. SPIE*, 4015, 544
Kemper, F., de Koter, A., Waters, L. B. F. M., *et al.* 2002, *A&A*, 384, 585
Kurayama, T., Sasao, T., & Kobayashi, H. 2005, *ApJ*, 627, L49
Kusuno, K., Asaki, Y., Imai, H., & Oyama, T. 2013, *ApJ*, 774, 107
Le Squeren, A. M., Baudry, A., Brillet, J., & Darchy, B. 1979, *A&A*, 72, 39
Lin, C. C., & Shu, F. H. 1964, *ApJ*, 140, 646
te Lintel Hekkert, P., Caswell, J. L., Habing, H. J., *et al.* 1991, *A&AS*, 90, 327
Matsuno, M., Nakagawa, A., Morita, A., *et al.* 2020, *PASJ*, 72, 56
Min, C., Matsumoto, N., Kim, M. K., *et al.* 2014, *PASJ*, 66, 38
Motogi, K., Sorai, K., Honma, M., *et al.* 2016, *PASJ*, 68, 69
Nakagawa, A., Tsushima, M., Ando, K., *et al.* 2008, *PASJ*, 60, 1013
Nakagawa, A., Omodaka, T., Handa, T., *et al.* 2014, *PASJ*, 66, 101
Nakagawa, A., Kurayama, T., Matsui, M., *et al.* 2016, *PASJ*, 68, 78
Nakagawa, A., Kurayama, T., Orosz, G., *et al.* 2018, in: A. Tarchi, M. J. Reid, & P. Castangia (eds.), *Astrophysical Masers: Unlocking the Mysteries of the Universe*, Proc. IAU Symp. No. 336 (Cambridge: Cambridge University Press), p. 365
Nakagawa, A., Kurayama, T., Orosz, G., *et al.* 2019, in: F. Kerschbaum, M. Groenewegen, & H. Olofsson (eds.), *Why Galaxies Care about AGB Stars: A Continuing Challenge through Cosmic Time*, Proc. IAU Symp. No. 343 (Cambridge: Cambridge University Press), p. 476
Nakagawa, A., Morita, A., Sakai, N., *et al.* 2023, *PASJ*, 75, 529
Nakagawa, A., Kurayama, T., Sudou, H., Orosz, G. 2023, *This work*
Nikzat, F., Ferreira Lopes, C. E., Catelan, M., *et al.* 2022, *A&A*, 660, A35
Nyman, L.-A., Hall, P. J., & Olofsson, H. 1998, *A&AS*, 127, 185
Nyu, D., Nakagawa, A., Matsui, M., *et al.* 2011, *PASJ*, 63, 63

Orosz, G., Imai, H., Dodson, R., *et al.* 2017, *AJ*, 153, 119
Oyama, T., Kono, Y., Suzuki, S., *et al.* 2016, *PASJ*, 68, 105
Reid, M. J., Menten, K. M., Brunthaler, A., *et al.* 2019, *ApJ*, 885, 131
Sellwood, J. A., & Carlberg, R. G. 1984, *ApJ*, 282, 61
Sparke, L. S., & Gallagher, J. S. 2000, *Galaxies in the Universe* (Cambridge: Cambridge University Press)
Sudou, H., Omodaka, T., Murakami, K., *et al.* 2019, *PASJ*, 71, 16
Takeuti, M., Nakagawa, A., Kurayama, T., *et al.* 2013, *PASJ*, 65, 60
Thompson, A. R., Moran, J. M., & Swenson, G. W. 2001, *Interferometry and Synthesis in Radio Astronomy*, 2nd Edition (New York: Wiley)
Urago, R., Yamaguchi, R., Omodaka, T., *et al.* 2020, *PASJ*, 72, 57
VERA Collaboration, Hirota, T., Nagayama, T., *et al.* 2020, *PASJ*, 72, 50
Vlemmings, W. H. T., van Langevelde, H. J., Diamond, P. J., Habing, H. J., & Schilizzi, R. T. 2003, *A&A*, 407, 213
Vlemmings, W. H. T., & van Langevelde, H. J. 2007, *A&A*, 472, 547
Vlemmings, W. H. T., Khouri, T., & Tafoya, D. 2021, *A&A*, 654, A18
Watson, C. L., Henden, A. A., & Price, A. 2006, in: B. D. Warner, J. Foote, D. A. Kenyon, & D. Mais (eds.), *Symposium on Telescope Science*, Proc. 25th Ann. Conf. of the Soc. for Astron. Sci. (Big Bear Lake: SAS), p. 47
Wood, P. R., Alcock, C., Allsman, R. A., *et al.* 1999, in: T. Le Bertre, A. Lebre, & C. Waelkens (eds.), *Asymptotic Giant Branch Stars*, Proc. IAU Symp. No. 191 (San Francisco: ASP), p. 151
Wright, E. L., Eisenhardt, P. R. M., Mainzer, A. K., *et al.* 2010, *AJ*, 140, 1868
Xu, S., Zhang, B., Reid, M. J., *et al.* 2018, *ApJ*, 859, 14
Zhang, B., Reid, M. J., Menten, K. M., Zheng, X. W., & Brunthaler, A. 2012, *A&A*, 544, A42
Zhang, B., Zheng, X., Reid, M. J., *et al.* 2017, *ApJ*, 849, 99

At the crossroads of astrophysics and cosmology: Period–luminosity relations in the 2020s
Proceedings IAU Symposium No. 376, 2024
R. de Grijs, P. A. Whitelock & M. Catelan, eds.
doi:10.1017/S1743921323003083

The structure of the Milky Way from period-luminosity relations

Dorota M. Skowron

Astronomical Observatory, University of Warsaw, Aleje Ujazdowskie 4, 00-478 Warsaw, Poland. email: dszczyg@astrouw.edu.pl

Abstract. Disentangling the structural components of the Milky Way requires knowledge of distances to various classes of objects, both young, which trace the Galactic disk, and old, which trace the Galactic bulge and halo. Variable stars that obey period-luminosity relations are perfect distance indicators for such studies. Here we discuss recent findings on the structure of our galaxy, inferred from period-luminosity relations for both young, old and intermediate-age variable stars, including Cepheids, RR Lyrae stars, δ Scuti stars, and long-period variables.

Keywords. Classical Cepheids, RR Lyrae, Delta Scuti, Miras, Long Period Variables, Milky Way Structure

1. Introduction

All of us astronomers would like to know what our own galaxy, the Milky Way, looks like. And we often see the artist impressions of the Milky Way disk showing the bright central overdensity and spiral arms full of star formation and prominent dust lanes. These are of course based on real data and models, and backed up by photographs of other galaxies in the Universe, but are nevertheless only schematic visualizations of what our galaxy might look like.

However, we agree by now, that we live in a spiral galaxy, with three main components: the bulge, the disk, and the halo. And in order to be able to constrain the structure of the Milky Way, we need to have tracers in each of these components, to which we can calculate reliable distances that will make it possible to investigate their three-dimensional distribution and thus the shape of our galaxy.

Fortunately, there are several groups of pulsating variable stars in the classical instability strips on the Hertzsprung-Russel diagram (see, e.g., Figure 1 in Kurtz 2022), both old and young, that can serve as tracers of the Galactic structure. The oldest group, i.e. the RR Lyrae stars, are ideal for learning about the bulge and the halo of the Galaxy. The youngest, i.e. the Classical Cepheids, have been successfully used to study the structure of the Milky Way disk. And finally, the intermediate group, including Type II Cepheids, δ Scuti, and Mira-type stars (which are in fact a mixture of the old, young, and intermediate populations), located all-sky, can help us understand each of the components of our galaxy.

Figure 1 shows the all-in-one period-luminosity diagram, or the Leavitt law, for all known classes of pulsating variable stars in the Large Magellanic Cloud (courtesy of Igor Soszyński), observed by the Optical Gravitational Lensing Experiment (OGLE) survey (Udalski *et al.* 2015). The fact that all these stars are located roughly at the same distance and that the reddening-free Wesenheit index was used as a measure of brightness allows us to construct such tight period-luminosity relations.

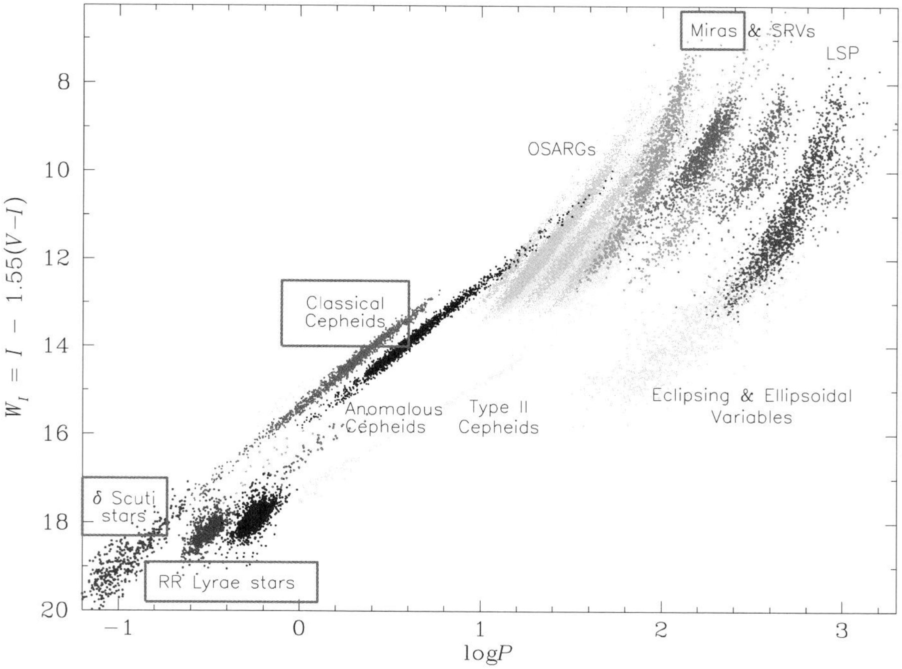

Figure 1. All-in-one period-Wesenheit relation for pulsating variable stars discovered by the OGLE survey in the Large Magellanic Cloud. The period P is in days and the Wesenheit index W_I is in magnitudes. Red frames mark variables that have been used to study the structure of the Milky Way galaxy. The top right part of the plot is occupied by long-period variable stars which are further subdivided into Miras, semi-regular variables (SRVs), OGLE small-amplitude red giants (OSARGs), long secondary period variables (LSP), eclipsing, and ellipsoidal variables. The plot was provided by Igor Soszyński.

In the case of our galaxy, stars are at different distances from us, so constructing period-luminosity relations is possible only for a subset of objects with distances known from other methods. Then, after calibrating the period-luminosity relation, it can be used to calculate distances to other stars of the same variable type, and as a result, to study the Galactic structure populated by this group of stars.

So far, four groups of pulsating variable stars in the Milky Was have been used to study its structure, and these groups are marked with red frames in Figure 1. They include RR Lyrae-type stars, Classical Cepheids, δ Scuti stars, and Miras.

2. Pulsating variable stars in numbers

During this Symposium Igor Soszyński told us about the revolution in the variable star research by the OGLE survey, which started in the very beginning of this century and continues until now, with over 1,088,000 variable stars classified so far. Gisella Clementini spoke about another revolution, by the *Gaia* mission, which constitutes over 9,976,000 variable stars released in the Data Release 3 (DR3), of which the majority are new discoveries. These numbers are overwhelming and provide an excellent database for numerous astrophysical studies. And while the majority of discoveries had been made by *Gaia* and OGLE, there are of course other important surveys that significantly contributed to this research area by providing data for the four variable star groups that have been used to study the Galactic structure. In an attempt to gather all these contributions in one place, we reference appropriate catalogs below, although the list is most certainly not complete, for which we apologize the reader.

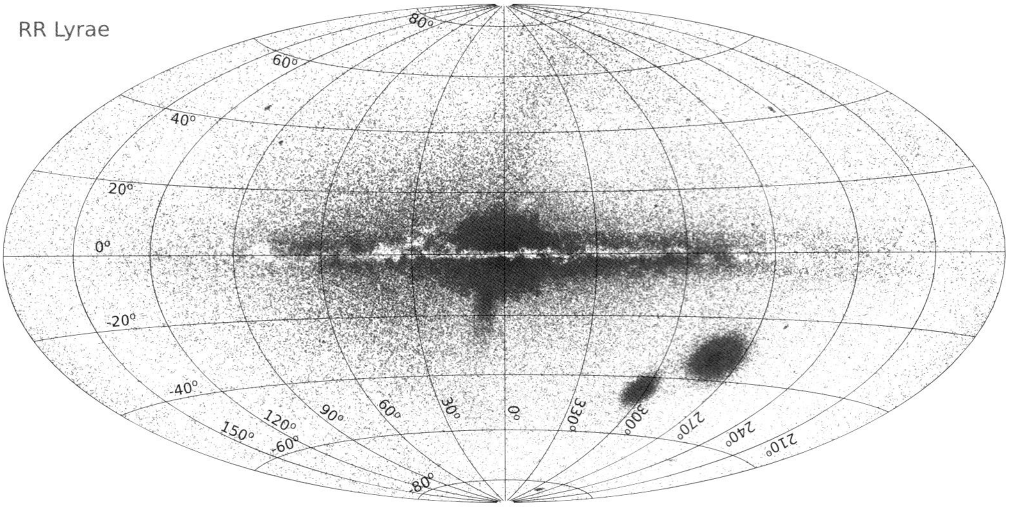

Figure 2. The all-sky distribution of RR Lyrae-type variable stars in Galactic coordinates.

2.1. *RR Lyrae-type stars*

The current number of all-sky RR Lyrae stars is around ~271,000, of which about ~223,900 are members of the Milky Way and the remaining ones belong mostly to the Magellanic Clouds. The main contributions to the RR Lyrae sample in our galaxy are from OGLE (Soszyński *et al.* 2011a, 2014, 2019) and *Gaia* (Clementini *et al.* 2023), but also from the All Sky Automated Survey (ASAS; Pojmanski 2002, 2003; Pojmanski & Maciejewski 2004, 2005; Pojmanski, Pilecki, & Szczygiel 2005), the All Sky Automated Survey for SuperNovae (ASAS-SN; Jayasinghe *et al.* 2019), the Catalina Sky Survey (CSS; Drake *et al.* 2013, 2014; Torrealba *et al.* 2015; Drake *et al.* 2017), the Lincoln Near-Earth Asteroid Research project (LINEAR; Palaversa *et al.* 2013; Sesar *et al.* 2013), the Panoramic Survey Telescope and Rapid Response System (Pan-STARRS1; Sesar et al. 2017), the Palomar Transient Factory (PTF; Sesar *et al.* 2014), the VISTA Variables in the Via Lactea survey (VVV; Contreras Ramos *et al.* 2018; Dékány *et al.* 2018), and the Zwicky Transient Facility (ZTF; Chen *et al.* 2020), as well as many other sources compiled in the General Catalog of Variable Stars (GCVS; Samus' *et al.* 2017).

Figure 2 shows the on-sky distribution of RR Lyrae-type variables from these catalogs.

2.2. *Classical Cepheids*

The current number of known Classical Cepheids in the Milky Way is 3,666, with the main contributions from OGLE (Pietrukowicz *et al.* 2013; Soszyński *et al.* 2017; Udalski *et al.* 2018; Pietrukowicz, Soszyński, & Udalski 2021) and from other surveys, as pictured in Figure 3, including ASAS (Pojmanski 2002, 2003; Pojmanski & Maciejewski 2004, 2005; Pojmanski, Pilecki, & Szczygiel 2005), ASAS-SN (Jayasinghe *et al.* 2019), the Asteroid Terrestrial-impact Last Alert System (ATLAS; Heinze *et al.* 2018), *Gaia* (Ripepi *et al.* 2023), and ZTF (Chen *et al.* 2020), as well as many other sources compiled in the GCVS (Samus' *et al.* 2017).

There are a number of Cepheids discovered by infrared (IR) surveys, e.g., VVV (Dékány *et al.* 2019), but here only optically confirmed Cepheids are included, because nearly sinusoidal light curves of Cepheids in the IR makes their classification unambiguous. Figure 4 shows the on-sky distribution of Classical Cepheids from these catalogs.

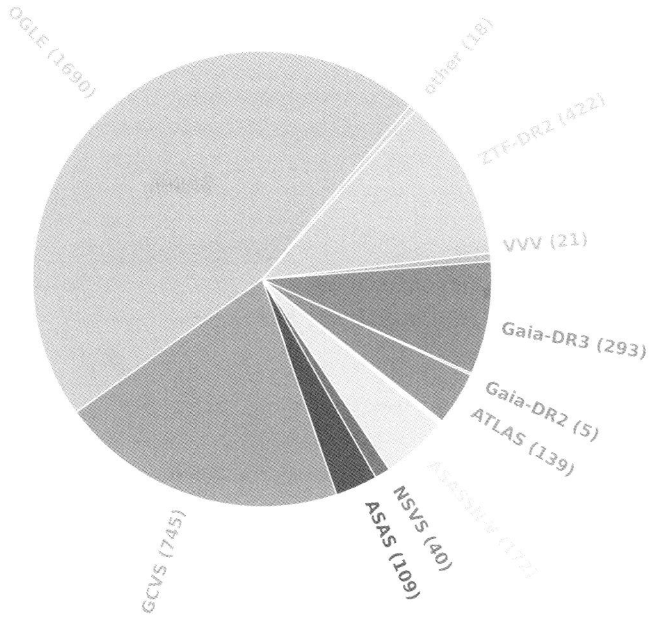

Figure 3. The number of Classical Cepheids by discovery survey and GCVS.

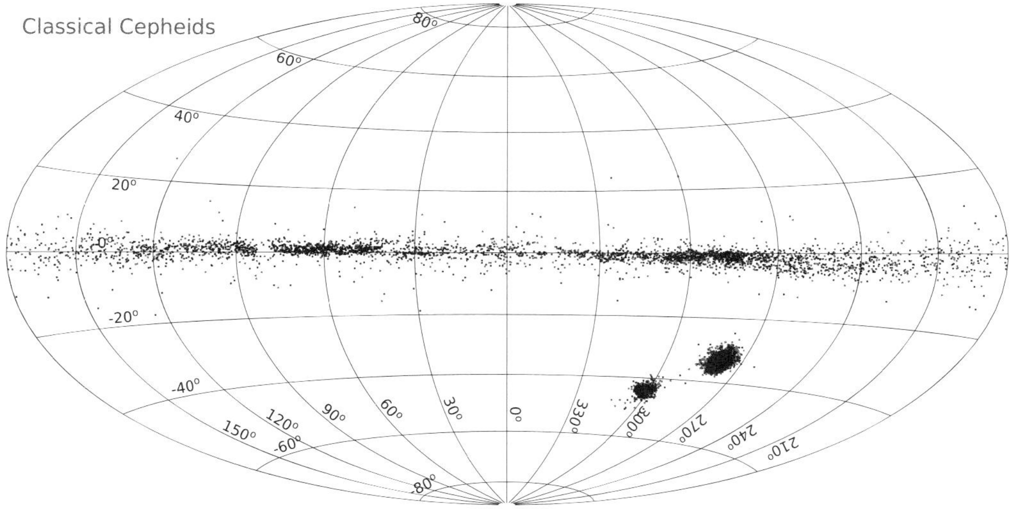

Figure 4. The all-sky distribution of Classical Cepheids in Galactic coordinates.

The up-to-date list of all genuine Galactic Classical Cepheids† is maintained by Paweł Pietrukowicz (Pietrukowicz, Soszyński, & Udalski 2021) and is being regularly updated as new discoveries are made.

2.3. *Mira-type stars*

Catalogs of Miras in the Milky Way are not as complete as in the case of RR Lyrae stars or Classical Cepheids. Nevertheless, recent years have seen much progress

† `http://www.astrouw.edu.pl/ogle/ogle4/OCVS/allGalCep.listID`

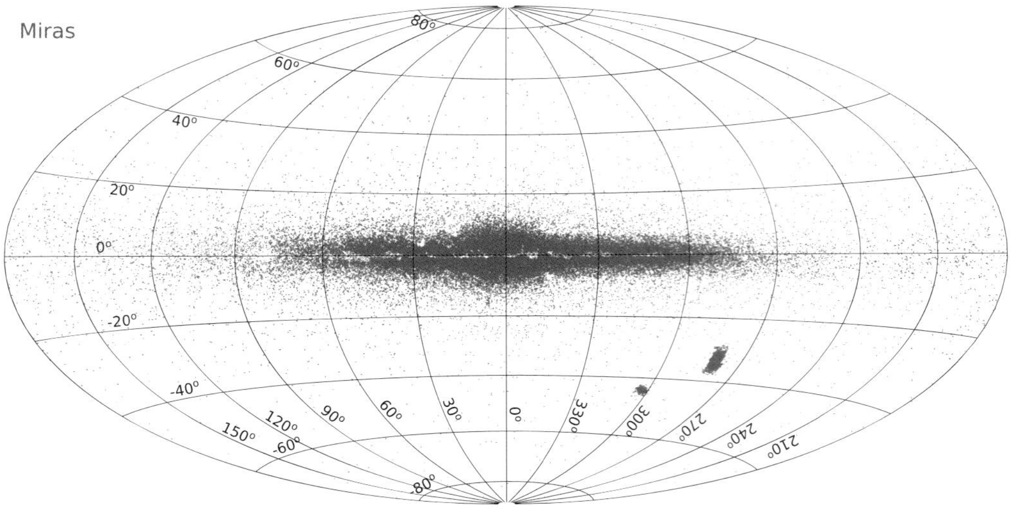

Figure 5. The all-sky distribution of Mira-type variable stars in Galactic coordinates.

and there are currently at least ~76,000 known Galactic Mira type stars. It is difficult to establish a more precise number, because some surveys only provide a "long-period variable" classification, without further subdivision into Miras, SRVs, OSARGs, etc. Catalogs of Miras have been published by OGLE (Soszyński *et al.* 2013; Iwanek *et al.* 2022), ASAS (Pojmanski 2002, 2003; Pojmanski & Maciejewski 2004, 2005; Pojmanski, Pilecki, & Szczygiel 2005), ASAS-SN (Jayasinghe *et al.* 2019), ATLAS (Heinze *et al.* 2018), CSS (Drake *et al.* 2014, 2017), *Gaia* (Lebzelter *et al.* 2023), VVV (Nikzat *et al.* 2022; Sanders *et al.* 2022), and ZTF (Chen *et al.* 2020), as well as many other sources compiled in GCVS (Samus' *et al.* 2017).

Figure 5 shows the on-sky distribution of Mira-type variables from these catalogs.

2.4. *δ Scuti-type stars*

There were 636 known δ Scuti stars two decades ago (see compilation by Rodríguez, López-González, & López de Coca 2000). This number has greatly increased and we now know over 33,000 δ Scuti stars in our galaxy, the majority of which were discovered by the OGLE survey (Pietrukowicz *et al.* 2020a, 2022; Soszyński *et al.* 2021) but also by ASAS-SN (Jayasinghe *et al.* 2019) and ZTF (Chen *et al.* 2020).

Figure 6 shows the on-sky distribution of δ Scuti-type variables from these catalogs.

3. The Milky Way structure

There are three main components of our galaxy: the bulge, the disk, and the halo. Vasily Belokurov gave an informative overview of the structure of the halo, so the focus here will be put on the remaining parts of the Milky Way.

3.1. *The bulge*

The bulge is composed mainly of old and intermediate-age stars, and thus most of the structural studies of the bulge, using pulsating variable stars, have been made with the use of RR Lyrae, Miras, and recently also δ Scuti-type stars.

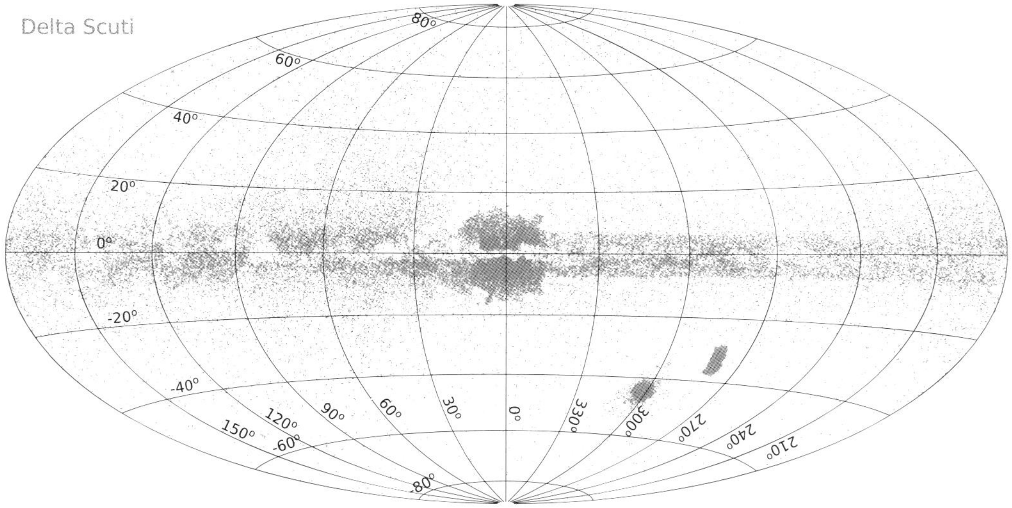

Figure 6. The all-sky distribution of δ Scuti-type variable stars in Galactic coordinates.

3.1.1. The bulge in RR Lyrae

The first study that used a large sample of bulge RR Lyrae stars was performed by Pietrukowicz *et al.* (2015) with over 27,000 objects from the OGLE-III survey. By modelling the RR Lyrae on-sky distribution and density profile, the authors found that the old bulge has the shape of a triaxial ellipsoid (see their Figures 6 and 8) with a major axis inclined to the line of sight by 20° within the Galactic plane. However, the X-shaped structure reported by studies of red clump giants in the bulge (McWilliam & Zoccali 2010) was not observed.

With the release of over 78,000 RR Lyrae in the bulge and disk area from OGLE-IV, Pietrukowicz *et al.* (2020b) investigated the properties of the old populations of the Milky Way's bulge, disk, and halo using both period-luminosity relations and photometric metallicity estimations. Their Figure 2 shows the distribution of RR Lyrae in these three components of our galaxy, while their Figure 3 presents metallicity distributions for each component, with the distribution maxima at $[Fe/H] = -1.2$ dex, -1.0 dex, and -0.6 dex for the halo, bulge, and disk, respectively. From the shape of the metallicity distributions, the authors conclude that halo stars are present deep in the bulge region.

Further analysis shows that about 25% of bulge RR Lyrae belong to the halo population (see Figure 4 of Pietrukowicz *et al.* 2020b), which is in excellent agreement with spectroscopy-based results by Kunder *et al.* (2020) based on the BRAVA-RR data, who found that 25% of bulge RR Lyrae are halo interlopers (see their Figure 6), but is slightly more than the 9% reported by Prudil *et al.* (2019) based on a kinematical study of bulge RR Lyrae. Kunder *et al.* (2020) also showed that, once the halo interlopers are removed from the bulge sample, we can clearly see two populations of RR Lyrae that are spatially and kinematically different. The existence of two bulge populations was also found by Pietrukowicz *et al.* (2020b) – their analysis of the period-amplitude diagram of bulge RR Lyrae revealed two sequences with likely different metallicities (see their Figures 6–8). The presence of multiple old populations in the Milky Way bulge was also observed in IR data (Dékány *et al.* 2018) and suggests that the bulge has been formed through merger events.

3.1.2. The bulge in Miras and δ Scuti

The old and intermediate-age bulge populations are represented by both Mira-type stars and δ Scuti-type stars, which have recently been used to study the Galactic bulge region.

A study of a clean sample of almost 7,500 δ Scuti stars from the OGLE catalogs is the first study of such a large sample of these variables in the bulge region (Deka, Deb, & Kurbah 2022). The authors determine the geometrical parameters of the bulge with the use of a fully Bayesian Markov Chain Monte Carlo analysis, and find that the δ Scuti distribution has the form of a triaxial ellipsoid (no X-shape morphology), where a bar-like structure is inclined at the angle of 22° to the line of sight, which is similar to results obtained by Pietrukowicz *et al.* (2015) for RR Lyrae stars.

The picture is not as consistent in the case of Mira-type variables. Both López-Corredoira (2017) and Chrobáková, López-Corredoira, & Garzón (2022) have shown that the population traced by these stars (with an average age estimated at ~ 9 Gyr) can be described as a boxy bulge, without the presence of an X-shaped structure, which is consistent with the results presented by Pietrukowicz *et al.* (2015) for RR Lyrae stars and Dékány *et al.* (2019) for Type II Cepheids. However, Catchpole *et al.* (2016) showed that the the bulge structure is a function of age, such that the younger Miras (age $\lesssim 5$ Gyr) follow a bar-like structure inclined to the line of sight, which may be part of an X-shaped structure, while the older ones follow a more spheroidal distribution (see their Figure 11).

The dependence of bulge morphology on age and metallicity has also been recently investigated by Grady, Belokurov, & Evans (2020), who found that the older Miras ($\sim 9 - 10$ Gyr) show little bar-like morphology, while the younger ones do (with a possible X-shaped structure), with the inclination angle of $\sim 21^\circ$ being similar to those of RR Lyrae, δ Scuti, and Type II Cepheids.

The most recent study of the largest by far sample of $\sim$66,000 Miras in the Milky Way bulge and disk was done by Iwanek *et al.* (2023). The authors used a much more complex model containing three barred components, and found that the X-shaped structure is present in the bulge (see their Figures 4–6). The analysis, however, did not include the age dependence.

3.2. *The disk*

The disk of the Milky Way is composed of stars, gas, and dust, and is an environment where most of the star formation is happening. For this reason, the youngest stars that are born in the disk are excellent tracers of its structure, as they have not had enough time to move away from their birthplaces.

3.2.1. The disk in Classical Cepheids

Classical Cepheids are typically younger than $300 - 400$ Myr, very bright, and obey a period-luminosity relation, which makes them ideal to study the shape of the Galactic disk. Recent years have seen exponential growth in the number of known Galactic Classical Cepheids (Figure 7), and we can safely say that the majority of these stars in the Milky Way have been discovered by now. This has led to a number of publications that used Cepheids to study the disk morphology (Skowron *et al.* 2019a,b; Chen *et al.* 2019; Dékány *et al.* 2019; Minniti *et al.* 2021; Poggio *et al.* 2021; Lemasle *et al.* 2022; Gaia Collaboration *et al.* 2023).

All of these studies are similar in the sense that they are based on more or less the same set of stars, and use period-luminosity relations to determine Cepheid distances. At the same time they differ, because different period-luminosity and period-age relations

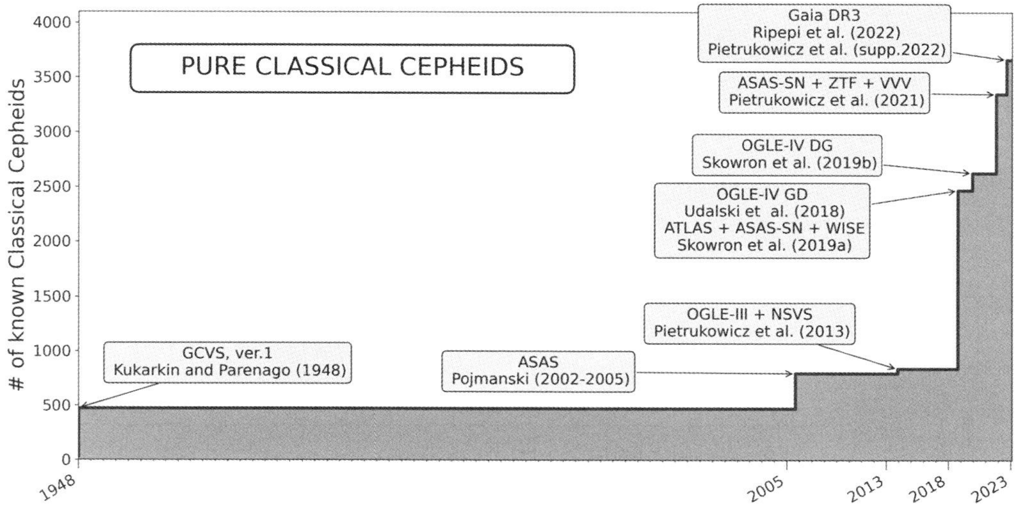

Figure 7. The change of the number of known Classical Cepheids in the Milky Way with time.

(to select the youngest Cepheids) are used in each study. For example, Skowron *et al.* (2019a,b) utilize mid-IR data from *Spitzer* and WISE for Cepheids and appropriate period-luminosity relations in the mid-IR, and correct for extinction with IR extinction maps; Gaia Collaboration *et al.* (2023) make use of period-Wesenheit-metallicity relations for *Gaia* bands and do not correct for extinction; while Lemasle *et al.* (2022) use period-Wesenheit relations in mid-IR WISE bands also without the extinction correction.

Thus, it is not surprising that Cepheid distances obtained in each study differ, which is illustrated in Figure 8, where the top panel shows how the distance difference between the study of Gaia Collaboration *et al.* (2023) and Skowron *et al.* (2019a,b) changes with Galactic longitude, while the bottom panel shows the same relation, but between the study of Lemasle *et al.* (2022) and Skowron *et al.* (2019a,b). Interestingly, the distance differences are largest toward the Galactic center, indicating that the treatment of extinction has a large impact on the final results. Using the reddening-free Wesenheit index, which by construction assumes a constant ratio of total-to-selective extinction, may not be appropriate in the highly extincted regions of the Milky Way, where the reddening law is not fully understood (Schlafly *et al.* 2016; Nataf *et al.* 2016). This is further supported by the fact that distances calculated with different methods are consistent toward the anti-center (around longitude 180°), where there is much less reddening.

Nevertheless, the distribution of Galactic Classical Cepheids has been successfully used to study the structure of the Milky Way disk. Figure 9 shows the top view of Cepheids in our galaxy based on the work of Skowron *et al.* (2019a,b). We see clear signs of spiral structure, although less evident in the outer disk, where the median age of Cepheids is larger (compare with Figure 3 in Skowron *et al.* 2019b). This is expected, because older Cepheids (150 – 250 Myr) have had enough time to move away from the spiral arm that they were born in, and as a result could be in the neighboring arm or the inter-arm region.

It has been long known that the Milky Way disk observed in neutral hydrogen is warped. Figure 10 (based on Skowron *et al.* 2019a) shows the the top view of the Galaxy in the middle (analogous to that in Figure 9), with individual Cepheids marked with colored dots. The disk is divided into 18 colored regions, and then the vertical distribution of Cepheids within each region is shown in 18 side panels with matching colors. The

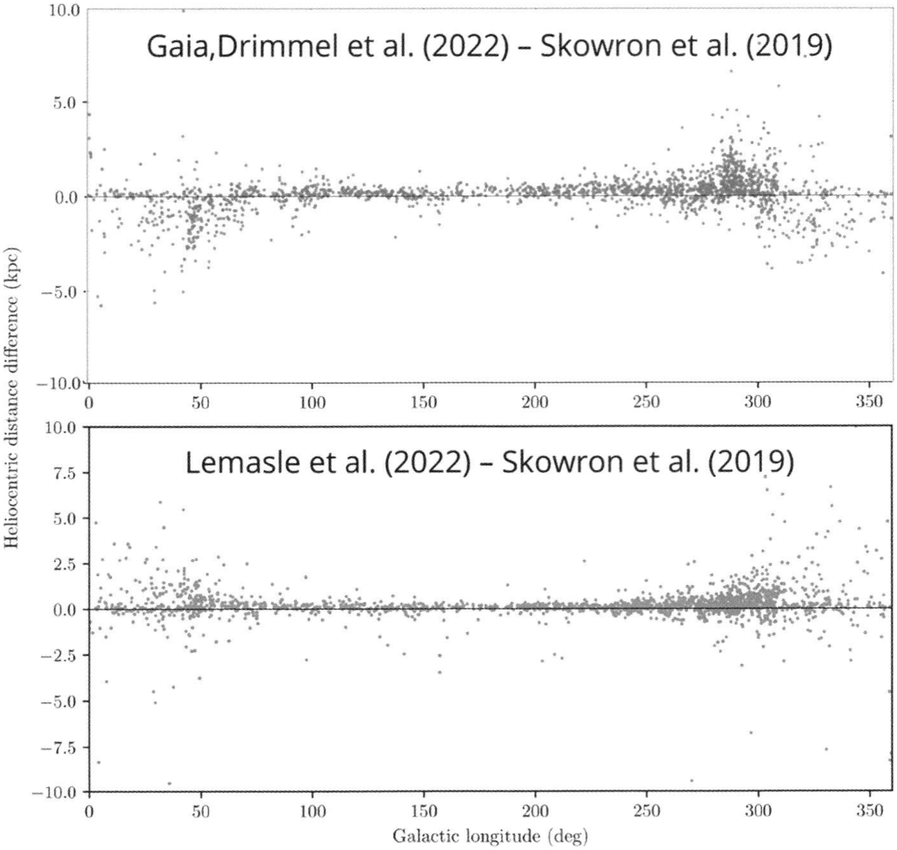

Figure 8. The difference between Classical Cepheid distances calculated with slightly different methods. Top: distance difference between the study of Gaia Collaboration *et al.* (2023) and Skowron *et al.* (2019a,b) vs. Galactic longitude. Bottom: distance difference between the study of Lemasle *et al.* (2022) and Skowron *et al.* (2019a,b) vs. Galactic longitude, provided by Bertrand Lemasle. The Galactic anti-center is at Galactic longitude 180°.

panels show the Galactocentric distance R plotted against distance from the Galactic plane Z, and the black lines show the intersection of the model surface fitted to the three-dimensional distribution of Cepheids. The agreement between the model and the observed distribution of Classical Cepheids is surprisingly good, even in the less populated sectors on the far side of the Galactic center, i.e., $140° < \phi < 160°$, $160° < \phi < 180°$, and $200° < \phi < 220°$.

Figure 10 also shows that the Milky Way disk is flared, that is the farther from the Galactic center, the thicker the disk is. The observed degree of flaring matches the one seen in neutral hydrogen.

There have been various attempts (Minniti *et al.* 2021; Poggio *et al.* 2021; Gaia Collaboration *et al.* 2023; Lemasle *et al.* 2022) to model the spiral structure of the Milky Way using the distribution of Classical Cepheids, or to match their distribution to existing spiral arm models based on neutral hydrogen (e.g. Levine, Blitz, & Heiles 2006) or high-mass star-forming regions (Reid *et al.* 2019). However, the results are not spectacular, most probably because Classical Cepheids are not good tracers of the spiral structure. Most importantly, they are not as young as high-mass star-forming regions, and so could have moved far enough from their birthplaces and currently reside in a

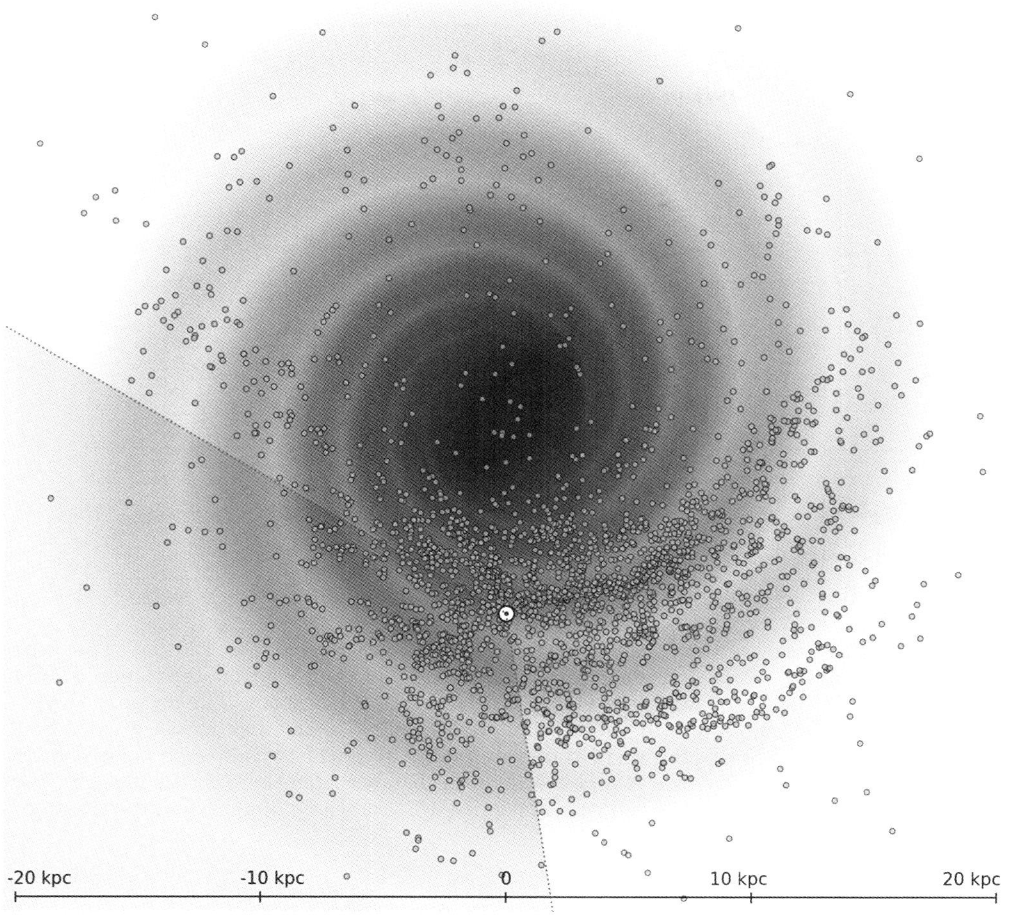

Figure 9. The top view of 2,390 Classical Cepheids in the Milky Way plotted over the spiral arm model consistent with the neutral hydrogen distribution (Levine, Blitz, & Heiles 2006). The image is based on data from Skowron *et al.* (2019a,b) and does not include Cepheids recently discovered from ASAS-SN, *Gaia*, and ZTF. The Sun is marked with a yellow dot. The gray area at the bottom left shows the region unavailable to the OGLE survey.

different arm or an inter-arm region (the rotational velocity of stars in the Galaxy differs from the speed of the spiral pattern, and depending on the distance from the Galactic center, stars are either faster or slower than the spiral pattern). Another uncertainty comes from the distance determination itself, as already shown in Figure 8, where distances determined by different groups differ by $2 - 3$ kpc in the highly extincted regions toward the Galactic center.

On the other hand, as we clearly see in Figure 9, Cepheids do form some overdensities which resemble the spiral arms. It is however important to realize that this does not have to be the same spiral structure as represented by the youngest tracers such as neutral hydrogen or high-mass star-forming regions. Lemasle *et al.* (2022) showed that Classical Cepheids form segments of the spiral arms rather than long spiral arms, and some segments are consistent with the spiral structure formed by the youngest tracers.

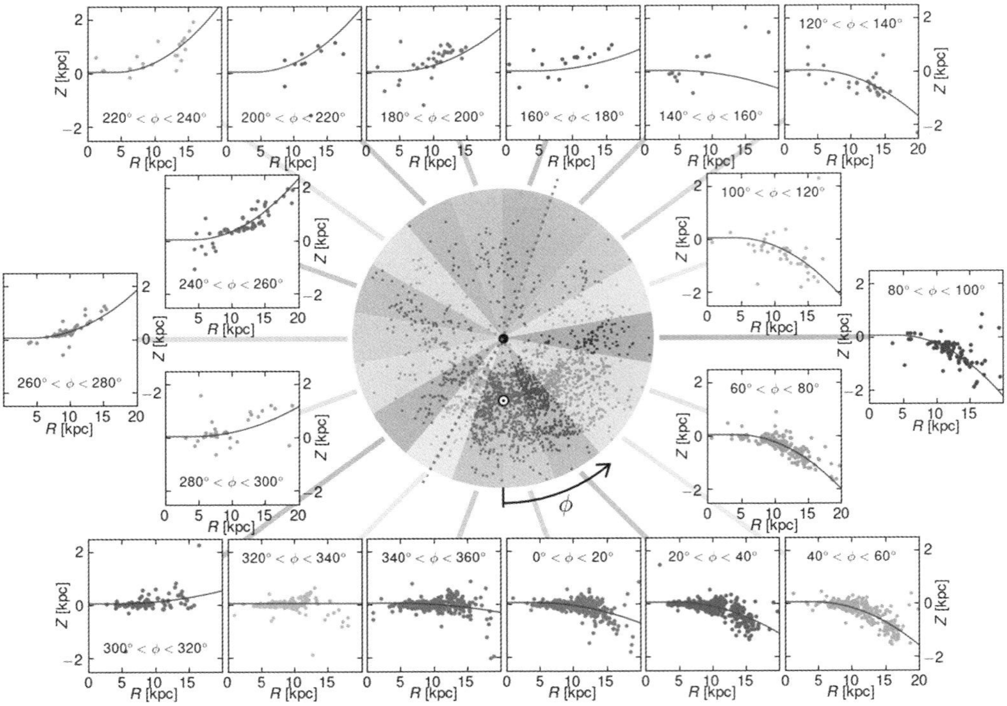

Figure 10. The model of the Galactic warp adapted from Skowron *et al.* (2019a). The central disk is divided into 18 colored regions in the Galactocentric polar coordinate system. The Sun is marked with a yellow dot while the Galactic center with a black dot. The azimuth $\phi = 0°$ is pointing toward the Sun, while the dotted line divides parts of the disk model warped South and North. The panels show the vertical distribution of Cepheids within each colored region, where the Galactocentric distance R is plotted against distance from the Galactic plane Z. Black lines show the intersection of the model surface fitted to the three-dimensional distribution of Cepheids.

Interestingly, the locations of these segments seem to have little dependence on the age cut used to select Cepheids (see their Figures 8, D1, D2, and D3).

When looking at Figure 9, we see a paucity of stars on the far side of the Galaxy. Cepheids residing in the area behind the Galactic center cannot be discovered by optical surveys, because of extremely high extinction, but can be found by IR surveys which see through the dust (Dékány *et al.* 2015, 2019; Minniti *et al.* 2020). The caveat is that IR light curves of Cepheids are less unique and often resemble other types of variable stars, making the unambiguous classification impossible. For example, Cepheids from the VVV IR data released by Dékány *et al.* (2019), that had counterparts in the OGLE optical data, were verified by Skowron *et al.* (2019a). It turned out that only about 55% were genuine Cepheids, whereas the remaining ones were eclipsing binaries or spotted stars.

However, in their recent study, Minniti *et al.* (2020) classified IR candidates for Classical Cepheids using high-quality IR spectra, from which they were able to derive radial velocities and metallicities, that allow to separate Classical Cepheids from other stars without relying solely on their IR light curves. Even though the sample had only 30 stars, it allowed Minniti *et al.* (2021) to investigate the structure of the far side of the Milky Way disk. The authors showed that there is no significant warping of the far side of the disk within 12 kpc from the Galactic center, and that their data are consistent with the warped structure observed by Skowron *et al.* (2019a,b).

3.2.2. The disk in Miras

Mira-type stars have been used as probes of Milky Way structure for over a century, but large catalogs of Miras became available only very recently (Iwanek *et al.* 2022). The first detailed three-dimensional map of the Galactic disk using Miras as tracers was presented by Iwanek *et al.* (2023), who showed that the disk seen in Miras is flared and possibly warped (see their Figure 7), although the latter is not as certain as in the case of Classical Cepheids, making it difficult to compare the degree of warping among different tracers. In their work on the Milky Way structure, Lemasle *et al.* (2022) showed that the degree of warping depends on the age of the tracer used, such that the older the tracer, the larger the degree of warping (see their Figure 3), meaning that the mean warp observed with Miras would be smaller that that of Classical Cepheids.

4. Summary

Period-luminosity relations are a powerful tool that can be used to study structures of galaxies, and in particular of our own Galaxy. So far, four groups of pulsating variable stars in the Milky Was have been successfully used to study the bulge, the disk, and the halo of the Milky Way: RR Lyrae-type stars, Classical Cepheids, δ Scuti stars, and Miras. These groups of stars are marked with red frames in Figure 1.

With the release of variable star catalogs from long-term surveys, it may seem that we have already discovered the majority of classical pulsators in the Milky Way, or at least, that the new discoveries will not significantly change the picture of our galaxy that had emerged from the analysis of the three-dimensional distribution of these variable stars. And while this may be true in the case of Classical Cepheids or RR Lyrae-type stars, there is still an unexplored wealth of long-period variables, among which only Miras have been partially examined so far. Beside Miras, long-period variables include OSARGs, SRVs, LSPs, eclipsing, and ellipsoidal variables (Wood *et al.* 1999; Soszyński *et al.* 2009, 2011b, 2013), all of which obey distinct period-luminosity relations (Figure 1), and may well serve as tracers of the Milky Way structure.

Acknowledgements. Funded by the European Union (ERC, LSP-MIST, 101040160). Views and opinions expressed are however those of the author only and do not necessarily reflect those of the European Union or the European Research Council. Neither the European Union nor the granting authority can be held responsible for them.

Discussion

Question (Ripepi): I'm the one that calculated the distance in the Gaia Collaboration *et al.* (2023) paper. Just wanted to ask how much larger would the difference in the Wesenheit index have to be to explain such a large difference in distances (Figure 8).

Answer: I didn't check this, but that's something that could be simulated. The differences in the Wesenheit index constant in the central regions of the Galaxy, as shown by Nataf *et al.* (2016), can vary from 1.5-1.6-1.9, so it would be good if someone checked how this translates to the distance difference.

Question (Eyer): I'm a bit surprised by your result about the Cepheid not tracing structures because we did a citizen science project where we asked people to classify objects and we have a projection of the Milky Way – the very nice picture done by people from NASA, it's very nice and I was finding always there is the Cepheid: you see that in the Galactic plane and you see it's near the arm, so that's the Cepheid.

Answer: Yes, that's the thing with Cepheids. I don't want to say that they don't trace the spiral structure – they do trace some structures. And thank you for this question, because now I can show you a movie†. This is the movie where we simulated star forming episodes in the Galaxy in different spiral arms at different moments; the younger stars here are blue and the older ones are red. In the end, the simulation is compared with observations. This is a toy model, where stars move according to the rotation curve and have random motions of about 8 km/s, and the spiral pattern moves with a fixed velocity. You will see stars forming in one arm, but as the galaxy rotates, some of them actually move to the neighboring arm just because of the different velocity of the spiral pattern and of the stars. Some end up in the inter-arm region. So the answer is that Cepheids do trace a structure but it's not necessarily consistent with the structure traced by younger traces, e.g. high-mass star-forming regions.

Question (Anderson): Did the preliminary comparison of Cepheid distances from different studies towards the bulge (Figure 8) consider information on color excess, e.g., $E(B-V)$, to test whether the increased scatter indeed can be traced to the inadequacies of the reddening law?

Answer: No, there is no sophisticated analysis here – just this simple comparison plot. But this is definitely something I will want to look into in more detail. We are at the point where we claim to have distances calculated with an accuracy better than 5% and we work on fine-tuning the period-luminosity relations. But in fact, the differences between distances calculated by different studies are much larger than that, which means that we really have to worry about other factors, for example about the reddening law.

Comment (Hajdu): About the extinction law implied by the Gaia Collaboration *et al.* (2023) results, the problem is that those are completely artificial and stem from incorrect zero point calibration in the VVV data that was done as a mistake by whoever did the calibration. It is shown both in my PhD thesis and in Hajdu *et al.* (2020). I told David Nataf that his results are completely artificial, and he agreed. You should not show this plot.

Answer: I wasn't aware of that, but maybe you're right.

Comment (Hajdu): It's a really unfortunate situation because it's not his fault that his results are incorrect.

Answer: It is definitely something to talk to David Nataf about, but it doesn't change what we see on the plot. So even if we acknowledge that David Nataf's work is wrong, it doesn't change the fact that there is a correlation with extinction.

Question (Riess): So the difference plot you were showing is all with near-infrared data (Figure 8)?

Answer: One comparison is between distances based on mid-infrared data and period-luminosity relations, in both studies. The other is between distances from a study based on mid-infrared and from a study based based on *Gaia* data and period-Wesenheit relations, so not infrared.

Comment (Riess): I don't understand how it could be a cause of large extinction. Or it could be extinction if both the results use either near- or mid-infrared.

Answer: Yes. So the blue plot is showing the comparison between studies where mid-infrared was used, but there was a different treatment of extinction. The red plot compares studies where one used mid-infrared and the other used *Gaia*. And I don't have an answer why.

Comment(Riess): OK, seems like more digging is required.

Answer: Yes, more checking is needed. I'm going over each step of distance calculations to check where we start getting these differences. Definitely not in the periods – I already checked those and they are the same in all studies.

† `http://www.astrouw.edu.pl/~jskowron/ogle/PR/milky_way_cepheids/cepheids_born_720.html`

References

Catchpole, R. M., Whitelock, P. A., Feast, M. W., Hughes, S. M. G., Irwin, M., & Alard, C. 2016, *MNRAS*, 455, 2216

Chen, X., Wang, S., Deng, L., de Grijs, R., Liu, C., & Tian, H. 2019, *New Astron.*, 3, 320

Chen, X., Wang, S., Deng, L., de Grijs, R., Yang, M., & Tian, H. 2020, *ApJS*, 249, 18

Chrobáková, Ž., López-Corredoira, M., & Garzón, F. 2022, *A&A*, 666, L13

Clementini, G., Ripepi, V., Garofalo, A., *et al.* 2023, *A&A*, 674, A18

Contreras Ramos, R., Minniti, D., Gran, F., *et al.* 2018, *ApJ*, 863, 79

Deka, M., Deb, S., & Kurbah, K. 2022, *MNRAS*, 514, 3984

Dékány, I., Hajdu, G., Grebel, E. K., & Catelan, M. 2019, *ApJ*, 883, 58

Dékány, I., Hajdu, G., Grebel, E. K., Catelan, M., Elorrieta, F., Eyheramendy, S., Majaess, D., & Jordán, A. 2018, *ApJ*, 857, 54

Dékány, I., Minniti, D., Majaess, D., *et al.* 2015, *ApJ*, 812, L29

Drake, A. J., Catelan, M., Djorgovski, S. G., *et al.* 2013, *ApJ*, 763, 32

Drake, A. J., Djorgovski, S. G., Catelan, M., *et al.* 2017, *MNRAS*, 469, 3688

Drake, A. J., Graham, M. J., Djorgovski, S. G., *et al.* 2014, *ApJS*, 213, 9

Gaia Collaboration, Drimmel, R., Romero-Gómez, M., *et al.* 2023, *A&A*, 674, A37

Grady, J., Belokurov, V., & Evans, N. W. 2020, *MNRAS*, 492, 3128

Hajdu, G., Dékány, I., Catelan, M., & Grebel, E. K. 2020, *Exp. Astron.*, 49, 217

Heinze, A. N., Tonry, J. L., Denneau, L., *et al.* 2018, *AJ*, 156, 241

Iwanek, P., Poleski, R., Kozłowski, S., *et al.* 2023, *ApJS*, 264, 20

Iwanek, P., Soszyński, I. Kozłowski, S., *et al.* 2022, *ApJS*, 260, 46

Jayasinghe, T., Stanek, K. Z., Kochanek, C. S., *et al.* 2019, *MNRAS*, 486, 1907

Kunder, A., Pérez-Villegas, A., Rich, R. M., *et al.* 2020, *AJ*, 159, 270

Kurtz, D. W. 2022, *ARAA*, 60, 31

Lebzelter, T., Mowlavi, N., Lecoeur-Taibi, I., *et al.* 2023, *A&A*, 674, A15

Lemasle, B., Lala, H. N., Kovtyukh, V., *et al.* 2022, *A&A*, 668, A40

Levine, E. S., Blitz, L., & Heiles, C. 2006, *Science*, 312, 1773

López-Corredoira, M. 2017, *ApJ*, 836, 218

Minniti, J. H., Sbordone, L., Rojas-Arriagada, A., *et al.* 2020, *A&A*, 640, A92

Minniti, J. H., Zoccali, M., Rojas-Arriagada, A., *et al.* 2021, *A&A*, 654, A138

McWilliam, A., & Zoccali, M. 2010, *ApJ*, 724, 1491

Nataf, D. M., Gonzalez, O. A., Casagrande, L., *et al.* 2016, *MNRAS*, 456, 2692

Nikzat, F., Ferreira Lopes, C. E., Catelan, M., *et al.* 2022, *A&A*, 660, A35

Palaversa, L., Ivezić, Ž., Eyer, L., *et al.* 2013, *AJ*, 146, 101

Pietrukowicz, P., Dziembowski, W. A., Mróz, P., *et al.* 2013, *AcA*, 63, 379

Pietrukowicz, P., Kozłowski, S., Skowron, J., *et al.* 2015, *ApJ*, 811, 113

Pietrukowicz, P., Ratajczak, M., Soszyński, I., *et al.* 2022, *AcA*, 72, 161

Pietrukowicz, P., Soszyński, I., Netzel, H., *et al.* 2020a, *AcA*, 70, 241

Pietrukowicz, P., Soszyński, I., & Udalski, A. 2021, *AcA*, 71, 205

Pietrukowicz, P., Udalski, A., Soszyński, I., *et al.* 2020b, *AcA*, 70, 121

Poggio, E., Drimmel, R., Cantat-Gaudin, T., *et al.* 2021, *A&A*, 651, A104

Pojmanski, G. 2002, *AcA*, 52, 397

Pojmanski, G. 2003, *AcA*, 53, 341

Pojmanski, G., & Maciejewski, G. 2004, *AcA*, 54, 153

Pojmanski, G., & Maciejewski, G. 2005, *AcA*, 55, 97

Pojmanski, G., Pilecki, B., & Szczygiel, D. 2005, *AcA*, 55, 275

Prudil, Z., Dékány, I., Grebel, E. K., Catelan, M., Skarka, M., & Smolec, R. 2019, *MNRAS*, 487, 3270

Reid, M. J., Menten, K. M., Brunthaler, A., *et al.* 2019, *ApJ*, 885, 131

Ripepi, V., Clementini, G., Molinaro, R., *et al.* 2023, *A&A*, 674, A17

Rodríguez, E., López-González, M. J., & López de Coca, P. 2000, *A&AS*, 144, 469

Samus', N. N., Kazarovets, E. V., Durlevich, O. V., Kireeva, N. N., & Pastukhova, E. N. 2017, *Astron. Rep.*, 61, 80

Sanders, J. L., Matsunaga, N., Kawata, D., Smith, L. C., Minniti, D., & Lucas, P. W. 2022, *MNRAS*, 517, 257
Schlafly, E. F., Meisner, A. M., Stutz, A. M., *et al.* 2016, *ApJ*, 821, 78
Sesar, B., Banholzer, S. R., Cohen, J. G., *et al.* 2014, *ApJ*, 793, 135
Sesar, B., Hernitschek, N., Mitrović, S., *et al.* 2017, *AJ*, 153, 204
Sesar, B., Ivezić, Ž., Stuart, J. S., *et al.* 2013, *AJ*, 146, 21
Skowron, D. M., Skowron, J., Mróz, P., *et al.* 2019a, *AcA*, 69, 305
Skowron, D. M., Skowron, J., Mróz, P., *et al.* 2019b, *Science*, 365, 478
Soszyński, I., Dziembowski, W., Udalski, A., *et al.* 2011a, *AcA*, 61, 1
Soszyński, I., Pietrukowicz, P., Skowron, J., *et al.* 2021, *AcA*, 71, 189
Soszyński, I., Udalski, A., Szymański, M. K., *et al.* 2009, *AcA*, 59, 239
Soszyński, I., Udalski, A., Szymański, M. K., *et al.* 2011b, *AcA*, 61, 217
Soszyński, I., Udalski, A., Szymański, M. K., *et al.* 2013, *AcA*, 63, 21
Soszyński, I., Udalski, A., Szymański, M. K., *et al.* 2014, *AcA*, 64, 177
Soszyński, I., Udalski, A., Szymański, M. K., *et al.* 2017, *AcA*, 67, 297
Soszyński, I., Udalski, A., Wrona, M., *et al.* 2019, *AcA*, 69, 321
Torrealba, G., Catelan, M., Drake, A. J., *et al.* 2015, *MNRAS*, 446, 2251
Udalski, A., Soszyński, I., Pietrukowicz, P., *et al.* 2018, *AcA*, 68, 315
Udalski, A., Szymański, M. K., & Szymański, G. 2015, *AcA*, 65, 1
Wood, P. R., Alcock, C., Allsman, R. A., *et al.* 1999, in: T. Le Bertre, A. Lebre, & C. Waelkens (eds.), *Asymptotic Giant Branch Stars*, Proc. IAU Symp. No. 191 (San Francisco: ASP), p. 151

At the crossroads of astrophysics and cosmology: Period–luminosity relations in the 2020s
Proceedings IAU Symposium No. 376, 2024
R. de Grijs, P. A. Whitelock & M. Catelan, eds.
doi:10.1017/S1743921323003319

Granulation in Red Supergiants: The Scaling Relations

Yi Ren[1], Zehao Zhang[2,3], Biwei Jiang[2,3], Igor Soszyński[4] and Tharindu Jayasinghe[5]

[1]College of Physics and Electronic Engineering, Qilu Normal University, Jinan 250200, China
email: yiren@qlnu.edu.cn

[2]Institute for Frontiers in Astronomy and Astrophysics, Beijing Normal University, Beijing 102206, China
email: bjiang@bnu.edu.cn

[3]Department of Astronomy, Beijing Normal University, Beijing 100875, China

[4]Astronomical Observatory, University of Warsaw, Al. Ujazdowskie 4, 00-478 Warszawa, Poland

[5]Department of Astronomy, University of California Berkeley, Berkeley CA 94720, USA
NASA Hubble Fellow

Abstract. The evolution of granulation is an important mechanism of the light variations of red supergiants (RSGs). Based on pure and complete samples of RSGs in the Magellanic Clouds, the mechanisms and characteristics of the granulation of RSGs are investigated based on time-series data. As predicted by the basic physical process of granulation and previous works, there are tight relations between granulation and stellar parameters of RSGs (i.e., the scaling relations). The scaling relations of RSGs provide a new method to infer stellar parameters by using the characteristic timescale and amplitude of granulations. Some faint sources deviate from the scaling relations, which may be due to the difference in the properties of the granulation of the RSGs before and after the blue loop or contamination by Mira variables. However, both of these possibilities suggest that the scaling relations of granulation is different among different types of stars.

Keywords. Red supergiant stars, Stellar pulsations, Stellar oscillations, Late-type stars

1. Introduction

Red supergiants (RSGs) are massive Population I stars in the core helium-burning stage. The initial mass of RSGs is generally considered to be between $7 - 30\,M_{\odot}$ (Yang *et al.* 2019; Ren *et al.* 2021a). They have relatively low effective temperatures, from $\sim 3000 - 4500$ K. The radii of RSGs can reach $\sim 1500\ R_{\odot}$. Hence, RSGs have the high luminosity of $3500 - 630{,}000\ L_{\odot}$ and low surface gravity.

RSGs exhibit various types of light variability; according to the characteristics of the light curves, the variations of RSGs are usually divided into three types: semi-regular variation, irregular variation, and variation with a long secondary period (LSP). Unfortunately, the mechanism producing LSPs remains unclear. The semi-regular variation exhibits the period-luminosity relations (PLR) theoretically and observationally (Guo & Li 2002; Kiss *et al.* 2006; Yang & Jiang 2011, 2012; Soraisam *et al.* 2018; Ren *et al.* 2019); this PLR makes RSGs a potential standard candle. In the case of irregular variations, the red noise ($1/f$ noise) found in the power spectra suggests that they are relates to stochastic mechanisms (i.e., granulation; Kiss *et al.* 2006; Yang & Jiang 2008). Three-dimensional radiative hydrodynamics (RHD) simulations and interferometric observations of Betelgeuse also proved that there are granulations on its surface (Chiavassa *et al.* 2009, 2010). However, direct imaging observation for distant RSGs is not possible, Betelgeuse is the only case for investigating the granulations of RSGs.

Using the time-series data from the All-Sky Automated Survey for SuperNovae (ASAS-SN; Shappee *et al.* 2014; Kochanek *et al.* 2017; Jayasinghe *et al.* 2020) and the intermediate Palomar Transient Factory (iPTF; Law *et al.* 2009; Rau *et al.* 2009) survey, and the sample of RSGs from Massey & Evans (2016), Yang *et al.* (2018, 2019), Ren *et al.* (2019), and Ren & Jiang (2020) investigated light variations of RSGs in the Small Magellanic Cloud (SMC), Large Magellanic Cloud (LMC), and M31, and determined the characteristic timescale and amplitude of granulations. They found tight relations (i.e., the scaling relations) between granulation parameters (timescale and amplitude) and stellar parameters (luminosity, mass, surface gravity, radius, effective temperature), which are in agreement with predictions from the basic physical process. They also compared their RSGs with red giant branch (RGB) stars and Betelgeuse. It is found that the relations fall close to the extrapolated relations followed by RGB stars. These results illustrate that the granulations constitute an important mechanism that contributes to the light variation of RSGs.

Moreover, for RSGs, the scaling relations provide a new method to infer stellar parameters using time-series data. This is very meaningful for the measurement of mass and surface gravity because there are few methods to determine the mass and surface gravity of RSGs. Therefore, we need more complete samples and time-series data to test whether the scaling relations are applicable to all RSGs. Using astrometric information from *Gaia* (Gaia Collaboration *et al.* 2018) and near-infrared photometry from the Two Micron All Sky Survey (2MASS; Skrutskie *et al.* 2006), Yang *et al.* (2020, 2021) and Ren *et al.* (2021b) removed foreground contaminations and established the most pure and complete samples of RSGs in the SMC and LMC. Based on the samples of RSGs and time-series data from the Optical Gravitational Lensing Experiment (OGLE; Udalski *et al.* 2015; Soszyński *et al.* 2015) and ASAS-SN, this work systematically investigates the scaling relations of RSGs.

2. Dynamic Properties of Granulations in Red Supergiants

The granulations in most of the RSGs evolve at an effective timescale ($\tau_{\rm eff}$) of several days to one year, with a characteristic amplitude ($\sigma_{\rm gran}$) of $10 - 1000$ mmag (Ren & Jiang 2020). $\tau_{\rm eff}$ and $\sigma_{\rm gran}$ are found to increase with metallicity, which may indicate the influence of metallicity on the granulation characteristics.

The most important dynamic property of granulations is the relation between granulation and stellar parameters, namely the scaling relation. Figure 1 presents the scaling relation of RSGs derived by Ren & Jiang (2020). RGB stars and some well-known stars (Betelgeuse, V1154 Cyg, the Sun) are included for comparison. Betelgeuse is an RSG, V1154 Cyg is a classical Cepheid, and the Sun is a G-type main-sequence star (G2V). In Figure 1, different types of stars almost follow the same scaling relation. Although RGB stars are offset towards smaller $\tau_{\rm eff}$ than suggested by the proposed scaling relation, it may be caused by different granulation parameters measurement methods. We can speculate that the scaling relation is applicable to all stars with granulations. If the scaling relation is universal for stars with granulations, then it would be a universal method for inferring stellar parameters that could be widely used.

In this work, we use the most complete and pure samples of RSGs to analyze the scaling relations in detail.

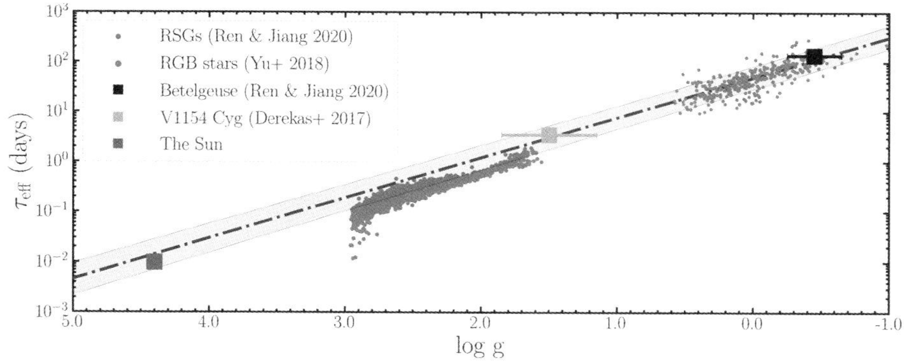

Figure 1. Relation of the granulation effective timescale $\tau_{\rm eff}$ with $\log g$. The results for RSGs (red dots) are taken from Ren & Jiang (2020), RGB stars (gray dots) are taken from Yu *et al.* (2018) and de Assis Peralta *et al.* (2018), V1154 Cyg (orange square) is taken from Derekas *et al.* (2017), and Betelgeuse (black square) is taken from Derekas *et al.* (2017) and Ren & Jiang (2020). The timescale of granulations in the Sun (green square) is $8-20$ minutes; we take the average value of 14 minutes in this figure. The blue dash-dotted line is the robust linear fit with the 95% confidence region (blue shaded region) based on RSGs.

3. Time-Series Data

ASAS-SN consists of 20 telescopes all around the globe that observed variable objects in the SMC and LMC at a cadence of $1-2$ days over a period time spanning ~ 2000 days. The magnitude limits are ~ 18 in the g band and ~ 17 in the V band. OGLE is a long-term photometric survey focused on variables, the brightness of sources in the SMC and LMC is sampled at a cadence of $4-8$ days, and OGLE has accumulated time-series data over 10 years (see Table 1).

Figure 2 shows the collections of g, V, and I-band light curves of two RSGs in the LMC. Figure 3 plots the 2MASS/K_0 distributions of sources along with ASAS-SN V-band and OGLE-IV I-band light curves, where K_0 are extinction corrected using reddening maps of MCs (Skowron *et al.* 2021). ASAS-SN mainly observes brighter sources, while OGLE mainly observes fainter sources.

3.1. *Time-Series Data Processing*

In order to exclude the influence of outliers in photometry on further analysis, we first calculate the first quartile (Q_1), the third quartile (Q_3), and the corresponding interquartile range (IQR; $\text{IQR} = Q_3 - Q_1$) according to the distribution of magnitudes for each light curve, and then the photometry points smaller than $Q_1 - 1.5 \times \text{IQR}$ or larger than $Q_3 + 1.5 \times \text{IQR}$ are removed from the light curve.

In this work, for ASAS-SN images, we perform forced photometry at the coordinates of the RSGs. Though the brightness of some faint sources is fainter than the detection threshold, ASAS-SN still generates a light curve. In addition, the amplitudes of granulations of some faint RSGs are relatively small. When observing at a longer wavelength (i.e., in the I band), their amplitudes will be even smaller. Therefore, it may be difficult to detect the signal of granulations in the light curves of some faint sources. In order to solve both these problems, we conduct a white noise test on the light curves. If the light curves are determined to be a white noise sequence, they are excluded from further analysis. Specifically, the autocorrelation functions (ACFs) of a light curve are calculated at 50 time lags. If less than 5% of the thus-computed ACFs meet the condition $|\text{ACFs}| < 1.96/\sqrt{n}$ (where n is the length of the time-series data), we accept the

Table 1. Number of Light Curves.

Galaxy	g band	V band	I band	At least one band	Containing gVI bands
SMC	886	452	1337	1924	83
LMC	2665	2123	2656	4471	451

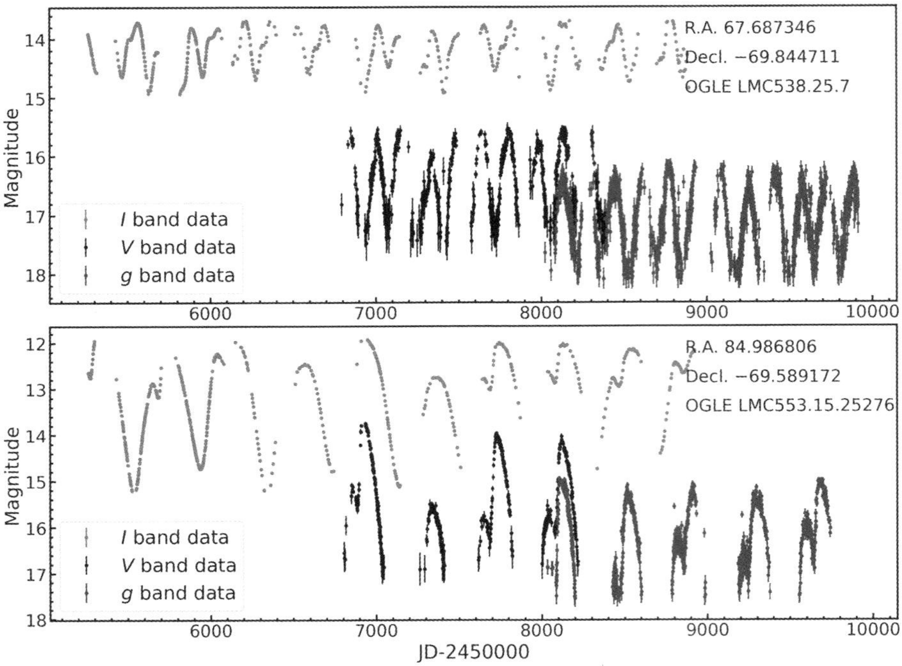

Figure 2. Examples of light curves of RSGs in the LMC after removing outliers. The ASAS-SN V and g band time-series data are shown in black and blue dots with error bars, respectively. I-band photometry is taken from OGLE-IV and is presented as red dots with error bars.

hypothesis that the time-series data constitutes a white noise sequence. Figure 4 gives an example of how we identify a white-noise process.

4. Granulation Parameters

Harvey (1985) proposed a famous model to describe the dynamic properties of granulations. The granulation evolution over time can be modeled as a stochastic process with an exponential decay ACF and variance $\sigma^2_{\rm gran}$. This model leads to a Lorentz profile power spectrum (hereafter Harvey function):

$$P(\nu) = \frac{4\sigma^2_{\rm gran}\tau_{\rm gran}}{1+(2\pi\nu\tau_{\rm gran})^2}, \tag{1}$$

in which $P(\nu)$ is the total power at frequency ν. The key parameters in Equation 1 are the characteristic timescale $\tau_{\rm gran}$ and characteristic amplitude $\sigma_{\rm gran}$. $\tau_{\rm gran}$ and $\sigma_{\rm gran}$ can thus be determined by fitting the power spectral density (PSD) using a Harvey-like function.

Ren & Jiang (2020) used the Continuous-time Auto Regressive Moving Average (CARMA) model (Kelly *et al.* 2014) to calculate the posterior PSDs of the light curves. Since the CARMA model is not affected by the noise of measurement, the Harvey-like

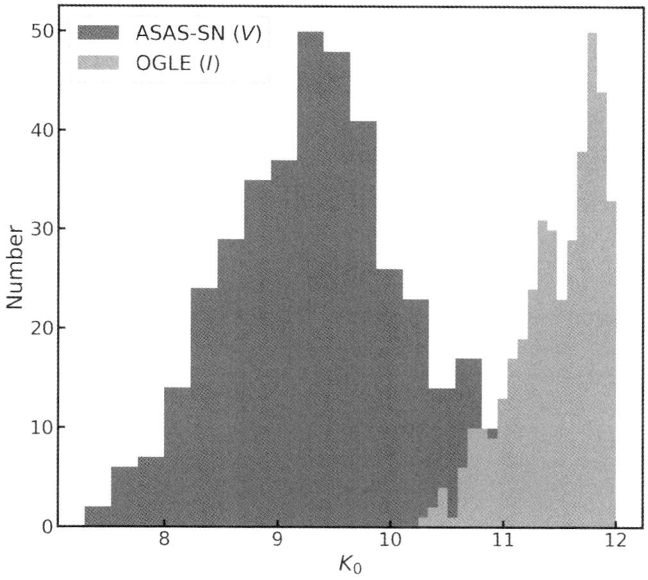

Figure 3. K_0 distributions of sources with ASAS-SN V-band light curves (blue histogram) and OGLE I-band light curves (orange histogram).

function is used to fit the posterior PSD. The Harvey-like function is the following:

$$P(\nu) = \frac{\xi \sigma_{\text{gran}}^2 \tau_{\text{gran}}}{1 + (2\pi\nu\tau_{\text{gran}})^{\alpha}}, \tag{2}$$

where ξ is a normalization factor related to α as

$$\xi = 2\alpha \sin(\pi/\alpha). \tag{3}$$

In this work, the PSD of a light curve is calculated using the Lomb-Scargle (LS) periodogram (Lomb 1976; Scargle 1982), as shown in Figure 5.

At the high-frequency end, the PSD approaches the measurement noise, so the Harvey-like function with white noise term is selected to fit the PSD,

$$P(\nu) = W + \frac{4\sigma^2 \tau_{\text{gran}}}{1 + (2\pi\nu\tau_{\text{gran}})^{\alpha}}, \tag{4}$$

where W represents the white noise component in the PSD. As shown in Figure 5, a total PSD consists of a granulation component and a white noise component. The granulation parameters are derived from the granulation component.

In order to compare the characteristic timescales resulting from different α values, the effective timescale τ_{eff} is introduced (Mathur *et al.* 2011). The ACF is the inverse Fourier transform of the PSD and is calculated from the latter numerically. Then τ_{eff} is the required time for the ACF to get reduced to $1/e$ of its initial value.

5. The Scaling Relations

Table 2 lists the methods for obtaining stellar parameters used in the scaling relations. For RSGs, there is no better way to obtain the mass and surface gravity. The M and $\log g$ are taken from the empirical mass-luminosity relation (Stothers & Leung 1971).

Using the measured granulation and stellar parameters of RSGs in the LMC, the obtained scaling relations are presented in Figure 6. The green points in Figure 6 use the light curves in the V band. Since the amplitudes of granulations are different in the V

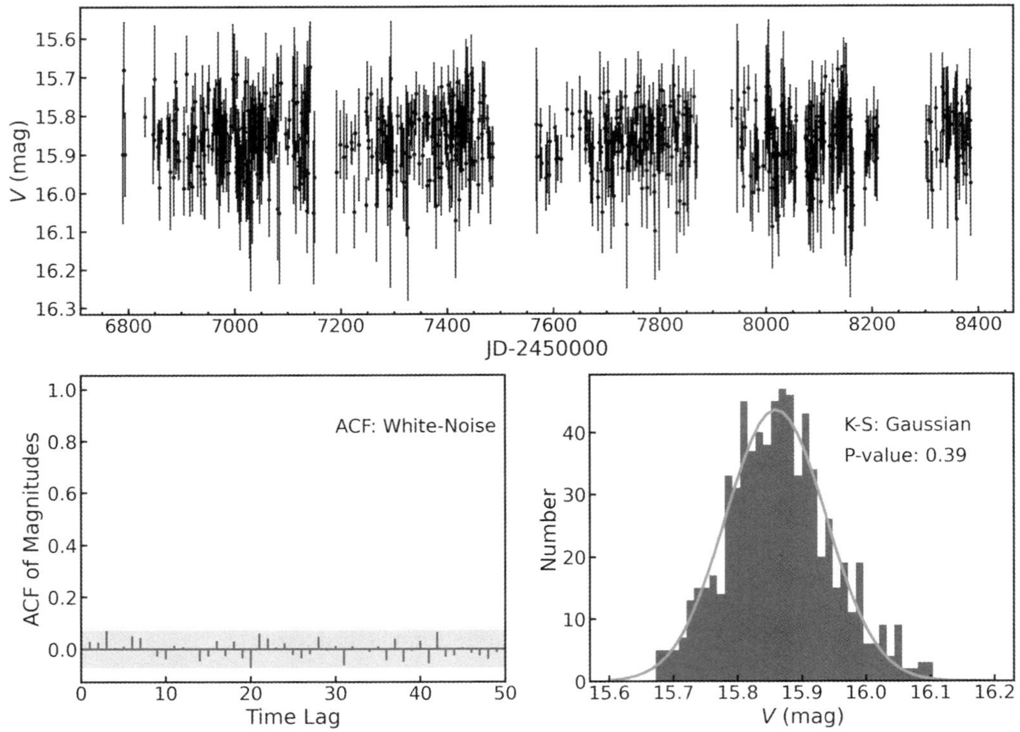

Figure 4. The upper panel shows an example of a light curve from ASAS-SN. This light curve constitutes a white noise sequence. The lower left panel indicates all the ACFs lay in the region assuming a white-noise process. In addition, we set up a null hypothesis that the distribution of brightness from the light curve obeys the normal distribution. Then a Kolmogorov-Smirnov (K-S) test is performed on the light curve, yielding a p-value of 0.39, which is greater than 0.05 (standard significance level). The p-value means we cannot reject the null hypothesis that the distribution of brightness from the light curve obeys the normal distribution (lower right panel). This confirms that the light curve is a Gaussian white noise sequence, which is thus excluded from further analysis.

and I bands, we multiply the amplitude in the I band by a factor of 1.231† to convert to the V-band amplitude for further analysis.

Compared to previous work (Ren & Jiang 2020), the scaling relations obtained include a group of fainter RSGs, which means that the sample covers a wider parameter space.

Most RSGs follow the scaling relations well ($\tau_{\rm eff}$ or $\sigma_{\rm gran}$ vs. L, $\log g$, M, etc). Figure 6 only displays the scaling relation between granulation parameters and $\log g$.

5.1. *Peculiar Sources*

As shown by the open circles in Figure 6, we note that some sources ("peculiar" sources) significantly deviate from the scaling relations. We speculate that there may be three possible reasons for these sources to deviate from the scaling relation.

(1) The light curves of these sources are rechecked, and it is found that some of them have relatively large amplitudes ($\Delta V > 0.3$ mag). Therefore, it is possible that the pulsating signal with large amplitude makes the measured granulation parameters larger. This is a technical explanation.

† The 29 RSGs in the LMC and 8 RSGs in the SMC have both I-band and V-band light curves. By fitting the granulation amplitudes $\sigma_{\rm gran}$ in these bands, we get $\sigma^V_{\rm gran} = 1.231\sigma^I_{\rm gran} + 0.006$. Thus, the amplitude ratio between the V and I band is 1.231.

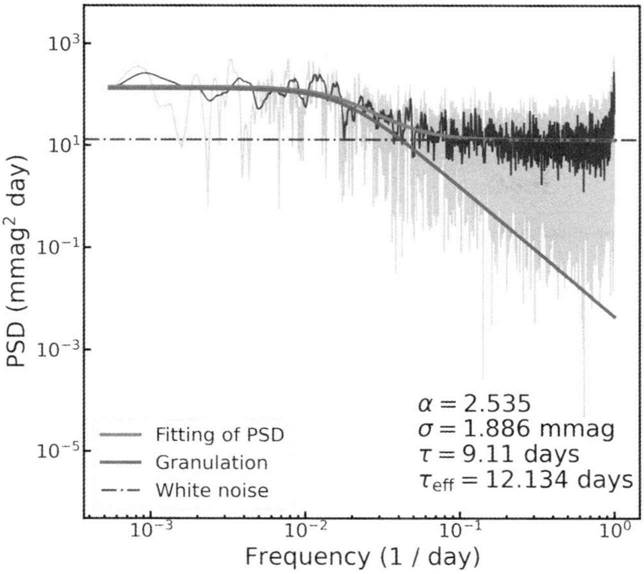

Figure 5. An example of fitting PSD using the Harvey-like function with white noise term. The gray line shows the PSD given by the LS algorithm, while the black line is the result of smoothing the gray line. The green solid line and blue dash-dotted line are the granulation component and measurement noise component, respectively.

(2) Another explanation is entirely physical. Some of these sources have very high light variation amplitudes ($\Delta V \sim 2$ mag), and their light variation characteristics are similar to those of Mira variables. Therefore, large-scale stellar phenomena may affect the dynamic properties of surface granulations.
(3) After calculating the mass of these sources, it is found that almost all sources with mass greater than 12 $M_\odot$ obey the scaling relations, while sources with mass less than 12 $M_\odot$ are divided into two groups. By comparing with Padova stellar evolution models (Chen *et al.* 2015; Pastorelli *et al.* 2020), it is found that an RSG with mass less than 12 $M_\odot$ will experience a blue loop, while an RSG with mass greater than 12 $M_\odot$ will not experience a blue loop. Therefore, the sources with less than 12 $M_\odot$ are divided into two groups, one of which represents the RSGs before the blue loop stage which follow the main branch, while the other group represents the RSGs after the blue loop stage which deviate from the main branch (minor branch).

The above three possibilities will be explored in a future paper through the stellar evolution model or data simulation method.

5.2. *Inferring Stellar Parameters Using the Scaling Relations*

The scaling relations provide a new method for inferring the stellar parameters of RSGs. In order to estimate the accuracy of this method, we perform linear fits (removing peculiar sources by eye-check) by taking granulation parameters as independent variables and the stellar parameters as dependent variables. The fitting form is as follows:

$$\log Y = k \times \log X + b, \tag{5}$$

where Y is one of g, $M/M_\odot$, $R/R\odot$, $L/L_\odot$, and $T_{\rm eff}$, and X is either $\tau_{\rm eff}$ or $\sigma_{\rm gran}$. Then an unbiased estimator, $\hat{\sigma}^{*2} = S_E^2/(n-2)$, where S_E is the residual sum of squares, is introduced to express the uncertainty of stellar parameters inferred by scaling relations.

Table 2. Methods for Obtaining Stellar Parameters.

Stellar Parameter	Method	Reference
$T_{\rm eff}$	$T_{\rm eff}-(J-K)_0$ relation^{*a}	—
L	$\log(L/L_\odot)=a+b\,(m_K-\mu)^{*b}$	Davies *et al.* (2013)
M	$(M/M_\odot)^\gamma=L/L_\odot,\ \gamma=4$	Stothers & Leung (1971)
R	$R=(4\pi\sigma)^{-0.5}L^{0.5}T_{\rm eff}^{-2}$	—
g	$g=GM/R^2$	—

*a The $T_{\rm eff}-(J-K)_0$ relation is obtained by performing linear fitting between $(J-K)_0$ from Ren *et al.* (2021b) and $T_{\rm eff}$ from APOGEE (Abdurro'uf *et al.* 2022).

*b $a=0.90$, $b=-0.40$, $\mu=18.72$ (18.26) for SMC (LMC).

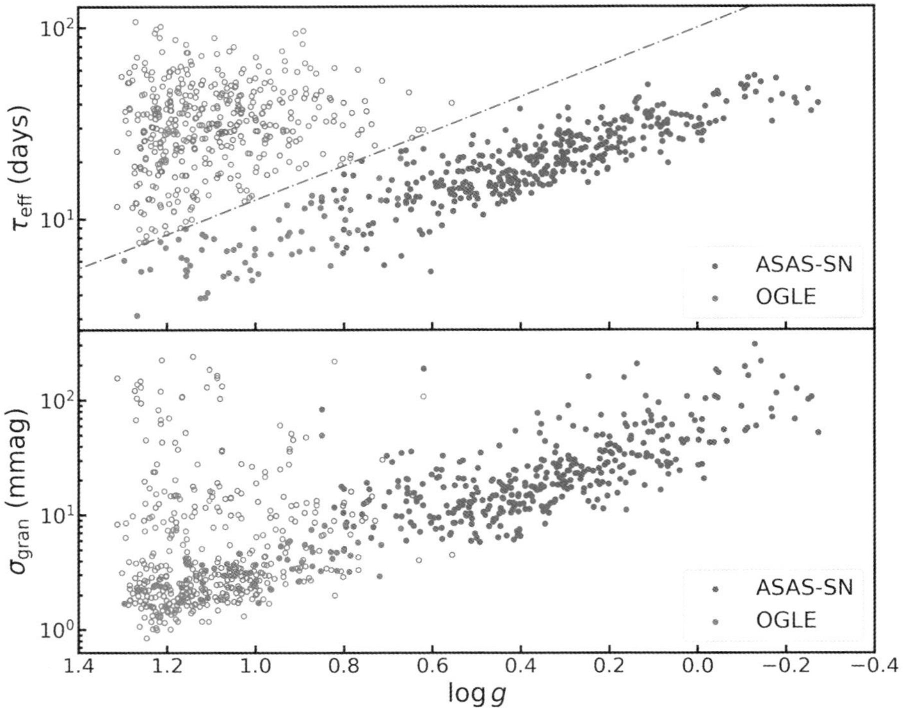

Figure 6. The scaling relation between effective timescale ($\tau_{\rm eff}$) and $\log g$ (top panel) and between granulation amplitude ($\sigma_{\rm gran}$) and $\log g$ (bottom panel) for RSGs in the LMC. Green and red dots represent RSGs, whose light curves are taken from ASAS-SN and OGLE, respectively. The blue dash-dotted line distinguishes between sources that follow the scaling relation and sources that deviate from the scaling relation. The open circles are sources that deviate from the $\tau_{\rm eff}-\log g$ relation.

$\hat{\sigma}^{*2}$ values of the fitting residuals of the scaling relations between five different stellar parameters and two granulation parameters are listed in Table 3.

As shown in Table 3, the scaling relations do indeed provide an accurate method to infer stellar parameters. For example, using the $\tau_{\rm eff}-\log g$ relation, the 1σ uncertainty in $\log g$ is 0.14.

6. Summary

Ren & Jiang (2020) proposed a method to measure the granulation parameters of RSGs using light curves and derived the scaling relations between granulation and stellar parameters for RSGs for the first time. They proved that the surfaces of RSGs are dominated by a few but huge granules, and that the evolution of granulations is an

Table 3. Uncertainty of Stellar Parameters Inferred by Scaling Relations.

Stellar Parameter	$\hat{\sigma}^{*2}$, stellar parameters from $\tau_{\rm eff}$	$\hat{\sigma}^{*2}$, stellar parameters from $\sigma_{\rm gran}$
$\log g$	0.14	0.18
$\log(M/M_\odot)$	0.04	0.06
$\log(R/R_\odot)$	0.09	0.12
$\log(L/L_\odot)$	0.17	0.22
$\log T_{\rm eff}$	0.01	0.01

important mechanism for modulating the light variation of RSGs. More importantly, the scaling relations provide a potential method for inferring their stellar parameters.

To test the universality of the scaling relations for RSGs, we investigated those based on new samples of RSGs in the SMC and LMC and time-series data from ASAS-SN and OGLE. The scaling relations obtained in this work cover a more expansive stellar parameter space. Most RSGs follow the scaling relations. However, some sources deviate from the latter. These sources generally have a large amplitude of light variation. The possible reasons may be the impact of large amplitude pulsations on measurements of granulation parameters, the impact of large amplitude pulsations on the physical properties of granulations, or the existence of two groups of sources corresponding to RSGs in the pre- or post- blue loop stages.

After removing sources that deviate from the scaling relations, we estimate the uncertainty of stellar parameters that are inferred from them. Excitingly, the scaling relations provide accurate inference of stellar parameters for RSGs.

Acknowledgements. This work is supported by the National Natural Science Foundation of China (NSFC) through grant Nos. 12133002 and 12203025, and Shandong Provincial Natural Science Foundation through project ZR2022QA064.

References

Abdurro'uf, Accetta, K., Aerts, C., *et al.* 2022, *ApJS*, 259, 35
Chen, Y., Bressan, A., Girardi, L., *et al.* 2015, *MNRAS*, 452, 1068
Chiavassa, A., Haubois, X., Young, J. S., *et al.* 2010, *A&A*, 515, A12
Chiavassa, A., Plez, B., Josselin, E., & Freytag, B. 2009, *A&A*, 506, 1351
Davies, B., Kudritzki, R.-P., Plez, B., *et al.* 2013, *ApJ*, 767, 3
de Assis Peralta, R., Samadi, R., & Michel, E. 2018, *AN*, 339, 134
Derekas, A., Plachy, E., Molnár, L., *et al.* 2017, *MNRAS*, 464, 1553
Gaia Collaboration, Brown, A. G. A., Vallenari, A., *et al.* 2018, *A&A*, 616, A1
Guo, J. H., & Li, Y. 2002, *ApJ*, 565, 559
Harvey, J. 1985, in ESA Special Publication, Vol. 235, Future Missions in Solar, Heliospheric & Space Plasma Physics, ed. E. Rolfe & B. Battrick, 199
Jayasinghe, T., Stanek, K. Z., Kochanek, C. S., *et al.* 2020, *MNRAS*, 491, 13
Kelly, B. C., Becker, A. C., Sobolewska, M., Siemiginowska, A., & Uttley, P. 2014, *ApJ*, 788, 33
Kiss, L. L., Szabó, G. M., & Bedding, T. R. 2006, *MNRAS*, 372, 1721
Kochanek, C. S., Shappee, B. J., Stanek, K. Z., *et al.* 2017, *PASP*, 129, 104502
Law, N. M., Kulkarni, S. R., Dekany, R. G., *et al.* 2009, *PASP*, 121, 1395
Lomb, N. R. 1976, *Ap&SS*, 39, 447
Massey, P., & Evans, K. A. 2016, *ApJ*, 826, 224
Mathur, S., Hekker, S., Trampedach, R., *et al.* 2011, *ApJ*, 741, 119
Pastorelli, G., Marigo, P., Girardi, L., *et al.* 2020, *MNRAS*, 498, 3283
Rau, A., Kulkarni, S. R., Law, N. M., *et al.* 2009, *PASP*, 121, 1334
Ren, Y., Jiang, B., Yang, M., *et al.* 2021a, *ApJ*, 907, 18
Ren, Y., Jiang, B., Yang, M., Wang, T., & Ren, T. 2021b, *ApJ*, 923, 232
Ren, Y., & Jiang, B.-W. 2020, *ApJ*, 898, 24
Ren, Y., Jiang, B.-W., Yang, M., & Gao, J. 2019, *ApJS*, 241, 35
Scargle, J. D. 1982, *ApJ*, 263, 835

Shappee, B. J., Prieto, J. L., Grupe, D., *et al.* 2014, *ApJ*, 788, 48
Skrutskie, M. F., Cutri, R. M., Stiening, R., *et al.* 2006, *AJ*, 131, 1163
Skowron, D. M., Skowron, J., Udalski, A., *et al.* 2021, *ApJS*, 252, 23
Soraisam, M. D., Bildsten, L., Drout, M. R., *et al.* 2018, *ApJ*, 859, 73
Soszyński, I., Udalski, A., Szymański, M. K., *et al.* 2015, *AcA*, 65, 297
Stothers, R., & Leung, K. C. 1971, *A&A*, 10, 290
Udalski, A., Szymański, M. K., & Szymański, G. 2015, *AcA*, 65, 1
Yang, M., & Jiang, B. W. 2008, in The Art of Modeling Stars in the 21st Century, ed. L. Deng & K. L. Chan, Vol. 252, 267
Yang, M., & Jiang, B. W. 2011, *ApJ*, 727, 53
Yang, M., & Jiang, B. W. 2012, *ApJ*, 754, 35
Yang, M., Bonanos, A. Z., Jiang, B.-W., *et al.* 2018, *A&A*, 616, A175
Yang, M., Bonanos, A. Z., Jiang, B.-W., *et al.* 2019, *A&A*, 629, A91
Yang, M., Bonanos, A. Z., Jiang, B.-W., *et al.* 2020, *A&A*, 639, A116
Yang, M., Bonanos, A. Z., Jiang, B., *et al.* 2021, *A&A*, 646, A141
Yu, J., Huber, D., Bedding, T. R., *et al.* 2018, *ApJS*, 236, 42

At the crossroads of astrophysics and cosmology: Period–luminosity relations in the 2020s
Proceedings IAU Symposium No. 376, 2024
R. de Grijs, P. A. Whitelock & M. Catelan, eds.
doi:10.1017/S1743921323003496

The distance scale of Type II Cepheids from near-infrared observations in the Magellanic Clouds

T. Sicignano[1], V. Ripepi[1], R. Molinaro[1], A. Bhardwaj[1], M. Marconi[1] and M.-R. L. Cioni[2]

[1] INAF-Osservatorio Astronomico di Capodimonte, Salita Moiariello 16, 80131, Naples, Italy.
email: teresa.sicignano@inaf.it

[2] Leibniz-Institut für Astrophysik Potsdam, An der Sternwarte 16, D-14482 Potsdam, Germany

Abstract. We employed data from the VISTA near-infrared YJK_s survey of the Magellanic System to analyse the light curves of Type II Cepheids (T2Cs) in the Large and Small Magellanic Clouds (LMC and SMC, respectively). Using the T2Cs identified by the OGLE IV survey and *Gaia* mission, we built up a sample of about 330 pulsators belonging to both galaxies. For all these objects we obtained accurate intensity-averaged magnitudes in the YJK_s bands by means of a template-fitting technique. We complemented our near-infrared data with optical photometry from the literature to calculate period-luminosity and period-Wesenheit relations for a variety of different bands and colour combinations and separately for the different T2C subclasses (BL Herculis, W Virginis, peculiar W Virginis, RV Tauri). These relations, calibrated with the LMC distance modulus, were tested using T2Cs belonging to Galactic globular clusters. We thus calculated the distances of 22 clusters and compared them with the literature values, mainly based on RR Lyrae stars, finding good agreement within 1 σ and dispersion of the order of $0.3 - 0.5$ kpc, depending on the adopted period-luminosity/period-Wesenheit relation.

Keywords. Variable stars, Type II Cepheids, Pulsations, Magellanic Clouds, Distance Scale.

1. Introduction

The ongoing Hubble tension is a hot topic in moden cosmology because of the 5σ discord in the Hubble constant (H_0) determination from the traditional extragalactic distance ladder and those from the Cosmic Microwave Background with the Λ cold dark matter theory (see Riess et al. 2022, and references therein). The extragalactic distance scale heavily relies on Classical Cepheids (CCs) as primary calibrators, but these stars belong to the young population and are therefore only present in late-type galaxies.

It is useful to check the presence of possible systematic errors in using CCs, which can be done using old standard candles such as RR Lyrae variables, Type II Cepheids (T2Cs) and the tip of the red giant branch (TRGB). These population II stars are present in almost all galaxies, including the early-type ones, unlike CCs.

In this framework, the Large and Small Magellanic Clouds (LMC and SMC, respectively), are very important objects, as they are anchors for the extragalactic distance scale. Moreover, by hosting CCs, T2Cs, and Anomalous Cepheids (ACs), as well as RR Lyrae stars, these galaxies provide an ideal laboratory for comparing the distance measurements obtained from different stellar standard candles.

In this study, we calculate the period-luminosity (PL) and the period-Wesenheit† (PW) relations for a variety of bands and colour combinations from the infrared towards the optical bands for the different T2C subclasses, namely BL Herculis (BL Her), W Virginis (W Vir), peculiar W Virginis (p W Vir), RV Tauri (RV Tau) (see Soszyński et al. 2018, for more details of the subclass separation and properties of each subclass). These relations will be extremely useful to compare the distance scales based on different standard candles.

2. Type II Cepheids in the VMC survey

To build a sample of T2Cs, we adopted both the Optical Gravitational Lensing Experiment IV survey (Soszyński et al. 2015, 2018) and the *Gaia* Data Releases 2 and 3 (DR2 and DR3, respectively Gaia Collaboration et al. 2016; Clementini et al. 2016, 2019; Ripepi et al. 2019, 2023; Vallenari et al. 2022) results. From these catalogues, we obtained the identification, position, and periods as well as V, I (from OGLE IV) and G, $G_{\rm BP}$, $G_{\rm RP}$ photometry (from the *Gaia* DR2 and DR3). The targets are distributed in the two Clouds as follows: 62 belong to the SMC and 277 to the LMC. Among these, 106 are BL Her, 121 W Vir, 33 p W Vir and 79 RV Tau variables.

The time-series photometry in the YJK_s bands has been obtained in the context of the VISTA† near-infrared $YJK_{\rm s}$ survey of the Magellanic System (VMC, see Cioni et al. 2011, for a description of the data and photometric reduction procedures). The data were retrieved from the Vista Science Archive (Cross et al. 2012) and the phased light curves were obtained following the same procedure as in our previous papers (e.g., Ripepi et al. 2012, 2015, 2012; Moretti et al. 2014). Examples of light curves for the different subclasses are shown in Figure 1.

3. Data analysis

3.1. *Template derivation*

We need to calculate accurate average magnitudes for each star to construct precise PL and PW relations for the T2Cs sample.

To this aim, we first fit the observed light curves with some analytical function, typically a truncated Fourier series, and then use the fitted curve to derive the intensity-averaged magnitude, which is the required quantity for pulsating stars (see Caputo et al. 1999). However, in our case, most of the light curves are too sparse to obtain good fitting with the Fourier series, especially in the Y and J bands for which we typically collected $5-6$ epochs. The $K_{\rm s}$ light curves are generally better sampled, with about 15 epochs on average. To overcome this problem, we adopted the template technique, which we already successfully used in our previous papers devoted to the study of CCs in the VMC survey (Ripepi et al. 2016, 2017, 2022). As in those papers, we built custom template light curves specific to T2Cs using the best light curves in each band.‡ The stars used to build the templates were chosen to span evenly the range of periods shown by the targets, and thus they represent all the light curve shapes observed in our sample. We refer interested readers to Ripepi et al. (2022, Section 3.1) for full details on template derivation.

† The Wesenheit functions were defined by Madore (1982) as reddening-free quantities, provided a chosen extinction law.

† Visible and Infrared Survey Telescope for Astronomy (Emerson et al. 2006)

‡ Note that about a quarter of the stars have twice as many epochs as the others because they fall in areas of overlap between adjacent fields.

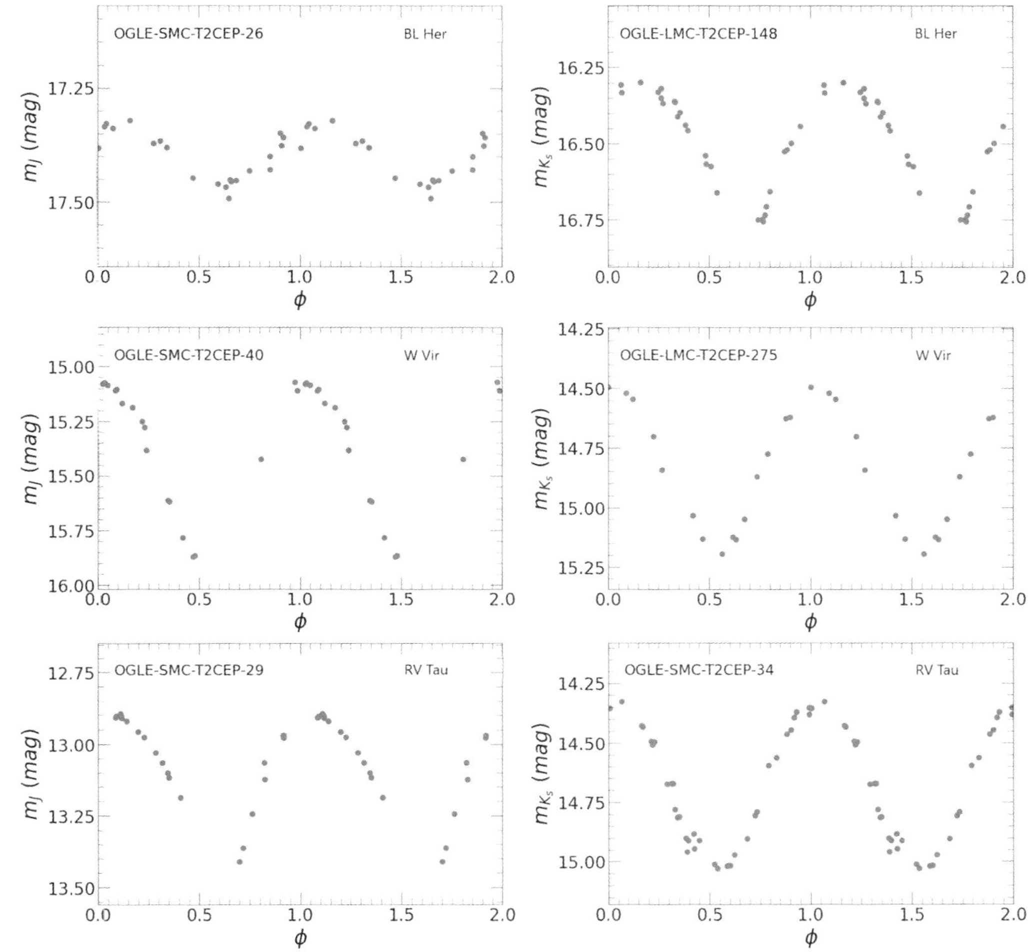

Figure 1. Some of the best observed light curves in our sample, used as a basis for the templates. To visualize the light curves, the observed magnitudes have been phased. From top left to bottom left in the J band: BL Her, W Vir, RV Tau of the SMC; from top right to bottom right in the K_s band: BL Her, W Vir of the SMC; RV Tau of the LMC.

3.2. *Fit to the data with templates*

We used our set of templates to derive the intensity-averaged magnitudes and amplitudes for each star in the sample following the well-established pipeline developed and tested for CCs in the aforementioned papers (in particular see Ripepi et al. 2022). In brief, the templates, which have zero average magnitude and unity peak-to-peak amplitude, were adjusted to the observed light curves by varying three parameters: (i) the magnitude to be added to the template, (ii) a scale factor (a) to adjust the template amplitude to match that of the observed light curves, and (iii) a phase shift to account for any misalignment between the starting points of the observed light curves and the templates. Conventionally, it is common to set the maximum of the light curve at phase 1 (or 0). The determination of these three unknown parameters for each template was achieved by minimizing a χ^2 function:

$$\chi^2 = \sum_{i}^{N_{\text{pts}}} \frac{[m_i - (a \cdot M_{\text{t}}(\phi_i + \delta\phi) + \delta M)]^2}{\sigma_i^2}. \tag{1}$$

In this equation, $N_{\rm pts}$ represents the number of epochs, whereas m_i, ϕ_i, and σ_i denote the observed magnitudes, corresponding phases, and uncertainties on the magnitudes, respectively. $\delta\phi$ and δM are the amounts to add to m_i and ϕ_i to better fit the template $M_{\rm t}(\phi_i)$. The procedure, with a weighting algorithm, does not favour the templates which require removing many epochs (points) to fit the light curve (see Ripepi et al. 2022, for more detailed information).

The outlined procedure successfully fitted all the available light curves, using the most appropriate template (see Figure 2 for a few examples), and provided the desired intensity-averaged magnitudes (and amplitudes) in the three bands. The uncertainties on the fitted parameters were determined using a Monte Carlo approach, ensuring robust estimates of the parameter errors.

3.3. *Reddening estimates for the target stars*

To continue our analysis, accurate reddening maps, published by Skowron et al. (2021), were utilized. However, these maps did not cover the extent of the Magellanic Clouds, so the authors supplemented them with the lower-resolution maps from Schlegel et al. (1998), referred to as SFD maps.

Intensity-averaged magnitudes have been corrected for reddening adopting Cardelli's law (as defined in Cardelli et al. 1989), assuming a value of $R_{\rm V} = 3.23$ as suggested by Inno et al. (2013). The same law is adopted to derive the color-term coefficients fot the PW relations. For the *Gaia* bands, coefficients published by Casagrande & VandenBerg (2018) were employed.

Examples of PL relations in J_0 and $K_{\rm s,0}$ for all the different T2C subclasses in the LMC are reported in the bottom panels of Figure 3.

The VMC near-infrared (NIR) data were supplemented with optical data from the literature, to investigate how the coefficients of the PL relations vary with wavelength. Specifically, optical data in the V and $I_{\rm C}$ bands were obtained from the OGLE IV survey (as mentioned earlier), while data in the G, $G_{\rm BP}$, and $G_{\rm RP}$ bands were acquired from the *Gaia* mission (*Gaia* DR2 and DR3).

By incorporating data from different wavelength ranges, a comprehensive analysis of the variation in coefficients for the PL and PW relations can be carried out, providing valuable insights into the properties and behaviour of the studied variable stars.

Examples of PL relations in G_0, $I_{C,0}$, J_0, $K_{\rm s,0}$ for all the different T2C subclasses in the LMC are reported in the top panels of Figure 3.

4. Results

A comparison of the panels of Figure 3, from the top to the bottom, allows us to make some considerations:

- The PL relations show a decreasing dispersion from the optical to the NIR bands.
- A unique slope, i.e. a unique PL relation, is followed by the different subclasses of T2Cs for bands redder than I_C.
- A greater dispersion characterises the RV Tau stars: this may be due to the possible mixed evolutionary status of this class of variables (Feast et al. 2008).

On this basis, we calculated a significant number of PL and PW relations separately for the LMC and SMC and for the different subclasses of T2Cs, adopting the following combinations: i) BL Her alone; ii) W Vir alone; iii) BL Her and W Vir; iv) BL Her, W Vir and p W Vir; v) BL Her, W Vir, p W Vir and RV Tau.

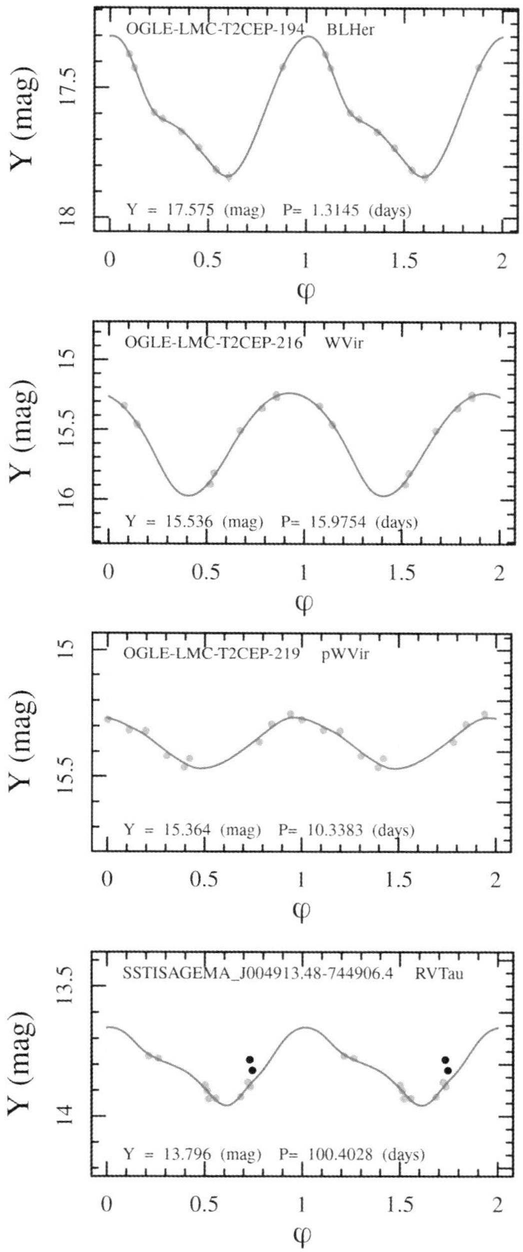

Figure 2. Some fitted light curves in the Y band. From the top: BL Her, W Vir, p W Vir, RV Tau. In the bottom panel, the black points are outliers.

To obtain the desired PL/PW relations, we fitted the following linear relations to the data:

$$m_{\lambda_0} = \alpha + \beta \cdot \log P\,; \quad \text{(PL)} \tag{2}$$

$$w(\lambda_1, \lambda_2) = \alpha + \beta \cdot \log P\,. \quad \text{(PW)} \tag{3}$$

Periods and dereddened intensity-averaged magnitudes or, in turn, Wesenheit functions are the observables while the α (intercept) and β (slope) coefficients are the unknowns.

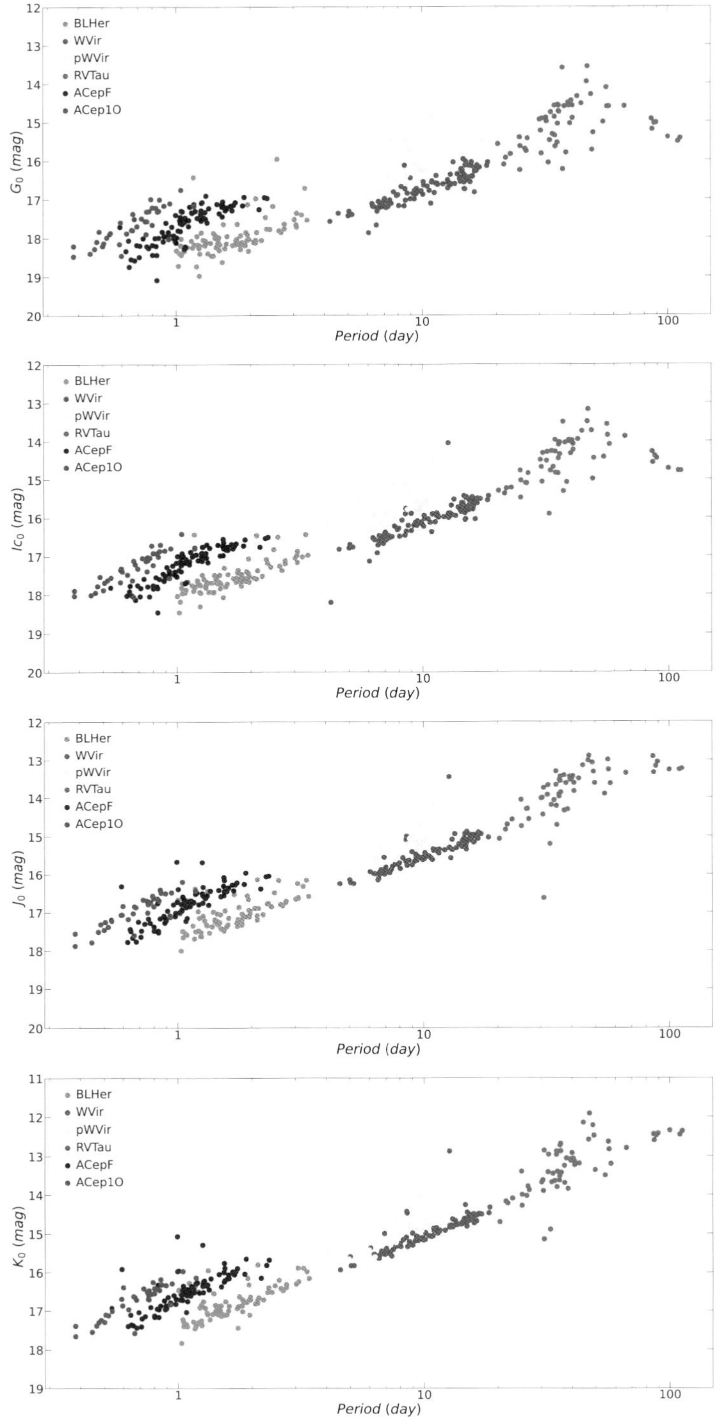

Figure 3. PL relations in different bands for T2Cs and ACs in the LMC.

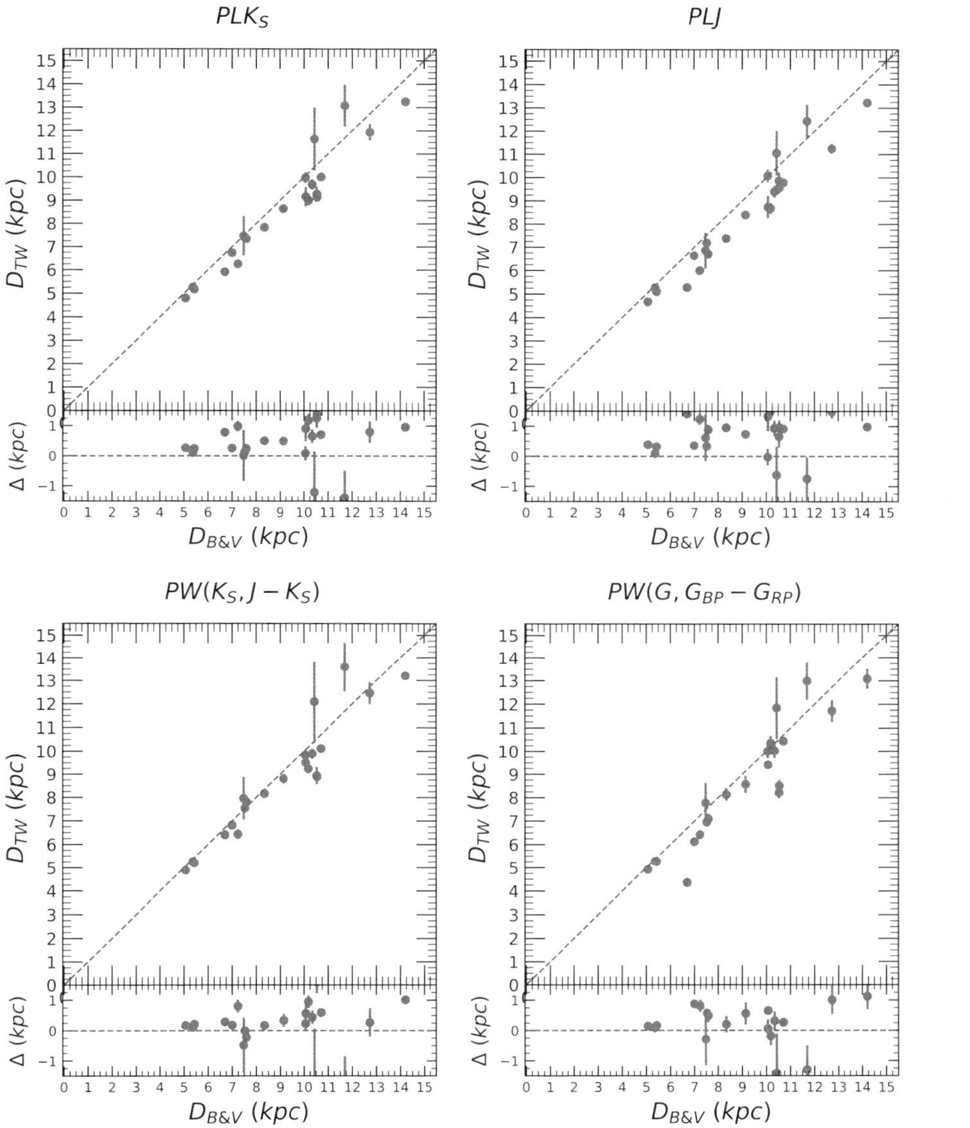

Figure 4. Comparison between our distances and those by Baumgardt & Vasiliev (2021).

To carry out the fitting of these relations, we adopted the `Python` package LTS (Least Trimmed Squares, Cappellari et al. 2013) which allows for a robust outlier rejection. In this way, we calculated more than 50 and 40 different PL and PW relations for the LMC and SMC, respectively. We do not report them here for brevity, but we discuss some results based on selected PL/PW relations in the next section.

5. Discussion

As a first application of our work, we selected the PL relations in the J, $K_{\rm s}$ bands and the PW relations in ($K_{\rm s}$, $J - K_{\rm s}$) and (G, $G_{\rm BP} - G_{\rm RP}$) combinations. Moreover, we restricted the choice to the BL Her + W Vir sample in the LMC.

To use these relations for calculating distances, we first have to calibrate their intercept to absolute magnitudes. To this end, we adopted the geometric distance modulus obtained by Pietrzyński et al. (2019), $\mu_{\rm LMC} = 18.477 \pm 0.027$ mag. In the following step, we used the calibrated relations to estimate the distances of a sample of Galactic globular clusters (GGCs) containing T2Cs. The aim is to validate the distances obtained from our relations for these objects, in comparison with the estimates from the literature. In particular, we adopted the recommended distances by Baumgardt & Vasiliev (2021), mainly based on independent methods.

In total, we analysed 58 T2Cs belonging to 22 different GGCs. The list of these pulsators, their periods and NIR intensity-averaged magnitudes were taken from Bhardwaj et al. (2017, 2021), Braga et al. (2020), Ngeow et al. (2022). Similar quantities in the *Gaia* bands were retrieved from the *Gaia* DR3 catalogue (Ripepi et al. 2023). As for the reddening, we used the Harris (2010) estimates.

These data allowed us to obtain distances of all GGCs hosting at least one T2C. In case of multiple stars hosted by the same GGC, the derived distances were averaged to obtain a unique distance for the specific GGC. Figure 4 shows the comparison between the distances calculated using our PL/PW relations and those obtained by Baumgardt & Vasiliev (2021). In the lower panels of each figure, Δ means literature minus this work, so the values of Δ underline that our distances are generally smaller than the literature ones.

The maximum discrepancy is found for the PL relation in the J band, while the minimum one is for the PW($K_{\rm s}$, $J - K_{\rm s}$) relation, also with a smaller dispersion of the residuals. It is possible to speculate that there might be some inhomogeneity in the J and $K_{\rm s}$ photometry for GGC data. The use of Wesenheit magnitudes seems to minimize the impact of any potential photometric inhomogeneities and provides more robust distance estimates for GGCs.

This first use of our relations seems to suggest that we find systematically lower distances by ~ 0.25 kpc for the GGCs compared with Baumgardt & Vasiliev (2021). This rather surprising occurrence will be discussed in detail in our future work.

References

Baumgardt, H., & Vasiliev, E. 2021, *MNRAS*, 505, 5957
Bhardwaj, A., Macri, L. M., Rejkuba, M., *et al.* 2017, *AJ*, 153, 154
Bhardwaj, A., Rejkuba, M., Sloan, G. C., *et al.* 2021, *ApJ*, 922, 20
Braga, V. F., Bono, G., Fiorentino, G., *et al.* 2020, *A&A*, 644, A95
Cappellari, M., McDermid, R. M., Alatalo, K., *et al.* 2013, *MNRAS*, 432, 1862.
Caputo, F., Marconi, M., & Ripepi, V. 1999, *ApJ*, 525, 784
Cardelli, J. A., Clayton, G. C., & Mathis, J. S. 1989, *ApJ*, 345, 245
Casagrande, L., & VandenBerg, D. A. 2018, *MNRAS*, 479, L102
Cioni, M.-R., Clementini, G., Girardi, L., *et al.* 2011, *Msngr*, 144, 25
Clementini, G., Ripepi, V., Leccia, S., *et al.* 2016, *A&A*, 595, A133
Clementini, G., Ripepi, V., Molinaro, R., *et al.* 2019, *A&A*, 622, A60
Cross, N. J. G., Collins, R. S., Mann, R. G., *et al.* 2012, *A&A*, 548, A119
Emerson, J., McPherson, A., & Sutherland, W. 2006, *Msngr*, 126, 41
Feast, M. W., Laney, C. D., Kinman, T. D., *et al.* 2008, *MNRAS*, 386, 2115
Gaia Collaboration, Prusti, T., de Bruijne, J. H. J., *et al.* 2016, *A&A*, 595, A1
Harris, W. E. 2010, *arXiv*:1012.3224
Inno, L., Matsunaga, N., Bono, G., *et al.* 2013, *ApJ*, 764, 84
Leroy, A. M., & Rousseeuw, P. J. 1987, *Robust Regression and Outlier Detection* (Wiley, New York)
Madore, B. F. 1982, *ApJ*, 253, 575
Moretti, M. I., Clementini, G., Muraveva, T., *et al.* 2014, *MNRAS*, 437, 2702
Ngeow, C.-C., Bhardwaj, A., Henderson, J.-Y., *et al.* 2022, *AJ*, 164, 154
Pietrzyński, G., Graczyk, D., Gallenne, A., *et al.* 2019, *Nature*, 567, 200

Riess, A. G., Yuan, W., Macri, L. M., *et al.* 2022, *ApJL*, 934, L7

Ripepi, V., Moretti, M. I., Clementini, G., *et al.* 2012, *Ap&SS*, 341, 51

Ripepi, V., Moretti, M. I., Marconi, M., *et al.* 2012, *MNRAS*, 424, 1807

Ripepi, V., Moretti, M. I., Marconi, M., *et al.* 2015, *MNRAS*, 446, 3034

Ripepi, V., Marconi, M., Moretti, M. I., *et al.* 2016, *ApJS*, 224, 21

Ripepi, V., Cioni, M.-R. L., Moretti, M. I., *et al.* 2017, *MNRAS*, 472, 808

Ripepi, V., Molinaro, R., Musella, I., *et al.* 2019, *A&A*, 625, A14

Ripepi, V., Catanzaro, G., Molinaro, R., *et al.* 2020, *A&A*, 642, A230

Ripepi, V., Chemin, L., Molinaro, R., *et al.* 2022, *MNRAS*, 512, 563

Ripepi, V., Clementini, G., Molinaro, R., *et al.* 2023, *A&A*, 674, A17

Skowron, D. M., Skowron, J., Udalski, A., *et al.* 2021, *ApJS*, 252, 23

Schlegel, D. J., Finkbeiner, D. P., & Davis, M. 1998, *ApJ*, 500, 525

Soszyński, I., Udalski, A., Szymański, M. K., *et al.* 2015, *MemSAIt*, 86, 257

Soszyński, I., Udalski, A., Szymański, M. K., *et al.* 2018, *AcA*, 68, 89

Vallenari, A., Arenou, F., Bellazzini, M., *et al.* 2022, *Gaia DR3 documentation*, European Space Agency; Gaia Data Processing and Analysis Consortium

At the crossroads of astrophysics and cosmology: Period–luminosity relations in the 2020s
Proceedings IAU Symposium No. 376, 2024
R. de Grijs, P. A. Whitelock & M. Catelan, eds.
doi:10.1017/S1743921323003447

Sub-percent binary star masses and distances from interferometric observations

Alexandre Gallenne

Universidad de Concepción, Departamento de Astronomía, Casilla 160-C, Concepción, Chile
French-Chilean Laboratory for Astronomy, IRL 3386, CNRS, Casilla 36-D, Santiago, Chile.
email: alexandre.gallenne@gmail.com

Abstract. Our knowledge of stellar evolution relies on constraints provided by measurements of the physical stellar properties such as the mass, effective temperature, and radii. The most fundamental parameter, the stellar mass, is rarely available or has a low accuracy, providing poor constraints on the stellar structure and evolution. Observing binary stars combining astrometry and spectroscopy offers the unique opportunity to measure very precise masses. In addition, double-lined spectroscopic binaries provide independent distance measurements with an extreme accuracy, allowing to test the *Gaia* parallaxes and the period-luminosity (P-L) relations. I will show that masses and distances with an accuracy level as high as 0.05% can be obtained by combining interferometric and spectroscopic observations for different types of binary systems, i.e. binary Cepheids, eclipsing and non-eclipsing binaries.

Keywords. Techniques: High angular resolution, Interferometric, Binaries (including multiple): Close, Stars: Fundamental parameters

1. Introduction

Detached binary systems are the only tool enabling direct and precise distance and mass measurements. When the spectral lines of both components can be detected, the orbital motion of the two stars can be measured via radial velocity (RV) measurements, which provides the spectroscopic orbit of the system. However, this only provides the mass ratio. With the apparent orbit from astrometry, we can measure the orbital inclination and obtain the individual masses of the system. Icing on the cake, spectroscopy provides the projected linear semi-major axis while astrometry gives its angular size, which directly provides the orbital parallax of the system. This is the only geometric and model-independent way of measuring masses and distances of stars (see, e.g., McAlister 1976; Torres et al. 2004; Pourbaix et al. 2004; Docobo and Andrade 2013; Torres et al. 2015; Gallenne et al. 2016).

Eclipsing binary (EB) systems are the main source of precise stellar parameters, combining RVs with photometric measurements during the eclipses (see, e.g., Andersen et al. 1984; Milone et al. 1992; Pietrzyński et al. 2009, 2013; Kirkby-Kent et al. 2016; Pietrzyński et al. 2019; Graczyk et al. 2020). Although a precision level of $1-3\,\%$ is routinely achieved, it is still subject to some modelling of the light curves, such as for limb darkening, oblateness, or stellar spots. To determine the distance, a surface brightness-colour relation is usually used to estimate the angular diameter of the stars, which is then combined with the linear value measured from the eclipses. EBs are a powerful tool and already provide precise masses and distances; however, this still depends on some modelling that needs to be well calibrated.

The combination of astrometry with RV is the optimum way to measure the basic stellar properties using a minimum of theoretical assumptions. An astrometric orbit can be measured using different observing techniques, from direct imaging to long-baseline interferometry (LBI). In this talk, I focused on the LBI of binary stars, which probes a different spatial scale, well below the diffraction limit of a single-dish telescope, and sensitive to orbital periods more suitable for monitoring observations (i.e., a few years at most, in contrast to direct imaging, which is sensitive to wide binaries, and therefore to systems with very long orbital periods). This technique has now proven its efficiency in terms of angular resolution and accuracy for close-in binary stars (see, e.g., Baron et al. 2012; Le Bouquin et al. 2013; Gallenne et al. 2013b, 2014b, 2015b; Pribulla et al. 2018; Gardner et al. 2018; Lester et al. 2020; Gardner et al. 2021), and it can now reach an astrometric precision of a few μas.

With the improvement of observing and analysis techniques, finer processes and stellar structures are being revealed (for instance, the limb-darkening effect has to be taken into account in the analysis of the photometric light curves of eclipsing binaries). Such observations provide a very precise determination of the stellar parameters, at $< 1\%$ level. Their comparison with predicted values from stellar evolution models start to show some discrepancies between models, which in turn provides a large range of possible ages when fitting isochrones. Models differ in various ways, as for instance in the input physics, the initial chemical compositions, the treatment of convective-core overshooting, the rotational mixing or the mixing length parameter (Marigo et al. 2017; Bressan et al. 2012). With high-precision mass measurements, evolutionary models can be tightly constrained and provide a better understanding of the stellar interior physics. We are now able to check, for instance, if the input physics is incomplete, the validity of the mixing length approximation in the convective atmosphere, or the efficiency of the convective core overshooting (Higl et al. 2018; Claret and Torres 2018).

When the primary star is a Cepheid, it makes the system even more interesting because Cepheids play a particularly important role as extragalactic distance indicators thanks to the empirical relation between their pulsation period and their luminosity. Binary Cepheids offer the opportunity to measure accurate distances and masses through the combination of astrometric and spectroscopic RV measurements. They will also help to make progress in resolving the Cepheid mass problem. For many years, stellar evolutionary models have predicted Cepheid masses larger than those derived from pulsation models (Bono et al. 1999, 2006; Keller 2008; Neilson et al. 2011).

2. Binary Cepheids

Measuring the mass and the distance of a Cepheid is a challenging task because of the small angular size of the orbits, which hampers astrometric measurements using single-dish telescopes, and the high contrast between the Cepheid and its companion, which complicates the detection of the companion's spectral lines and the measurement of its RV. A number of companions of Galactic Cepheids have been detected from RV measurements and the variability of the systemic velocity of the Cepheids (see, e.g., Szabados 1989, 1991). An ultraviolet spectroscopic survey was also carried out with the International Ultraviolet Explorer (IUE), where companions were detected from low- and high-resolution spectra (Böhm-Vitense and Proffitt 1985; Evans 1992), providing us with a range of spectral types. From an evolutionary timescale point of view, most of the companions should be stars close to the main sequence, and because of the Cepheid's brightness, only bright (and hence massive) companions can be detected by photometric or spectroscopic surveys. Fainter (and hence less massive) companions have a small effect on the Cepheid's astrometry, and might be detected from high-precision radial velocity measurements. However, because of non-symmetric lines from the Cepheid's atmosphere,

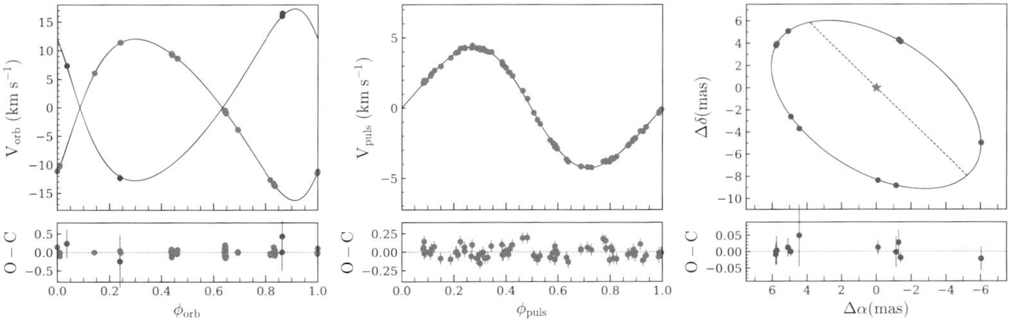

Figure 1. Left: fitted (solid lines) and extracted primary (blue dots) and secondary (red dots) orbital velocities. Middle: fitted (solid line) and extracted (blue dots) pulsation velocity. Right: modelled (black line) and measured (red points) relative astrometric orbit of V1334 Cyg Ab (Gallenne et al. 2018a).

a precision of the order of 1 m s^{-1} is not possible. So far, all binary Cepheids are single-line spectroscopic binaries (SB1) for which masses and distances are degenerate parameters.

LBI has an important impact by spatially resolving the orbiting companion and measuring its astrometric position relative to the Cepheid. First detections were reported by Gallenne et al. (2013a, 2014b,a) who combined for the first time astrometric measurements with the RVs of the Cepheids. However, only lower limits on the Cepheid mass and distance can be determined this way. In 2018, we reported the most precise mass for the Galactic Cepheid V1334 Cyg and the most precise distance to a Cepheid (Gallenne et al. 2018a) by combining astrometry, ground-based RVs for the Cepheid and new space-based RVs for the companion (Proffitt et al. 2017). We measured the mass and distance of the Cepheid at the 3 % and 1 % accuracy level, respectively. The orbital fit is shown in Figure 1.

With such precise measurements, we can do two things. First, we can compare the measured mass of the Cepheid with the predicted value from evolutionary models. Evans et al. (2018) showed that the mass is significantly lower than the prediction by fitting to evolutionary tracks in the Hertzsprung-Russell diagram, regardless of the code or rotation. Second, our measured distance can be compared to period-luminosity relations and the *Gaia* parallax (Ripepi et al. 2023). As we can see in Figure 2, only the newest relations agree with the measured value, even after correcting for the flux contamination from the companion. The latest *Gaia* measurements (Gaia Collaboration 2023) are more than 1σ away, but it is likely due to known problems with bright and pulsating stars (Gaia Collaboration 2023), which should be corrected in the next data releases.

Finally, *Gaia* will provide a unique opportunity to study binary Cepheids for which we have an astrometric orbit but no measured velocities for the companions. Assuming the distance, we will be able to measure the mass of SB1 binary systems. An example of such a system is presented in Figure 3, showing the astrometric observations of AW Per obtained during several years and combined with RVs of the Cepheid taken from the literature.

3. Binary stars: general case

Binary Cepheids are challenging systems, but in general, a lot more binary stars with a more favourable contrast enable easier spectroscopic and interferometric observations. Accurate mass determinations in binary systems in different evolutionary stages serve as benchmarks for calibrating theoretical stellar models and improving our understanding of the physical processes occurring within stars. In addition, they can be used as benchmark

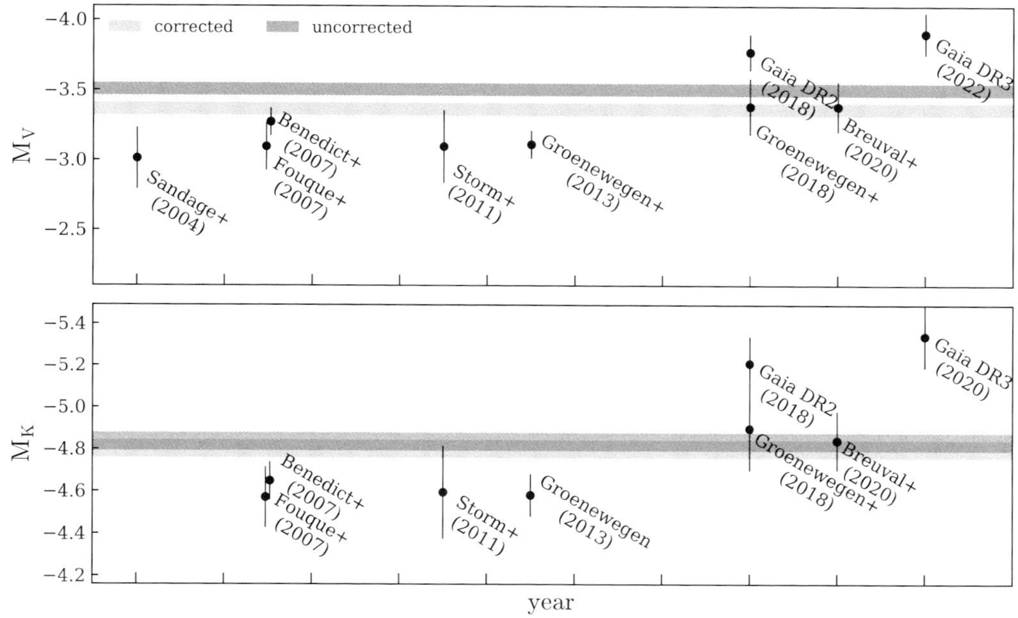

Figure 2. Comparison between the absolute magnitudes of V1334 Cyg predicted from literature P-L relations (black dots) and the present distance measurement (red and blue areas) in the V and K photometric bands (upper and bottom panels, respectively). The blue and red areas represent the measured absolute magnitude with and without the subtraction of the companion's flux contamination, respectively. In the K band, these two regions overlap as the contribution of the companion is smaller (updated from Gallenne et al. 2018a).

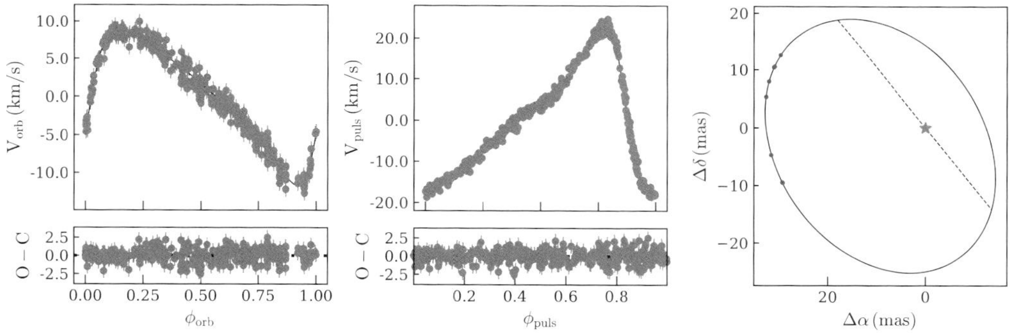

Figure 3. Same as Figure 1 but for the Cepheid AW Per. RVs are taken from Griffin et al. (2016) and the CHARA/MIRC interferometric astrometry from Gallenne et al. (2015b, updated).

systems to check or validate the *Gaia* parallaxes. Previous works show that combining spectroscopy and interferometry provides very precise and accurate measurements (see, e.g., Hummel et al. 2001; Konacki et al. 2010). However, to constrain stellar interior parameters and double-check *Gaia*, it requires an observational precision at a sub-percent level, which is rarely achieved (e.g., *Gaia* provides parallaxes at the μas level).

Using the PIONIER instrument (Precision Integrated-Optics Near-infrared Imaging ExpeRiment, Le Bouquin et al. 2011) at the Very Large Telescope Interferometer (Haguenauer et al. 2010), we collected interferometric measurements of the eclipsing binary system TZ For with an average astrometric precision of 50μas. Combining the astrometry with new precise radial velocities, we were able to measure the mass of each

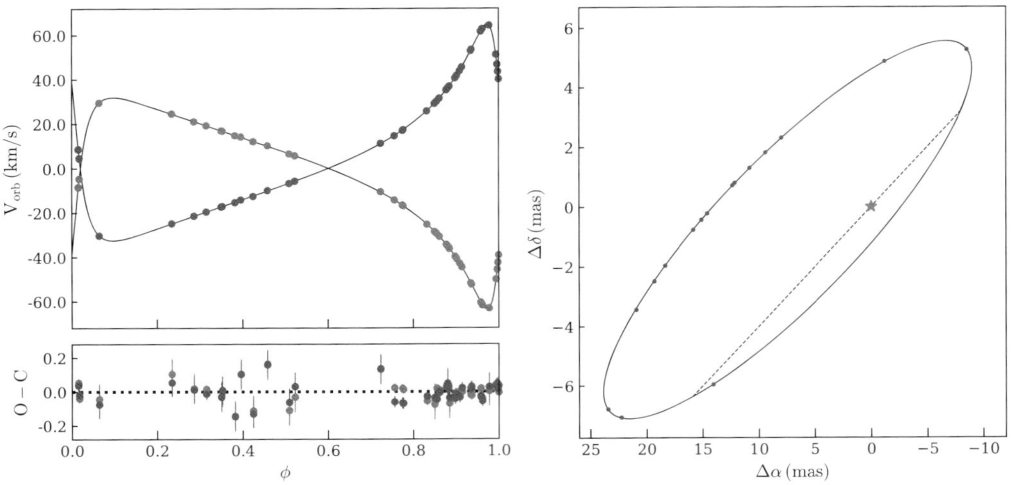

Figure 4. Combined fit of HD188088. Upper left: radial velocities of the primary (blue) and the secondary (red) star. Lower left: Residuals of the velocities. Right: modelled (black) and measured (red) GRAVITY astrometric orbit. The shaded grey area represents the 1σ orbit (Gallenne et al. 2023).

component with a precision and an accuracy of 0.05 % and the distance to the system at 0.4 % precision and accuracy levels (Gallenne et al. 2016). These are the most precise and accurate masses available in the literature. We then observed additional eclipsing systems for which we obtained the same level of precision and accuracy (Gallenne et al. 2019). However, due to the internal calibration of the instrument wavelength scale, we had a limited accuracy on any dimensional measurements of ~ 0.4 % (Gallenne et al. 2018b).

The purpose of observing eclipsing binaries with PIONIER was to check the calibration of the surface brightness-colour relation (Pietrzyński et al. 2019; Gallenne et al. 2018b), but it also resulted in very accurate mass and distance determinations. In light of these unprecedented results, I decided to start a new project for any type of "simple" detached binary stars with the interferometric instrument GRAVITY (Eisenhauer et al. 2011). This instrument provides a much better accuracy of the wavelength calibration thanks to a dedicated internal reference laser source. The goal is to have a large sample of benchmark stars with high-accuracy mass and distance measurements to test stellar evolutionary models and the *Gaia* parallaxes. First results were published in Gallenne et al. (2023), where we demonstrated that extremely precise and accurate (i.e., at the 0.03 % level) dynamical masses and orbital parallaxes can be determined. We display an example of one fitted orbit in Figure 4, for which we obtained an astrometric residual scatter of $\sim 20\,\mu$as. For this project, we combined our new interferometric measurements with new spectroscopic observations from VLT/UVES (UV Echelle Spectrograph, Dekker et al. 2000) in order to achieve the best possible accuracy.

We confronted our measurements with four stellar evolution models and we showed that with this level of precision on individual masses theory is clearly deficient for most of the systems when simultaneously fitting stellar parameters ($T_{\rm eff}$, R, L, and M). This stresses the importance of precise stellar measurements for stellar evolution modelling and its calibration. Unfortunately, with precomputed isochrones (and predefined evolution parameters) and without more precise and accurate measurements of the metallicity and temperature, it is impossible to conclude which specific evolution model is more appropriate.

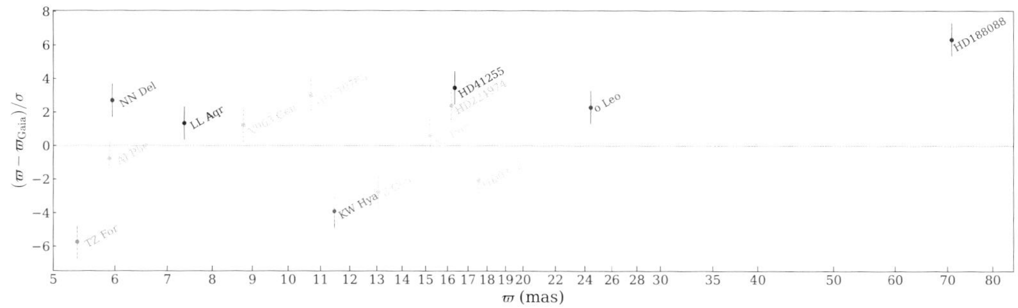

Figure 5. Comparison of our orbital parallaxes with the *Gaia* data release 3 measurements (Gallenne et al. 2023).

The very precise orbital parallaxes also provide a stringent test of *Gaia* measurements. As displayed in Figure 5, we found that 50 % (8/16 stars, including our previous works) of our sample is $> 1\sigma$ away from the *Gaia* parallax, and within the "nominal" Renormalised Unit Weight Error (RUWE) range $1 - 1.4$ (the RUWE is an astrometric quality indicator†). This can be problematic for stars with unresolved companions, which would bias the parallax. We did not see any correlation between the RUWE metric and the relative parallax difference (i.e., the relative difference between *Gaia* and our measured parallaxes), but we did confirm a correlation with the semi-major axis of the photocenter. Finally, all systems are within the bright-star magnitude range of *Gaia*, and therefore may be subject to additional systematic errors (see, e.g., Cantat-Gaudin and Brandt 2021; Lunz et al. 2022).

4. Concusions

I presented a new generation of interferometric and spectroscopic observations of binary stars, including Cepheids. I simultaneously fit the interferometric astrometry and RVs to measure stellar masses and distances with a high precision and accuracy.

In the case of binary Cepheids, I am able to measure the astrometric orbit of the close companions with a very good accuracy. When combined with Space Telescope Imaging Spectrograph (STIS) spectra on board the *Hubble Space Telescope* (HST), the mass of the Cepheid and its distance can be determined with an accuracy level better than 3 %, as demonstrated for the Cepheid V1334 Cyg. The comparison of the measured mass with prediction from evolutionary models shows that the dynamical mass is smaller than expected. The causes are still unknown, but the discrepancy may be related to a mass-loss process, the system being the product of a binary merger, or issues with evolution models. Direct and independent distance measurements of Cepheids are important and provide a stringent test of existing P-L relations and *Gaia* parallaxes. We showed that most of the relations are in marginal agreement with our value but the latest ones seem to provide a good estimate. The *Gaia* parallax is also $\sim 2 - 3\sigma$ away from our measured value, but there are known issues with bright and pulsating stars, so this is somewhat expected. Hopefully this will be solved in the next data release.

Observations of low-contrast binary systems offer a unique opportunity of measuring masses and distances with a very high precision and accuracy. New spectroscopic and interferometric instruments can now provide very precise measurements, with RVs of the order of a few m/s and interferometric astrometry as precise as a few μas. Our recent

† https://gea.esac.esa.int/archive/documentation/GDR2/Gaia_archive/chap_datamodel/sec_dm_main_tables/ssec_dm_ruwe.html

works showed that uncertainties as small as 0.03 % on the distances and masses are possible. A comparison with previous studies and different datasets demonstrated that our measurements are both precise and accurate. This is possible thanks to the precision and sensitivity of the GRAVITY instrument, which provides exquisite differential astrometry ($10 - 20\,\mu$as). The comparison of our orbital parallaxes with *Gaia* showed differences at a $1 - 4\sigma$ level. A possible explanation is that all stars are in the bright-star magnitude range of the detector, and therefore possibly affected by additional systematics. Our measured masses were also used to test four isochrone models. We showed that theory is still deficient for most of the systems in fitting all observables, especially with such precision level on the masses. To reconcile the models, it is likely that a fine-tuning of the models of each star in a system is necessary, as was done by Graczyk et al. (2016). With such precise masses, stellar interior parameters such as the mixing length and envelope overshooting can now be better constrained and lead to an improved calibration of stellar evolution models.

References

Andersen, J. & Vaz, L. P. R. 1984, *A&A*, 130, 102

Baron, F., Monnier, J. D., Pedretti, E., *et al.* 2012, *ApJ*, 752, 20

Böhm-Vitense, E. & Proffitt, C. 1985, *ApJ*, 296, 175

Bono, G., Caputo, F. & Castellani, V. 2006, *Mem. Soc. Astron. Italiana*, 77, 207

Bono, G., Marconi, M., & Stellingwerf, R. F. 1999, *ApJS*, 122, 167

Bressan, A., Marigo, P., Girardi, L., *et al.* 2012, *MNRAS*, 427, 127

Cantat-Gaudin, T. & Brandt, T. D. 2021, *A&A*, 649, A124

Claret, A. & Torres, G. 2018, *ApJ*, 859, 100

Dekker, H., D'Odorico, S., Kaufer, *et al.* 2000, *SPIE Conference Series*, ed. M. Iye & A. F. Moorwood, Vol. 4008, 534

Docobo, J. A. & Andrade, M. 2013, *MNRAS*, 428, 321

Eisenhauer, F., Perrin, G., Brandner, W., *et al.* 2011, *The Messenger*, 143, 16

Evans, N. R. 1992, *ApJ*, 384, 220

Evans, N. R., Karovska, M., Bond, H. E., *et al.* 2018, *ApJ*, 863, 187

Gaia Collaboration, Vallenari, A., Brown, A. G. A., *et al.* 2023, A&A, 674, A1

Gallenne, A., 2015, *in RMxAC Conference Series: VI Reunión de Astronomía Dinámica en Latinoamérica* Vol. 46, 32

Gallenne, A., Kervella, P., Evans, N. R., *et al.* 2018a, *ApJ*, 867, 121.

Gallenne, A., Kervella, P., Mérand, A., *et al.* 2013a, *in EAS Publications Series*, ed. Pavlovski K., Tkachenko, A. & Torres, G., Vol. 64, 197–204.

Gallenne, A., Kervella, P. & Mérand, A., 2014a, *IAU Symposium*, ed. Guzik, J. A., Chaplin, W. J., Handler, G., & Pigulski, Vol. 301, 411

Gallenne, A., Mérand, A., & Kervella, P., 2015a, *The Physics of Evolved Stars: A Conference Dedicated to the Memory of Olivier Chesneau*, ed. E. Lagadec, F. M. & Lanz, T., Vol. 71, 187

Gallenne, A., Mérand, A., Kervella, P., *et al.* 2014b, *A&A*, 561, L3

Gallenne, A., Mérand, A., Kervella, P., *et al.* 2023, *A&A*, 672, A119

Gallenne, A., Mérand, A., Kervella, P., *et al.* 2015b, *A&A*, 579, A68

Gallenne, A., Monnier, J. D., Mérand, A., *et al.* 2013b, *A&A*, 552, A21

Gallenne, A., Pietrzyński, G., Graczyk, D., *et al.* 2016, *A&A*, 586, A35.

Gallenne, A., Pietrzyński, G., Graczyk, D., *et al.* 2018b, *A&A*, 616, A68

Gallenne, A., Pietrzyński, G., Graczyk, D., *et al.* 2019, *A&A*, 632, A31.

Gardner, T., Monnier, J. D., Fekel, F. C., *et al.* 2021, *AJ*, 161, 40

Gardner, T., Monnier, J. D., Fekel, F. C., *et al.* 2018, *ApJ*, 855, 1

Graczyk, D., Pietrzyński, G., Thompson, I. B., *et al.* 2020, *ApJ*, 904, 13

Graczyk, D., Smolec, R., Pavlovski, K., *et al.* 2016, *A&A*, 594, A92

Griffin, R. F. 2016, *The Observatory*, 136, 209

Haguenauer, P., Alonso, J., Bourget, P., *et al.* 2010, *in SPIE Conference Series*, ed. Danchi, W. C, Delplancke, F., Rajagopal, J. K., Vol. 7734
Higl, J., Siess, L., Weiss, A., *et al.* 2018, *A&A*, 617, A36
Hummel, C. A., Carquillat, J. M., Ginestet, N., *et al.* 2001, *AJ*, 121, 1623
Keller, S. C. 2008, *ApJ*, 677, 483
Kirkby-Kent, J. A., Maxted, P. F. L., Serenelli, A. M., *et al.* 2016, *A&A*, 591, A124
Konacki, M., Muterspaugh, M. W., Kulkarni, S. R., *et al.* 2010, *ApJ*, 719, 1293
Le Bouquin, J.-B., Berger, J.-P., Lazareff, B., *et al.* 2011, *A&A*, 535, A67
Le Bouquin, J.-B., Beust, H., Duvert, G., *et al.* 2013, *A&A*, 551, A121
Lester, K. V., Fekel, F. C., Muterspaugh, M., *et al.* 2020, *AJ*, 160, 58
Lunz, S., Anderson, J., Xu, M., *et al.* 2022, *in European VLBI Network Mini-Symposium and Users' Meeting 2021*, Vol. 2021, 32
Marigo, P., Girardi, L., Bressan, A., *et al.* 2017, *ApJ*, 835, 77
McAlister, H. A. 1976, *PASP*, 88, 957
Milone, E. F., Stagg, C. R., & Kurucz, R. L. 1992, *ApJS*, 79, 123
Neilson, H. R., Cantiello, M., & Langer, N. 2011, *A&A*, 529, L9
Pietrzyński, G., Graczyk, D., Gallenne, A., *et al.* 2019, *Nature*, 567, 200
Pietrzyński, G., Graczyk, D., Gieren, W., *et al.* 2013, *Nature*, 495, 76
Pietrzyński, G., Thompson, I. B., Graczyk, D., *et al.* 2009, *ApJ*, 697, 862
Pourbaix, D., Tokovinin, A. A., Batten, A. H., *et al.* 2004, *A&A*, 424, 727
Pribulla, T., Mérand, A., Kervella, P., *et al.* 2018, *A&A*, 616, A49
Proffitt, C. R., Evans, N. R., Winston, E. M., *et al.* 2017, *in European Physical Journal Web of Conferences*, Vol.152, 04003
Ripepi, V., Clementini, G., Molinaro, R., *et al.* 2023, A&A, in press [e-prints: 2206.06212]
Torres, G. 2004, *in Astronomical Society of the Pacific Conference Series*, ed. R. W. Hilditch, H. Hensberge, & K. Pavlovski, Vol. 318, 123
Torres, G., Claret, A., Pavlovski, K., *et al.* 2015, *ApJ*, 807, 26
Szabados, L. 1989, *Commun. of the Konkoly Observatory Hungary*, 94, 1
Szabados, L. 1991, *Commun. of the Konkoly Observatory Hungary*, 96, 123

At the crossroads of astrophysics and cosmology: Period–luminosity relations in the 2020s
Proceedings IAU Symposium No. 376, 2024
R. de Grijs, P. A. Whitelock & M. Catelan, eds.
doi:10.1017/S1743921323002880

Globular cluster metallicities and distances from disentangling their RR Lyrae light curves

A. Arellano Ferro

Instituto de Astronomía, Universidad Nacional Autónoma de México, Ciudad Universitaria, 04510, Ciudad de México, México. email: armando@astro.unam.mx

Abstract. We present mean horizontal branch absolute magnitudes and iron abundances for a sample of 39 globular clusters. These quantities were calculated in an unprecedented homogeneous fashion based on Fourier decomposition of ligt curves of RR Lyrae cluster members. Zero points for the luminosity calibrations are discussed. Our photometrically derived metallicities and distances compare very well with spectroscopic determinations of [Fe/H] and accurate distances obtained using *Gaia* and *Hubble Space Telescope* data. The need to distinguish between the results for RRab and RRc stars for a correct evaluation of the M_V–[Fe/H] relation is discussed. For RRab stars, the relation is non-linear, and the horizontal branch structure plays a significant role. For RRc stars, the relation remains linear and tight, and the slope is very shallow. Hence, the RRc stars seem better indicators of the parental cluster distances. Systematic time-series CCD imaging performed over the last 20 years enabled to discover and classify 330 variables in our sample of globular clusters.

Keywords. Globular clusters metallicities, Horizontal branch luminosity, stars: RR Lyrae stars, stars: Variable stars in clusters.

1. Introduction

The Fourier light curve decomposition technique is based on the representation of the observed light curve by a series of harmonics of adequate amplitudes and displacements, A_k and ϕ_k, in an equation of the form,

$$m(t) = A_0 + \sum_{k=1}^{N} A_k \cos\left(\frac{2\pi k}{P}(t-E) + \phi_k\right), \tag{1}$$

where $m(t)$ is the magnitude at time t, P is the pulsation period, and E is the epoch, typically selected as the time of maximum brightness. A linear minimization routine is used to derive the best-fitting values for the amplitudes A_k and phases ϕ_k of the sinusoidal components. The Fourier parameters, defined as $\phi_{ij} = j\phi_i - i\phi_j$ and $R_{ij} = A_i/A_j$, are computed from the amplitudes and phases of the harmonics in eq. 1. In principle, a proper combination of some of these Fourier parameters, to be determined on theoretical and/or empirical grounds, can yield appropriate physical stellar parameters. The demonstration of this fact and its practical application has taken a few decades, and it traces its origins to the pioneering work of van Albada & Baker (1971) and Simon (1988a,b). Simon employed hydrodynamic pulsation models to construct light curves and experimented with Fourier decomposition in search of useful correlations, eventually expressing that "Finally, we argue that if a reliable hydrodynamic pulsation code were available, the Fourier technique would be capable of determining masses, luminosities and temperatures of individual stars, in many cases using observations already in hand." And, in fact, hydrodynamical models were developed in another early work by Simon & Clement (1993), where the significance of the ϕ_{31} parameter in the estimation of the stellar mass and luminosity of RRc stars was demonstrated.

Towards the end of the 20th century, a program aimed at calculating accurate empirical relations for the physical parameters of RR Lyrae stars was developed at the Konkoly Observatory. Soon, robust and simple empirical calibrations emerged for the determination of [Fe/H] (Jurcsik & Kovács 1996) and the absolute magnitude (Kovács & Jurcsik 1996) in terms of the pulsation period, yielding the Fourier A_1 and ϕ_{31} parameters for a large sample of RRab stars in the globular clusters of the Sculptor dwarf galaxy. The then-recent set of iron abundances (Suntzeff, Kraft, & Kinman 1994; Layden 1994) and photometric data (Kovács & Zsoldos 1995; Lub 1977) were employed.

Soon after, it was argued that some hydrodynamical pulsation models did not follow the empirical light curves (Kovács 1998; Kovács & Kanbur 1998), and tso empirical calibrations were preferred (Kovács & Walker 2001).

A thorough [Fe/H] calibration for RRc stars (Morgan, Wahl, & Wieckhorst 2007) employed a sample of 106 calibrator stars in globular clusters and rendered useful formulae for the iron abundance estimation on the Zinn–West metallicity scale (Zinn & West 1984). More recently, a further effort to improve the accuracy of the calibrations, using high precision *Kepler* photometry and detailed spectroscopic iron abundances for RRab and RRc stars, was performed (Nemec et al. 2013). Twenty-six calibrators were included in the RRab calibration (nine with Blazhko modulations) and 101 calibrators for the RRc calibration.

For nearly 20 years our research group has been interested in performing extensive time-series CCD photometry in the *VI* bands for a sample of Galactic globular clusters. We aimed at estimating cluster mean metallicities and distances from their member RR Lyrae stars via the Fourier decomposition technique, and from other variables in a cluster when available, e.g. SX Phe via their period–luminosity (P–L) relation, or red giant semi-periodic variables near the tip of the red giant branch (RGB). Meanwhile, other aspects of the clusters under study have been addressed, such as their reddening, stellar membership and a census of their variable star populations; their periods and classifications have also been updated. As a natural by-product we discovered and classified about 330 new variables.

Numerous RR Lyrae light curve decompositions and their implied physical parameters can be found in the literature, but often different calibrations and zero points to estimate [Fe/H] and M_V are used. For the determination of these quantities, in earlier investigations, our group also used some different equations, since we were also learning what were probably the best and most reliable calibrations. Hence, we decided to reconsider all our data and determine Fourier fits to pursue homogeneous calculations with the selected calibrations, for our entire sample of globular clusters (Arellano Ferro 2022). The specific calibrations employed will be presented in Section 3.

The present paper is organized as follows. In Section 4, we summarize the resulting values of reddening, [Fe/H] and M_V (and, hence, distance) for our sample of globular clusters and compare them with spectroscopically determined iron abundance values, while comparing the distances with well established mean distances based on *Gaia* and *Hubble Space Telescope* (*HST*) data. In Section 5 we will revisit the M_V–[Fe/H] correlation in light of the homogeneous set of Fourier [Fe/H] and M_V values. We highlight the need to split the discussion of results obtained from RRab stars and those from RRc stars, as they are of an obviously different nature. In Section 6 we introduce the role of the HB structure on the M_V–[Fe/H] relation and discuss its results. Section 7 recounts briefly the census of new variables discovered in this project. Finally, in Section 8 we summarize our results and conclusions.

2. CCD Photometry based on the DIA approach

All our CCD photometry obtained over the years has been performed based on Difference Image Analysis (DIA) through the pipeline DanDIA (Bramich 2008; Bramich et al. 2013, 2015). A detailed description of the method and its caveats can be found in Bramich et al. (2011). A good overview of all our procedures applied to the analysis of the individual clusters is found in Deras et al. (2019) and Yepez et al. (2020).

3. The [Fe/H] and M_V calibrations employed

3.1. *The [Fe/H] calibrations*

For the calculation of [Fe/H] we adopted the following calibrations:

$$[\mathrm{Fe/H}]_{\mathrm{J}} = -5.038 - 5.394P + 1.345\phi_{31}^{(s)}, \tag{2}$$

and

$$[\mathrm{Fe/H}]_{\mathrm{ZW}} = 52.466P^2 - 30.075P + 0.131\phi_{31}^{(c)2} - 0.982\phi_{31}^{(c)} - 4.198\phi_{31}^{(c)}P + 2.424, \tag{3}$$

from Jurcsik & Kovács (1996) and Morgan, Wahl, & Wieckhorst (2007) for RRab and RRc stars, respectively. The phases $\phi^{(c)}$ and $\phi^{(s)}$ are calculated either based on a cosine or a sine series, respectively, and they are correlated via $\phi^{(s)} = \phi^{(c)} - \pi$. The iron abundance on the Jurcsik & Kovács (1996) scale, $[\mathrm{Fe/H}]_{\mathrm{J}}$, can be converted to the Zinn & West (1984) scale via $[\mathrm{Fe/H}]_{\mathrm{J}} = 1.431\ [\mathrm{Fe/H}]_{\mathrm{ZW}} + 0.88$ (Jurcsik 1995). All $[\mathrm{Fe/H}]_{\mathrm{ZW}}$ values can be converted to the UVES spectroscopic scale of Carretta et al. (2009) via $[\mathrm{Fe/H}]_{\mathrm{UV}} = -0.413 + 0.130\ [\mathrm{Fe/H}]_{\mathrm{ZW}} - 0.356\ [\mathrm{Fe/H}]^2_{\mathrm{ZW}}$.

We have also considered the more recent non-linear calibrations calculated by Nemec et al. (2013) for both RRab and RRc stars. These calibrations made use of high-precision *Kepler* photometry and spectroscopic iron abundances for 26 RRab and 101 RRc calibrators. Their resulting equations are, respectively,

$$[\mathrm{Fe/H}]_{\mathrm{N}} = -8.65 - 40.12P + 5.96\phi_{31}^{(s)}(K) + 6.27\phi_{31}^{(s)}(K)P - 0.72\phi_{31}^{(s)}(K)^2, \tag{4}$$

where $\phi_{31}^{(s)}(K) = \phi_{31}^{(s)} + 0.151$ is given on the *Kepler* scale (Nemec et al. 2013), and

$$[\mathrm{Fe/H}]_{\mathrm{N}} = 1.70 - 15.67P + 0.20\phi_{31}^{(c)} - 2.41\phi_{31}^{(c)}P + 18.0P^2 + 0.17\phi_{31}^{(c)^2}. \tag{5}$$

The $[\mathrm{Fe/H}]_{\mathrm{N}}$ values resulting from these calibrations, calibrated with respect to spectroscopic iron abundances, should produce iron abundances on the spectroscopic UVES scale, $[\mathrm{Fe/H}]_{\mathrm{UV}}$.

3.2. *The M_V calibrations*

For the calculation of M_V we adopted the calibrations:

$$M_V = -1.876\ \log P - 1.158A_1 + 0.821A_3 + 0.41, \tag{6}$$

$$M_V = -0.961P - 0.044\phi_{21}^{(s)} - 4.447A_4 + 1.061, \tag{7}$$

from Kovács & Walker (2001) and Kovács (1998) for the RRab and RRc stars, respectively. We call attention to the zero points in the above two equations, which do not correspond to those in the original publications. The zero points of eqs. 6 and 7 have been calculated to scale the luminosities of RRab and RRc stars to the distance modulus of 18.5 mag for the Large Magellanic Cloud (LMC) (Freedman et al. 2001; Pietrzyński et al. 2013; de Grijs et al. 2014), assuming a mean magnitude for the RR Lyrae stars in the LMC of $V_0 = 19.064 \pm 0.064$ mag (Clementini et al. 2003), and a mean absolute magnitude for the star RR Lyrae of $M_V = 0.61 \pm 0.10$ mag (Benedict et al. 2002). Other discussions as to these zero points can be found in Kinman (2002) and Cacciari, Corwin, & Carney (2005). We refer the reader to section 4.2 of Arellano Ferro, Giridhar, & Bramich (2010) for a detailed discussion of the above zero point values that we finally adopted.

4. The Fourier [Fe/H] and distance values for globular clusters

4.1. *Comparison of Fourier* [Fe/H] *with spectroscopic values*

At this point, we would like to compare our photometric calculations of the iron abundances with independent, solidly sustained values. We have chosen to compare with the values given by Carretta et al. (2009) on the spectroscopic scale $[Fe/H]_{Carr}$, for the clusters in Table 1. In Fig. 1, the photometric values $[Fe/H]_{UV}$ and $[Fe/H]_{N}$, obtained as described in the previous section, are plotted versus $[Fe/H]_{Carr}$. Given that the results for RRab and RRc stars come from different specific calibrations, the comparison is done separately for both pulsation modes. The top two panels, *a* and *b*, show the comparison of $[Fe/H]_{UV}$, obtained based on the calibrations of eqs. 2 and 3, duly transformed to the UVES scale. These plots show good agreement between the photometric iron abundance values obtained from the RRab and the RRc stars and the spectroscopic values. There may be a slight hint that the RRc calibration might need a small correction (~ -0.2 dex) for the most metal-poor clusters ($[Fe/H]_{UV} \leq -1.8$ dex). The bottom two panels, *c* and *d*, show the comparison of the values $[Fe/H]_{N}$, obtained from the calibrations of Nemec et al. (2013) (eqs. 4 and 5). For the RRc star, the comparison is satisfactory and comparable to the case in panel *b*. However, for the RRab stars it is obvious from panel *c* that the metallicities obtained from eq. 4 are largely overestimated. We do not encourage the use of that calibration for the RRab stars. The same conclusion was reached by Varga et al. in a poster delivered at this IAU Symposium. See also the discussion at the end of the present paper.

4.2. *Comparison of Fourier distances with Gaia and HST data based distances*

We now proceed to compare the resulting distances from the absolute magnitude determinations resulting from eqs. 6 and 7 described in the previous section. The calculation of the distance involves a value of $E(B-V)$; the adopted values are listed in Table 1. We perform our comparison with accurate critical mean distances from Baumgardt & Vasiliev (2021, BV21), calculated for a large sample of globular clusters using data from *Gaia* early data release 3 (EDR3) parallaxes, line-of-sight velocity dispersion profiles from *Gaia* and *HST*-based proper motions. The top panel of Fig. 2 shows that our distances compare very well with those of BV21. The distance differences are always < 1.9 kpc and the standard deviation is 0.7 kpc. The smaller bottom panels show the run of distance differences with $[Fe/H]_{UV}$ and with the horizontal branch (HB) structural parameter HBt. In no case seems there to be a dependence of the distances on either of these quantities.

Individual cluster distances are listed in Table 2. As a complement we have estimated the cluster distances from the P–L relation of SX Phe stars. We have taken into consideration the relations derived by Arellano Ferro et al. (2011, AF11) and by Cohen & Sarajedini (2012, CS12). A close look at the table corroborates that these distances are consistent with the determinations from the RR Lyrae stars calculated via the Fourier light curve decomposition.

5. The M_V–[Fe/H] relation

Once we have proven that our photometric determinations of the iron abundances and the absolute magnitudes (hence distances) compare satisfactorily with sound and well-respected independent values, we are in a position to revisit the M_V–[Fe/H] relation as implied by our results. Before we carry on, let us recall that traditionally, the M_V–[Fe/H]

Table 1. Mean values of [Fe/H], on three different scales, and M_V from a homogeneous Fourier decomposition of the light curves of cluster member RR Lyrae stars.[1,2]

GC	Oo	[Fe/H]$_{ZW}$ dex	[Fe/H]$_{UV}$ dex	[Fe/H]$_{N}$ dex	M_V mag	N	[Fe/H]$_{ZW}$ dex	[Fe/H]$_{UV}$ dex	[Fe/H]$_{N}$ dex	M_V mag	N	$E(B-V)$ mag	*HBt*
NGC (M)		RRab					RRc						
1261	I	−1.48±0.05	−1.38	−1.27	0.59±0.04	6	−1.51±0.13	−1.38	−1.41	0.55±0.02	4	0.01	−0.71
1851	I	−1.44±0.10	−1.33	−1.18	0.54±0.03	10	−1.40±0.13	−1.28	−1.28	0.59±0.02	5	0.02	−0.36
3201	I	−1.49±0.10	−1.39	−1.29	0.60±0.04	19	−1.47±0.08	−1.37	−1.36	0.58±0.01	2	diff.	+0.08
4147	I	–	–	–	–	–	−1.72±0.26	−1.68	−1.66	0.57±0.05	6	0.01	+0.55
5272 (M3)	I	−1.56±0.16	−1.46	−1.46	0.59±0.05	59	−1.65±0.14	−1.57	−1.56	0.56±0.06	23	0.01	+0.08
5904 (M5)	I	−1.44±0.09	−1.33	−1.19	0.57±0.08	35	−1.49±0.11	−1.39	−1.38	0.58±0.03	22	0.03	+0.31
6171 (M107)	I	−1.33±0.12	−1.22	−0.98	0.62±0.04	6	−1.02±0.18	−0.90	−0.88	0.59±0.03	4	0.33	−0.73
6229	I	−1.42±0.07	−1.32	−1.13	0.61±0.06	12	−1.45±0.19	−1.32	−1.58	0.53±0.10	8	0.01	+0.24
6266[7] (M62)	I	−1.31±0.11	−1.19	–	0.63±0.03	40	−1.23±0.09	−1.11	–	0.51±0.03	21	diff	+0.55
6362	I	−1.25±0.06	−1.13	−0.83	0.62±0.01	2	−1.21±0.15	−1.09	−1.10	0.59±0.05	6	0.06	−0.58
6366	I	−0.84	−0.77	−0.31	0.71	1	–	–	–	–		0.80	−0.97
6401	I	−1.36±0.09	−1.24	−1.04	0.60±0.07	19	−1.27±0.23	−1.09	−1.16	0.58±0.03	9	diff	+0.13
6712	I	−1.25±0.06	−1.13	−0.82	0.55±0.03	6	−1.10±0.04	−0.95	−0.96	0.57±0.18	3	0.35	−0.62
6934	I	−1.56±0.14	−1.48	−1.49	0.58±0.05	15	−1.53±0.12	−1.41	−1.50	0.59±0.03	5	0.10	+0.25
6981 (M72)	I	−1.48±0.11	−1.37	−1.28	0.63±0.02	12	−1.66±0.08	−1.60	−1.55	0.57±0.04	4	0.06	+0.14
7006	I	−1.51±0.13	−1.40	−1.36	0.61±0.03	31	−1.53	−1.44	−1.43	0.55	1	0.08	−0.28
Pal 2	I	−1.39±0.55	−1.09	−1.20	0.52±0.08	11	–	–	–	–		0.93	–
Pal 13	I	−1.64±0.15	−1.56	−1.67	0.65±0.05	4	–	–	–	–	–	0.10	−0.30
288	II	−1.64	−1.58	−1.42	0.38	1	−1.59	−1.52	−1.54	0.58	1	0.03	+0.98
1904 (M79)	II	−1.63±0.14	−1.55	−1.47	0.41±0.05	5	−1.71	−1.66	−1.69	0.58	1	diff	+0.74
4590 (M68)	II	−2.07±0.09[4]	−2.21	−2.01	0.49±0.07	5	−2.09±0.03	−2.24	−2.23	0.53±0.01	15	0.05	+0.17
5024 (M53)	II	−1.94±0.06[4]	−2.00	−1.68	0.45±0.05	18	−1.84±0.13	−1.85	−1.85	0.52±0.06	3	0.02	+0.81
5053	II	−2.05±0.14[4]	−2.18	−2.07	0.46±0.08	3	−2.00±0.18	−2.05	−2.06	0.55±0.05	4	0.18	+0.50
5286[7]	II	−1.68±0.15	−1.64	–	0.52±0.04	59	−1.71±0.23	−1.68	–	0.57±0.04	23	0.24	+0.80

Table 1. Continued.

GC	Oo	$[Fe/H]_{ZW}$ dex	$[Fe/H]_{UV}$ dex	$[Fe/H]_N$ dex	M_V mag	N	$[Fe/H]_{ZW}$ dex	$[Fe/H]_{UV}$ dex	$[Fe/H]_N$ dex	M_V mag	N	$E(B-V)$ mag	*HBt*
NGC (M)		RRab					RRc						
5466	II	−2.04±0.14[4]	−2.16	−2.01	0.44±0.09	8	−1.90±0.21	−1.89	−1.96	0.53±0.06	5	0.00	+0.58
6139	II	−1.63±0.02	−1.57	−1.33	0.49±0.06	2	−1.62±0.39	−1.55	−1.47	0.56±0.10	2	diff	–
6205 (M13)	II	−1.60	−1.54	−1.00	0.38	1	−1.70±0.20	−1.63	−1.71	0.59±0.05	3	0.02	+0.97
6254 (M10)	II?	–	–	–	–	–	−1.59	−1.52	−1.52	0.52	1	0.25	+1.00
6333 (M9)	II	−1.91±0.13[4]	−1.96	−1.72	0.47±0.04	7	−1.71±0.23	−1.66	−1.66	0.55±0.04	6	diff	+0.87
6341 (M92)	II	−2.12±0.18[4]	−2.16[5]	−2.26	0.45±0.03	9	−2.01±0.11	−2.11	−2.17	0.53±0.06	3	0.02	+0.91
6809[7]	II	−1.61±0.20	−1.55	–	0.53±0.09	59	–	–	–	–	–	0.08	+0.87
7078 (M15)	II	−2.22±0.19[4]	−2.46	−2.65	0.51±0.04	9	−2.10±0.07	−2.24	−2.27	0.52±0.03	8	0.08	+0.67
7089 (M2)	II	−1.60±0.18	−1.51	−1.25	0.53±0.13	10	−1.76±0.16	−1.73	−1.76	0.51±0.08	2	0.06	+0.38[5]
7099 (M30)	II	−2.07±0.05[4]	−2.21	−1.88	0.40±0.04	3	−2.03	−2.14	−2.07	0.54	1	0.03	+0.89
7492[6]	II	−1.89[4,5]	−1.93	−0.83	0.37	1	–	–	–	–	–	0.00	+0.76
6402 (M14)	Int	−1.44±0.17	−1.32	−1.17	0.53±0.07	24	−1.23±0.21	−1.12	−1.12	0.58±0.05	36	0.57	+0.65
6779 (M56)	Int	−1.97[4]	−2.05	−1.74	0.53	1	−1.96	−2.03	−2.05	0.51	1	0.26	+0.98
6388	III	−1.35±0.05	−1.23	−1.00	0.53±0.04	2	−0.67±0.24	−0.64	−0.56	0.61±0.07	6	0.40	−1.00
6441	III	−1.35±0.17	−1.23	−0.80	0.43±0.08	7	−1.02±0.34	−0.82	−1.00	0.55±0.08	8	0.51	−0.73

Notes: [1] Original light curves and Fourier parameters can be found in the papers long list in table 1 of Arellano Ferro (2022). **Unfortunately the table 1 as published in Arellano Ferro (2022) was an older version and exhibits some small differences compared with the present table, which in fact is used for the production of all figures and discussion in the present paper. The present table supersedes that in Arellano Ferro (2022)** [2] Quoted uncertainties are 1σ errors calculated from the scatter in the data for each cluster. The number of stars considered in the calculations is given by N. [3] The only RRL V1 is probably not a cluster member. [4] This value has −0.21 dex added; see Arellano Ferro (2022) for a discussion. [5] Our calculation. [6] Based on a single light curve that was not fully covered. [7] Metallicity and M_V taken from the compilation of Contreras et al. (2010).

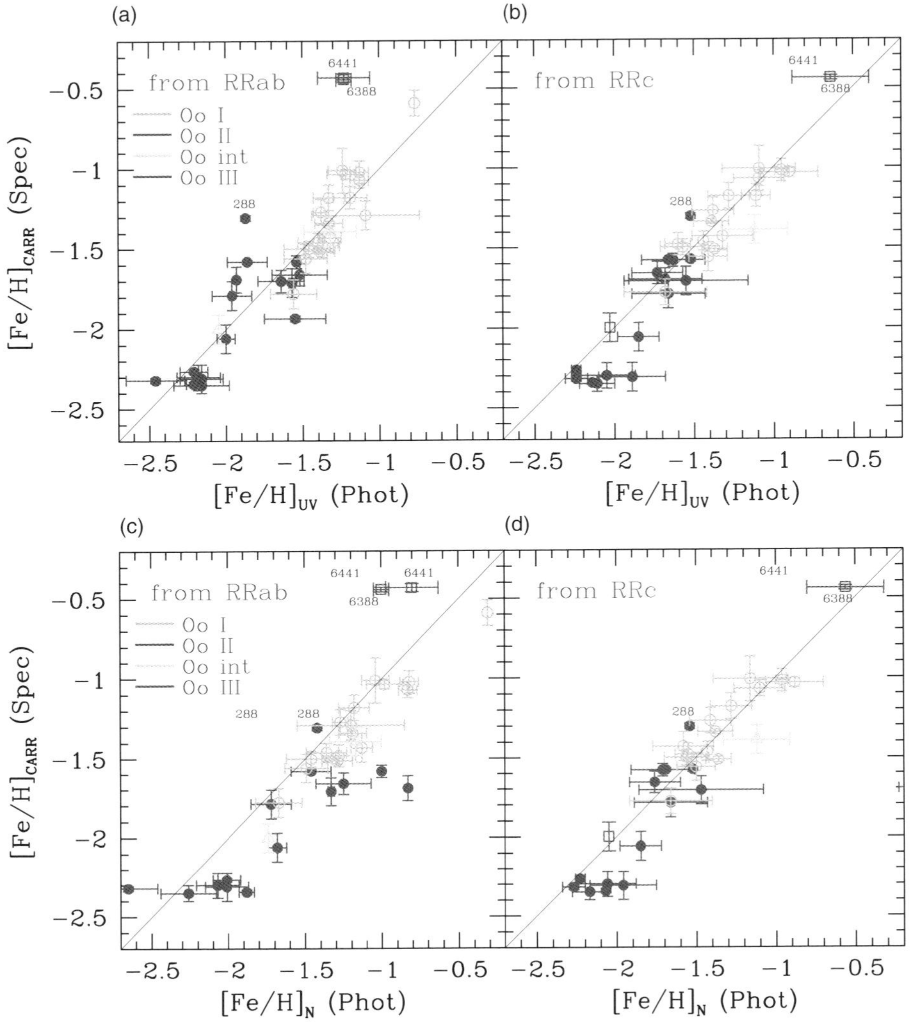

Figure 1. Comparison of the photometric, Fourier decomposition-based $[Fe/H]_{UV}$ and $[Fe/H]_N$ (Table 1), with equivalent values from high-resolution spectroscopy, $[Fe/H]_{Carr}$ (Carretta et al. 2009). (Updated from figure 1 in Arellano Ferro 2022, Rev. Mex. A&A, 58, 257). The color figure can be viewed online.

relation has been considered linear, of the form $M_V = \mu$ [Fe/H] + γ, and that the values of the slope μ found in the literature range between 0.09 and 0.30 depending on the authors and approach (e.g. Demarque et al. 2000).

We stress at this point that the values of M_V and [Fe/H], obtained from the calibrations for RRab and RRc stars, should not be mixed when displaying the M_V–[Fe/H] relation since, in doing so, the correlation will turn very noisy and become meaningless, as is obvious from Fig. 3 and the following discussion. In the left panels of Fig. 3, we show the resulting M_V–[Fe/H] relation based on the results for RRab stars. From top to bottom, panels *a*, *b* and *c*, we display the values of $[Fe/H]_{ZW}$, $[Fe/H]_{UV}$ and $[Fe/H]_N$, respectively, versus the corresponding M_V values from eq. 6. Panel *a* shows a linear run between $[Fe/H]_{ZW}$ and M_V, but the scatter is substantial. Interestingly enough, when $[Fe/H]_{ZW}$ is converted to the spectroscopic scale, $[Fe/H]_{UV}$, panel *b*, the correlation becomes less dispersed and, most remarkably, it suggests a non-linear trend, very similar in fact to

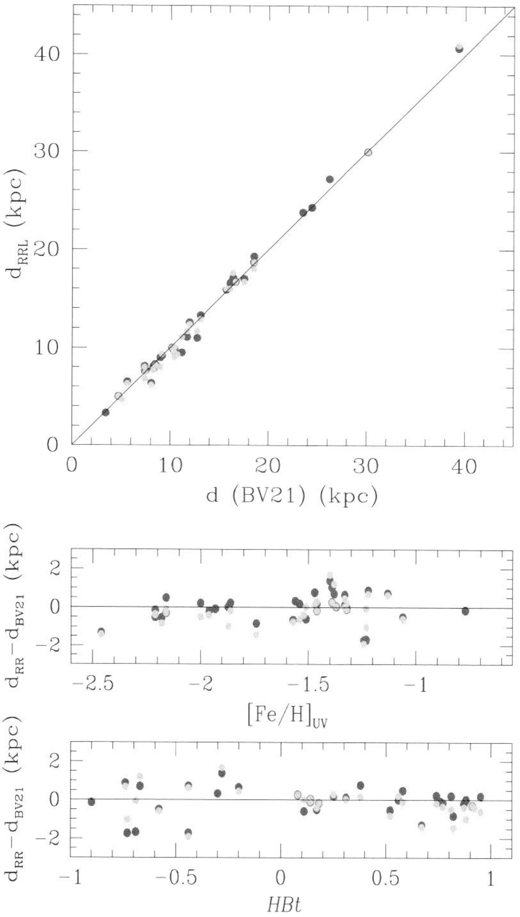

Figure 2. Comparison of distances obtained from the RR Lyrae light curve Fourier decomposition and those of BV21. Blue and green symbols are for distances derived from the calibrations for RRab and RRc, respectively. All distance differences are contained within ±1.9 kpc of each other and the standard deviation is 0.7 kpc. (Updated from figure 4 in Arellano Ferro 2022, Rev. Mex. A&A, 58, 257). The color figure can be viewed online.

the theoretical predictions of Cassisi et al. (1999) and VandenBerg et al. (2000), shown as segmented gray curves in the figure. To the best of our knowledge this is the first empirical M_V–[Fe/H] relation that follows these theoretical results, at least for RRab results. We will return to this point in the Discussion. In panel *c* the metallicities from Nemec et al. (2013) are considered. In this case, and due to the fact noted in Section 4.1 that this calibration overestimates the metallicities, the resulting M_V–$[Fe/H]_N$ relation turns out to be very scattered and with the RRab results systematically too large.

The quadratic fit in panel *b* was calculated while excluding the obvious outliers (NGC 288, NGC 7492, M13 based on a single star, and the OoIII type clusters NGC 6388 and NGC 6441), and including M15, and weighing by the number of RRab stars considered in the calculation for each cluster. The correlation is of the form,

$$M_V = +0.081(\pm 0.060)[Fe/H]^2_{UV} + 0.428(\pm 0.207)[Fe/H]_{UV} + 1.016(\pm 0.170), \qquad (8)$$

with r.m.s. = 0.060 mag.

Let us turn to the case of the RRc stars, i.e., panels *d*, *e* and *f* on the right of Fig. 3. Now the three panels display linear, tight and very similar correlations. The linear fit in

Table 2. Individual distances for a sample of globular clusters estimated homogeneously from the RR Lyrae stars' light curve Fourier decompositions and two versions of the SX Phe P–L relation.

GC NGC(M)	*d (kpc)* (RRab)	*d (kpc)* (RRc)	*d* (kpc) (SX Phe) P–L AF11	No. of SX Phe	*d* (kpc) P–L CS12	*E(B-V)* mag	d (kpc) BV21
288	9.0±0.2	8.0	8.8±0.4	6	9.4±0.6	0.03	8.988
1261	17.1±0.4	17.6±0.7	–	–	–	0.01	16.400
1851	12.6±0.2	12.4±0.2	–	–	–	0.02	11.951
1904 (M79)	13.3±0.4	12.9	–	–	–	0.01	13.078
3201	5.0±0.2	5.0±0.1	4.9±0.3	16	5.2±0.4	dif	4.737
4147	19.3	18.7±0.5	–	–	–	0.02	18.535
4590 (M68)	9.9±0.3	10.0±0.2	9.8±0.5	6	–	0.05	10.404
5024 (M53)	18.7±0.4	18.0±0.5	18.7±0.6	13	20.0±0.8	0.02	18.498
5053	17.0±0.4	16.7±0.4	17.1±1.1	12	17.7±1.2	0.02	17.537
5272	10.0±0.2	10.0±0.4	–	–	–	0.01	10.175
5466	16.6±0.2	16.0±0.6	15.4±1.3	5	16.4±1.3	0.00	16.120
5904 (M5)	7.6±0.2	7.5±0.3	6.7±0.5	3	7.5±0.2	0.03	7.479
6139	9.7±0.7	9.6±0.6	–	–	–	dif.	10.35
6171	6.5±0.3	6.3±0.2	–	–	–	0.33	5.631
6205 (M13)	7.6	6.8±0.3	7.2±0.7	4	–	0.02	7.419
6229	30.0±1.5	30.0±1.1	27.9	1	28.9	0.01	30.106
6254 (M10)	–	4.7	5.2±0.3	15	5.6±0.3	0.25	5.067
6333 (M9)	8.1±0.2	7.9±0.3	–	–	–	dif	8.300
6341 (M92)	8.2±0.2	8.2±0.4	–	–	–	0.02	8.501
6362	7.8±0.1	7.7±0.2	7.1±0.2	6	7.6±0.2	0.09	8.300
6366	3.3	–	–	–	–	0.80	3.444
6388	9.5±1.2	11.1±1.1	–	–	–	0.40	11.171
6401	6.35±0.7	6.15±1.4	–	–	–	dif	8.064
6402 (M14)	9.1±0.9	9.3±0.5	–	–	–	0.57	9.144
6441	11.0±1.8	11.7±1.0	–	–	–	0.51	12.728
6712	8.1±0.2	8.0±0.3	–	–	–	0.35	7.382
6779 (M56)	9.6	9.0	–	–	–	0.26	10.430
6934	15.9±0.4	16.0±0.6	15.8	1	18.0	0.10	15.716
6981 (M72)	16.7±0.4	16.7±0.4	16.8±1.6	3	18.0±1.0	0.06	16.661
7006	40.7±1.6	41.0±1.6	–	–	–	0.08	39.318
7078 (M15)	9.4±0.4	9.3±0.6	–	–	–	0.08	10.709
7089 (M2)	11.1±0.6	11.7±0.02	–	–	–	0.06	11.693
7099 (M30)	8.32±0.3	8.1	8.0	1	8.3	0.03	8.458
7492	24.3	–	22.1±3.2	2	24.1±3.7	0.00	24.390
Pal2	27.2±1.8	–	–	–	–	0.93	26.174
Pal 13	23.8±0.6	–	–	–	–	0.10	23.475

panel *e* is of the form,

$$M_V = 0.034(\pm 0.009)[\mathrm{Fe/H}]_{\mathrm{UV}} + 0.601(\pm 0.015), \tag{9}$$

with r.m.s. = 0.022 mag.

Note that the slope in the above correlation is shallower than the shallowest M_V–[Fe/H] correlations found in the literature, but it is tight and meaningful. It is also worth to stress that, for the RRc stars, Nemec et al. (2013)'s calibration is not affected by the bias noticed in their calibration of the RRab stars, and it produces [Fe/H] values that are similar to those of the calibration of Morgan, Wahl, & Wieckhorst (2007) (panels *d* and *e*).

It shall be obvious now that mixing the RRab and RRc stars in a single M_V–[Fe/H] relation will lead to a noisy and rather meaningless relation. While this requires some explanation, we shall defer our speculations to the Conclusions.

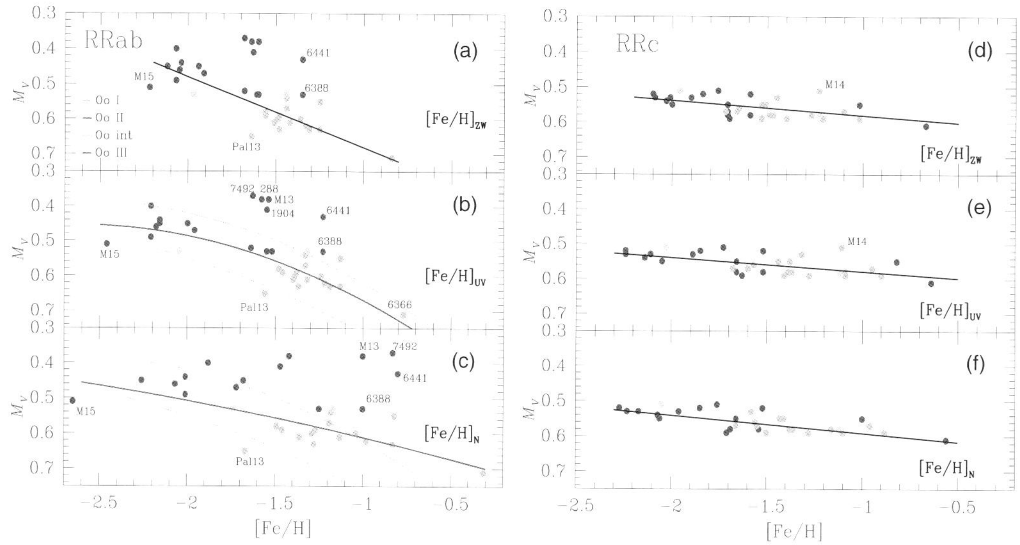

Figure 3. The [Fe/H] versus M_V correlations for RRab and RRc stars. The metallicity scales used are, from the top to bottom panels, $[\mathrm{Fe/H}]_{\mathrm{ZW}}$, $[\mathrm{Fe/H}]_{\mathrm{UV}}$ and $[\mathrm{Fe/H}]_{\mathrm{N}}$. Cluster type is color coded in the legend in panel *(a)*. All fits have been weighted by the number of RR Lyrae stars included in each cluster. In the panel *(b)*, the gray curves are the theoretical predictions of Cassisi et al. (1999) (long dashes) and VandenBerg et al. (2000) (short dashes), which are remarkably similar to the photometric solution. The color figure can be viewed online.

6. The role of the HB structure parameter

In a theoretical study, Demarque et al. (2000) concluded that the M_V–[Fe/H] relation is anything but universal, that the slope is also a function of the metallicity range considered, and that for a given metallicity, the luminosity depends on the HB morphology. Following Demarque et al. (2000), we have adopted the HB type parameter defined as $HBt \equiv (B-R)/(B+V+R)$ (Lee, Demarque & Zinn 1994), where B, V and R are the numbers of HB stars to the blue of the instability strip, RR Lyrae, and those to the red of the instability strip, respectively. We have estimated the value of HBt from the resulting color–magnitude diagrams (CMDs) for almost all clusters in Table 1. Before counting stars we performed a membership analysis using the proper motions, radial velocities and positions available in the different *Gaia* data releases (Gaia Collaboration et al. 2016, 2022) and the membership analysis approach of Bustos Fierro & Caldrón (2019). For a few clusters for which we have no CMD, we adopted HBt from Torelli et al. (2019), who also applied membership considerations.

In Fig. 4 the well-known dependence of HBt on metallicity is shown in the left panels for the RRab and the RRc iron abundance solutions. In the right panels, the run of the absolute magnitude M_V with HBt is displayed, again separately for the two pulsation modes. The striking feature of these plots is the quadratic correlation between M_V and HBt for the RRab solutions, plus the obvious linear correlation of these two quantities for the RRc solutions. The fact that the three parameters $[\mathrm{Fe/H}]_{\mathrm{UV}}$, M_V and HBt are interrelated is clear. A multiple regression analysis leads to the following correlations for the RRab and the RRc stars, respectively:

$$M_V = A + B\ [\mathrm{Fe/H}]_{\mathrm{UV}} + C\ [\mathrm{Fe/H}]^2_{\mathrm{UV}} + D\ HBt + E\ HBt^2, \tag{10}$$

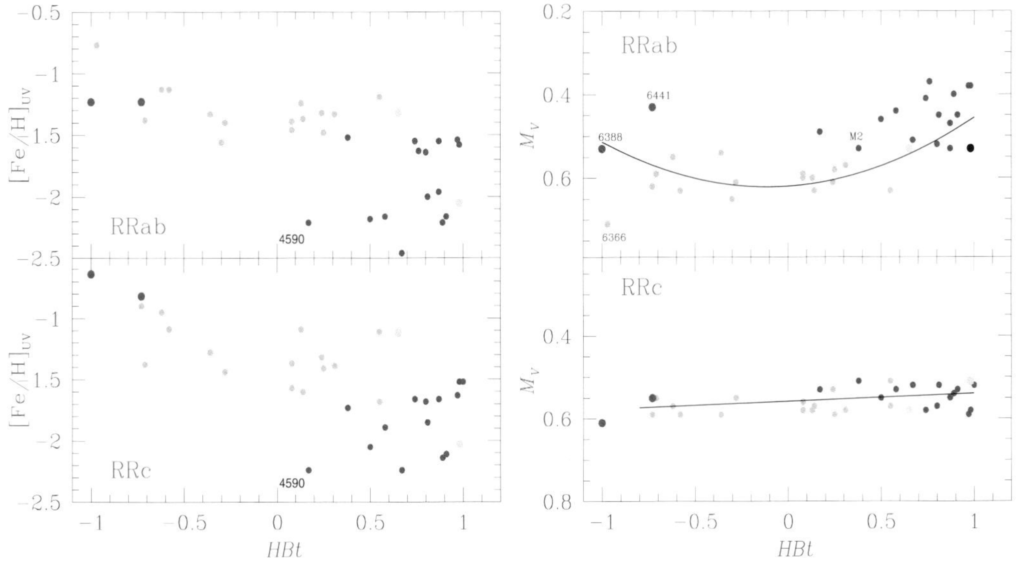

Figure 4. Interdependencies of M_V, $[\mathrm{Fe/H}]_{\mathrm{UV}}$ and the HBt parameter from the solutions for the RRab and the RRc stars. The color figure can be viewed online.

with $A = +1.096(\pm0.141)$, $B = +0.519(\pm0.172)$, $C = +0.119(\pm0.050)$, $D = +0.006(\pm0.014)$, $E = -0.111(\pm0.029)$, and r.m.s. = 0.053 mag.

$$M_V = +0.609(\pm0.016) + 0.032(\pm0.009)\ [\mathrm{Fe/H}]_{\mathrm{UV}} + 0.015(\pm0.011)\ HBt, \qquad (11)$$

with r.m.s. = 0.024 mag.

We note that in eq. 10 the coefficient D is not significant and that term can therefore be neglected. However, E is significant, hence the quadratic dependence of the calibration on HBt as well as on $[\mathrm{Fe/H}]_{\mathrm{UV}}$. In eq. 11 the coefficient of HBt is not significant, implying a linear correlation between M_V and $[\mathrm{Fe/H}]_{\mathrm{UV}}$ for the RRc stars' solutions. If this latter term is ignored, we can see that eqs. 11 and 9 are virtually identical.

The results exposed in this section are in a way the empirical confirmation of the theoretical argumentation of Demarque et al. (2000), i.e., that a simple linear M_V–[Fe/H] relation for RR Lyrae in globular clusters is not sufficient but that, for a given metallicity, the luminosity depends on the HB structure, which can be described by the HBt parameter. To this statement we add that this seems indeed the case when the HB luminosities are calculated from RRab stars, whereas the relation remains simple and linear if the luminosity indicators are the overtone RRc stars. If the analysis is performed without distinguishing between both pulsation modes, the presence of RRab stars will contribute to the scatter and non-linearity between M_V and [Fe/H]. If RR Lyrae stars are to be used as distance indicators, one should be inclined to prefer the simple RRc stars, and a proper instrument for this purpose could be the M_V–[Fe/H] relation offered in this work in the form of eq. 9.

7. Variable Stars in our sample of Globular Clusters

The exercise of obtaining accurate photometry in the field of our sample of globular clusters via the DIA, for a time-series of images, as described in Section 2, produces light curves of many point sources in the field of view. Typically between 1000 and 15,000 stars can be isolated and measured, depending on the cluster distance, size, reddening and the prevailing sky conditions during our observations. Exploring all light curves, employing different methods, became a parallel routine in the project. This allowed us to recover the light curves of all known variables and, very often, to find variables not detected previously. The analysis of the light curve shape, period, and the position of the star in the corresponding CMD, generally enables a proper classification of the variable type.

A noteworthy example of the discovery of unexpected variable stars is the case of Palomar 2, well known since a long time as a cluster devoid of variables. However a detailed analysis of our data led to the discovery of 20 RRab stars and 1 RRc. A revision of *Gaia* DR3 data enabled us to identify 10 additional variables. It has been shown that 16 RRab, 1 RRc and 1 RGB are cluster member stars (Arellano Ferro et al. 2023).

Table 3 summarizes the number of variables and their types found during the development of our project. All pertinent details for each cluster, such as their light curves, ephemerides, Fourier fits, classifications and other discussions as to cluster membership and specific peculiarities, can be found in the many original papers referenced and properly coded in table 3 of Arellano Ferro (2022). The present Table 3 has been updated and edited, and it supersedes table 3 of Arellano Ferro (2022). In Table 3 we have not included any variable star listed as such in the *Gaia* DR3 in the corresponding field of each cluster. Many of the *Gaia* variables coincide with previously known variables and/or may be field stars. Therefore, while we are not aware of a dedicated analysis of these variables, as for example for the case of Palomar 2 (Arellano Ferro et al. 2023) or NGC 6139 (Yepez et al. 2023), we have based our star counts on the present edition of the Catalog of Variable Stars in Globular Clusters (Clement et al. 2001) and our own findings.

8. Summary and Conclusions

For nearly two decades, systematic observational scrutiny of the variable star populations in a sample of 39 globular clusters by means of time-series CCD imaging in the *VI* passbands, and their treatment by the DIA, enabled to collect numerous light curves of variable stars, of many types, contained in these stellar systems. Of particular interest to the present work were the light curves of RR Lyrae stars of both pulsation modes, the fundamental RRab and the first overtone RRc groups. We have performed Fourier decompositions of these light curves and the resulting Fourier parameters have been employed in carefully selected empirical calibrations to calculate, in an unprecedentedly homogeneous manner, the atmospheric iron abundances, [Fe/H], and absolute magnitudess, M_V, via this photometric approach. Attention was paid to the zero points of the absolute magnitude calibrations.

The mean M_V values and estimates of the interstellar reddening led to photometric distances calculated for our sample 39 globular clusters and reported here. The mean iron abundance for a given cluster was calculated on the scale of Zinn & West (1984) $[Fe/H]_{ZW}$, transformed to the UVES spectroscopic scale of Carretta et al. (2009) $[Fe/H]_{UV}$, and on the scale of Nemec et al. (2013) $[Fe/H]_{N}$.

The photometric $[Fe/H]_{UV}$ and the spectroscopic $[Fe/H]_{CARR}$ values taken from Carretta et al. (2009) were found to be in very good agreement for both the solutions from the RRab and the RRc stars. Nemec et al. (2013)'s calibration for the RRc stars are also in good agreement with both $[Fe/H]_{UV}$ and $[Fe/H]_{CARR}$. It was found, however, that the photometric values from the calibration for RRab stars of Nemec et al. (2013) are systematically overestimated relative to the spectroscopic values. This result becomes of particular relevance since the calibration of Nemec et al. (2013) was used by *Gaia* to calculate the iron abundances from the photometry of RR Lyrae stars. It is worth noting that a similar result has been reported in a poster presented at this meeting by Varga et al. See also the Q&A section below for further comments on this issue.

Table 3. Number of presently known variables per type and per cluster for the most common variable types, in the globular clusters studied by our group†.

GC NGC (M)	RRab	RRc	RRd	SX Phe	Binaries	CW-(AC) -RV	SR, L, M	unclass others ∗	Total per cluster
288	0/1	0/1	0/0	0/6	0/1	0/0	0/1	–	0/10
1261	0/16	0/6	0/0	0/3	0/1	0/0	0/3	–	0/29
1904 (M79)	0/6	1/5	0/0	0/5	0/1	0/1	0/14	0/1	1/32
3201	0/72	0/7	0/0	3/24	0/11	0/0	0/8	0/7	3/122
4147	0/5	0/19	0/1	0/0	0/14	0/0	2/2	0/3	2/41
4590 (M68)	0/14	0/16	0/12	4/6	0/0	0/0	0/0	1/2	4/48
5024 (M53)	0/29	2/35	0/0	13/28	0/0	0/0	1/12	–	16/104
5053	0/6	0/4	0/0	0/5	0/0	0/0	0/0	–	0/15
5466	0/13	0/8	0/0	0/9	0/3	0/1	0/0	2/2	0/34
5904 (M5)	2/ 89	1/40	0/0	1/6	1/3	0/2	11/12	0/1	16/152
6171 (M107)	0/15	0/6	0/0	0/1	0/0	0/0	2/3	0/3	2/25
6139	0/4	4/5	0/0	0/0	1/1	0/0	4/7		9/17
6205 (M13)	0/1	1/7	2/2	2/6	1/3	0/3	3/22	0/4	9/44
6229	10/42	5/15	0/0	1/1	0/0	2/5	6/6	0/1	24/69
6254 (M10)	0/0	0/1	0/0	1/15	2/10	0/3	0/5	0/2	3/34
6333 (M9)	0/8	2/10	1/1	0/0	3/4	1/1	5/6	3/4	12/30
6341 (M92)	0/9	0/5	1/1	1/6	0/0	0/1	1/1	0/6	3/23
6362	0/16	0/15	1/3	0/6	0/12	0/0	0/0	0/22	1/52
6366	0/1	0/0	0/0	1/1	1/1	1/1	3/4	–	6/8
6388	1/14	2/23	0/0	0/1	0/10	1/11	42/58	–	46/117
6397	0/0	0/0	0/0	0/5	0/15	0/0	0/1	0/13	0/21
6401	6/23	6/11	0/0	0/0	0/14	0/1	3/3	14/14	15/52
6402 (M14)	0/55	3/56	1/1	1/1	0/3	0/6	18/32		23/154
6441	2/50	0/28	0/1	0/0	0/17	2/9	43/82	0/10	47/187
6528	1/1	1/1	0/0	0/0	1/1	0/0	4/4	–	7/7
6638	3/10	2/18	0/0	0/0	0/0	0/0	3/9	0/25	8/37
6652	0/3	0/1	0/0	0/0	1/2	0/1	0/2	1/5	1/9
6712	0/10	0/4	0/0	0/0	2/2	0/0	5/11	0/8	7/27
6779 (M56)	0/1	0/2	0/0	1/1	3/3	0/2	0/3	1/6	4/12
6934	3/68	0/12	0/0	3/4	0/0	2/3	3/5	1/6	11/92
6981 (M72)	8/37	3/7	0/0	3/3	0/0	0/0	0/1		14/48
7078 (M15)	0/65	0/67	0/32	0/4	0/3	0/2	0/3	0/11	0/176
7089 (M2)	5/23	3/15	0/0	0/2	0/0	0/4	0/0	0/12	8/44
7099 (M30)	1/4	2/2	0/0	2/2	1/6	0/0	0/0	0/3	6/14
7492	0/1	0/2	0/0	2/2	0/0	0/0	1/2	–	3/7
Pal 2	16/16	1/1	0/0	0/0	0/0	0/0	1/1	–	18/18
Pal 13	0/4	0/0	0/0	0/0	0/0	0/0	1/1		1/5
Total per type	58/732	41/454	6/54	39/153	17/141	9/57	162/324	*23/171* *	330/1916

†. The variable star types have been adopted from the General Catalog of Variable Stars (Kazarovets et al. 2009; Samus et al. 2009). Entries expressed as M/N indicate the M variables found or reclassified by our program and the total number N of presently known variables. Relevant papers on individual clusters can be found properly coded in table 3 of Arellano Ferro (2022).

∗. Numbers from this column are not considered in the totals, since they include unclassified or likely field variables.

Similarly, we have compared our photometric determinations of the accurate mean cluster distances derived by Baumgardt & Vasiliev (2021). The agreement is outstanding, yielding distance differences within ±1.9 kpc for the entire distance range up to 40 kpc.

The resulting M_V–[Fe/H] relation for the RRab stars looks non-linear and scattered. However, based on the RRc stars the relation remains linear and tight. The latter relation shows only a mild but significant dependence of the luminosity on the metallicity. It is clear that if the M_V and [Fe/H] results from RRab and RRc stars are mixed, the resulting M_V–[Fe/H] relation will be apparently linear, very scattered and steeper. We find that it is important to segregate the RRab and RRc results for a more meaningful interpretation of the M_V–[Fe/H] relation.

As suggested on theoretical grounds (Demarque et al. 2000), we have empirically shown that the HB type parameter, HBt, plays a significant role in the M_V–[Fe/H] relation if calculated using the RRab stars as tracers, in which case the relation becomes quadratic in [Fe/H] and HBt. No dependence on HBt was found for the RRc stars. These results strengthen the notion that RRc stars may be more trusted distance indicators than their more complex RRab counterparts.

While I really do not have an in-depth explanation for the differences in the luminosity dependence on the metallicity for RRab and from RRc stars, it may be relevant to mention that RRab stars carry a greater degree of complexity; they have larger amplitudes, many of them have (undetected) Blazhko modulations, their light curves are asymmetrical and hence prone to more inaccurate Fourier decompositions; as their periods are often longer, their light curves are not fully or rather unevenly covered. Their evolutionary tracks toward the RGB are spread more widely with small variations in age, inner mass structure and chemical differences. The RRc stars, on the other hand, are simpler, easier to decompose and their luminosities are less sensitive to age, mass or chemistry. However I emphasize that the remarks in the present work may require further theoretical input.

Systematic time-series CCD imaging and differential image analysis performed over the last nearly 20 years has enabled us to report and classify 330 newly detected variables in our sample of globular clusters.

9. Acknowledgments

I am indebted to all my friends and colleagues who have accompanied me on this globular cluster journey that has lasted nearly two decades. They number too many to specifically name them all here; their names are all included as co-authors of our papers; to all of them, my gratitude for having shared with me and taught me their expertise. I am particularly thankful to the Time Allocation Committees of several observatories for having generously granted the time needed to push forward our observing strategies. For many years, the project has received support from the Institute of Astronomy and DGAPA, of the National Autonomous University of Mexico, through a range of grant numbers.

References

Arellano Ferro A., 2022, Rev. Mex. A&A, 58, 257

Arellano Ferro, A., Bustos Fierro, I., Muneer, S., Giridhar, S. 2023, Rev. Mex. A&A, 59, 3

Arellano Ferro, A., Figuera Jaimes, R., Giridhar, S., Bramich, D. M., Hernández Santisteban, J. V., Kuppuswamy, K. 2011, MNRAS, 416, 2265 (AF11)

Arellano Ferro A., Giridhar S., Bramich D. M., 2010, MNRAS, 402, 226

Baumgardt, H., Vasiliev, E. 2021, MNRAS, 505, 5957 (BV21)

Benedict G. F., McArthur B. E., Fredrick L. W., Harrison T. E., Lee J., Slesnick C. L., Rhee J., et al., 2002, AJ, 123, 473

Bramich D.M. 2008, MNRAS, 386, L77

Bramich, D. M., Horne, K., Albrow, M. D., Tsapras, Y., Snodgrass, C., Street, R. A., Hundertmark, M., Kains, N., Arellano Ferro, A., Figuera, J. R., Giridhar, S. 2013, MNRAS, 428, 2275

Bramich D. M., Figuera Jaimes R., Giridhar S., Arellano Ferro A., 2011, MNRAS, 413, 1275

Bramich, D. M., Bachelet, E., Alsubai, K. A., Mislis, D., Parley, N. 2015, A&A, 577, A108

Bustos Fierro I. H., Calderón J. H., 2019, MNRAS, 488, 3024

Cacciari C., Corwin T. M., Carney B. W., 2005, AJ, 129, 267

Carretta E., Bragaglia A., Gratton R., D'Orazi V., Lucatello S., 2009, A&A, 508, 695

Carretta E., Gratton R. G., 1997, A&AS, 121, 95

Cassisi, S., Castellani, V., deglInnocenti, S., Salaris, M., & Weiss, A. 1999, A&AS, 134, 103

Clement, C.M, Muzzin,A., Dufton, Q., et al., 2001, AJ, 122, 2587
Clementini G., Gratton R., Bragaglia A., Carretta E., Di Fabrizio L., Maio M., 2003, AJ, 125, 1309
Cohen, R. E., Sarajedini, A. 2012, MNRAS, 419, 342 (CS12)
Contreras, R., Catelan, M., Smith, H. A., Pritzl, B. J., Borissova, J., Kuehnet, C. A. 2010, AJ, 140, 1766
de Grijs, R., Wicker, J.E., Bono, G. 2014, AJ, 147, 122
Deras, D., Arellano Ferro, A., Lázaro, C., Bustos Fierro, I. H., Calderón, J. H., Muneer, S., Giridhar, S. 2019, MNRAS, 626
Demarque P., Zinn, R., Lee, Y-W, Yi, A.S. 2000, ApJ 119, 1398
Freedman, W. L., Madore, B. F., Gibson, B. K. et al., 2001, ApJ, 553, 47
Gaia collaboration, Prusti T., de Bruijne J. H. J., et al. 2016, A&A, 595, A1
Gaia collaboration Vallenari, A., Brown, A. G. A., Prusti, T. et al. 2022, in press (arXiv:2208.00211)
Jurcsik J., 1995, AcA, 45, 653
Jurcsik, J., Kovács, G. 1996, A&A 312, 111
Kazarovets E. V., Samus' N. N., Durlevich O. V., Kireeva N. N., Pastukhova E. N., Pojmanski G., 2009, ARep, 53, 1013
Kinman T. D., 2002, IBVS, 5354, 1
Kovács G., 1998, Mem. Soc. Astron. It., 69, 49
Kovács G., Jurcsik J., 1996, ApJL, 466, L17
Kovács G., Kanbur S. M., 1998, MNRAS, 295, 834
Kovács G., Walker A. R., 2001, A&A, 371, 579
Kovács, G., Zsoldos, E. 1995, A&A 293, L57
Layden A. C., 1994, AJ, 108, 1016
Lee, Y.-W., Demarque, P., & Zinn, R. 1994, ApJ, 423, 248
Lub J. 1977, A&AS, 29, 345
Morgan S. M., Wahl J. N., Wieckhorst R. M., 2007, MNRAS, 374, 1421
Nemec J. M., Cohen J. G., Ripepi V., Derekas A., Moskalik P., Sesar B., Chadid M., et al., 2013, ApJ, 773, 181
Pietrzyński, G., Graczyk, D., Gieren, W. et al. 2013, Nature, 495, 76
Samus N. N., Kazarovets E. V., Pastukhova E. N., Tsvetkova T. M., Durlevich O. V., 2009, PASP, 121, 1378
Simon N. R., 1988a, ApJ, 328, 747
Simon, N. R., 1988b, *Pulsation and Mass Loss in Stars* 148, 27
Simon, N. R., Clement, M. C. 1993, ApJ, 410, 526
Suntzeff N. B., Kraft R. P., Kinman T. D. 1994, ApJS, 93, 271
Torelli, M., Iannicola, G., Stetson, P. B., et al. 2019, A&A, 629, A53
van Albada T. S., Baker N., 1971, ApJ, 169, 311
VandenBerg, D. A., Swenson, F. J., Rogers, F. J., Iglesias, C. A., & Alexander, D. R. 2000, ApJ, 532, 430
Yepez M. A., Arellano Ferro A., Deras D., 2020, MNRAS, 494, 3212
Yepez M. A., Arellano Ferro A., Bustos Fierro, I, Luna, A., 2023, MNRAS, submitted
Zinn R., West M. J., 1984, ApJS, 55, 45

Questions and Answers

Clara Martínez-Vázquez: Thanks for your talk and the interesting comments offered! I would like to bring to your attention a set of recently published new P–ϕ_{31}–[Fe/H] relationships for RRab and RRc types. They have been calibrated based on a very large and homogeneous sample of spectroscopic metallicities. The testing of the relations has proven to be very efficient. Here are the papers in case you want to check them: https://iopscience.iop.org/article/10.3847/1538-4357/abefd4 (P–ϕ_{31}–[Fe/H] for RRab) and https://iopscience.iop.org/article/10.3847/1538-4357/ac67ee (P–ϕ_{31}–[Fe/H] for RRc).

AAF: Thank you so much for the references. I will check them carefully.

Héctor Vázquez Ramló: Thanks for the interesting talk. A technical question: if I understood you correctly, you use differential imaging to obtain the light curves of the RR Lyrae in globular clusters. That implies soundly coping with PSF variations from image to image (at least if observations were obtained from the ground, I didn't get if that is the case). Could you expand a little on this, please? Which package do you use?

AAF: We use Difference Image Analysis (DIA) through the pipeline DanDIA (Bramich 2008, MNRAS, 386, L77; Bramich et al. 2013 MNRAS, 428, 2275). DanDIA is constructed based on the DanIDL library of IDL routines available at http://www.danidl.co.uk. Our use is described in detail in Bramich et al. (2011 MNRAS, 413, 1275). A recent paper where you can gain an overall view of all our procedures is perhaps Yepez et al. (2020, MNRAS, 494, 3212). Regarding the PSF, DanDIA constructs a reference image from a combination of the best images in our collection. Then, it isolates up to 400 stars in the image using some isolation criteria, and it uses those individual PSFs to search for possible trends across the chip, fitting high-order polynomials. The PSF kernel is then used for the individual differential images.

Héctor Vázquez Ramló: Another question regarding the statement of RRc being more suitable for distance measurements, that was an interesting one. It is clear that RRab are more widely used due to their much higher abundance and their higher pulsation amplitude (easier to identify) in comparison with RRc. Out of curiosity, I really don't know: are there studies exclusively focusing on RRc for estimating distances? Is there a general consensus in the community on that issue?

AAF: There are of course papers dedicated exclusively to RRc stars, for instance one on spectral metallicity determinations by Sneden et al., (2018, AJ, 155, 45), but not on the premise that they may be better distance indicators, I think. This is, in fact, one of the highlights among our results; this, and the fact that absolute magnitudes of mixed samples of RRab and RRc stars will lead to larger uncertainties and mislead calibrations of the famous M_V–[Fe/H] correlation. I do not think there is, at present, a general consensus on these points, or not just yet. The community has become more interested in the presence of secondary frequencies caused by non-radial modes in RRc stars, I believe, the ones with ratios relative to the overtone frequency (f_1) known as the $f_{61}/f_1 \sim 0.61$–0.65.

Gergely Hajdu: Thank you very much for the update on globular cluster RR Lyrae. I would like to discourage the use of the Nemec et al. (2013) relations for metallicity estimation of RRab stars. They come from a very small sample of variables, and the authors overfit their data with parameters that are highly correlated (especially their three parameters that all contain ϕ_{31}), as shown by the large formal errors (4.64 dex on the intercept!). Unfortunately, the baseline *Gaia* DR3 [Fe/H] estimates are based on this formula, leading to clearly artificial features appearing when the Bailey diagram is colored according to that metallicity estimate (i.e., fig. 35 of https://ui.adsabs.harvard.edu/abs/2022arXiv220606278C/abstract).

AAF: Thanks for your comments. As you may have noticed, in our comparison of the photometric estimates of $[Fe/H]_{UV}$ vs. the spectroscopic values, $[Fe/H]_{CARR}$, we also noticed that something does not work well for the calibration of Nemec et al. (2013) for the RRab stars, since their calibration clearly overestimates the iron abundance values.

For the RRc stars, it seems as good as the calibration of Morgan et al. (2007). Thus, it is interesting to note your comments and recommendation against that calibration. By the way, there is a poster at this Symposium by Varga et al.; the authors reached the same conclusion as regards Nemec et al.'s calibration for the RRab and, apparently, they have corrected it.

Vázsony Varga: I also recommend comparision with Fabrizio et al.'s (2021) ΔS Bailey diagram metallicity distribution. I'd also like to add that the anomalous metal-rich regions in the current *Gaia* Bailey diagram are likely caused by the small parameter space covered by the calibration data set.

AAF: Thank you for your comments. I have noticed that you also highlight that the calibration of Nemec et al. (2013) overestimates the iron abundances for RRab stars.

At the crossroads of astrophysics and cosmology: Period–luminosity relations in the 2020s
Proceedings IAU Symposium No. 376, 2024
R. de Grijs, P. A. Whitelock & M. Catelan, eds.
doi:10.1017/S1743921323003368

The PL diagram for δ Sct stars: back in business as distance estimators

Antonio García Hernández[1], Javier Pascual-Granado[2], Mariel Lares-Martiz[2], Giovanni M. Mirouh[1], Juan Carlos Suárez[1], Sebastiá Barceló Forteza[1] and Andrés Moya[3]

[1]Departamento de Física Teórica y del Cosmos, Universidad de Granada, Campus de Fuentenueva s/n, 18071, Granada, Spain. email: agh@ugr.es

[2]Instituto de Astrofísica de Andalucía (CSIC). Glorieta de la Astronomía s/n. 18008, Granada, Spain

[3]Departament d'Astronomia i Astrofísica, Universitat de València, C. Dr. Moliner 50, 46100, Burjassot, Spain

Abstract. In this work, we focus on the period-luminosity relation (PLR) of δ Sct stars, in which mode excitation and selection mechanisms are still poorly constrained, and whose structure and oscillations are affected by rotation. We review the PLRs in the recent literature, and add a new inference from a large sample of δ Sct. We highlight the difficulty in identifying the fundamental mode and show that rotation-induced surface effects can impact the measured luminosities, explaining the broadening of the PLR. We derive a tight relation between the low-order large separation and the fundamental radial mode frequency (F0) that holds for rotating stars, thus paving the way towards mode identification. We show that the PLRs we obtain for different samples are compatible with each other and with the recent literature, and with most observed δ Sct stars when taking rotation effects into account. We also find that the highest-amplitude peak in the frequency spectrum corresponds to the fundamental modein most δ Sct, thus shedding some light on their elusive mode selection mechanism.

Keywords. Period-luminosity relations, δ Sct stars, Stellar pulsations, Stellar rotation, Mode identification

1. Introduction

Period-luminosity relations (PLRs) have been studied since the early twentieth century. The discovery of such a relation for classical Cepheids (Leavitt 1908) allowed to estimate distances to other galaxies and changed our vision of the scale of the Universe. Inspired by this success, complementary PLRs have been searched for other pulsating stars, mainly those in the Cepheid instability strip (see, e.g., Balona 2010; Owens *et al.* 2022, and references therein). In particular, a PLR for δ Sct stars has been established since Fernie (1964).

δ Sct stars offer several observational advantages compared to Cepheids. One is their shorter periods, which arise from their smaller radii. Furthermore, as pointed out by Breger (1979), δ Sct stars constitute the second most abundant group of pulsators in the Galaxy, following the pulsating white dwarfs. Having a tight PLR, along with these favourable characteristics, would enable us to compare between observations and pulsation theory. Additionally, it would facilitate the determination of stellar luminosities and distances to stars, open and globular clusters, nearby galaxies, and the Galactic centre.

However, there are several difficulties in the interpretation of the pulsation spectra of δ Sct stars. They have spectral types from late A to early F, with masses in the range 1.5 $M_{\odot}$ to 2.5 $M_{\odot}$, and can be found either on the main sequence or at the H-shell burning phase. They show low-order radial and non-radial modes near the fundamental radial mode (whose frequency and period we denote as F0 and P0, respectively), although the latter is not always the highest amplitude peak in the spectrum (Aerts *et al.* 2010). This is one of the main difficulties to get a proper PLR, which requires the order of the radial mode to be properly identified. Moreover, other difficulties are related to their usually rapid rotation (Royer *et al.* 2007) and the poor understanding of their excitation and selection mechanisms (Aerts *et al.* 2010).

1.1. *Recent advances in understanding the δ Sct pulsation spectrum*

The pulsation spectrum of solar-like pulsators, including red giants, is easily understood thanks to the periodic patterns they exhibit because their pressure (p) modes appear in the asymptotic regime ($n \gg \ell$, see, for example, Aerts *et al.* 2010; Corsaro *et al.* 2012). The most important pattern is known as large separation, $\Delta\nu$, a spacing between consecutive radial orders of the same spherical degree that is a direct measurement of the stellar mean density.

δ Sct stars do not pulsate in the asymptotic regime but in the low-order regime of p modes. Nonetheless, prior to the launch of space missions dedicated to ultra-precise photometric time-series, several efforts were made to find a large separation structure in the pulsation spectrum of these stars (Handler *et al.* 1997; Breger *et al.* 1999). With the advent of space missions like Microvariability and Oscillations of STars (*MOST*; Walker *et al.* 2003), Convection, Rotation and planetary Transits (*CoRoT*; Auvergne *et al.* 2009), and *Kepler* (Koch *et al.* 2010), it became clear that a similar pattern in the low-order regime could be found in the pulsation spectra of δ Scuti stars (García Hernández *et al.* 2009, 2013; Zwintz *et al.* 2011; Paparó *et al.* 2016; Barceló Forteza *et al.* 2017; Michel *et al.* 2017; Bedding *et al.* 2020; Hasanzadeh *et al.* 2021).

Theoretical predictions for these patterns were also made (Reese *et al.* 2008; Ouazzani *et al.* 2015) and were found to be consistent with a large separation scaling with the mean density of the star ($\Delta\nu_{\rm low} - \langle\rho\rangle$, Suárez *et al.* 2014). This scaling relation differs with respect to the solar-like one because of the different regimes where the modes appear in both pulsating classes. Empirical confirmation of this scaling law was subsequently provided by García Hernández *et al.* (2015, 2017) using eclipsing binary systems with a δ Scuti component. Interestingly, Mirouh *et al.* (2019) employed a convolutional neural network to identify island modes in 2D pulsation models, and obtained a similar $\Delta\nu_{\rm low} - \langle\rho\rangle$ scaling relation for fast-rotating stars. The most recent theoretical relation was derived by Rodríguez-Martin *et al.* (2020) using synthetic spectra of 1D rotating models. The independence of the scaling relation with rotation makes $\Delta\nu_{\rm low}$ a promising seismic index to shed light on the pulsation behaviour of δ Sct stars.

1.2. *Previous PLRs for δ Sct stars*

The PLR of δ Sct stars has been extensively studied in previous works. McNamara (2011) investigated the PLR using a sample of 26 high-amplitude δ Scuti (HADS) stars. Cohen & Sarajedini (2012) examined 77 SX Phoenicis (SX Phe) stars, which are generally considered the Population II analogues of δ Sct stars, to explore the PLR in that class. Ziaali *et al.* (2019) analysed a larger dataset consisting of 1352 *Kepler* δ Sct stars and utilised additional information from Rodríguez & Breger (2001).

Jayasinghe *et al.* (2020) conducted a study on approximately 4000 δ Sct stars sourced from the *ASAS-SN* catalogue (Jayasinghe *et al.* 2018). Poro *et al.* (2021) focused on 27 δ Sct from the Kourovka Planet Search (KPS) project. Barac *et al.* (2022) investigated 434 δ Sct stars, comprising 301 from Rodríguez & Breger (2001) and 133 from Chang *et al.* (2013), using data obtained from the Transiting Exoplanet Survey Satellite (*TESS*; Ricker *et al.* 2014).

De Ridder *et al.* (2022) studied a significant sample size of 6511 *Gaia* δ Sct stars and found evidence of a broken PLR within this population. Martínez-Vázquez *et al.* (2022) examined 3799 extragalactic δ Sct from various sources, including the Super MAssive Compact Halo Object (*SuperMACHO*; Garg *et al.* 2010) and the Optical Gravitational Lensing Experiment-III (*OGLE-III*; Poleski *et al.* 2010) projects. They also observed a broken PLR for these extragalactic objects.

Furthermore, Deka *et al.* (2022) studied 3202 δ Sct stars in the Galactic bulge from the *OGLE-IV* catalogue (Soszyński *et al.* 2021) and the Large Magellanic Cloud (LMC) from *OGLE-III*. Their analysis revealed a broken PLR for the bulge δ Sct stars, but not for those in the LMC.

Despite the numerous studies conducted, many unanswered questions persist, and a definitive PLR for δ Sct stars is yet to be established. This study aims to introduce a novel PLR and utilises rotating equilibrium and pulsation models to demonstrate that rotation (through seismic and gravity darkening effects) contributes to the solution of the unresolved issues affecting the δ Scuti PLR.

2. The samples

In our study, we utilised various samples of δ Sct stars to investigate the PLR. We employed approximately 3700 δ Sct stars from the *ASAS-SN* catalogue, as described by Jayasinghe *et al.* (2020). To ensure these objects were δ Sct, we selected only those stars that fell within the effective temperature range of $T_{\rm eff} = [6700, 10000]$ K and surface gravity range of $\log g = [3.5, 4.2]$ dex, using the parameters derived from their own analysis. Additionally, we incorporated around 3800 δ Sct, γ Dor, and hybrid stars from Sódor et al. (in prep.), obtained from the *Kepler* mission. We ended up discarding the pure γ Dor stars because they are not relevant for this work.

We also considered a sample of 17 HADS stars with *Gaia* luminosities from data release 3 (DR3, Fouesneau *et al.* 2023) based on the work of McNamara (2011). We added a new sample of HADS composed of 11 HADS stars with non-distorted lightcurves from Lares-Martiz (2022) and 36 HADS stars introduced in this work for the first time, collectively forming what we refer to as Lares-Martiz's HADS sample. These 47 lightcurves were acquired from the *TESS* satellite. The Best Parent Method (Lares-Martiz *et al.* 2020) was employed to determine their F0 frequency, which is usually that of the dominant peak for HADS. By fitting not only the main peaks but also their possible combinations, this method ensures a robust and more precise determination of F0. The list and F0 values of the stars in this sample are shown in Table 1.

Finally, we have used 60 young δ Sct stars identified by Bedding *et al.* (2020) in our analysis. This sample has the advantage of showing clear periodicities, so $\Delta\nu_{\rm low}$ was identified.

To ensure consistency and obtain luminosity information, we cross-matched all the aforementioned catalogues with *Gaia* DR3 (Fouesneau *et al.* 2023). The resulting PL diagram is shown in Figure 1. The combined *ASAS-SN* and *Kepler* sample is shown as grey points, whereas McNamara's, Lares-Martiz's, and Bedding's samples are depicted with red squares, blue circles, and purple triangles, respectively. Uncertainties both in luminosity and period are also shown for the smaller sample (for the sake of clarity), but

Table 1. List of HADS comprising the Lares-Martiz sample. The 11 HADS taken from Lares-Martiz (2022) are indicated by a star symbol preceding their *TESS* Input Catalog (TIC) ID, while the remaining 36 are presented here for the first time. F0 was extracted from *TESS* data using the Best Parent Method (Lares-Martiz *et al.* 2020).

TIC ID	F0 [d^{-1}]	TIC ID	F0 [d^{-1}]	TIC ID	F0 [d^{-1}]
17153995	8.54481	181087723	7.72287	334436767	7.47803
35920894	5.64856	⋆ 183532876	5.98212	339675323	7.44527
43173526	11.93649	200624064	5.2087	342500794	6.5115
46937596	7.88982	210548440	5.42033	⋆ 355547586	10.51244
78850814	9.11811	⋆ 224285325	18.19356	⋆ 355687188	8.19196
⋆ 51991595	10.91561	⋆ 231632224	6.91544	⋆ 358502706	11.67333
80781425	6.5968	241787384	11.0103	362384415	7.21626
118080601	7.54064	241843363	5.98816	374753270	7.24599
⋆ 121731704	14.22167	242302902	7.63725	383604347	9.49098
⋆ 139845816	6.80218	262652067	5.63467	393420032	6.01012
⋆ 144309524	8.97501	293110952	14.20631	396424970	7.1762
145372195	7.74984	298112357	9.84591	447363101	5.22441
⋆ 166979292	17.217331	299948201	7.07711	448892817	13.43279
168384036	9.25355	304196197	8.82052	454665792	8.21272
178463456	4.94822	308396022	13.20399	464401254	8.33771
178616716	9.76366	321528595	5.57497		

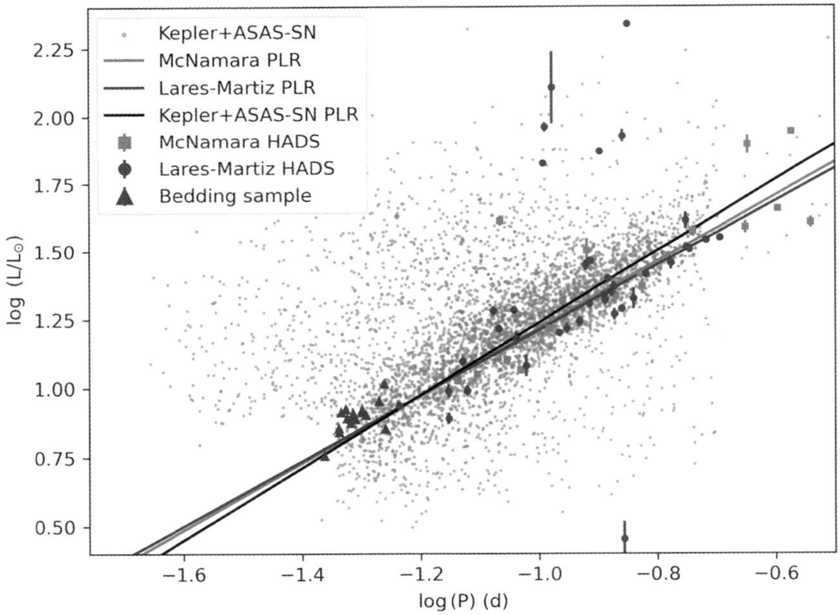

Figure 1. PLRs of the samples used in this study after removing outliers: δ Scuti, γ Dor and hybrid stars from *Kepler* combined with *ASAS-SN* catalogue; McNamara's HADS; Lares-Martiz's HADS; and stars from Bedding *et al.* (2020). Each sample is represented with a different colour and symbol. Three different PLRs are shown corresponding to the first three samples.

they are usually smaller than the symbol size. The entire δ Sct sample spans approximately $\log P \simeq [-1.6, -0.7]$ and $\log(L/L_\odot) \simeq [0.75, 2.00]$. Within this region, a denser band was observed, primarily consisting of stars where the fundamental mode (of period P0) also exhibited the highest amplitude (Jayasinghe *et al.* 2020). This denser band was used to calculate the PLR, as described in Section 3.

3. Getting the PLR

We have obtained three different PLRs with the previous samples that are depicted in Figure 1. We obtain a PLR using Lares-Martiz's sample of HADS. We avoid outliers in the fit (more than 2σ of deviation from the fit). Some outliers seem to be binary systems, for example, TIC 299948201, identified as a double or multiple-star system in Simbad (Wenger *et al.* 2000). Other outliers suspected to be binaries are TIC 383604347 and TIC 396424970. Initially, these were identified as δ Sct stars by Chang *et al.* (2013), but recently Barac *et al.* (2022) identified them as non-δ Sct since binary systems without a pulsating component can also look like HADS. This can explain the high luminosity of TIC 383604347 ($\log(L/L_\odot) > 1.75$, see Figure 1) and low luminosity ($\log(L/L_\odot) < 0.5$) of TIC 396424970.

We also calculated the PLR for McNamara's sample of HADS. We did not take into account AE UMa, the star with $\log(L/L_\odot) \sim 1.5$ and $\log P \sim -1.05$, because its luminosity differs from that found by other authors (for example, Xue *et al.* 2022 find $\log(L/L_\odot)$ between 1.00 and 1.24).

Finally, to get the PLR from the combined *Kepler* and *ASAS-SN* sample, we used the denser band of stars in the PL diagram (see Section 2). We divided the set in equal bins of $\log P$ to get 20 bins. This assured a statistically significant number of stars per bin. Then, we created a histogram in $\log(L/L_\odot)$ of every bin and calculated the Gaussian kernel density estimator to obtain its maximum value. With that maximum and the centre of the bin, we get the points to be fitted and derive the PLR.

The relations found in the way described above give, respectively:

$$\log(L/L_\odot)_{\text{Lares-Martiz}} = (1.18 \pm 0.10) \cdot \log P + (2.39 \pm 0.09), \tag{1a}$$

$$\log(L/L_\odot)_{\text{ASAS-SN}} = (1.21 \pm 0.15) \cdot \log P + (2.43 \pm 0.13), \tag{1b}$$

$$\log(L/L_\odot)_{\text{McNamara}} = (1.31 \pm 0.03) \cdot \log P + (2.55 \pm 0.03). \tag{1c}$$

These relations, in which the period is expressed in days, are all in agreement, within the uncertainties at 1σ. The higher deviation comes from the independent term in the case of equation (1a), corresponding to Lares-Martiz's sample. This is probably due to the fact that most of the stars appear in the lower part of the broad region of the *ASAS-SN* sample in the PL diagram. For the rest of the work, we will use equation (1b), corresponding to the *ASAS-SN* sample, because the corresponding PLR is obtained for a larger number of stars. This statistically more robust sample allows avoiding certain problems and errors, such as wrong luminosity values. It is not surprising that this relation has the lowest error bars.

We also compared our findings to previous PLRs. Figure 2 shows several previous studies† together with our new relations. As can be seen, the relations obtained with the different samples in this work are essentially the same as in previous studies. Small deviations are expected due to differences in the samples used to get them.

4. Gravity darkening effects due to rapid rotation

δ Sct stars are rapid rotators (see, e.g., Royer *et al.* 2007). Due to their high rotation speeds, these stars become oblate, deviating from the shape of an ideal sphere. This deformation gives rise to distinct equilibrium conditions on their surfaces, resulting in higher temperatures and brightness at the poles, while the equator experiences lower temperatures and fainter luminosity. This phenomenon is commonly referred to as gravity darkening, and it holds significant importance when considering observed luminosities.

† Because we just used the previous PLRs for comparison reasons, we transform the visual absolute magnitudes given by the authors into bolometric ones assuming that most δ Sct are A-type stars for which the bolometric correction is 0.

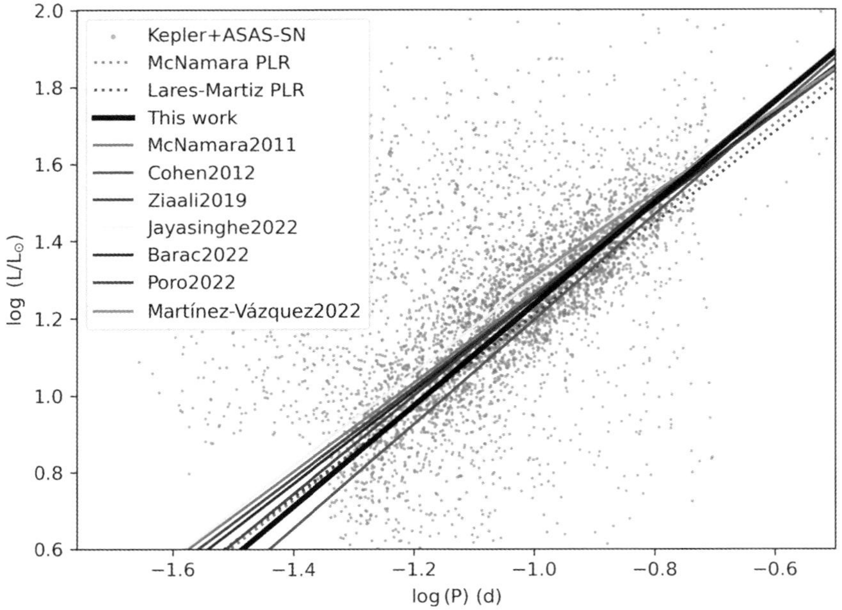

Figure 2. Comparison between PLRs found in previous studies and this work (corresponding to equation (1b)). The PLRs of HADS stars from the McNamara's and Lares-Martiz's samples are also shown as a guide. Notice how similar the slopes of the different fits are. The short-period segment of the Martínez-Vázquez *et al.* (2022) relation is not depicted because of the broken relation they found, we only include the segment most coherent with the other PLRs.

Rotation effects have commonly been incorporated into 1D codes as a correction factor for the stellar equations, as done, for instance, in the Modules for Experiments in Stellar Astrophysics (MESA; Paxton *et al.* 2019). In this code, after calculating the stellar structure for a specific rotation rate, a gravity darkening model derived from 2D solutions is applied (Espinosa Lara & Rieutord 2011). Inclination also affects the observed physical parameters in this case, as the relative visibility of the pole versus the equator plays a role. Under extreme conditions, the combined influence of stellar deformation and inclination can lead to a luminosity increase of up to 50% and a temperature variation of 2.5%. Consequently, this effect can significantly impact a PLR.

To show the possible impact of not taking into account the gravity darkening, we have used the two limiting cases, i.e., the highest rotation rate possible but with the line of sight at 0° (pole-on) or 90° (equator-on). In these cases, the luminosity can be higher by up to 50% or lower by up to 20%, respectively (Paxton *et al.* 2019). The angle at which the projected luminosity is equal to the real luminosity is approximately 55°, meaning that it is slightly more probable to find a star with higher projected luminosity (supposing randomly distributed inclinations). All this is shown in Figure 3. We plot our PLR and also the limits of the projected luminosity, $C_L \equiv L_{\rm proj}/L$, over the diagram with the samples. Clearly, the gravity darkening can explain the broader region observed in our large sample of more than 4000 stars. Moreover, if we estimate the first overtone period (P1) based on P1/P0 $\sim$ 0.77, then we also see in the figure that there exists overlapping between the pole-on case and that overtone, depicted as a dashed green line. This translates into difficulty in getting a reliable mode identification, even for the fundamental mode. The only way to overcome this problem is to derive the unprojected stellar rotational velocity, something hard to get even with asteroseismology (see, for example, Ramón-Ballesta *et al.* 2021).

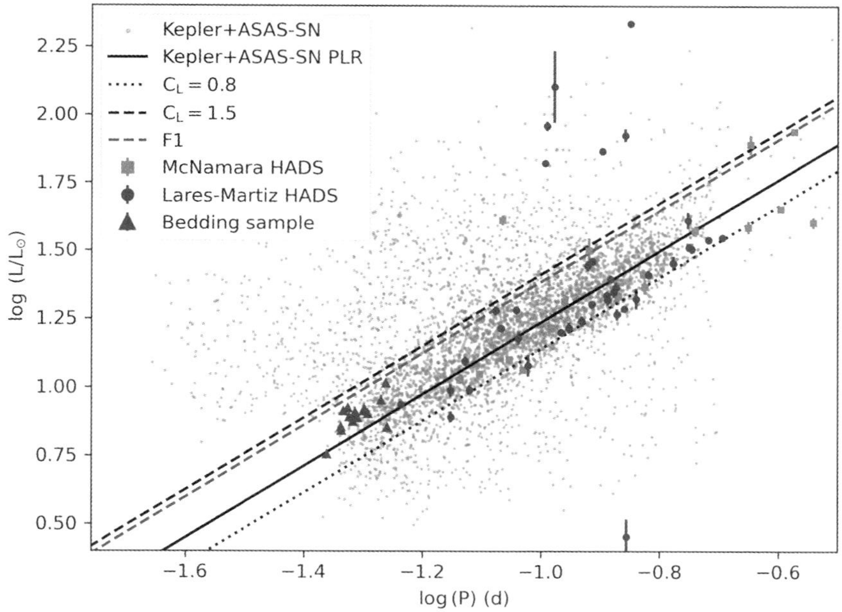

Figure 3. The PLR found in this work is shown with the upper and lower limits of projected luminosity C_L, the first overtone F1, and all samples of stars analysed.

4.1. *The* $F0 - \Delta\nu_{low}$ *relation*

Apart from dealing with gravity darkening, positioning a particular star in the PLR means a correct identification of the fundamental radial mode. Because both the low-order large separation and P0 are directly related to the stellar mean density (García Hernández *et al.* 2015, 2017), a direct relation between both quantities is also likely. To check that, we have used the rotating equilibrium and pulsation models by Rodríguez-Martin *et al.* (2020) to look for such a relation. Figure 4 shows the results. In this figure, we have plotted the fundamental mode frequency (F0) as a function of the low-order large separation ($\Delta\nu_{\rm low}$) for the models in the observational instability strip of δ Sct stars, as found by Murphy *et al.* (2019). A tight linear fit is clearly seen, with some scatter in the most evolved stages that correspond to the lowest values of F0 and $\Delta\nu_{\rm low}$. The deviation from the relation is somewhat expected in the models since the stars at those stages are quite evolved, probably indicating that they are not even δ Sct pulsators. A closer inspection of these models with non-adiabatic calculations of the oscillation frequencies would be necessary to confirm their mode stability.

The result of the fit in $\rm d^{-1}$ is:

$$\rm F0 = 3.022 \cdot \Delta\nu_{low} + 0.603, \tag{2}$$

with $r^2 = 0.995$ and a standard deviation of the residuals equal to $\sigma = 0.311\,\rm d^{-1}$. There have been previous experimental determinations of a $\rm F0 - \Delta\nu_{low}$ relation (Jayasinghe *et al.* 2020; Bedding *et al.* 2020). Our fit is not very different from others, although it is much more robust. The difficulties in determining the correct fundamental mode and $\Delta\nu_{\rm low}$ have hampered the use of this relation in the past.

4.2. *Towards a fundamental mode identification*

One may consider using the relations we present here to derive P0 knowing $\Delta\nu_{\rm low}$, but the large separation has only been determined for a handful of δ Sct. The larger sample

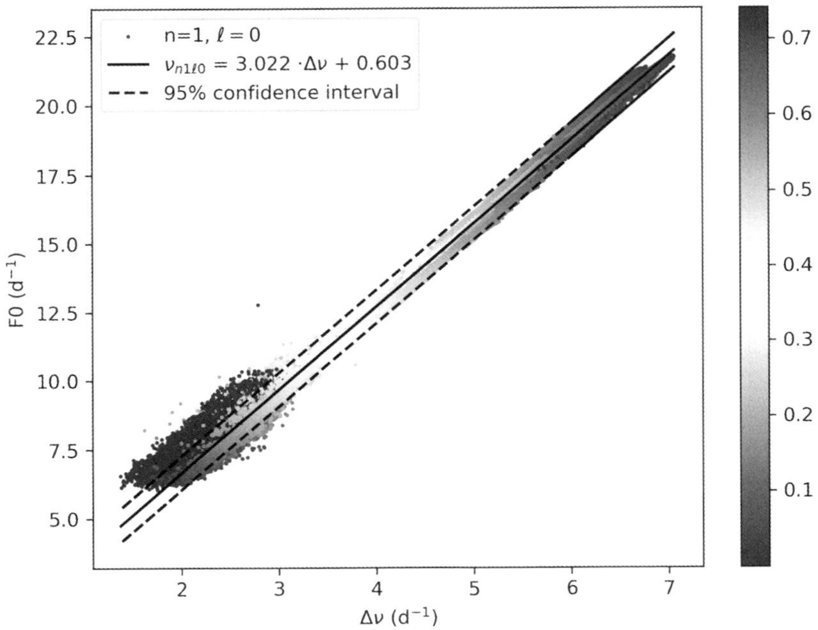

Figure 4. $F0 - \Delta\nu_{\rm low}$ relation. The fundamental mode ($n = 1$, $\ell = 0$) and the low-order large separation ($\Delta\nu_{\rm low}$) for rotating equilibrium and pulsation models in the observational instability strip of δ Sct stars are shown. The plot is colour-coded according to the central hydrogen abundance, Xc, illustrating the evolutionary stage of each model. A fit to this relation is shown as a continuous line, with the associated 95% confidence interval shown as dashed lines.

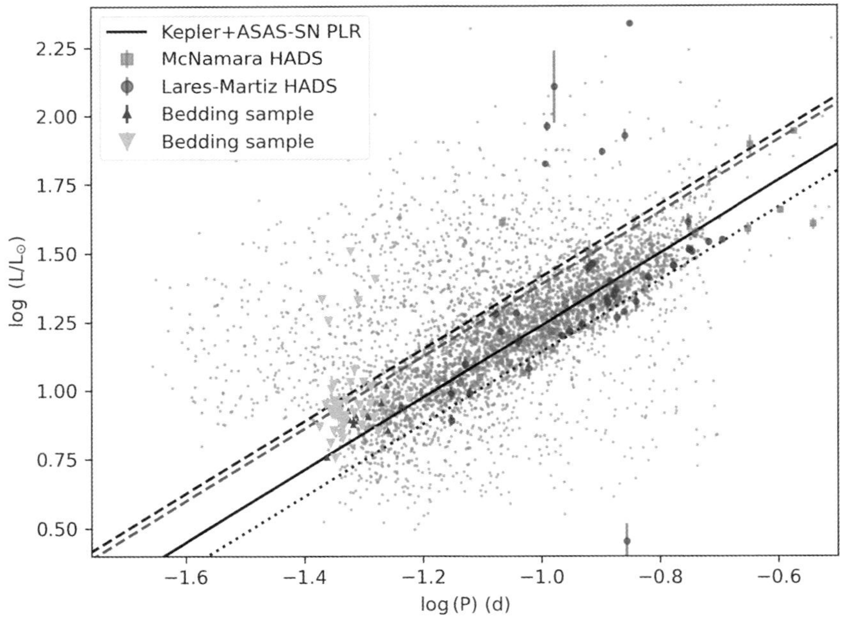

Figure 5. The PLR found in this work is shown with the upper and lower limits of the projected luminosity C_L (dashed and dotted black lines, respectively), the first overtone F1 (green dashed line), and all samples of stars analysed. F0 computed from the $F0 - \Delta\nu_{\rm low}$ relation given in equation (2) for the Bedding *et al.* (2020) stars is shown with downward yellow triangles.

to date is that of Bedding *et al.* (2020). They were able to find $\Delta\nu_{\rm low}$ for 60 young δ Sct stars, for which the pattern of the large separation is clearer. They also identified the fundamental mode for some of them, which are the stars depicted in Figure 1 as purple triangles. We tried to check the validity of their mode identification and complete it for those stars without an F0 determination in Bedding *et al.* (2020).

Using equation (2), we computed new F0 values based on the provided $\Delta\nu_{\rm low}$. Figure 5 shows a similar plot to Figure 3, but with the computed F0 as yellow triangles. Most of our determined fundamental modes are in agreement with those found by Bedding *et al.* (2020), with some slight differences coming from a non-perfect fit (included in the uncertainties). However, for some stars showing higher luminosities, F0 does not seem to be correctly determined: a complete description thus requires a better understanding of the selection mechanism. While our models predict fundamental modes corresponding to luminosities above and below the PLR, observations only populate the region close to it. These results point to the impossibility of populating the PLR with new, isolated stars using just $\Delta\nu_{\rm low}$. However, it appears that only stars in which the fundamental mode is actually excited lie close to the PLR: if this holds true, a detailed study of the stars with a detectable fundamental mode at F0 might be of interest to shed light on the selection mechanism, which has remained evasive even with the advances permitted by recent space missions.

5. Conclusions

We have derived a new PLR for δ Sct stars by utilizing a sample of around 4000 such stars from *ASAS-SN* and *Kepler* data. Our relation is consistent with previous determinations. We have also brought out that gravity darkening plays a significant role in causing the broadening effect within the PLR. This broadening includes the predicted position of the first overtone, hampering the identification of both F0 and F1.

We have derived a theoretical $\mathrm{F0} - \Delta\nu_{\rm low}$ relation that is compatible with previous observations, demonstrating its robustness even in the presence of rotation. Using this relation to guess F0 does not help to get a more constrained PLR. Only in the cases when the fundamental mode is excited (i.e., detected in the frequency spectrum) does it place the pulsator close to the PLR. This observational fact provides valuable insights into the excitation and mode selection processes.

There also seems to be an internal process related to mode selection that favours the high-amplitude peak being associated with the fundamental mode in the majority of δ Sct stars, as suggested by the broader region of the PL diagram that forms the PLR. This might offer valuable clues for further investigations into their excitation mechanisms and mode selection processes.

Acknowledgements. The authors acknowledge T. Bedding for allowing the use of the F0 values from his 2020 publication. AGH acknowledges funding support from 'FEDER/junta de Andalucía-Consejería de Economía y Conocimiento' under project E-FQM-041-UGR18 by Universidad de Granada. JPG and MLM acknowledge financial support from project PID2019-107061GB-C63 from the 'Programas Estatales de Generación de Conocimiento y Fortalecimiento Científico y Tecnológico del Sistema de I+D+i y de I+D+i Orientada a los Retos de la Sociedad' and from the grant CEX2021-001131-S funded by MCIN/AEI/10.13039/501100011033. AGH, JCS, GMM and SBF acknowledge funding support from the Spanish State Research Agency (AEI) project PID2019-107061GB-064. Funding for open access charge: Universidad de Granada.

References

Aerts, C., Christensen-Dalsgaard, J., & Kurtz, D. W. 2010, *Asteroseismology* (Dordrecht: Springer)

Auvergne, M., Bodin, P., Boisnard, L., *et al.* 2009, *A&A*, 506, 411

Balona, L. A. 2010, *Challenges in Stellar Pulsation* (Sharjah: Bentham Science Publishers)
Barac, N., Bedding, T. R., Murphy, S. J., & Hey, D. R. 2022, *MNRAS*, 516, 2080
Barceló Forteza, S., Roca Cortés, T., García Hernández, A., & García, R. A. 2017, *A&A*, 601, A57
Barceló Forteza, S., Roca Cortés, T., & García, R. A. 2018, *A&A*, 614, A46
Barceló Forteza, S., Moya, A., Barrado, D., *et al.* 2020, *A&A*, 638, A59
Bedding, T. R., Murphy, S. J., Hey, D. R., *et al.* 2020, *Nature*, 581, 147
Breger, M. 1979, *PASP*, 91, 5
Breger, M., Pamyatnykh, A. A., Pikall, H., & Garrido, R. 1999, *A&A*, 341, 151
Chang, S.-W., Protopapas, P., Kim, D.-W., & Byun, Y.-I. 2013, *AJ*, 145, 132
Cohen, R. E., & Sarajedini, A. 2012, *MNRAS*, 419, 342
Corsaro, E., Stello, D., Huber, D., *et al.* 2012, *ApJ*, 757, 190
Deka, M., Kanbur, S. M., Deb, S., *et al.* 2022, *MNRAS*, 517, 2251
De Franciscis, S., Pascual-Granado, J., Suárez, J. C., *et al.* 2018, *MNRAS*, 481, 4637
De Ridder, J., Ripepi, V., Aerts, C., *et al.* 2022, *A&A*, 50
Espinosa Lara, F., & Rieutord, M. 2011, *A&A*, 533, A43
Fernie, J. D. 1964, *ApJ*, 140, 1482
Fouesneau, M., Frémat, Y., Andrae, R., *et al.* 2023, *A&A*, 674, A28
García Hernández, A., Moya, A., Michel, E., *et al.* 2009, *A&A*, 506, 79
García Hernández, A., Moya, A., Michel, E., *et al.* 2013, *A&A*, 559, A63
García Hernández, A., Martín-Ruiz, S., Monteiro, M. J. P. F. G., *et al.* 2015, *ApJ*, 811, L29
García Hernández, A., Suárez, J. C., Moya, A., *et al.* 2017, *MNRAS*, 471, L140
Garg, A., Cook, K. H., Nikolaev, S., *et al.* 2010, *AJ*, 140, 328
Handler, G., Pikall, H., O'Donoghue, D., *et al.* 1997, *MNRAS*, 286, 303
Hasanzadeh, A., Safari, H., & Ghasemi, H. 2021, *MNRAS*, 505, 1476
Jayasinghe, T., Kochanek, C. S., Stanek, K. Z., *et al.* 2018, *MNRAS*, 477, 3145
Jayasinghe, T., Stanek, K. Z., Kochanek, C. S., *et al.* 2020, *MNRAS*, 493, 4186
Koch, D. G., Borucki, W. J., Basri, G., *et al.* 2010, *ApJ*, 713, L79
Leavitt, H. 1908, *Annals of Harvard College Observatory*, 60, 87
Lares-Martiz, M., Garrido, R., & Pascual-Granado, J. 2020, *MNRAS*, 498, 1194
Lares-Martiz, M. 2022, *Frontiers in Astronomy and Space Sciences*, 9, 301
Martínez-Vázquez, C. E., Salinas, R., Vivas, A. K., & Catelan, M. 2022, *ApJ*, 940, L25
McNamara, D. H. 2011, *AJ*, 142, 110
Michel, E., Dupret, M.-A., Reese, D. R., *et al.* 2017, *EPJ Web of Conferences*, 160, 03001
Mirouh, G. M., Angelou, G. C., Reese, D. R., & Costa, G. 2019, *MNRAS*, 483, L28
Moya, A., Suárez, J. C., García Hernández, A., & Mendoza, M. A. 2017, *MNRAS*, 471, 2491
Murphy, S. J., Hey, D. R., Van Reeth, T., & Bedding, T. R. 2019, *MNRAS*, 485, 2380
Ouazzani, R.-M., Roxburgh, I. W., & Dupret, M.-A. 2015, *A&A*, 579, A116
Owens, K. A., Freedman, W. L., Madore, B. F., & Lee, A. J. 2022, *ApJ*, 927, 8
Paparó, M., Benkő, J. M., Hareter, M., & Guzik, J. A. 2016, *ApJS*, 224, 41
Pascual-Granado, J., Garrido, R., & Suárez, J. C. 2015, *A&A*, 581, A89
Pascual-Granado, J., Suárez, J. C., Garrido, R., *et al.* 2018, *A&A*, 614, A40
Paxton, B., Smolec, R., Schwab, J., *et al.* 2019, *ApJS*, 243, 10
Poleski, R., Soszyński, I., Udalski, A., *et al.* 2010, *AcA*, 60, 179
Poro, A., Paki, E., Mazhari, G., *et al.* 2021, *PASP*, 133, 084201
Ramón-Ballesta, A., García Hernández, A., Suárez, J. C., *et al.*, 2021 *MNRAS*, 505, 6217
Reese, D. R., Lignières, F., & Rieutord, M. 2008, *A&A*, 481, 449
Ricker, G. R., Winn, J. N., Vanderspek, R., *et al.* 2014, *Journal of Astronomical Telescopes, Instruments, and Systems*, 1, 014003
Rodríguez, E., & Breger, M. 2001, *A&A*, 366, 178
Rodríguez-Martín, J. E., García Hernández, A., Suárez, J. C. & Rodón, J. R. 2020, *MNRAS*, 498, 1700
Royer, F., Zorec, J., & Gómez, A. E. 2007, *A&A*, 463, 671
Soszyński, I., Pietrukowicz, P., Skowron, J., *et al.* 2021, *AcA*, 71, 189

Suárez, J. C., García Hernández, A., Moya, A., *et al.* 2014, *A&A*, 563, A7
Walker, G., Matthews, J., Kuschnig, R., *et al.* 2003, *PASP*, 115, 1023
Wenger, M., Ochsenbein, F., Egret, D., *et al.* 2000, *A&AS*, 143, 9
Xue, H.-F., Niu, J.-S., & Fu J.-N. 2022, *Research in Astronomy and Astrophysics*, 22, 105006
Ziaali, E., Bedding, T. R., Murphy, S. J., *et al.* 2019, *MNRAS*, 486, 4348
Zwintz, K., Lenz, P., Breger, M., *et al.* 2011, *A&A*, 533, A133

At the crossroads of astrophysics and cosmology: Period–luminosity relations in the 2020s
Proceedings IAU Symposium No. 376, 2024
R. de Grijs, P. A. Whitelock & M. Catelan, eds.
doi:10.1017/S1743921323002776

Period–Luminosity–Metallicity relations for Classical Pulsators at Near-infrared Wavelengths

Anupam Bhardwaj

INAF - Osservatorio Astronomico di Capodimonte, Via Moiariello 16, 80131, Napoli, Italy.
email: anupam.bhardwajj@gmail.com

Abstract. Classical pulsating stars such as Cepheid and RR Lyrae variables exhibit well-defined Period–Luminosity relations at near-infrared wavelengths. Despite their extensive use as stellar standard candles, the effects of metallicity on Period–Luminosity relations for these pulsating variables, and in turn, on possible biases in distance determinations, are not well understood. We present ongoing efforts in determining accurate and precise metallicity coefficients of Period–Luminosity-Metallicity relations for classical pulsators at near-infrared wavelengths. For Cepheids, it is crucial to obtain a homogeneous sample of photometric light curves and high-resolution spectra for a wide range of metallicities to empirically determine metallicity coefficient and reconcile differences with the predictions of the theoretical models. For RR Lyrae variables, using their host globular clusters covering a wide range of metallicities, we determined the most precise metallicity coefficient at near-infrared wavelengths, which is in excellent agreement with the predictions of the horizontal branch evolution and stellar pulsation models.

Keywords. Variable Stars - Cepheids, RR Lyrae, Pulsations, Distance Scale

1. Introduction

Cepheid and RR Lyrae variables are the most popular subclasses of radially pulsating stars and are often referred as classical pulsators. These variables are robust stellar standard candles for astronomical distance measurements because their luminosities are strongly correlated with their pulsation periods (see the reviews by Subramanian et al. 2017; Bhardwaj 2020). The Period-Luminosity (PL) relation or the "Leavitt Law" is at the basis of the extragalactic distance ladder used to determine the Hubble constant or the present expansion rate of the Universe (Freedman et al. 2001; Riess et al. 2022). At present, there is a discord between the Hubble constant values based on the traditional Classical Cepheid-Supernova distance ladder (Riess et al. 2022) and those from the *Planck* mission (Planck Collaboration et al. 2020). The discrepancy in these two measurements of the expansion rate at two different ends of the Universe hint at missing or new physics in the standard cosmological model. Since the PL relation of Classical Cepheid variables plays a vital role in the local determination of the Hubble constant, it is now crucial to explore all possible sources of systematic uncertainties in the calibration of these relations and improve the accuracy and the precision of the extragalactic distance ladder.

Despite the extensive use of PL relations for distance measurements, their universality in different stellar environments has not been established and is often debated in the literature (Bhardwaj 2020). When approaching a percent-level precision, the impact of chemical compositions, topology of the instability strip, and the mass-luminosity relations obeyed by classical pulsators on PL relations can also become significant (Bono et al. 2010; Marconi et al. 2015). Some of these evolutionary effects become negligible at

Table 1. Metallicity coefficients of RR Lyrae PLZ relations in NIR bands.

	$N_{\rm RRL}$	Δ[Fe/H]	J	H	K_s	Ref.
GC + MW	1310	2.0	0.20 ± 0.02	0.19 ± 0.01	0.18 ± 0.01	Bhardwaj et al. (2023)
MW	448	2.8	—	—	0.10 ± 0.03	Muhie et al. (2021)
GC + MW	403	2.7	—	—	0.17 ± 0.01	Bhardwaj et al. (2021a)
MW	55	2.5	0.20 ± 0.03	0.17 ± 0.03	0.17 ± 0.03	Neeley et al. (2019)
ω Cen	55	1.0	0.15 ± 0.02	0.13 ± 0.02	0.15 ± 0.02	Braga et al. (2018)
ω Cen	64	1.0	0.15 ± 0.03	-	0.14 ± 0.02	Navarrete et al. (2017)
Theory	—	$--$	0.18 ± 0.01	0.18 ± 0.01	0.18 ± 0.01	Marconi et al. (2015)
MW	23	2.6	—	—	0.07 ± 0.04	Muraveva et al. (2015)
LMC	70	1.5	—	—	0.03 ± 0.07	Muraveva et al. (2015)
MW	403	2.0	—	—	0.09 ± 0.03	Dambis et al. (2013)
LMC	50	1.3	—	—	0.05 ± 0.07	Borissova et al. (2009)
GC	538	2.0	—	—	0.08 ± 0.11	Sollima et al. (2008)
Theory	—	$--$	$0.19 \pm -$	$0.18 \pm -$	$0.18 \pm -$	Catelan et al. (2004)
MW	27	2.0	—	—	0.23 ± 0.01	Bono et al. (2003)
Theory	—	$--$	—	—	$0.17 \pm -$	Bono et al. (2001)

Notes: [1] RR Lyrae host stellar systems in column 1 are: GC - globular clusters, MW- Milky Way, LMC - Large Magellanic Cloud.

near-infrared (NIR) wavelengths because of the narrow width of the instability strip and small temperature variations. NIR observations also offer advantages over optical bands for distance measurements because the PL relations are tighter at these longer wavelengths and extinction effects are significantly smaller. Furthermore, smaller variability amplitudes in NIR lead to more accurate mean magnitude determinations even with a small number of observations or sparsely sampled light curves. The absolute calibration of NIR PL relations for classical pulsators are now crucial in view of the upcoming large ground-based observational facilities and space telescopes, which will predominantly operate at these longer wavelengths.

Even at NIR wavelengths, one of the main sources of residual systematic uncertainty on the Classical Cepheid based distance scale is due to metallicity effects on PL relations (Ripepi et al. 2021; Riess et al. 2022). While there is no consensus on the exact dependence of metallicity, it is expected to affect both the slope and zero-point of the PL relations (Bono et al. 2010). Breuval et al. (2022) provided a compilation of theoretical and empirical estimates of the metallicity coefficient (commonly referred as γ in the literature) of Classical Cepheid PL relations over the past two decades. It is interesting to note that the metallicity term of the K_s-band Period-Luminosity-Metallicity (PLZ) relation varies significantly between -0.45 and $+0.02$ mag/dex in different empirical studies, and thus, even the sign of the metallicity dependence is debated. However, more recently the values of γ are converging between -0.4 and -0.2 mag/dex in NIR bands. In the case of Type II Cepheids, both empirical and theoretical calibrations predict a small or negligible dependence in limited studies at NIR wavelengths (Matsunaga et al. 2006; Das et al. 2021). Recently, Wielgórski et al. (2022) found a relatively larger dependence of the order of -0.2 mag/dex in JHK_s band PL relations albeit using a small sample of Type II Cepheid variables. Similar to Cepheids, the distance measurements based on RR Lyrae stars will also result in large systematic biases if the metallicity term on their PL relation is not taken into account (Catelan et al. 2004; Marconi et al. 2015; Bhardwaj et al. 2023). Table 1 lists the empirical and theoretical estimates of metallicity coefficient in NIR filters in different stellar environments. It is evident that the sign of metallicity dependence is well established for RR Lyrae stars such that metal-rich stars are fainter. Moreover, the metallicity coefficient in NIR bands seems to converge around 0.2 mag/dex in both empirical and theoretical studies.

In the following subsections, we will discuss our recent and ongoing efforts to improve the quantification of metallicity dependence on PL relations for Cepheid and RR Lyrae stars.

2. Classical or Type I Cepheid variables

Until a few years back, a precise determination of metallicity coefficient of Classical Cepheid PLZ relations was hampered by the lack of accurate distances to Galactic Cepheids. High precision parallaxes for thousands of Classical Cepheids are now being provided by the *Gaia* astrometric mission along with their photometric light curves (Ripepi et al. 2022). However, *Gaia* does not provide high-resolution spectroscopic abundances and NIR light curves of classical pulsators, which are equally important for deriving precise PLZ relations.

2.1. *High-resolution spectroscopic metallicities*

High-resolution spectroscopic metallicities of Classical Cepheid variables are available in the literature for small samples of stars in a single study, which were then used to homogenize larger samples of literature [Fe/H] measurements in different metallicity scales (e.g., Genovali et al. 2014; Luck 2018). However, Classical Cepheids with metallicities measured using high-resolution spectra were mainly restricted to the solar neighbourhood and there are a very few variables with low metallicities ([Fe/H]<-0.4 dex). Since the solar neighbourhood Classical Cepheids exhibit a rather narrow range of [Fe/H] ($\sigma = 0.3$ dex), it is difficult to properly quantify the metallicity term of the PLZ relations. In the latest determination of the Hubble constant, 75 Milky Way Cepheid standards were used to calibrate the PLZ relation for the extragalactic distance ladder along with independent anchors such as LMC, and NGC 4258 (Riess et al. 2022). These 75 Milky Way Cepheid standards were carefully selected to be in less crowded and less extincted regions in the Galaxy and exhibit a [Fe/H] range of 0.7 dex. The spectroscopic [Fe/H] values for these Cepheids were adopted from Groenewegen (2018) based on the compilation of Genovali et al. (2015). Homogeneous high-resolution spectroscopy of these 75 Milky Way Cepheid standards is crucial to properly quantify the metallicity term and measure its impact on the distance scale and the Hubble constant.

We are obtaining high-resolution spectra of Milky Way Cepheid standards using the ESPaDOnS instrument ($R \sim 81,000$) at the Canada-France-Hawaii Telescope (CFHT). In addition to optical spectra, we are also collecting high-resolution ($R \sim 45,000$) HK_s band spectra for these Classical Cepheids for the first-time using IGRINS instrument at Gemini-South Telescope. Multi-epoch spectra for 42 out of 75 Classical Cepheids have already been obtained and analysed to derive [Fe/H] values. Fig. 1 displays the spectra of two Cepheid standards at two different phases along their pulsation cycle. The iron abundances were measured using the same methodology as discussed in Ripepi et al. (2021) and the interested readers are referred to this paper for more details. The preliminary analysis with derived [Fe/H] values suggests an excellent agreement with the [Fe/H] values from Luck (2018), which are also based on multi-epoch spectra of Classical Cepheid variables. Our spectroscopic [Fe/H] values for 42 Milky Way Cepheids were combined with the remaining 35 Cepheids having metallicities from Luck (2018) to obtain a homogenized sample of 75 stars with new [Fe/H] measurements. Multiband photometric light curves for majority of these 75 Cepheids were analysed by Bhardwaj et al. (2015, 2017b) using a Fourier decomposition technique. Therefore, accurate mean magnitudes in $VIJHK_s$ bands are available, and together with high-resolution [Fe/H] measurements and precise *Gaia* parallaxes, our sample will be used to derive PLZ relations for these Milky Way standard Classical Cepheid variables.

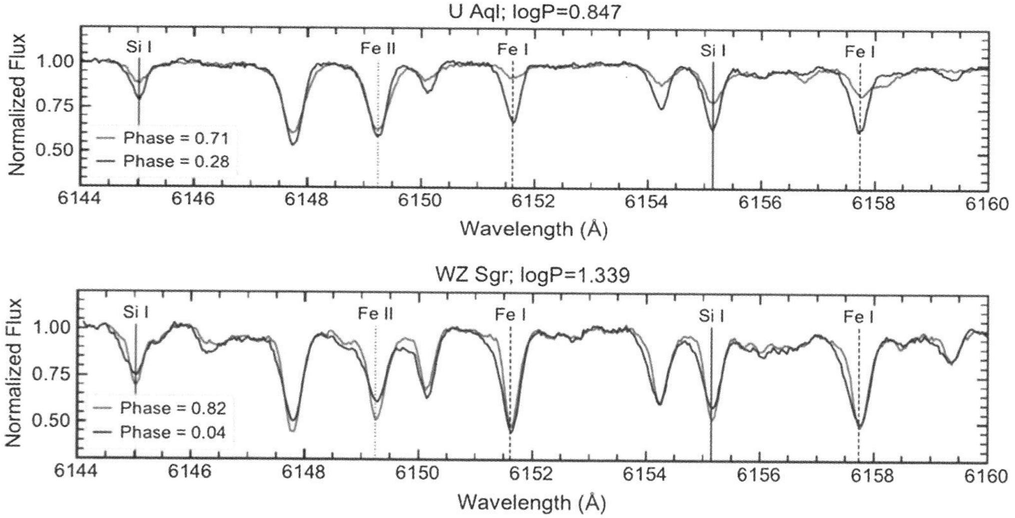

Figure 1. Optical spectra of a short-period (top) and a long-period (bottom) Milky Way Cepheid standard at two different phases along their pulsation cycles. Iron lines are marked as dotted lines that are used to determine spectroscopic metallicities ([Fe/H]).

At the same time, high-resolution spectroscopic metallicities for more metal-poor Classical Cepheids are being obtained as part of the Cepheid Metallicity in the Leavitt Law (C-MetaLL) survey (Ripepi et al. 2021). This program aims to enlarge the metallicity range of Classical Cepheids by observing more distant targets in the Galactic disk and anti-center directions using HARPS instrument at Telescopio Nazionale Galileo (TNG) and UVES instrument at Very Large Telescope (VLT). Ripepi et al. (2021) found a rather large metallicity coefficient in NIR bands of the order of ~ -0.4 mag/dex. This is in contradiction to Riess et al. (2021) value of ~ -0.2 mag in the H-band Wesenheit relation, which was determined using 75 Milky Way Cepheids in the solar vicinity. Note that the Wesenheit magnitudes (see Madore 1982; Inno et al. 2015; Bhardwaj et al. 2016, for details) are by construction reddening-independent. Most studies which either used a small sample of Milky Way Cepheids in the solar neighbourhood or combined those with Classical Cepheids in the Magellanic Clouds to enlarge the metallicity range, have found a metallicity coefficient of the order of -0.2 mag/dex in NIR bands (e.g. Gieren et al. 2018; Breuval et al. 2022). Therefore, increasing the metallicity range using more than 110 published C-MetaLL target Classical Cepheids is crucial to better constrain the metallicity coefficient of the PLZ relation. However, since these targets are more distant, there are no NIR light curves for these stars in the literature and their latest *Gaia* parallaxes also exhibit larger uncertainties.

2.2. *Homogeneous optical and NIR photometric light curves*

Multiband light curves of Classical Cepheid variables are useful to not only obtain precise and accurate mean magnitudes, but also determine their reddening values. Optical light curves for Classical Cepheids in traditional Johnson-Cousins filters are available in the literature from large variability surveys (e.g., VI in OGLE or V in ASAS-SN) or from photoelectic observations (e.g. Berdnikov 2008, in $UBVRI$). However, no homogeneous light curves in Sloan (optical) and NIR filters are available for most of our target Cepheid variables with spectroscopic metallicities. The calibration of Cepheid PLZ relations in the Sloan-like and NIR filters will be particularly useful in the era of Vera C. Rubin

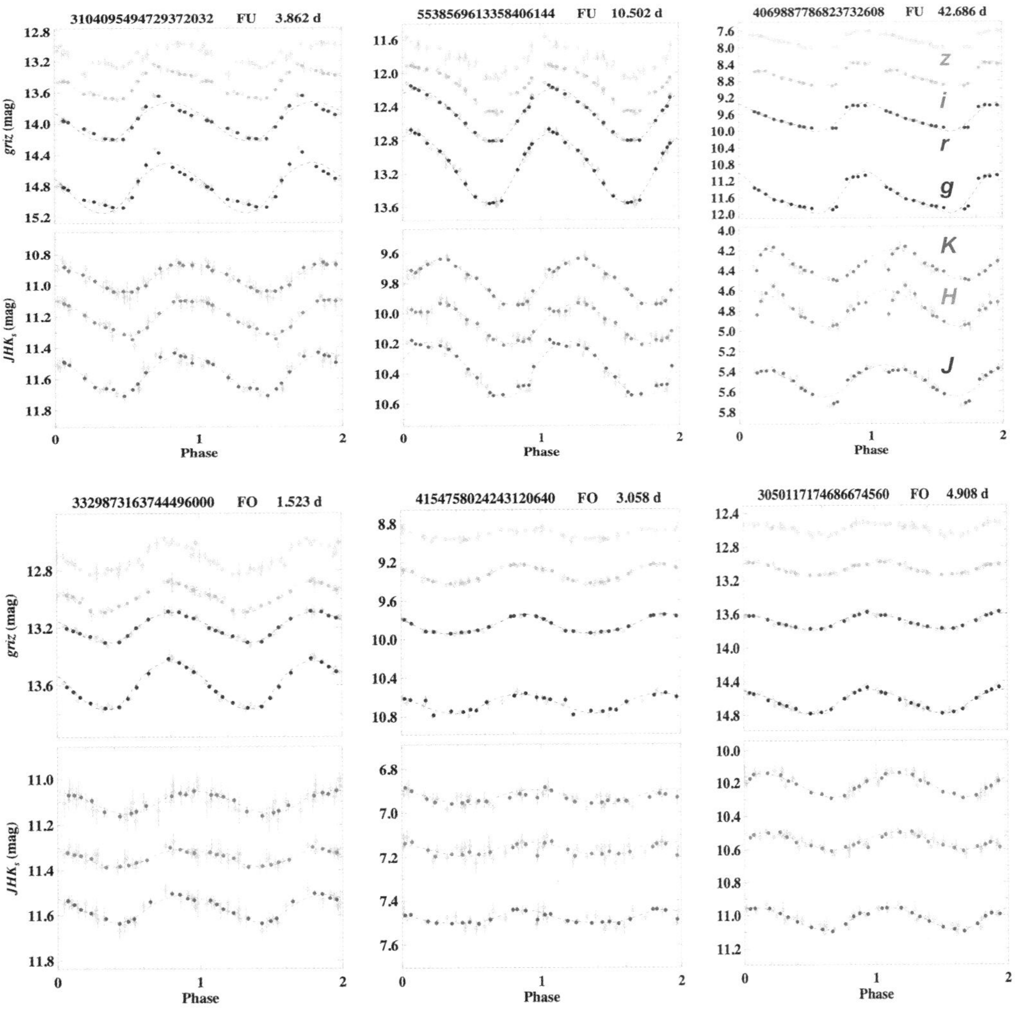

Figure 2. Example light curves of fundamental (FU) and first-overtone (FO) mode Cepheids with different periods in our sample. The grey points in each panel show all the data points while filled circles represent phase averaged light curves. The best-fitted sinusoidal templates are also overplotted. The *Gaia* DR3 Source ID, pulsation mode, and periods are listed on the top of each panel.

Observatory's Legacy Survey of Space and Time (LSST) and upcoming thirty meter class telescopes operating at infrared wavelengths, respectively.

We are obtaining multiband ($grizJHK_s$) light curves of Cepheids using the Rapid Eye Mount (REM) telescope. The observations are collected simultaneously in all filters using two different instruments - ROSS (optical) and REMIR (NIR). The observations were scheduled nightly for a given target, if visible, for up to 60 nights within a semester. We have already obtained imaging data for 80 Cepheids and reduced the images using DAOPHOT/ALLFRAME suite of softwares (Stetson 1996). The light curves were calibrated in Sloan Digital Sky Survey (SDSS) filters using *Gaia* synthetic photometry and in NIR filters using Two Micron All Sky Survey (2MASS, Skrutskie et al. 2006) observations. Fig. 2 displays preliminary light curves of fundamental (FU) and first-overtone (FO) mode Cepheids in our sample. The number of data points per light curve range between 10 and 60, and there are on average 25 epochs. Template-fitted light curves were used to obtain mean-magnitudes for these Classical Cepheid variables and derive preliminary PL relations. Our analysis suggests a large scatter in the PL relations due to larger parallax uncertainties for these distant targets, and will be presented in detail in Bhardwaj et al. (2023, in prep.). Our new homogeneous spectra and multiband photometry of Galactic Classical Cepheids will be useful to derive accurate and precise PLZ relations.

3. Type II Cepheids

Type II Cepheids have never been used as extensively as Classical Cepheids for distance measurements primarily because these are fainter and less abundant. These old and low-mass stars are further separated in three distinct subclasses of BL Herculis (BL Her), W Virginis (W Vir), and RV Tauri (RV Tau) stars based on their different evolutionary status (see the review by Wallerstein 2002; Bhardwaj 2022). Observationally, the separation of Type II Cepheids in three different subclasses is based on their pulsation periods (Soszyński et al. 2018). While these population II stars also follow a tight PL relation in NIR bands, the metallicity dependence on these relations has not been explored in detail in the literature. Similar to Classical Cepheids, the primary reason for not having a well-calibrated PLZ relation is the lack of a homogeneous sample of Type II Cepheids with both high-resolution spectroscopic metallicities and multiband photometry. Empirically, the best constraints on the metallicity coefficient of Type II Cepheid PL relations in JHK_s bands were provided by Matsunaga et al. (2006) using variables in globular clusters. The authors assumed mean-metallicity of the cluster for individual Type II Cepheids and adopted the distances to the clusters based on horizontal branch stars, and combined those with mean JHK_s magnitudes to find a small or negligible metallicity coefficient. Theoretical pulsation models have also predicted no statistically significant metallicity dependence of Type II Cepheid PL relations at NIR wavelengths (Di Criscienzo et al. 2007; Das et al. 2021). The lack of accurate parallaxes in the pre-*Gaia* era precluded such empirical studies for Milky Way field Type II Cepheids. More recently, Wielgórski et al. (2022) provided NIR time-series photometry for 21 Galactic field Type II Cepheids and obtained a larger metallicity coefficient in NIR filters using 8 stars having high-resolution spectroscopic metallicities. In contrast, Ngeow et al. (2022) found no metallicity dependence on NIR PL relations for Type II Cepheids in globular clusters.

3.1. *Spectroscopic metallicities of Type II Cepheids from Gaia*

Homogeneous spectroscopic metallicities and NIR time-series photometry for a statistically significant sample of Type II Cepheids is crucial for the empirical calibration of their PLZ relations in NIR bands. This is particularly important now because accurate parallaxes for more than 1600 Type II Cepheids are available in the latest *Gaia* data release (Ripepi et al. 2022). However, high-resolution spectroscopic metallicities are still limited to small samples of 10-20 Type II Cepheids variables (Maas et al. 2007; Kovtyukh et al. 2018). In addition to precision astrometry, *Gaia* also provides medium resolution spectra for bright sources using Radial Velocity Spectrometer (RVS) instrument. We explored *Gaia*-RVS spectroscopic metallicities for Type II Cepheids and found 166 variables with both optical light curves and [Fe/H] measurements. This represents the largest sample of Milky Way field Type II Cepheids with homogeneous [Fe/H] determinations.

Fig. 3 displays *Gaia* parallaxes and photometry for 166 Type II Cepheids with RVS spectroscopic metallicities. The top panel clearly shows that the parallaxes for RV Tau stars are the most uncertain while BL Her have most accurate and precise parallaxes. This is also evident in the scatter in the Period-Wesenheit (PW) relation in *Gaia* bands

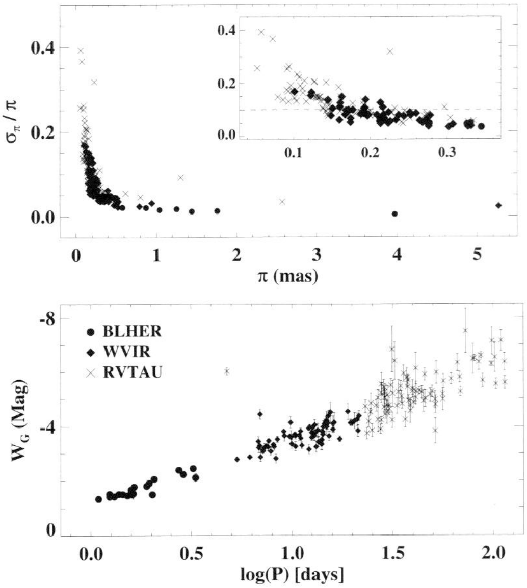

Figure 3. *Top:* Parallaxes of 166 Type II Cepheids in *Gaia* with spectroscopic metallicities from RVS. It is evident that parallaxes for RV Tau stars exhibit large uncertainties. *Bottom:* The Wesenheit magnitudes in *Gaia* bands showing tight PW relation for BL Her and W Vir stars and larger scatter for RV Tau variables.

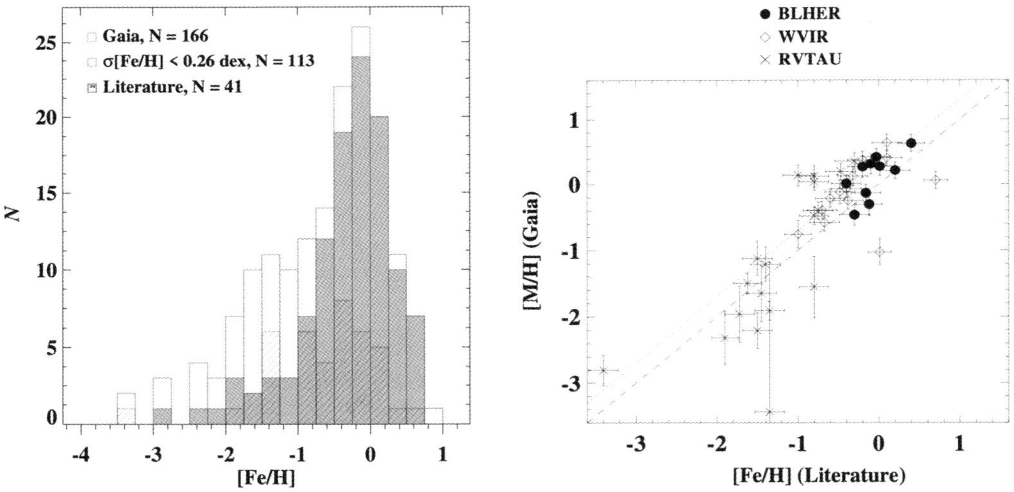

Figure 4. *Left:* Histograms of spectroscopic metallicities for *Gaia* Type II Cepheids. *Right:* Comparison of high-resolution spectroscopic metallicities from *Gaia*-RVS with those in the literature for common stars.

in the bottom panels. Fig. 4 displays the histogram of spectroscopic metallicities for all 166 variables. The *Gaia*-RVS metallicities have been assigned different quality flags (see Recio-Blanco et al. 2022, for details), and accordingly, we adopted a minimum uncertainty of 0.13 dex corresponding to 1σ systematics. We considered 113 *Gaia*-RVS measurements with acceptable quality flags with a maximum uncertainty of $2\sigma \sim 0.26$ dex. All the *Gaia*-RVS measurements were corrected using the recipe suggested by Recio-Blanco et al. (2022). For a relative comparison, we compiled 41 common Type II Cepheids with high-resolution spectroscopic metallicities in the literature. The right panel of Fig. 4 shows a

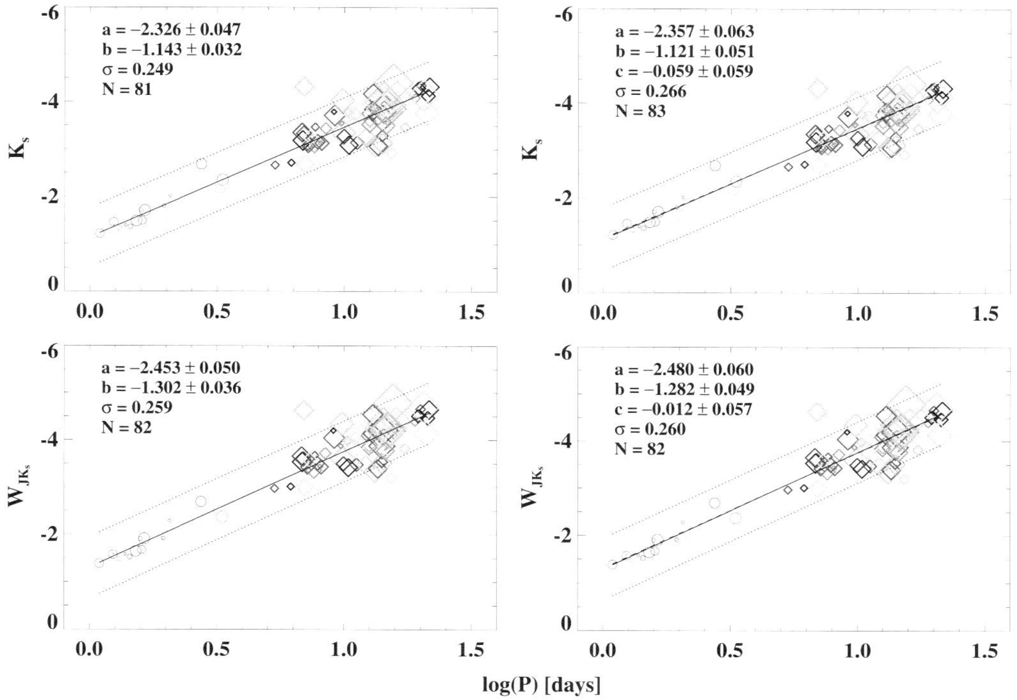

Figure 5. *Top:* K_s-band PL relation (left) and PLZ relation (right) for the combined sample of BL Her and W Vir stars. *Bottom:* Same as the top panels but for W_{JK_s} Wesenheit magnitudes. The slope (a), zero-point (b), metallicity coefficient (c), dispersion (σ) and the number of Type II Cepheids in the final fit (N) are mentioned in each panel. Note that the symbol size represents the error in parallaxes while colors represent variation in [Fe/H] values. The solid lines show the best-fitting linear regressions and the dashed lines represent $\pm 3\sigma$ scatter.

comparison of *Gaia*-RVS spectroscopic metallicities with those available in the literature. We found a systematic offset of ~ 0.3 dex between these two sets of [Fe/H] values with measurements for RV Tau stars exhibiting the largest scatter.

3.2. *NIR Period-Luminosity-Metallicity relations*

Other than spectroscopic metallicities and accurate *Gaia* parallaxes, NIR light curves of Type II Cepheids are also needed to derive precise PLZ relations. However, NIR light curves are not available for our sample of 166 Type II Cepheids and are being obtained using the REM telescope, as are those of Classical Cepheids. Nevertheless, we can do a preliminary analysis using the random-phase 2MASS observations. Given the large uncertainties in parallaxes for RV Tau stars, we restricted our sample to BL Her and W Vir variables. It is known that the PL relations of RV Tau do not follow their short-period counterparts (Ripepi et al. 2014; Bhardwaj et al. 2017a). Therefore, we excluded RV Tau stars and utilized a combined sample of 85 BL Her and W Vir stars for PLZ analysis.

Fig. 5 displays the PL and PW relations for Type II Cepheids in the left panels. The extinction corrections were applied assuming the reddening law of Cardelli et al. (1989). The precise parallaxes of BL Her stars result in tighter PL/PW relations, but the scatter increases for W Vir stars. Note that the dispersion in these relations is nearly two times larger than the scatter in the PL relations for Type II Cepheids in globular clusters (Matsunaga et al. 2006; Bhardwaj 2022; Ngeow et al. 2022). This increase in scatter can be attributed to random-phase 2MASS magnitudes and larger uncertainties in the parallaxes of W Vir stars. For these field stars with periods smaller than 10 days, the scatter in the PL relations is comparable to cluster variables. The right panels display PLZ relations after including the metallicity term. It is evident that the metallicity coefficient is statistically consistent with zero and this dependence is in agreement with the pulsation model predictions (Das et al. 2021). Our ongoing optical ($griz$) and NIR (JHK_s) observations of Type II Cepheids using REM will allow us to constrain better the empirical metallicity coefficients at multiple wavelengths.

4. RR Lyrae variables

Similar to Classical Cepheids, one of the main issues in RR Lyrae-based distance determinations is the impact of metallicity dependence on their PL relations. While for Cepheids, these effects had previously been predicted to be smaller, the luminosities of RR Lyrae are known to depend significantly on their metallicities. Catelan et al. (2004) predicted a metallicity coefficient of ~ 0.2 mag/dex in JHK_s PL relations using models of horizontal branch evolution, and Marconi et al. (2015) found similar dependence using RR Lyrae pulsation models. However, several empirical studies in the recent past predicted a small or negligible dependence of metallicity in NIR PL relations (see Table 1). To resolve these inconsistencies in empirical calibrations, we observed RR Lyrae variables in several globular clusters of different mean-metallicities at these longer wavelengths.

4.1. *NIR monitoring of RR Lyrae variables in globular clusters*

RR Lyrae variables in globular clusters provide a unique opportunity to study the variation in their pulsation properties as a function of metallicity, assuming that the clusters under consideration are mono-metallic or do not exhibit a significant metallicity spread. For this purpose, we started an observational program to monitor several globular clusters in JHK_s bands using the WIRCam instrument at the Canada-France-Hawaii Telescope. To increase the metallicity baseline, we are also observing several metal-rich bulge clusters using the Flamingos 2 instrument at the Gemini-South Telescope. The observational program consisted of typically 10-15 epochs per globular clusters with each epoch separated by at least 2 hours. This ensured a full phase coverage for RR Lyrae light-curves in NIR bands. The goal was to quantify the metallicity term by comparing cluster RR Lyrae stars with metallicity differences of up to Δ[Fe/H]~ 2 dex, when the effect of metallicity difference on their luminosities ($\Delta M \sim 0.4$ mag) will be quite evident assuming the predictions of theoretical models.

Fig. 6 displays example light curves of RR Lyrae variables with different periods in two clusters having different number of epochs per light curve. The left panels represent variables in M53 (Bhardwaj et al. 2021a), which are the best sampled with more than 20 epochs per light curve. The right panels display M15 variables with sparsely sampled light curves derived from archival imaging data (Bhardwaj et al. 2021b). More epochs are now available for this cluster, which was observed with both CFHT and Gemini Telescopes in our observational program. The NIR light curves were fitted using the templates derived by Braga et al. (2019). While these templates were based on NIR light curves of RR Lyrae in ω Cen and were divided in three separate period bins for fundamental mode variables (RRab), they do not necessarily fit our light curves in other clusters in the given period-bin. This is expected since the period distributions of RR Lyrae stars vary in different clusters depending on their metallicities. We, therefore, fitted all three sets of templates for RRab stars to observed NIR light curves. For overtone mode (RRc) and mixed-mode (RRd) stars, the single set of template from Braga et al. (2019) works well for all periods. Fig. 6 shows the best-fitted light curve templates for RR Lyrae stars

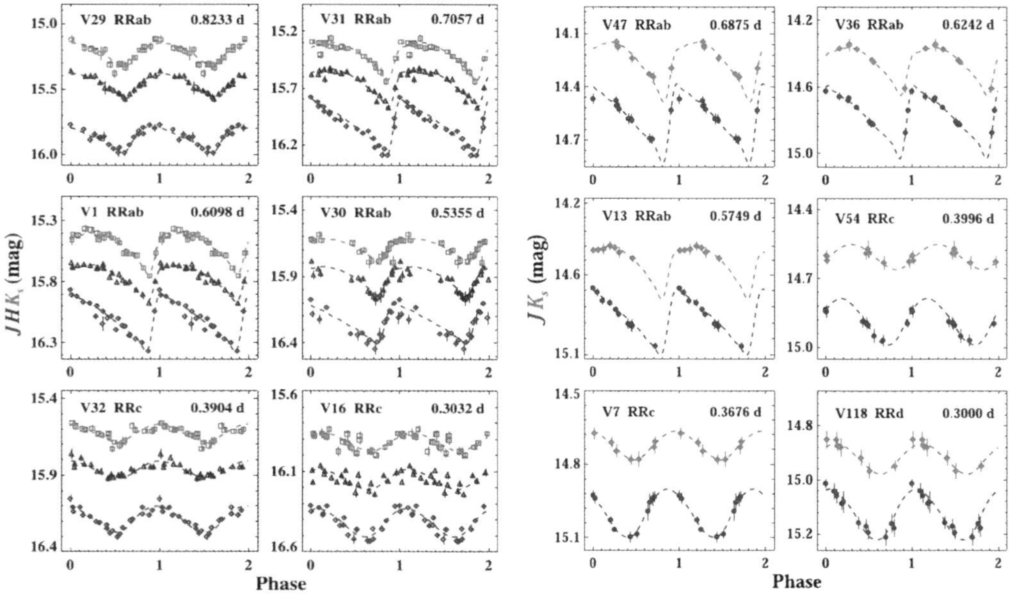

Figure 6. Example light curves for RR Lyrae stars in two different clusters - M53 (left) and M15 (right) with different light curve sampling (Bhardwaj et al. 2021a,b). The best-fitted templates are also shown as dashed lines. The magnitudes in different filters are slightly offset for visualization purposes.

with different periods. For each RR Lyrae in a given cluster, we also obtained the mean-metallicity of the cluster from Carretta et al. (2009) and the reddening value from the catalog of Harris (2010).

4.2. *Color–magnitude diagrams and period-luminosity relations*

Template-fitted light curves of RR Lyrae variables in different globular clusters were used to determine their accurate and precise mean-magnitudes and colors. The extinction corrections were adopted using the Cardelli et al. (1989) reddening law together with $E(B-V)$ values from Harris (2010). Fig. 7 displays color–magnitude diagrams for three different globular clusters with varying mean-metallicities. It is evident that these clusters are well-populated with RR Lyrae stars. NGC 6441 is a peculiar metal-rich cluster with red horizontal branch and hosts RR Lyrae stars. RR Lyrae variables in this cluster have unusually long periods and are considered to be helium enhanced (Pritzl et al. 2003; Catelan 2009). However, our NIR observations of RR Lyrae variables in NGC 6441 for the first time showed that their NIR magnitudes and colors are more in agreement with stellar pulsation models adopting canonical helium content (Bhardwaj et al. 2022). RR Lyrae models with helium enhancement predict bluer colors and brighter magnitudes than observations. In the case of metal-intermediate M3 and metal-poor M53 (see Bhardwaj 2020; Bhardwaj et al. 2021a, for details), the magnitudes and colors of RR Lyrae stars fall within the predicted boundaries of the instability strip (Marconi et al. 2015) at NIR wavelengths.

We investigated PL relations for RR Lyrae stars in globular clusters with different metallicities. For this purpose, we utilized our homogeneous photometric data in several clusters: NGC 6441, NGC 6121, NGC 6402, NGC 5272, NGC 6934, NGC 7089, NGC 5024, and NGC 7078. We supplemented our clusters with already published data in the literature for NGC 5904, NGC 5139, and NGC 6341. Interested readers are referred

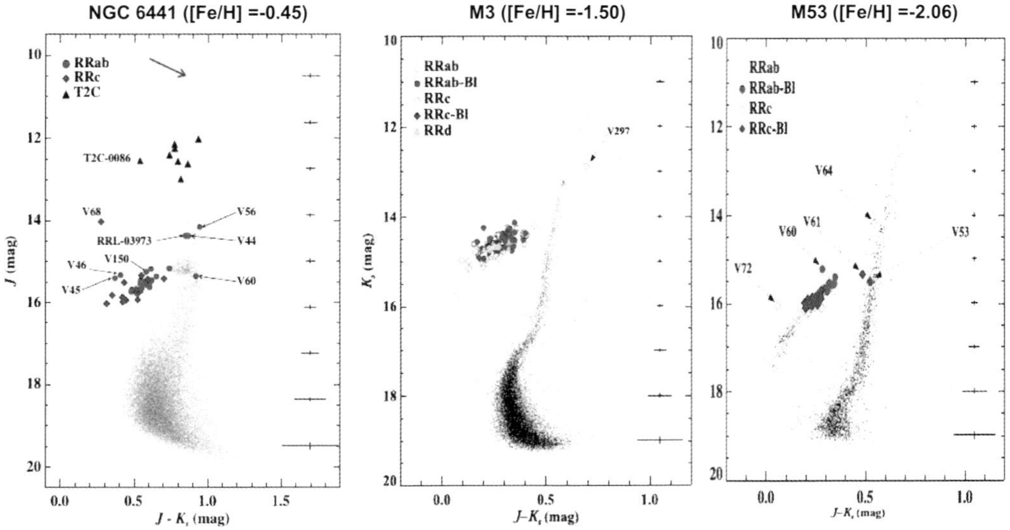

Figure 7. NIR color–magnitude diagrams for RR Lyrae variables in different globular clusters.

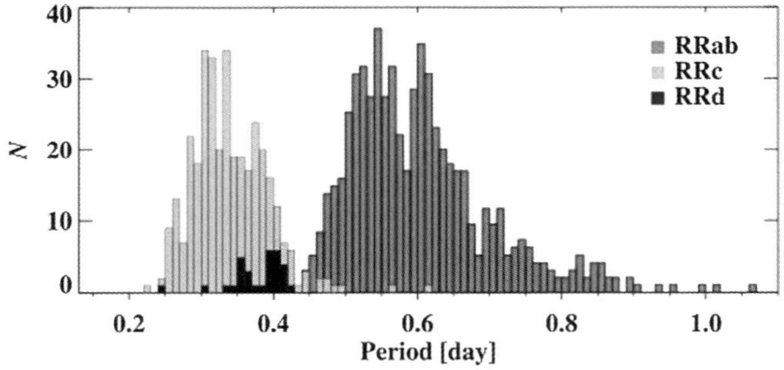

Figure 8. Histogram of period distributions of 964 globular cluster RR Lyrae variables.

to Bhardwaj et al. (2023) for the details regarding these datasets. Fig. 8 displays the histograms of period distributions of all 964 cluster RR Lyrae variables in our sample. We derived PL relations for different subtypes of RR Lyrae stars and the combined sample of RRab+RRc+RRd stars by fundamentalizing the periods of RRc/RRd stars using: $\log(P_f) = \log(P_{fo}) + 0.127$ (Braga et al. 2022). We assume that the RRd stars exhibit the first-overtone mode as the dominant mode of pulsation.

Fig. 9 displays K_s-band PL relations for RR Lyrae variables in different globular clusters, and we have similar relations in JH bands as well (Bhardwaj et al. 2023). In the case of metal-rich NGC 6441, NIR PL relations are tight (with dispersion of ~ 0.05 mag) and fully consistent with the predictions of the pulsation models with canonical helium content from Marconi & Minniti (2018). The predicted PL relations with higher helium content result in systematically brighter zero-points even though the slopes remain statistically consistent with observations. Therefore, we concluded that RR Lyrae variables in NGC 6441 are either not significantly helium enhanced, as previously thought, or the impact of such enhancement in NIR bands is much smaller than the predictions of the pulsation codes (Bhardwaj et al. 2022). In the case of normal metal-intermediate and metal-poor clusters, the PL relations in NIR agree well with theoretically predicted relations by Marconi et al. (2015). For high latitude clusters with less extinction and field

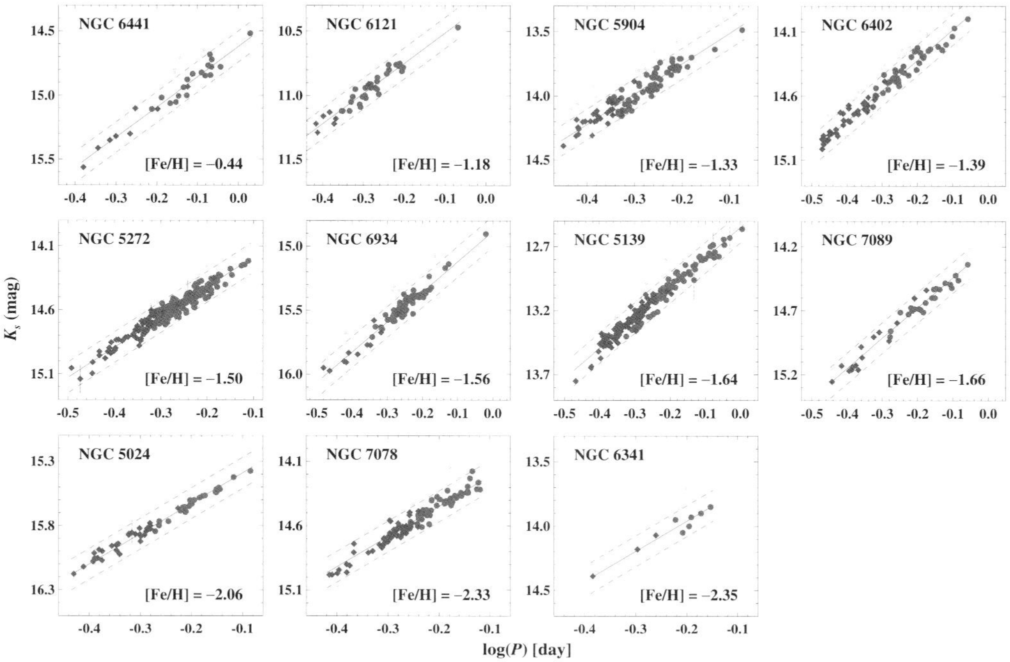

Figure 9. K_s-band PL relations for RR Lyrae variables in globular clusters with different mean-metallicities. The red circles and blue squares represent RRab and RRc stars respectively. The best-fitted linear regression is shown in solid lines while the dashed lines display 3σ scatter in the underlying relation. The small grey dots represent the outliers.

contamination, the PL relations for RR Lyrae stars are very tight. For example, the dispersion of ~ 0.02 mag for RRab in M53 is at the limit of photometric uncertainties. The bulge clusters with severe crowding and differential extinction result in relatively larger scatter (with dispersion of ~ 0.07 mag) in the PL relations. Nevertheless, the slopes of PL relations in different clusters are statistically consistent for a given sample of RR Lyrae subtype. The PL relations for the combined sample of all RR Lyrae stars are best constrained thanks to the larger statistics and longer period range under consideration.

4.3. *NIR period-luminosity-metallicity relations*

We used individual PL relations in different globular clusters to investigate and quantify the dependence of metallicity at NIR wavelengths. Previously, Sollima et al. (2006) used random-phase K_s-band photometry for more than 500 RR Lyrae variables in 16 globular clusters covering a metallicity range of ~ 1.2 dex to derive PLZ relation. For our analysis, homogeneous NIR photometry for 964 RR Lyrae variables in 11 globular clusters covering Δ[Fe/H]~ 2 dex was combined with NIR photometry of 346 RR Lyrae stars in the Milky Way field. The sample of field RR Lyrae variables was adopted from Muraveva et al. (2018) with metallicities and reddening estimates. NIR photometry for these variables was adopted from 2MASS catalog as discussed in Bhardwaj et al. (2023). These field stars with accurate parallaxes from *Gaia* allowed us to constrain an absolute zero-point of the PLZ relation in JHK_s bands. We derived a PLZ relation for the combined sample of 1310 field and globular cluster RR Lyrae variables by simultaneously solving for a slope, zero-point, metallicity coefficient, and individual distances to globular clusters. This approach does not depend on prior assumption of distance to a given globular cluster.

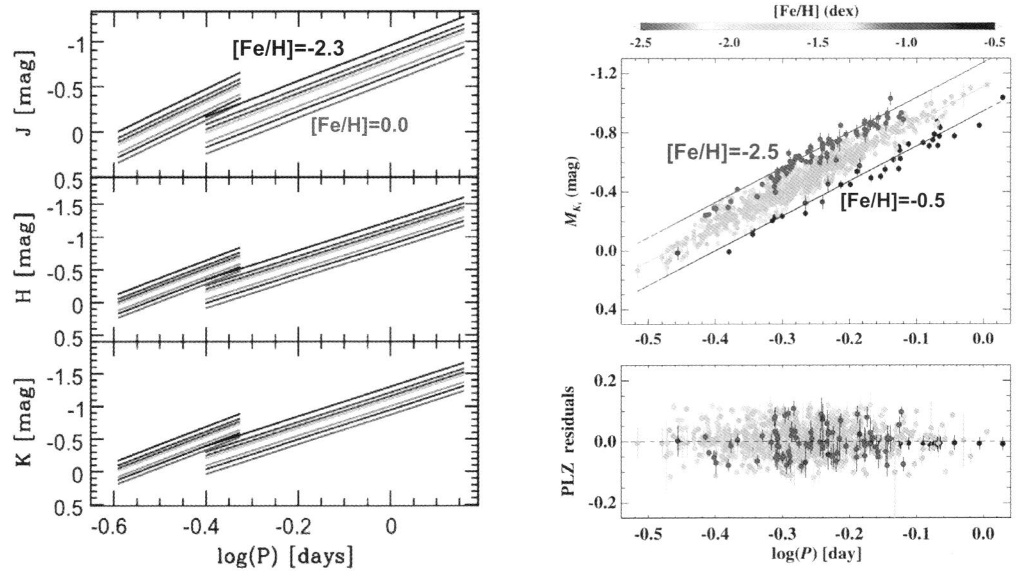

Figure 10. *Left:* NIR PLZ relation for RR Lyrae variables predicted by Marconi et al. (2015) based on stellar pulsation models. The figure is adopted and slightly modified from the original Figure 14 of Marconi et al. (2015). *Right:* Empirical PL relations for RR Lyrae stars in different globular clusters as a function of metallicity (top). The bottom panel displays the residuals of PLZ relation after taking into account the metallicity term. It is evident that the trend in the residuals as a function of metallicity, as seen in the top panels, vanishes in the lower panel.

The left panels of Fig. 10 display theoretical PLZ relation in JHK_s band for RR Lyrae variables as predicted by the pulsation models. The right panel displays our empirical relation in K_s-band and similar relations are also derived in JH filters as well. The combined sample of all RR Lyrae stars was used in deriving these PLZ relations. The variation in the PL relations as a function of metallicity is clearly evident in both the theoretical and empirical relations. We find an excellent agreement between the metallicity coefficients from the theoretical models and the observations. To investigate the possible effects of different metallicity scales and reddening laws, we also adopted recently published spectroscopic metallicities of field stars by Crestani et al. (2021). For the globular clusters, we experimented with different metallicity scales of Carretta et al. (2009), Harris (2010), and Dias et al. (2016). No statistically significant variation was seen in the PLZ relations for different metallicity scales. Similarly, extinction corrections based on different reddening maps (Schlafly & Finkbeiner 2011; Green et al. 2019) resulted in small or negligible variations in PLZ coefficients within their associated uncertainties.

Our recommended PLZ relations in NIR filters are:

$$M_J = -0.44\ (\pm 0.03) - 1.83\ (\pm 0.02) \log(P_\mathrm{f}) + 0.20\ (\pm 0.02)\ [\mathrm{Fe/H}] \quad (\sigma = 0.05\ \mathrm{mag}),$$
$$M_H = -0.74\ (\pm 0.02) - 2.29\ (\pm 0.02) \log(P_\mathrm{f}) + 0.19\ (\pm 0.01)\ [\mathrm{Fe/H}] \quad (\sigma = 0.05\ \mathrm{mag}),$$
$$M_{K_s} = -0.80\ (\pm 0.02) - 2.37\ (\pm 0.02) \log(P_\mathrm{f}) + 0.18\ (\pm 0.01)\ [\mathrm{Fe/H}] \quad (\sigma = 0.05\ \mathrm{mag}),$$

where M_λ represents the absolute magnitude in the JHK_s filters, and P_f represents the periods of RRab stars or the fundamentalized periods of RRc/RRd stars. These empirical relations are fully consistent with the theoretical relations provided by Marconi et al. (2015). Our empirical calibrations will be used to determine robust RR Lyrae based distances to nearby dwarf spheroidal galaxies like Draco, Sextans, and Ursa Minor, for which NIR photometry using WIRCam at CFHT has already been obtained. These precise and accurate RR Lyrae distances will then allow us to provide an independent calibration of other population II distance indicators such as the tip of the red giant branch for the population II distance scale.

Since we solved for the individual distances to our sample of globular clusters, we can compare those with independent distances in the literature. We adopted globular cluster distances from the catalog of Baumgardt & Vasiliev (2021) which provides average distances based on several independent determinations. A direct comparison of our distance estimates with those provided by Baumgardt & Vasiliev (2021) suggests a small offset of -0.03 ± 0.03 mag such that our distances are systematically smaller albeit within 1σ errors. This difference can be explained due to overcorrection of parallax zero-point offset, because a smaller parallax correction than Lindegren et al. (2021) will result in brighter absolute zero-points (Bhardwaj et al. 2021a) and in turn a larger distance to our sample of globular clusters. Nevertheless, our independently derived RR Lyrae based distances also agree well with those available in the literature within their quoted uncertainties.

5. Summary

Classical pulsators, in particular, Classical Cepheids, Type II Cepheids, and RR Lyrae variables are stellar standard candles that are used to obtain accurate and precise distances to their host galaxies. The advent and the increase of NIR detectors in the past two decades have improved the accuracy and precision of these variables as distance indicators since their PL relations are tighter at these longer wavelengths. Despite the extensive use and improvements in the PL relations, their universality in different stellar systems is not well-established. In particular, the metallicity effects on NIR PL relations for Cepheid and RR Lyrae variables are not properly quantified empirically and there are inconsistencies between observations and theoretical model predictions.

We discussed our ongoing efforts to properly quantify the metallicity effects on the PL relations for these classical pulsators by utilizing the increasingly accurate *Gaia* parallaxes. For the precise calibration of these relations, we also need high-resolution spectroscopic metallicities and multiband time-series photometry for a statistically significant sample of variables. Our observational programs are obtaining homogeneously both the spectra and the light curves for these classical pulsators to fully utilize current and future *Gaia* parallaxes. We presented preliminary results on the high-resolution spectroscopic metallicities of Milky Way Cepheid standards that have been used in the latest local determination of the Hubble constant. Our new spectroscopic metallicities together with time-series photometry and *Gaia* parallaxes will provide a new calibration for the distance scale. At the same time, the C-MetaLL survey is obtaining spectra for most metal-poor Classical Cepheids in our Galaxy to increase the metallicity baseline for a better quantification of metallicity effects. For Type II Cepheids, we show that NIR light curves are equally important when deriving PLZ relations even with a homogeneous sample of spectroscopic metallicities and accurate *Gaia* parallaxes. For RR Lyrae stars, we presented the most precise quantification of metallicity dependence to date using variables in globular clusters of different mean-metallicities. Our latest results and ongoing efforts to improve the accuracy and precision of PL calibrations will be useful for distance scale studies in the near-future. These empirical calibrations will be particularly important in the era of upcoming large ground-based telescopes and space telescopes, which will operate mostly at longer wavelengths.

This project has received funding from the European Union's Horizon 2020 research and innovation program under the Marie Sklodowska-Curie grant agreement No. 886298.

References

Baumgardt H., Vasiliev E., 2021, MNRAS, 505, 5957
Berdnikov L. N., 2008, VizieR Online Data Catalog, 2285, 0
Bhardwaj A., 2020, Journal of Astrophysics and Astronomy, 41, 23
Bhardwaj A., 2022, Universe, 8, 122
Bhardwaj A., et al., 2015, MNRAS, 447, 3342
Bhardwaj A., et al., 2016, AJ, 151, 88
Bhardwaj A., et al., 2017a, AJ, 153, 154
Bhardwaj A., et al., 2017b, MNRAS, 466, 2805
Bhardwaj A., et al., 2021a, ApJ, 909, 200
Bhardwaj A., et al., 2021b, ApJ, 922, 20
Bhardwaj A., et al., 2022, A&A, 668, A59
Bhardwaj A., et al., 2023, ApJL, 944, L51
Bono G., et al., 2001, ApJ, 563, 319
Bono G., et al., 2003, MNRAS, 344, 1097
Bono G., et al., 2010, ApJ, 715, 277
Borissova J., et al., 2009, A&A, 502, 505
Braga V. F., et al., 2018, AJ, 155, 137
Braga V. F., et al., 2019, A&A, 625, A1
Braga V. F., et al., 2022, MNRAS, 517, 5368
Breuval L., et al., 2022, ApJ, 939, 89
Cardelli J. A., Clayton G. C., Mathis J. S., 1989, ApJ, 345, 245
Carretta E., et al., 2009, A&A, 508, 695
Catelan M., 2009, AP&SS, 320, 261
Catelan M., Pritzl B. J., Smith H. A., 2004, ApJS, 154, 633
Crestani J., et al., 2021, ApJ, 914, 10
Dambis A. K., et al., 2013, MNRAS, 435, 3206
Das S., et al., 2021, MNRAS, 501, 875
Di Criscienzo M., et al., 2007, A&A, 471, 893
Dias B., et al., 2016, A&A, 590, A9
Freedman W. L., et al., 2001, ApJ, 553, 47
Genovali K., et al., 2014, A&A, 566, A37
Genovali K., et al., 2015, A&A, 580, A17
Gieren W., et al., 2018, A&A, 620, A99
Green G. M., et al., 2019, ApJ, 887, 93
Groenewegen M. A. T., 2018, A&A, 619, A8
Harris W. E., 2010, arXiv:1012.3224, p. arXiv:1012.3224
Inno L., et al., 2015, A&A, 576, A30
Kovtyukh V., et al., 2018, MNRAS, 477, 2276
Lindegren L., et al., 2021, A&A, 649, A2
Luck R. E., 2018, AJ, 156, 171
Maas T., Giridhar S., Lambert D. L., 2007, ApJ, 666, 378
Madore B. F., 1982, ApJ, 253, 575
Marconi M., Minniti D., 2018, ApJL, 853, L20
Marconi M., et al., 2015, ApJ, 808, 50
Matsunaga N., et al., 2006, MNRAS, 370, 1979
Muhie T. D., et al., 2021, MNRAS, 502, 4074
Muraveva T., et al., 2015, ApJ, 807, 127
Muraveva T., et al., 2018, MNRAS, 481, 1195
Navarrete C., et al., 2017, A&A, 604, A120
Neeley J. R., et al., 2019, MNRAS, 490, 4254
Ngeow C.-C., et al., 2022, AJ, 164, 154
Planck Collaboration et al., 2020, A&A, 641, A6
Pritzl B. J., et al., 2003, AJ, 126, 1381

Recio-Blanco A., et al., 2022, arXiv e-prints, p. arXiv:2206.05541
Riess A. G., et al., 2021, ApJL, 908, L6
Riess A. G., et al., 2022, ApJL, 934, L7
Ripepi V., et al., 2014, MNRAS, 437, 2307
Ripepi V., et al., 2021, MNRAS, 508, 4047
Ripepi V., et al., 2022, arXiv e-prints, p. arXiv:2206.06212
Schlafly E. F., Finkbeiner D. P., 2011, ApJ, 737, 103
Skrutskie M. F., et al., 2006, AJ, 131, 1163
Sollima A., Cacciari C., Valenti E., 2006, MNRAS, 372, 1675
Sollima A., et al., 2008, MNRAS, 384, 1583
Soszyński I., et al., 2018, Acta Astron., 68, 89
Stetson P. B., 1996, PASP, 108, 851
Subramanian S., et al., 2017, Space Science Reviews, 212, 1817
Wallerstein G., 2002, PASP, 114, 689
Wielgórski P., et al., 2022, ApJ, 927, 89

Discussion

Question (Braga): Did you check what happens to the matrix solution when removing ω Cen RR Lyrae, since for them, you adopted the average [Fe/H] value of ω Cen, but this globular cluster is known for having a large iron abundance spread, even among its RR Lyrae stars.
Answer: Thanks. We did check this in Bhardwaj et al. (2021a) by adopting different metallicities of ω Cen RR Lyrae. For this work, I have not checked this, but I doubt this will make any difference since the overall metallicity range is significantly larger.

Question (Riess): Very nice talk. What did you conclude from, or did you, compare RR Lyrae distances for globular clusters to *Gaia* distances? Did you sort of average stars in *Gaia*?
Answer: We didn't compare directly with *Gaia* because some of these clusters are quite distant. We have a few of these clusters at 15-19 kpc where *Gaia* parallaxes are unlikely to be that good. But we did compare with some independent measurements based on, for example, the compilation by Baumgardt & Vasiliev (2021) where we did find a small offset, which is somewhat statistically consistent with zero.

Question (Anderson): Given that Classical Cepheids and RR Lyrae stars occupy the same instability strip and share many similarities, what could be the origin of the different sign of the metallicity term for RR Lyrae and Cepheid PL-relations? For context, all individual PL relations and all PW relations of RR Lyrae stars shown in your talk have positive signs. Conversely, the observed metallicity effect sign for Cepheids is negative in both PL and PW relations, and models predict a negative sign for PW relations (and Geneva models also for individual bands). I wonder what physics this could point to.
Answer: I have no idea. I think first we have to be sure that for Cepheids, it goes the other way. From the models, we see something else; from the observations we see something else. At least for RR Lyrae stars, I am now confident that we have consistency between models and observations. Having thought about it, RR Lyrae and Cepheids have different evolutionary parameters, mass-luminosity relations, compositions, and it will be difficult to speculate what might be contributing to different metallicity dependence.

Comment (Hocdé): Just to answer the question, I have a curious idea that needs to be checked - but this metallicity effect difference with Cepheids could come from circumstellar envelopes of Cepheids, if the emission of the envelopes depends on the metallicity. That could possibly be the physical effect that causes this. I can give details later.
Answer: I think then one has to ask how much these effects are there at longer wavelengths or if there is a variation in the metallicity coefficients going from optical to infrared. I think this effect gets minimized as you go for longer wavelengths.

Comment (Hocdé): For example, there is an interesting phenomenon which has to do with the opacity of the negative hydrogen ion which in fact depends on the quantity of the metals in the gas. I can also give details later. We have to study the physics of the negative hydrogen opacity.
Answer: Sure, thanks.

Question (Martínez-Vázquez): Very interesting talk. Have you considered (or tried already) to obtain period-$\phi31$-[Fe/H] relations in the different infrared bands you have?
Answer: Not yet, but certainly in the list things to do. This is something I thought about when the RRab templates from Braga et al. (2019) did not work well based on their period bins, since the shape of RRab light curves varies at different periods in different metallicity clusters.

At the crossroads of astrophysics and cosmology: Period–luminosity relations in the 2020s
Proceedings IAU Symposium No. 376, 2024
R. de Grijs, P. A. Whitelock & M. Catelan, eds.
doi:10.1017/S1743921323003563

A multiphase study of classical Cepheids in the Magellanic Clouds- Models and Observations

Kerdaris Kurbah[1], Shashi M. Kanbur[2], Sukanta Deb[1], Susmita Das[3], Mami Deka[1], Anupam Bhardwaj[4], Hugh Riley Randal[2] and Selim Kalici[2]

[1] Department of Physics, Cotton University, Guwahati, Assam, 781001, India.
email: kerdarisbahkur@gmail.com

[2] Department of Physics, State University of New York, Oswego, NY 23126, USA

[3] Konkoly Observatory, Research Centre for Astronomy and Earth Sciences, Eötvös Loránd Research Network (ELKH), Konkoly-Thege Miklós út 15-17, H-1121, Budapest, Hungary

[4] INAF-Osservatorio Astronomico di Capodimonte, Via Moiariello 16, I-801301, Napoli, Italy

Abstract. This work presents the study of multiphase relations of classical Cepheids in the Magellanic Clouds for short periods ($\log P < 1$) and long periods ($\log P > 1$). From the analysis, it has been found that the multiphase relations obtained using the models as well as observations are highly dynamic with pulsational phase. The multiphase relations for short and long periods are found to display contrasting behaviour for both LMC and SMC. It has been observed that the multiphase relations obtained using the models agree better with the observations in the PC plane in most phases in comparison to the PL plane. Multiphase relations obtained using the models display a clear distinction among different convection sets in most phases. Comparison of models and observations in the multiphase plane is one way to test the models with the observations and to constrain the theory of stellar pulsation.

Keywords. stars: variable: Cepheids-galaxies: Magellanic Clouds-methods: data analysis-methods: statistical

1. Introduction

Theoretical pulsation properties of Cepheid variables have been investigated by Bono and his group in the 1990s (Bono et al. 1998, 1999, 2000; Caputo et al. 2000a,b). The comparison between the observed and theoretical period-color (PC), amplitude-color (AC) and period-luminosity (PL) relations for Cepheids, RR Lyraes, and δ Scuti stars in the Galaxy and the Magellanic Clouds in the literature were mostly done at maximum, minimum, and mean light (Kanbur et al. 2004; Kanbur and Ngeow 2006; Kanbur et al. 2007; Das et al. 2020; De Somma et al. 2020; Deka et al. 2022; De Somma et al. 2022). However, the relation at mean light is the average of the relations at all phases, hence, investigating the Cepheid PL/PC relations at other phases of the stars provides a better understanding of the pulsational properties of the stars. A multiphase study of Cepheid PL/PC/AC relations has been done by Ngeow and Kanbur (2006) and the comparative study of the same with the models has been carried out by Kanbur et al. (2010). In this study, we investigate the theoretical PL/PC relations of Cepheids at multiple phases during a pulsation cycle, using the radial stellar pulsation (RSP) code in Modules for

Experiments in Stellar Astrophysics (MESA, Paxton et al. (2011, 2013, 2015, 2018, 2019) and the same for the observations using OGLE-IV database.

2. Methodology

Theoretical light curves of Cepheids are generated using MESA-RSP (Smolec and Moskalik 2008; Paxton et al. 2019) in all four sets of convection parameters A, B, C and D. A brief explanation of the parameters in each convection set is outlined in Paxton et al. (2019). The chemical composition appropriate for the LMC ($Z = 0.008, X = 0.742$) and SMC ($Z = 0.004, X = 0.746$) are adopted from Bono et al. (1998). For each chemical composition and mass in the range 3.6 $M_\odot$ − 6.8 $M_\odot$, the luminosity is calculated using the rotation-averaged mass-luminosity (ML) relation of Anderson et al. (2014). The models cover a wide range of effective temperatures between ~ 4000 K to 8000 K. It is to be noted that the Cepheid light curves obtained using the models pulsate in the fundamental mode (FU) and fulfilled the condition of full-amplitude and stable pulsation state where the amplitude of radius variation (ΔR), the pulsation period computed on a cycle to cycle basis (P), and the fractional growth of the kinetic energy per pulsation period (Γ) vary by less than 0.01 in the last 100 pulsation cycles of the total pulsation cycles computed (Das et al. 2021). The period range of the light curves generated using the models is between $3 < P(\mathrm{d}) < 32$.

Light curves for FU Cepheids in the optical bands (VI) in the Magellanic Clouds (MCs) are obtained from the OGLE-IV catalogue (Soszyński et al. 2015). Stars listed in 'remarks.text' file provided by OGLE-IV are rejected. Light curves consisting of more than 30 data points and within period range of $3 < P(\mathrm{d}) < 32$ are selected from both the bands leaving 1530 and 703 stars in the LMC and SMC, respectively. The apparent magnitudes are corrected for interstellar extinction using Skowron et al. (2021) reddening map, the extinction values are adopted using the reddening law of Cardelli et al. (1989). Furthermore, the same period range of $3 < P(\mathrm{d}) < 32$ for models and observations is considered throughout the analysis.

The light curves obtained from the observations and the models are fitted using a cosine Fourier decomposition (Deb and Singh 2009, and reference therein)

$$m(\Phi) = A_0 + \sum_{i=1}^{N} A_i \cos(2\pi i\Phi + \phi_i) \tag{1}$$

Φ is the pulsation phase ranges from 0 to 1 given by

$$\Phi = \frac{(t - t_0)}{P} - \mathrm{Int}\left[\frac{(t - t_0)}{P}\right], \tag{2}$$

where t is the time of observation in days, t_0 is the epoch of maximum light and P denotes the period in days.

To investigate the behaviour of the PL/PC relation as a function of phase, equation (1) is interpolated in Φ from 0 to 1 in steps of 0.02 on the Fourier-fitted data points. The magnitudes corresponding to the 50 pulsation phases were extracted from the interpolated light curves for both the models and observations. The multiphase study was carried out for short periods ($\log P < 1$) and long periods ($\log P > 1.0$) in both the LMC and SMC. The number of light curves used in the analysis for both LMC and SMC are listed in Table 1.

3. Results

The coefficients of the multiphase period-luminosity (MPPL) relations in both VI-bands for both the models and observations are displayed in Figures- 1, 2, 3, 4. A distance

Table 1. Summary of light curves used in the analysis.

		Set A	Set B	Set C	Set D	Observations
LMC	N_s	169	108	115	66	1407
	N_l	138	129	142	147	123
SMC	N_s	203	149	155	87	612
	N_l	90	85	85	84	91

N_s: Short period light curves N_l: Long period light curves

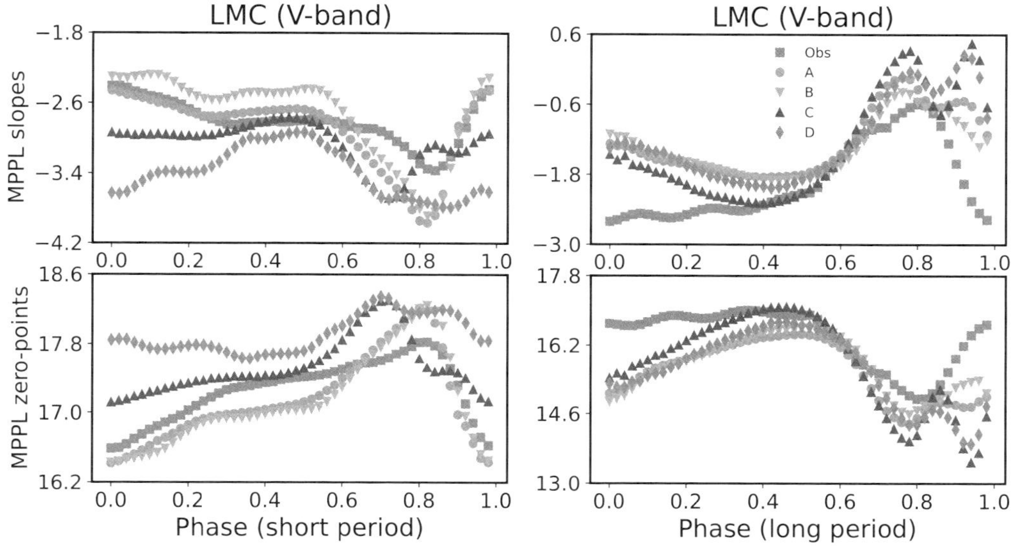

Figure 1. Plots of the MPPL relations obtained from Cepheid models and observations for the LMC in V-band. Orange, green, red and purple denote the relations obtained using convection sets A, B, C and D, respectively. Blue represent observed relations. Left/right panels represent the relations obtained using short/long period light curves.

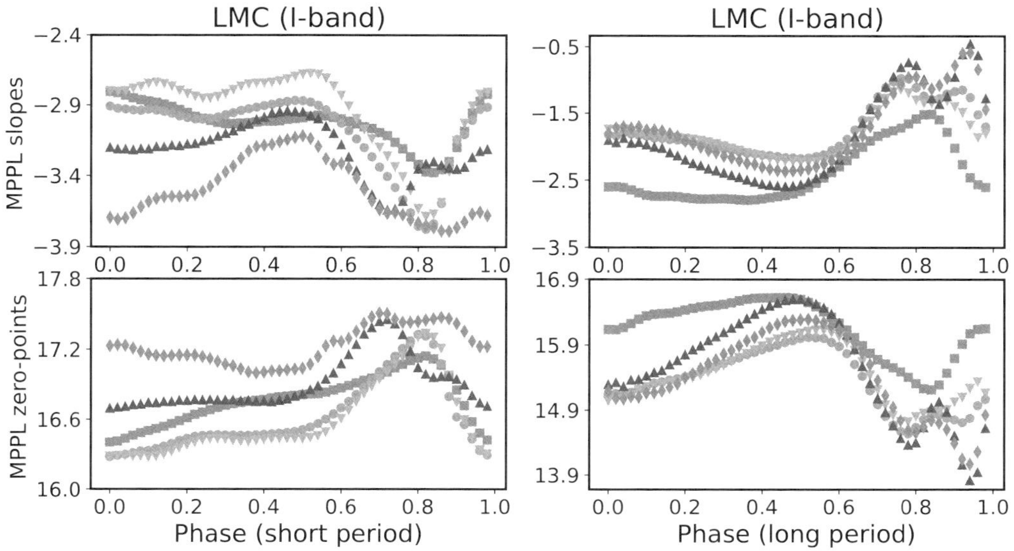

Figure 2. Same as Figure 1 but for LMC in I-band.

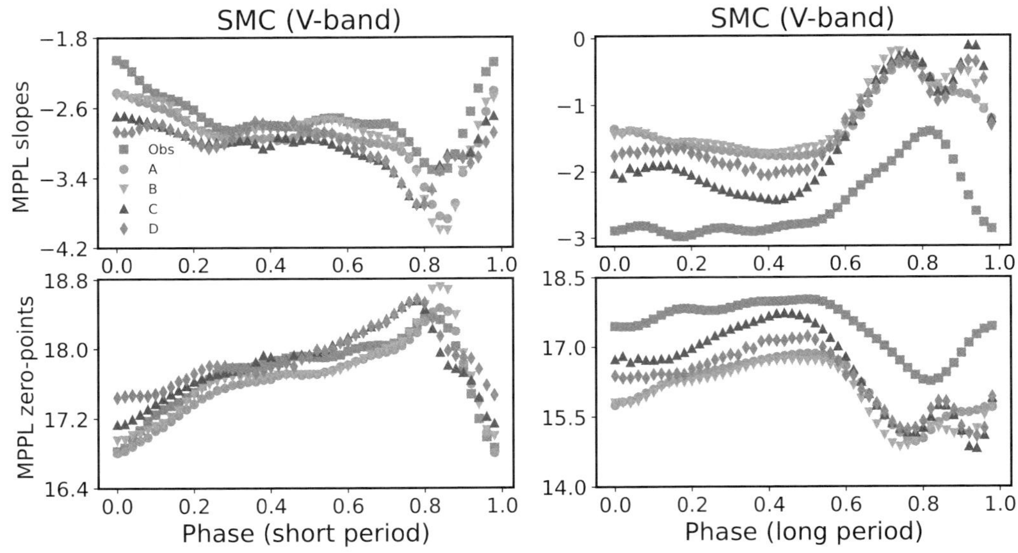

Figure 3. Same as Figure 1 but for SMC.

modulus of $\mu_{\rm LMC} = 18.49 \pm 0.008$(statistical) $\pm\, 0.047$(systematic) mag (Pietrzyński et al. 2019) and $\mu_{\rm SMC} = 18.97 \pm 0.016 \pm 0.028$ mag (Grazyk et al. 2020) are added to the intercept of the theoretical MPPL relations obtained for LMC and SMC, respectively, for comparison with the observed relations. Similarly the coefficients of the multiphase period-color (MPPC) relations from the models and observation are shown in Figures- 5 and 6. The results obtained in the study are discussed below:

(1) MPPL Relations: The MPPL slopes/zero-points obtained from the observations display minimum/maximum values at phase $\Phi \sim 0.85$ for short periods for both LMC and SMC Cepheids in both the bands. However, long period Cepheids in VI-bands exhibit maximum slopes and minimum zero-points at $\Phi \sim 0.8$ for both LMC and SMC. For LMC, sets A & B exhibit minimum slopes and maximum zero-points at $\Phi \sim 0.8$, and at $\Phi \sim 0.7$ using sets C & D for short periods. In case of the SMC, the same was found at $\Phi \sim 0.85$ using sets A & B, while those using sets C and D at $\Phi \sim 0.8$. However, for long periods, all the four convection sets exhibit maximum/minimum slopes/zero-points at $\Phi \sim 0.8$ for both LMC and SMC. For both the LMC and SMC, the MPPL relations obtained using convection set A is more consistent with set B, whereas the relations obtained using set C is found to be more consistent with set D. There is a discrepancy between the models and the observations in the MPPL plane. The discrepancy may be because there are fewer models than observations.

(2) MPPC Relations: For both LMC and SMC, the results of MPPC slopes/zero-points using the observations display maximum/minimum points at $\Phi \sim 0.85$ for short periods. However, for long periods, the largest variation of the slopes/zero-points was found at $\Phi \sim 0.8$. At long periods, the LMC models (sets A, B, C, D) exhibit maximum/minimum slopes/zero-points at $\Phi \sim 0.8$ and the SMC models at $\Phi \sim 0.75$. For LMC models at short periods, convection sets A, B display minimum/maximum slopes at $\Phi \sim 0.85$, and sets C & D at $\Phi \sim 0.7$. In case of the SMC, all the four convection sets display maximum/minimum slopes/zero-points at $\Phi \sim 0.7$.

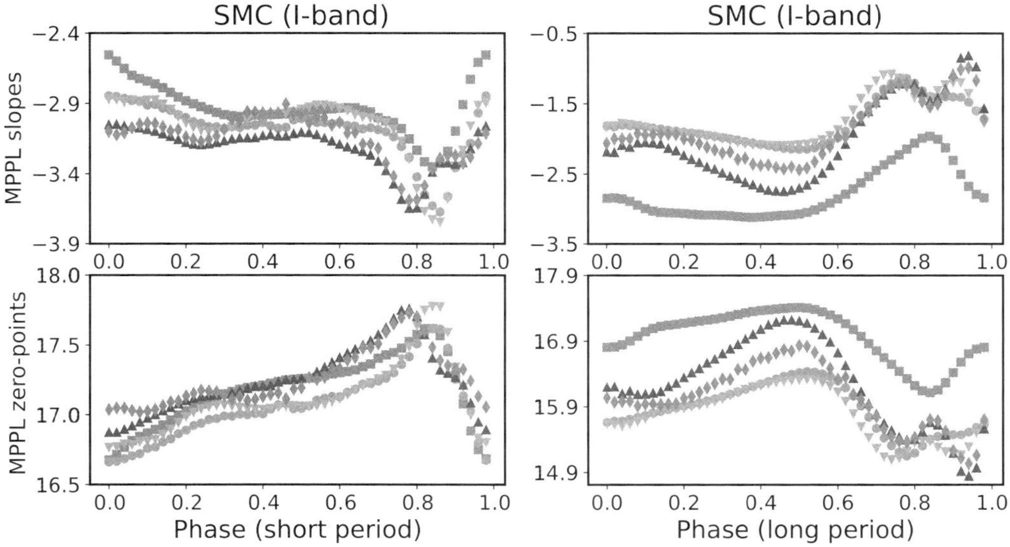

Figure 4. Same as Figure 1 but for SMC in *I*-band.

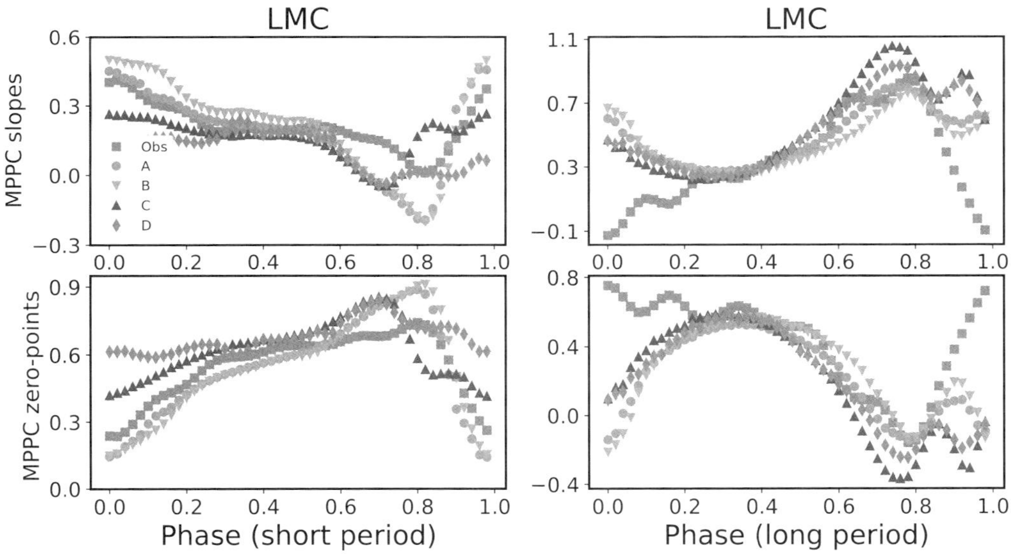

Figure 5. Same as Figure 1 but for MPPC in the LMC.

4. Discussion and Conclusion

Analysis of the light curves of Cepheid models generated using MESA-RSP and observations was carried for the LMC and SMC in both *V*- and *I*-bands. The comparison of the results obtained using the models with the observed ones has been done in the multiphase and Fourier parameter planes. The MPPL/MPPC relations obtained from observations and models are found to be highly dynamic in nature, with the effect more pronounced at $\Phi \sim 0.7 - 0.85$ for short and long periods, in both the LMC and SMC. This is consistent with the results obtained by Ngeow and Kanbur (2006). The significantly smaller number of theoretical light curves compared to observations seem to have a role in the discrepancy between the models and observations in the MPPL plane. The availability of a larger set of non-linear models that cover the whole range of the observation

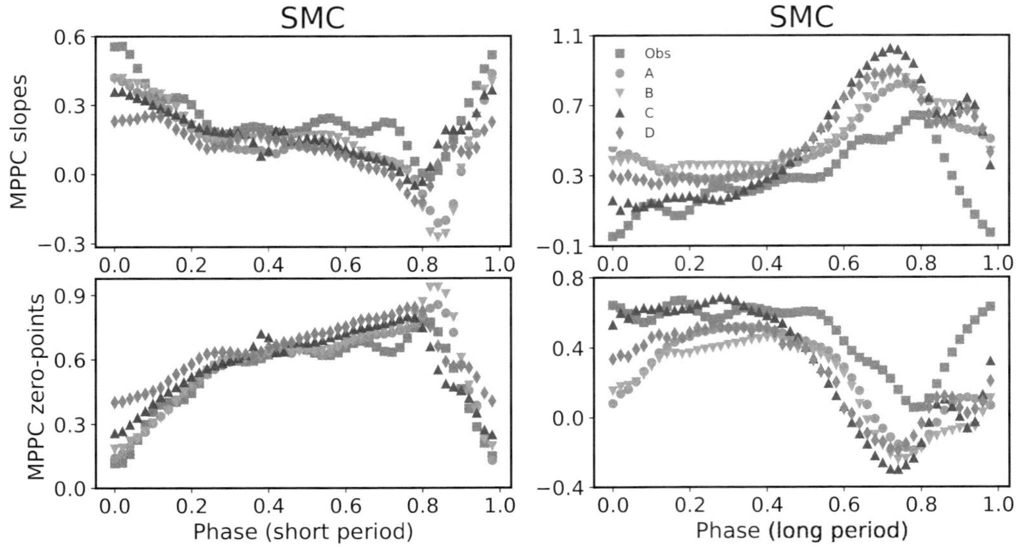

Figure 6. Same as Figure 1 but for MPPC in the SMC.

data will provide more details of the apparent discrepancy between the models and the observations.

In case of the LMC, a discrepancy between the modeled and observed light curves at long periods can be seen at phases between $\Phi \sim 0.0 - 0.4$ and $\Phi \sim 0.9 - 1.0$ in the MPPL plane in VI- bands, and between $\Phi \sim 0.0 - 0.2$ and $\Phi \sim 0.9 - 1.0$ in the MPPC plane. The discrepancy between models and observations at long periods in the MPPL plane is comparatively larger in the SMC compared to the LMC. Similarly, the same can be seen in the MPPC plane. However, from Table- 1, it can be seen that the number of long period modeled light curves are comparatively similar that of the observations for both LMC and SMC. From these results, we demonstrate that a thorough investigation of the ensemble of the models, the effects of the $MLTZX$ combinations and the convective parameters on the multiphase relations for long period Cepheids must be considered.

It is worth mentioning here that the MPPC relations obtained using the four convection sets are found to agree well with the observations for most of the phases in both LMC and SMC. The agreement was observed for both short and long periods Cepheids.

Interestingly, it has been observed that the nature of the MPPL/MPPC observed relations for short/long period Cepheids exhibit contrasting behaviour for both LMC and SMC, and this behaviour is supported by models as well. This is an interesting result revealed using the multiphase approach and is worthy of a thorough and detailed inspection.

It can be seen that different convection sets are distinctive from each other in the multiphase plane for most of the phases when compared to the FP plane. Figure 7 displays the Fourier phase and amplitude parameters as a function of period, obtained using the models and the observations for both LMC and SMC in V- band. From the figure, an overlap of the parameters among the four convection sets can be seen. Moreover, the comparison in the multiphase planes requires the ensemble of models must match the ensemble of observations. It may be stated that in this study, interesting results are obtained using the multiphase approach. A deep and detailed investigation considering the shortcomings mentioned in the present study can lead to a better way to test the models and to constrain the pulsation theories.

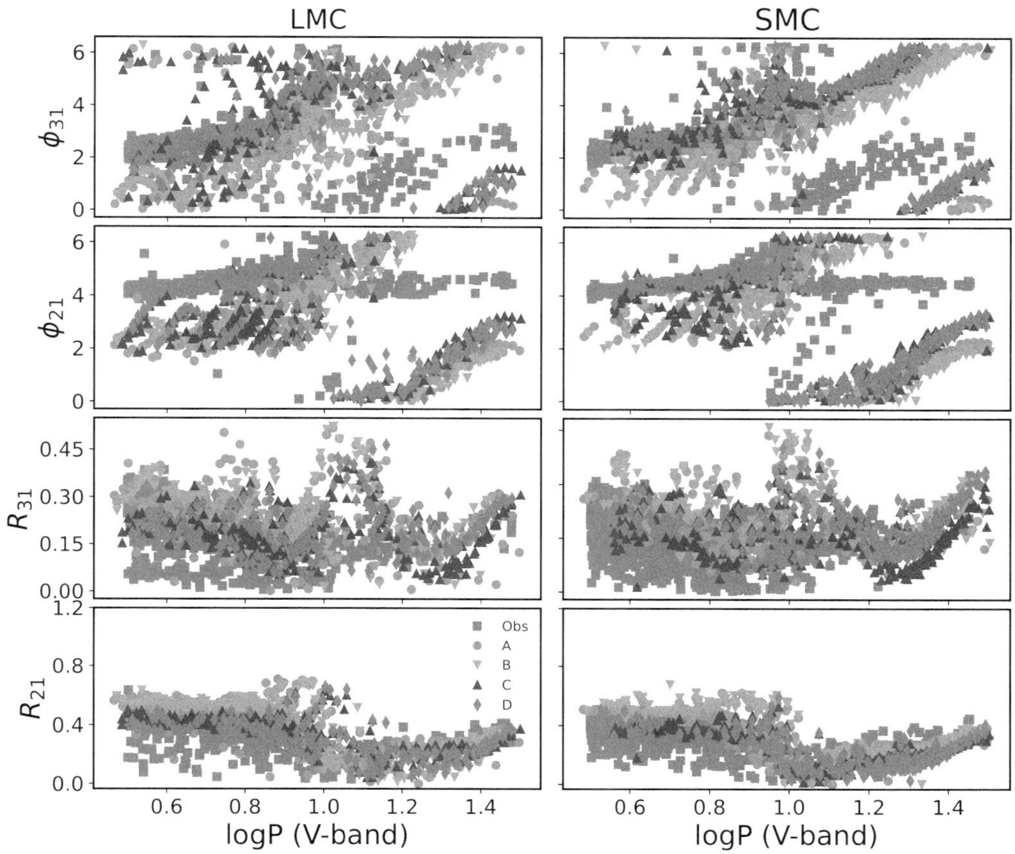

Figure 7. Comparison of Fourier parameters obtained using convection sets A, B, C, D with the observed ones for the LMC (left panel) and SMC (right panel) in V-band.

5. Acknowledgements

KK thanks the Council of Scientific and Industrial Research (CSIR), Govt. of India for the Senior Research Fellowship (SRF). KK acknowledges the travel support provided by the International Astronomical Union (IAU) to attend the IAU Symposium 376, "At the cross-roads of astrophysics and cosmology: Periodluminosity relations in the 2020s". SD thanks Council of Scientific and Industrial Research (CSIR), Govt. of India, New Delhi for a financial support through the research grant "03(1425)/18/EMR-II". SMK thanks State University of New York, Oswego, NY 23126, USA and Cotton University, Guwahati, Assam for the support. MD thanks CSIR for Junior Research Fellowship provided through CSIR-NET under the project. S. Das acknowledges the support of the KKP-137523 'SeismoLab' Élvonal grant of the Hungarian Research, Development and Innovation Office (NKFIH). AB acknowledges funding from the European Union's Horizon 2020 research and innovation programme under the Marie Skłodowska-Curie grant agreement No.886298. HRR and SK thanks State University of New York, Oswego, NY 23126, USA. The authors acknowledge IUCAA, Pune for the use of High Performance Computing facility Pegasus.

References

Anderson, Richard I. and Ekström, Sylvia and Georgy, et al., 2014, A& A, 564, 100
Bono, Giuseppe and Caputo, Filippina and Marconi, M., et al., 1998, ApJ, 497, L43

Bono, Giuseppe and Caputo, et al., 1999, ApJ, 512, 711
Bono, G. and Marconi, M. and Stellingwerf, R. F., 2000, A& A, 360, 245
Caputo, F. and Marconi, M. and Musella, I. and Santolamazza, P., 2000, A& A, 359, 1059
Caputo, F. and Marconi, M. and Musella, I., 2000, A& A, 354, 610
Cardelli, Jason A. and Clayton, Geoffrey C. and Mathis, John S., 1989, ApJ, 345, 245
Deb, S. and Singh, H. P. 2009, A& A, 507, 1729
Das, Susmita and Kanbur, Shashi M. and Bellinger, Earl P., et al., 2020, MNRAS, 493, 29
Das, S., Kanbur, S. M., Smolec, R., et al. 2021, MNRAS, 501, 875
De Somma, G., Marconi, M. Molinaro, R., et al. 2020, ApJS, 247, 30
De Somma, G., Marconi, M., Molinaro, R., et al. 2022, ApJS, 262, 25
Deka, Mami and Kanbur, Shashi M. and Deb, Sukanta, et al., 2022, MNRAS, 517, 2251
Graczyk, Dariusz and Pietrzyński, Grzegorz and Thompson, Ian B. and Gieren, et al., 2020, ApJ, 904, 13
Kanbur, Shashi M. and Ngeow, Chow-Choong and Buchler, J. Robert, 2004, MNRAS, 354, 212
Kanbur, Shashi M. and Ngeow, Chow-Choong, 2006, MNRAS, 369, 705
Kanbur, Shashi M. and Ngeow, Chow-Choong and Feiden, Greg, 2007, MNRAS, 380, 819
Kanbur, Shashi M. and Marconi, M., et al., 2010, MNRAS, 408, 695
Ngeow, Chow-Choong and Kanbur, Shashi M., 2006, MNRAS, 369, 723
Paxton, Bill and Bildsten, Lars and Dotter, Aaron, et al., 2011, ApJS, 192, 3
Paxton, Bill and Cantiello, Matteo and Arras, et al., 2013, ApJS, 208, 4
Paxton, Bill and Marchant, Pablo and Schwab, et al., 2015, ApJS, 220, 15
Paxton, Bill and Schwab, Josiah and Bauer, Evan B., et al., 2018, ApJS, 234, 34
Paxton, Bill and Smolec, R. and Schwab, Josiah, et al., 2019, ApJS, 243, 10
Pietrzyński, G. and Graczyk, D. and Gallenne, A., et al., 2019, Nature, 567, 200
Skowron, D. M. and Skowron, J. and Udalski, A., et al., 2021, ApJS, 252, 23
Smolec, R., & Moskalik, P. 2008, AcA, 58, 193
Soszyński, I., and Udalski, A., et al., 2015, AcA, 65, 297

At the crossroads of astrophysics and cosmology: Period–luminosity relations in the 2020s
Proceedings IAU Symposium No. 376, 2024
R. de Grijs, P. A. Whitelock & M. Catelan, eds.
doi:10.1017/S1743921323003186

Double-mode RR Lyrae star — robust distance and metallicity indicators

Shu Wang[1,3], Xiaodian Chen[1,2,3], Jianxing Zhang[1,2] and Licai Deng[1,2,3]

[1]CAS Key Laboratory of Optical Astronomy, National Astronomical Observatories, Chinese Academy of Sciences, Beijing 100101, China
email: shuwang@nao.cas.cn

[2]School of Astronomy and Space Science, University of the Chinese Academy of Sciences, Beijing, 100049, China

[3]Department of Astronomy, China West Normal University, Nanchong, 637009, China

Abstract. RR Lyrae (RR Lyr) stars are a well-known and useful distance indicator for old stellar populations such as globular clusters and dwarf galaxies. Fundamental-mode RR Lyr (RRab) stars are commonly used to measure distances, and the accuracy of the determined distance is strongly constrained by metallicity. Here, we investigate the metallicity dependence in the period–luminosity (PL) relation of double-mode RR Lyr (RRd) stars. We find and establish a linear relation between metallicity and period or period ratio for RRd stars. This relation can predict the metallicity as accurately as the low-resolution spectra. Based on this relation, we establish a metallicity-independent PL relation for RRd stars. Combining the distance of the Large Magellanic Cloud and Gaia parallaxes, we calibrate the zero point of the derived PL relation to an error of 0.022 mag. Using RRd stars, we measure the distances of globular clusters and dwarf galaxies with an accuracy of 2-3% and 1-2%, respectively. In the future, RRd stars could anchor galaxy distances to an accuracy of 1.0% and become an independent distance ladder in the Local Group.

Keywords. RR Lyrae variable stars, RRd variable stars, distance indicators, metallicity, dwarf galaxies

1. Introduction

RR Lyrae (RR Lyr) stars are low-mass stars that have evolved to the horizontal branch stage and begin to pulsate as they move into the instability strip. The pulsation periods of RR Lyr stars are about 0.2 to 1.0 days. According to the pulsation mode, RR Lyr stars are classified into RRab (fundamental mode), RRc (first-overtone mode) and RRd (double or multiple mode) stars. In the Hertzsprung–Russell diagram, RRd stars are located in the intersection region of the instability strips of RRab and RRc stars.

Most RRd stars are clustered in a sequence on the Petersen diagram (Petersen 1973) and are called classical RRd stars. There are also anomalous and peculiar RRd stars. The anomalous or peculiar RRd stars are located above or below the sequence of classical RRd stars on the Petersen diagram (Nemec & Moskalik 2021). The dominant mode of the anomalous or peculiar RRd stars is the fundamental mode, while the dominant mode of classical RRd stars is the first-overtone mode. In addition, anomalous or peculiar RRd stars usually have long-term amplitude modulation. In this paper, the default RRd star is the classical RRd star.

RR Lyr stars were searched for by a large number of time domain survey, including the Optical Gravitational Lensing Experiment (OGLE, Soszyński et al. 2009; Soszynśki et al. 2016, 2019), the Gaia (Clementini et al. 2022), the Zwicky Transient Facility (ZTF, Chen et al. 2020), the All-Sky Automated Survey for Supernovae (ASASSN, Jayasinghe et al. 2018), the Asteroid Terrestrial-impact Last Alert System (ATLAS, Heinze et al. 2018), the Wide-field Infrared Survey Explorer (WISE, Chen et al. 2018), and the PanSTARRS1(Sesar et al. 2017).

RR Lyr stars are good distance tracers for the old population. They satisfy the metallicity–luminosity (LZ) relation in the V band and the period–metallicity–luminosity (PLZ) relation in infrared bands. Since the two periods of RRd stars also satisfy the PLZ relation of RRab or RRc, respectively, RRd stars also have LZ and PLZ relations. Usually, RRd stars were not adopted to measure distances because the determined periods of RRd stars are not as accurate as those of RRab stars.

For RRd stars, both theory (Popielski et al. 2000; Marconi et al. 2015) and observations (Catelan & Smith 2015; Soszynśki et al. 2019) suggest that compared to metal-poor RRd stars, metal-rich RRd stars have shorter fundamental periods and smaller period ratios. There seems to be a good linear relation between the period ratio and metallicity (Braga et al. 2022). In this work, we investigate the metallicity dependence in the period–luminosity (PL) relation of RRd stars and establish their PL relation (Chen et al. 2023).

2. Period–Metallicity Relation

We collect RRd stars mainly based on Gaia and OGLE databases. From Gaia DR3, we obtain 1021 Galactic RRd stars. From OGLE, we collect 2083 and 674 RRd stars belonging to the Large Magellanic Cloud (LMC) and the Small Magellanic Cloud (SMC), respectively. Metallicity information for these stars is taken from low-resolution spectra of the Sloan Digital Sky Survey (SDSS, Eisenstein et al. 2011) and the Large Sky Area Multi-Object Fiber Spectroscopic Telescope (LAMOST, Cui et al. 2012). After cross-matching, we obtain 100 RRd stars with SDSS or LAMOST metallicities. In addition, we supplement another 203 RRd stars with SDSS or LAMOST metallicity based on the ZTF data. The final sample contains 207 and 96 RRd stars with metallicity measurements from SDSS and LAMOST, respectively.

$$\begin{aligned}
[\mathrm{Fe/H}]_{\mathrm{SDSS}} =& -(173\pm 38)\left(\log\frac{P_{1\mathrm{O}}}{P_{\mathrm{F}}}-\log 0.745\right)\\
&-(6.62\pm 0.88)(\log P_{1\mathrm{O}}-\log 0.37)-(1.76\pm 0.01),\\
[\mathrm{Fe/H}]_{\mathrm{LAMOST}} =& -(184\pm 70)\left(\log\frac{P_{1\mathrm{O}}}{P_{\mathrm{F}}}-\log 0.745\right)\\
&-(5.69\pm 1.50)(\log P_{1\mathrm{O}}-\log 0.37)-(1.68\pm 0.03),
\end{aligned} \tag{1}$$

Based on the selected RRd stars, we establish the period–period ratio–metallicity relation. The results are shown in equation 1. For the SDSS metallicity, the scatter σ is 0.16 dex and the correlation coefficient R^2 is 0.74. For the LAMOST metallicity, the scatter σ is 0.21 dex and the correlation coefficient R^2 is 0.60. The scatter of this relation decreases when a tighter criterion on metallicity error is adopted. For example, when we restrict the metallicity error to less than 0.04, with the rest of 56 RRd, the scatter of this relation will decrease to 0.13 dex. By comparing the scatter of the period–period ratio–metallicity relation with the external error of metallicity, we find that the metallicity estimated from the period and the period ratio of RRd stars can be as accurate as the low-resolution spectra.

With metallicities from high-resolution spectra, the scatter of the period–period ratio–metallicity relation could be greatly reduced. For example, for Cepheids, the scatter of

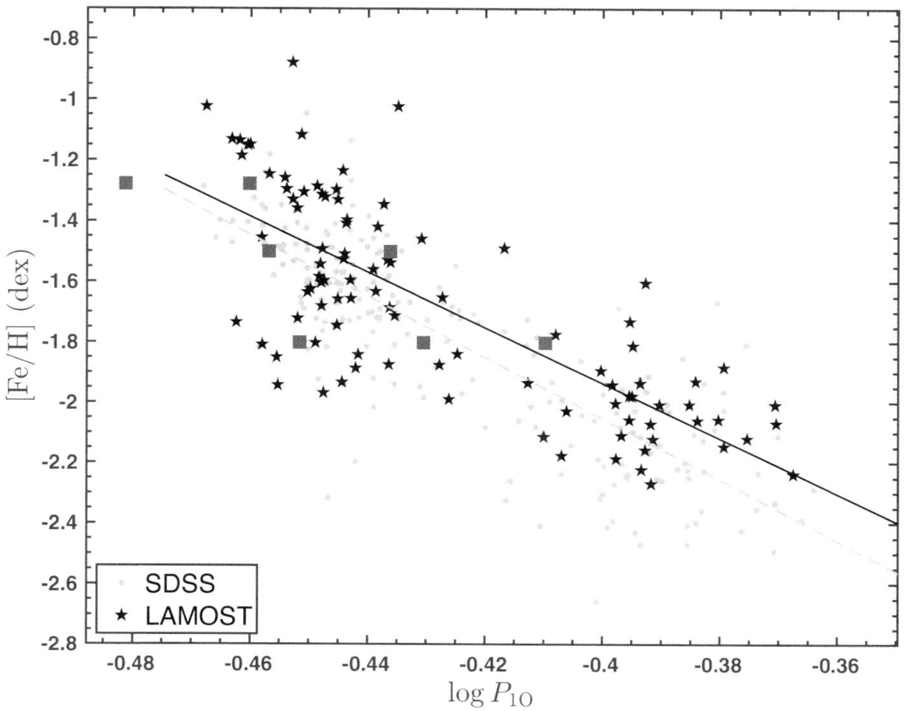

Figure 1. The Period–metallicity diagram of RRd stars with metallicities from SDSS (green dots) and LAMOST (black pentagrams). The green dashed line and the black solid line are linear fit lines. Red squares are the grid points of the theoretical model.

the relation is only 0.033 dex (Kovtyukh et al. 2016). In the future, with metallicities from high-resolution spectra, the period–period ratio–metallicity relation of RRd stars could also have a small scatter. Kovtyukh et al. (2016) found that for double–mode Cepheids, their metallicity depend on period ratio more than the period. Based on our sample, we find that RRd metallicity is more dependent on the period than the period ratio.

Figure 1 shows the distribution of RRd stars in the period–metallicity (PZ) diagram. The green dots and black pentagrams are RRd stars with metallicities from SDSS and LAMOST, respectively. Red squares are the grid points of the theoretical model (Marconi et al. 2015). There is an intuitive linear relationship between the metallicity and the period. The green dashed line and the black solid line are linear fit lines. The determined PZ relations for RRd stars with SDSS metallicities are listed in equation 2.

$$\begin{aligned}
[\mathrm{Fe/H}]_{\mathrm{SDSS}} &= (-10.33 \pm 0.47)\log P_{\mathrm{F}} + (-4.87 \pm 0.14), \sigma = 0.17\,\mathrm{dex}, R^2 = 0.71, \\
[\mathrm{Fe/H}]_{\mathrm{SDSS}} &= (-10.15 \pm 0.46)\log P_{1\mathrm{O}} + (-6.11 \pm 0.19), \sigma = 0.17\,\mathrm{dex}, R^2 = 0.71
\end{aligned} \tag{2}$$

3. Period–Luminosity Relation

Combining the PLZ relation with the period–period ratio–metallicity relation, the period–period ratio–luminosity relation is derived. We determined the period–period ratio–luminosity relation based on OGLE LMC RRd stars. We find that there is an approximate linear relation between the period and the period ratio, with a correlation coefficient of 0.7 and a dispersion of 0.0003. Therefore, we try to simplify the period–period ratio–luminosity relation to the period–luminosity (PL) relation. We find that the absolute magnitude difference predicted by the PL relation and the period–period ratio–luminosity relation is small. The scatter in magnitude difference is 0.01 mag and

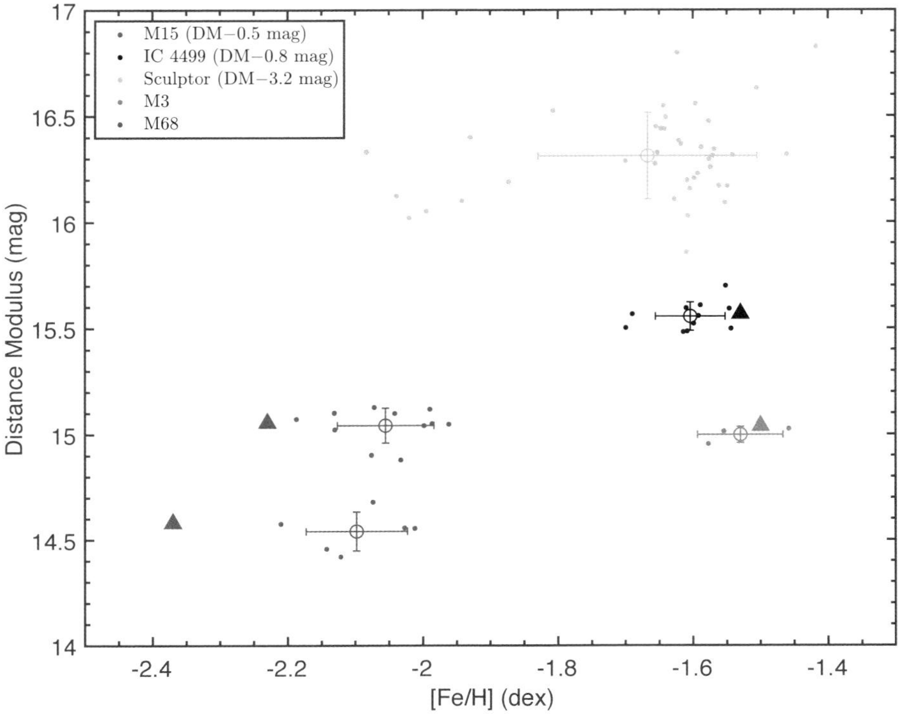

Figure 2. Metallicity and distance modulus (DM) determinations based on RRd stars for M15 (red dots), IC 4499 (black dots), Sculptor (green dots), M3 (magenta dots), and M68 (blue dots). The mean values of RRd stars in five targets are shown as open circles with 1σ internal errors. Solid triangles indicate the parameters of the four globular clusters from the Harris's globular cluster catalog. To make the figure clearer, the distance modulus of M15, IC 4499, and Sculptor are reduced by 0.5 mag, 0.8 mag, and 3.2 mag, respectively.

the maximum magnitude difference is only 0.06 mag. In addition, the scatter of the PL relation and the period–period ratio–luminosity relation is the same. Since the metallicity is more dependent on period, we ultimately simplify the period–period ratio–luminosity relation to the PL relation (see equation 3). Here, we adopt the Wesenheit magnitude (Madore 1982) i.e. $W_{VI} = I - 1.55(V - I)$ and $W_{G,BP,RP} = G - 1.90(BP - RP)$.

$$\begin{aligned}
M_{W_{VI}} &= (-4.523 \pm 0.156) \log P_{\mathrm{F}} + (16.620 \pm 0.048) - \mathrm{DM_{LMC}}, \sigma = 0.132\,\mathrm{mag}, \\
M_{W_{VI}} &= (-4.434 \pm 0.153) \log P_{1\mathrm{O}} + (16.079 \pm 0.067) - \mathrm{DM_{LMC}}, \sigma = 0.132\,\mathrm{mag}, \\
M_{W_{G,BP,RP}} &= (-3.623 \pm 0.229) \log P_{\mathrm{F}} + (17.042 \pm 0.071) - \mathrm{DM_{LMC}}, \sigma = 0.159\,\mathrm{mag}, \\
M_{W_{G,BP,RP}} &= (-3.557 \pm 0.225) \log P_{1\mathrm{O}} + (16.606 \pm 0.098) - \mathrm{DM_{LMC}}, \sigma = 0.159\,\mathrm{mag}
\end{aligned} \tag{3}$$

We combine the LMC distance (Pietrzyński et al. 2019) and Gaia parallaxes to calibrate the zero points of the PL relations as $M_{W_{G,BP,RP}}(P_{1\mathrm{O}} = 0.37\mathrm{d}) = -0.348 \pm 0.022$ mag and $M_{W_{VI}}(P_{1\mathrm{O}} = 0.37\mathrm{d}) = -0.496 \pm 0.022$ mag. Meanwhile, we derive the zero point of the Gaia DR3 parallax, which is 13.4 ± 5.8 μas. This parallax zero point agrees well with that determined by classical Cepheids (Riess et al. 2021).

4. Distance and Metallicity Measurements

The established PL relation and PZ relation of RRd stars can be used to determine the distances and metallicities of distant objects. We measure the distances and metallicities of four globular clusters (IC 4499, M15, M3, and M68) and a dwarf galaxy (Sculptor) based on RRd stars with Gaia data. Figure 2 shows the derived metallicities and distance moduli. The different colored dots indicate individual metallicity and distance modulus determinations based on RRd stars in M15 (red), IC 4499 (black), Sculptor (green), M3 (magenta), and M68 (blue). Solid triangles indicate the results for the four globular clusters from the globular cluster catalog of Harris (1996). The mean values of RRd stars in five targets are shown as open circles with 1σ internal errors. For dwarf galaxy Sculptor, the distance accuracy is about 1.8%, and the metallicity dispersion is 0.2 dex. For globular clusters, the distance accuracy is about 2-3%, and the metallicity dispersion is 0.05 dex. Recently, we use RRd stars to measure the distances and metallicities of other dwarf galaxies. For example, we measure the distances and metallicities of five dwarf galaxies in M31, where the distance accuracy is about 1.8%. We also determine the distance of the Fornax dwarf galaxy with an accuracy of 1.2% based on ~ 130 RRd stars.

There are several advantages of adopting RRd stars as distance indicators. Since the RRd star PL relation is not affected by metallicity, the galaxy distance error can be optimized to 1.2% based on ~ 100 RRd stars. This accuracy is as good as the distance accuracy of the LMC. RRd stars are not rare, and they account for 3-5% of RR Lyr stars. The number of RRd stars in the SMC, LMC, and the Galactic halo is comparable to the total number of classical Cepheids in these galaxies. Moreover, distance measurements based on RRd stars are less affected by extinction and binarity than the classical Cepheids. RRd stars are the best distance indicators for dwarf galaxies. They can also provide metallicity information for distance measurements of RRab stars. In the future, the China Space Station Telescope and the Vera C. Rubin Observatory Legacy Survey of Space and Time will discover more RRd stars. The RRd star can be an independent distance ladder in the Local Group, used to check the Cepheid (Riess et al. 2022) and the tip of the red giant branch (Freedman et al. 2019) distance ladders.

5. Summary

We discover a period–(period ratio)–metallicity relation for RRd stars and find it can predict metallicity as accurately as the low-resolution spectra. We propose that the PL relation of RRd stars is not affected by the metallicity and derive the PL relation of RRd stars. The zero point of the PL relation is determined with an error of 0.022 mag (1.0% error in the distance). We find that RRd stars can anchor galaxy distances with an accuracy of 1-2% (dwarf galaxies) and 2-3% (globular clusters). In the Future, RRd stars will be established as an independent distance ladder in the Local Group.

Acknowledgements

This work is supported by the National Natural Science Foundation of China (NSFC) through the projects 12003046, 12173047, 11903045, 12133002, and 11973001. S.W. and X.C. acknowledge support from the Youth Innovation Promotion Association of the CAS (grant No. 2023065 and 2022055).

References

Braga, V. F., Fiorentino, G., Bono, G., et al. 2022, *Monthly Notices of the Royal Astronomical Society*, 517, 5368

Catelan, M., & Smith, H. A. 2015, Pulsating Stars. (Wiley-VCH)

Chen, X., Wang, S., Deng, L., et al. 2018, *ApJS*, 237, 28

Chen, X., Wang, S., Deng, L., et al. 2020, *ApJS*, 249, 18

Chen, X., Zhang, J., Wang, S., & Deng, L. 2023, *Nat. Astron.*, arXiv:2306.10708

Clementini, G., et al., et al. 2022, *arXiv*, 2206.06278

Cui, X.-Q., et al. 2012, *Res. Astron. Astrophys.*, 12, 1197

Eisenstein, D. J., et al. 2011, *Astron. J.*, 142, 72
Freedman, W. L., et al. 2019, *ApJ*, 882, 34
Harris, W. 1996, *ApJ*, 112, 1487
Heinze, A. N. Tonry J. L., Denneau L., et al. 2018, *ApJ*, 156, 241
Jayasinghe, T., et al. 2018, *Monthly Notices of the Royal Astronomical Society*, 477, 3145
Kovtyukh, V., et al. 2016, *Monthly Notices of the Royal Astronomical Society*, 460, 2077
Madore, B. F. 1982, *ApJ*, 253, 575
Marconi, M., et al. 2015, *ApJ*, 808, 50
Nemec, J. M., & Moskalik, P. 2021, *MNRAS*, 507, 781
Petersen J. O. 1973, *A&A*, 27, 89
Pietrzyński, G., et al. 2019, *Nature*, 567, 200
Popielski, B. L., Dziembowski, W. A., Cassisi, S. 2000, *Acta Astron.*, 50, 491
Riess, A. G., et al. 2021, *ApJL*, 908, L6
Riess, A. G., et al. 2022, *ApJL*, 934, L7
Sesar, B. et al. 2017, *ApJ*, 153, 204
Soszyński, I. et al. 2009, *Acta Astron.*, 59, 1
Soszynśki, I., et al. 2016, *Acta Astron.*, 66, 131
Soszynśki, I., et al. 2019, *Acta Astron.*, 69, 321

At the crossroads of astrophysics and cosmology: Period–luminosity relations in the 2020s
Proceedings IAU Symposium No. 376, 2024
R. de Grijs, P. A. Whitelock & M. Catelan, eds.
doi:10.1017/S1743921323002673

Empirical constraints for the instability strip from the analysis of LMC Cepheids

F. Espinoza-Arancibia and B. Pilecki

Nicolaus Copernicus Astronomical Center, Polish Academy of Sciences, Bartycka 18, 00-716 Warsaw, Poland. email: fespinoza@camk.edu.pl

Abstract. The instability strip (IS) of classical Cepheids has been extensively studied theoretically. Comparison of the theoretical IS edges with those obtained empirically, using the most recent Cepheids catalogs available, can provide us with insights into the physical processes that determine the position of the IS boundaries. We investigate the empirical positions of the IS of the classical Cepheids in the Large Magellanic Cloud (LMC) using data of classical fundamental-mode and first-overtone LMC Cepheids from the OGLE-IV variable star catalog, together with a recent high-resolution reddening map from the literature. We studied their position on the Hertzsprung-Russell diagram and determined the IS borders by tracing the edges of the color distribution along the strip. We obtain the blue and red edges of the IS in V- and I-photometric bands, in addition to $\log T_{\rm eff}$ and $\log L_{\odot}$. The results obtained show a break located at the Cepheids' period of about 3 days, which was not reported before. This phenomenon is most likely explained by the depopulation of second and third crossing classical Cepheids in the faint part of the IS, since blue loops of evolutionary tracks in this mass range do not extend blueward enough to cross the IS at the LMC metallicity. Furthermore, our empirical borders show good agreement with theoretical ones published in the literature. This proves that our empirical IS is a useful tool to put constraints on theoretical models.

Keywords. Stars: variables: Cepheids – Stars: evolution – Magellanic Clouds

1. Introduction

During their evolution, intermediate- and high-mass stars can radially pulsate as Classical Cepheids (hereafter Cepheids) when they cross a defined region in the Hertzsprung-Russel diagram (HRD), called the instability strip (IS). Cepheids span a typical mass range between 3 and 13 $M_{\odot}$ (see, e.g. Bono et al. 2000b; Anderson et al. 2016), though Cepheids with masses higher than 13 $M_{\odot}$ are also known (Musella 2022, and references therein). Such stars first cross the IS during the H-shell burning phase, after they leave the main sequence (MS). Later, the star crosses the IS again during the helium-core burning phase, called the blue loop. The time scale of the first crossing is more than 10 times shorter than that of subsequent crossings. Thus, it is assumed that most of the observed Cepheids are burning helium during the second or third crossing of the IS.

The IS has shown not to be really a "strip" but rather to have a "wedge" structure. The structure of the IS was reported originally using Galactic Cepheids by Pel & Lub (1978) and Fernie (1990), while in the Magellanic Clouds it was noted by Martin et al. (1979), and Caldwell & Laney (1991). A large number of theoretical studies have addressed this issue, among them Bono et al. (2000a) showed how nonlinear theoretical models that consider the coupling between pulsation and convection produce an IS much less steep than linear models. Fiorentino et al. (2002), and more recently De Somma et al. (2022), concluded that the IS location depends on both the metallicity and helium abundance, moving to lower effective temperatures as the metal content increases. Petroni et al. (2003) studied the effect of using different equations of state on the IS topology, finding a weak dependence. Fiorentino et al. (2007) explored the effects of varying the mixing-length parameter l/H_P, finding that the IS becomes narrower as l/H_P increases, these results agree with the recent work done by De Somma et al. (2022). Anderson et al. (2016) conducted a study of the effects of rotation on Cepheid models, finding that the blue edge of the IS does not depend significantly on rotation, while the red edge is affected by metallicity and rotation.

On the other hand, empirical studies have been conducted to study the properties of the IS. Turner (2001) mapped the IS with a sample of 293 Cepheids in the Milky Way with reddening measurements, finding agreement with the results of Pel & Lub (1978), but disagreement with those presented by Fernie (1990). Tammann et al. (2003) obtained period-color and period-luminosity (P-L) relations of a sample of 321 Galactic Cepheids, in addition to 314 and 486 Cepheids from the Large Magellanic Cloud (LMC) and the Small Magellanic Cloud (SMC), respectively. Using these relations, they obtained the IS of the Galaxy and the Magellanic Clouds, finding differences in the slopes of the ISs between the three galaxies. Sandage et al. (2004, 2009) presented a detailed analysis of the samples of 593 and 460 Cepheids from the LMC and SMC, respectively. For both, they found breaks in the P-L relations, as well as in the IS edges. Recently, Narloch et al. (2019) used a sample of over 3200 LMC Cepheids to obtain an empirical IS using a step-detection technique, to find candidates for non-pulsating stars lying inside the IS.

Remarkably, the Optical Gravitational Lensing Experiment (OGLE) project presented an almost complete census of Cepheids in the Magellanic Clouds (Soszyński et al. 2015). In addition, numerous reddening maps have been published in the literature (see, e.g. Inno et al. 2016; Górski et al. 2020; Skowron et al. 2021; Chen et al. 2022). In this context, we aim to obtain an empirical intrinsic IS for the Cepheids in the LMC using the most recent Cepheids catalogs available. The comparison of theoretical IS edges with those obtained empirically can provide us with insights into the physical processes that determine the position of the IS boundaries. For this purpose, we use data of classical fundamental-mode (F) and first-overtone (1O) LMC Cepheids from the OGLE variable star catalog, together with recent high-resolution reddening maps from the literature. We study the position of the Cepheids on the color-magnitude diagram (CMD) and determine the IS borders by tracing their color distribution along the strip. In addition, we compare our empirical borders with theoretical ones presented in the literature.

The outline of this paper is as follows: Sect. 2 describes the sample selection procedure; Sect. 3 describes the method used to obtain the IS borders; Sect. 4 presents a discussion of our results, including a comparison theoretical ISs published in the literature. Finally, Sect. 5 presents our conclusions.

2. Sample selection

We use data of F and 1O Cepheids in the LMC from the OGLE-IV variable stars catalog (Soszyński et al. 2015) and OGLE-III Shallow Survey (Ulaczyk et al. 2013). The data is affected by reddening, produced by the interstellar extinction along the line of sight. To correct for this effect we use the high-resolution (1.7 arcmin $\times$ 1.7 arcmin) reddening map of Skowron et al. (2021), which is based on red clump stars from the Magellanic Clouds, selected from the OGLE-IV catalog. This map has shown good agreement with previous and other recent reddening maps (e.g. Inno et al. 2016; Chen et al. 2022), and yields a small dispersion compared to other maps in the literature. This last feature is important when trying to recover the intrinsic IS, as discussed in Sect. 3.

In order to clean our sample, we discarded 164 objects deviating more than 3 sigma from the reddening-free period-luminosity relationship (also known as period-Wesenheit or P-W relation). A significant fraction of them are objects located significantly above the period-luminosity (P-L) relationship, which were identified recently to be mostly binary Cepheids with giant companions (Pilecki et al. 2021, 2022; Pilecki 2022). Moreover, we discarded 112 Cepheids that presented remarks in the OGLE-IV catalog, to avoid objects that have uncertain characterization. In addition, following the method described in Madore et al. (2017), 113 Cepheids were not considered due to their high vertical deviation from the relation between the magnitude residuals of the I-band P-L relation and the corresponding residual of the P-W relation. These deviations are possibly due to errors in the individual adopted extinctions. Since intrinsic color uncertainties are dominated by reddening uncertainties, as a final cleaning process, we did not consider Cepheids with reddening errors greater than the 95th percentile of the reddening error distribution. Consequently, 109 F Cepheids and 73 1O Cepheids were discarded.

Distances to the stars, used to calculate their I-band absolute magnitudes, were computed using the method described in Jacyszyn-Dobrzeniecka et al. (2016). We adopted their P-L relations for the full sample of F and 1O LMC Cepheids, specifically, we did not consider their P-L relation with a break. Furthermore, we considered the recent determination of the geometrical distance to the LMC by Pietrzyński et al. (2019).

3. Instability Strip borders

The intrinsic width of the IS must be corrected for any effect that could change it, to be able to compare the strip with theoretical models. As we already removed from the sample all clear outliers (also those due to luminous companions) and objects with uncertain measurements, the major factor that is left is the effect of not perfectly known reddening. The non-zero uncertainties of the applied reddening values mean that the scatter of the Cepheid colors is higher and the IS width larger. The uncertainty of the photometry itself has a similar effect but to a much lesser extent. We performed the following steps to obtain an intrinsic IS that is free of this effect. We binned the complete sample, including F and 1O Cepheids, with respect to absolute magnitude in the I-band M_I. The central bins contain 233 stars, while the faintest and brightest bins contain around 100 stars. Firstly, we determine the initials blue and red edges of the IS of each bin, by locating the 1$^{\rm st}$ and 99$^{\rm th}$ percentile of the intrinsic color distribution, namely $B_{\rm 1st}$ and $R_{\rm 99th}$, respectively. We then generate and add random errors to the intrinsic color of each Cepheid, making the distribution of color wider. Subsequently, we count the number of stars that lie outside the initial edges, in other words, stars bluer and redder than $B_{\rm 1st}$ and $R_{\rm 99th}$, respectively. We repeated this process 10000 times and computed the median of the distribution of the number of stars outside the initial edges, namely $n_{\rm blue}$, $n_{\rm red}$.

The final blue and red IS positions of each bin, namely, $B_{\rm final}$ and $R_{\rm final}$, were obtained by moving the initial edges "inside" the IS by $n_{\rm blue}$ and $n_{\rm red}$ stars, respectively. In this way, we subtract the widening effect due to reddening uncertainties and obtain a better approximation to the intrinsic IS. Note that only such corrected IS can be compared directly with theoretical models that are not affected by reddening. The interior and exterior uncertainties in the intrinsic color of the IS were obtained by moving $B_{\rm 1st}$ and $R_{\rm 99th}$ by the 16$^{\rm th}$ and 84$^{\rm th}$ percentile of the distribution of the number of stars that lie outside these initial edges. The positions in M_I of each bin of the IS edges were considered as the median of the M_I distribution of each bin. The computed IS boundaries for the complete sample are shown in Fig. 1. In this figure, the periods of the 1O Cepheids were fundamentalized using equation 1 of Pilecki et al. (2021). To obtain the overplotted constant period lines for 1, 3, and 10 days we used the period-luminosity-color

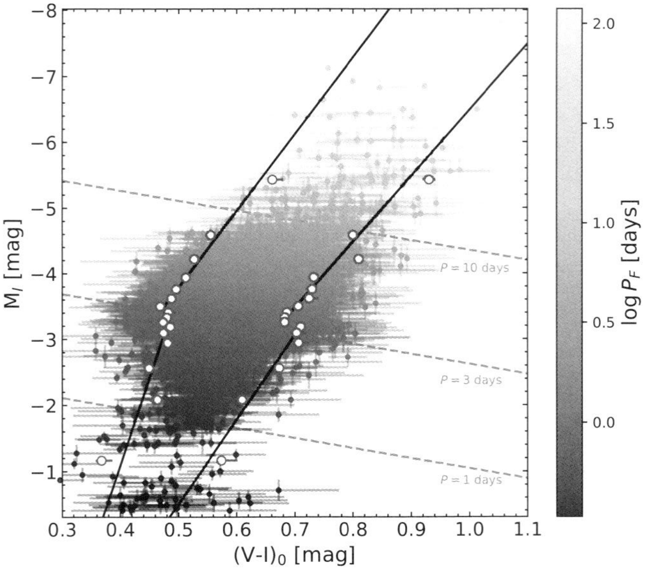

Figure 1. CMD of F and 1O LMC Cepheids. The boundaries of our empirical IS are shown as red points. Fits for the upper and lower part of the red and blue edges are shown as black solid lines. Periods for these stars are shown with a color gradient. For 1O Cepheids, periods were fundamentalized. Dashed lines of constant periods are overplotted.

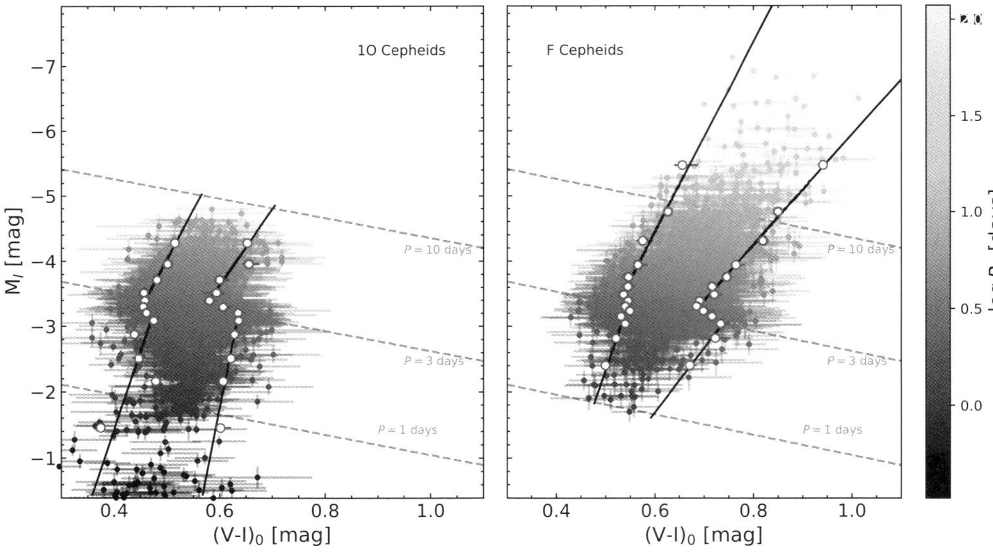

Figure 2. Same as Fig. 1, for 1O (left panel) and F (right panel) independently.

(PLC) relation computed using the previously calculated I-band absolute magnitude and intrinsic color, and the periods provided in the OGLE catalog. We repeated the IS edge determination process for F and 1O Cepheids separately, using bins that contain 162 and 109 stars, respectively. The obtained IS edges are shown in Fig. 2.

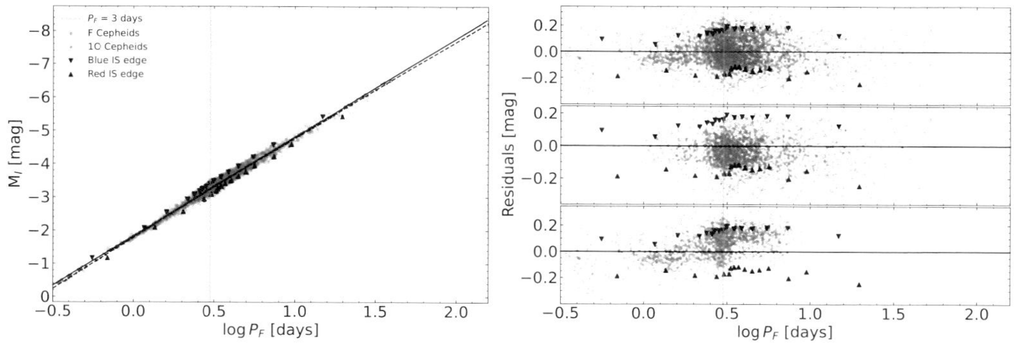

Figure 3. P-L relation of F (blue points) and 1O (red points) LMC Cepheids (left panel). The 1O Cepheid periods were fundamentalized. The boundaries of the IS are displayed as black points. At $P = 2.5$ days, a vertical gray line is traced. An average fit of the P-L relation considering the entire sample of stars is shown as a solid line. The differences between the fit of P-L and the data are presented in the right panel, for the entire sample (upper subpanel), only F Cepheids (middle subpanel), and only 1O Cepheids (lower subpanel).

4. Discussion

4.1. *Break in IS boundaries in $(V - I)$ color*

Previously in the literature, breaks in the P-L relation of F and 1O Cepheis have been reported in Magellanic clouds. Bauer et al. (1999); Sharpee et al. (2002) and Sandage et al. (2009) found a slope change of the P-L relation for the SMC F Cepheids with periods shorter than 2 days. On the other hand, Sandage et al. (2004) and Ripepi et al. (2022) describe a break in the PL relation for LMC Cepheids with periods of $P_F = 10$ d and $P_{1O} = 0.58$ d (the equivalent of $P_F = 0.81$ d), respectively. Bhardwaj et al. (2016, and references therein) performed an analysis of possible non-linearities in the P-L relations of Cepheids in the Magellanic clouds, confirming the statistical significance of the aforementioned breaks.

Bauer et al. (1999) proposed several possible scenarios to explain the break of the P-L relation at 2 days for the SMC. One of these scenarios, also mentioned by Ripepi et al. (2022), is the depopulation of second and third-crossing Cepheids in the faint part of the IS, due to the fact that the extension of the blue loops decreases as the mass and period of Cepheids decrease, hence these stars would spend less time inside the IS, or they would not enter at all. This implies that Cepheids fainter than the break are likely on their first crossing of the IS. This scenario should be noticeable in the CMD and could have an impact on the shape of the IS.

By analyzing the obtained IS edges of the entire sample, shown in Fig. 1, we observe that there is a break in the position of the IS boundaries, characterized by a shift to hotter temperatures above $P \sim 3$ days. This effect is also present in the IS edges of the F and 1O Cepheids subsamples (most noticeable in the red edge) presented in Fig. 2. Motivated by the detection of these breaks, using the Python package `emcee`, we performed a linear fit with two parts for our empirical IS borders. This separation considers the change of slope and discontinuity in the boundaries of the IS, between faint and bright Cepheids. In addition, a simplified wedge-shaped IS version was also computed.

In Fig. 3 we show an average P-L relation of our complete sample (considering F and 1O Cepheids), the periods of the 1O Cepheids were fundamentalized. Using the PLC relation we converted our IS to the period-luminosity plane and marked its edges with black triangles. In the right panel, we show residuals from the subtraction of the fit of the P-L relation from the data. For periods longer than $P \sim 3$ days, there is a sudden increase

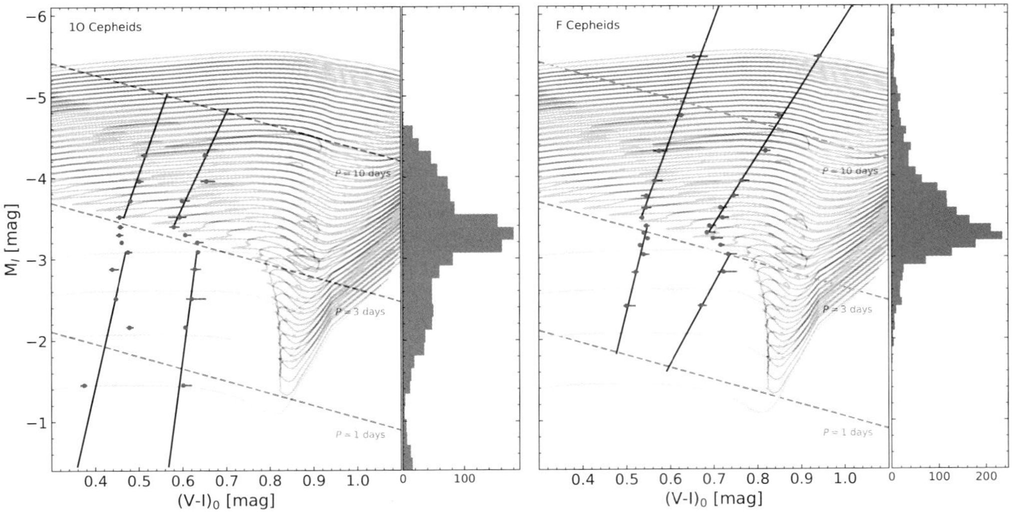

Figure 4. CMD showing the empirical IS limits (red points), separately for 1O (left panel) and F-mode (right panel) LMC Cepheids. Fits for the blue and red edge, considering a break at $P \sim 3$ days, are shown as solid blue and red lines. Blue loops of evolutionary tracks of 3 to 7 $M_\odot$ in steps of 0.1 $M_\odot$ are presented as gray lines. The subgiant phase is displayed for tracks of 3 to 4.5 $M_\odot$ in steps of 0.5 $M_\odot$. The I-band absolute magnitude distributions of the 1O and F LMC Cepheids are shown on the right side of each panel.

in the number of Cepheids. Stars with periods shorter than 2.5 days or longer than about 10 days tend to lay mostly below the P-L relation. The 10-day limit corresponds to the break reported in the literature, while below 0.81 days (a break from Ripepi et al.) we can only see another significant decrease in the number of Cepheids. Regarding the IS edges, some change can be seen at these two period values but the statistics are too low to quantify it reliably.

From a theoretical point of view, blue loops are very sensitive to metallicity and the adopted input physics, such as convective core overshooting and nuclear reactions (Xu & Li 2004a,b; Walmswell et al. 2015). Several previous works in the literature have shown that the general trend for the blue loop extension is to decrease with stellar mass. In particular, for stars with $M < 4\ M_\odot$ loops do not enter the IS (see, e.g., Anderson et al. 2016; De Somma et al. 2022). This supports the scenario proposed by Bauer et al. (1999) and Ripepi et al. (2022). To further test theoretical models and the depopulation of the faint part of the IS, using the stellar evolution code Modules for Experiments in Stellar Astrophysics (MESA; Paxton et al. 2011), we obtained evolutionary tracks for non-rotating stars covering the mass range from 3 to $7M_\odot$ in steps of $0.1M_\odot$, with metallicity $Z = 0.008$, and compared them with our IS positions. These tracks consider exponential overshooting, semi-convection, and mass loss. The comparison between the evolutionary tracks and our empirical IS borders is shown in Fig. 4. Evolutionary tracks with $M \lesssim 4.2M_\odot$ show blue loops that do not pass through the red edge of the IS. As a consequence, in this mass range, we expect an increasingly (toward shorter P) large contribution of Cepheids at the first crossing through the IS, where they evolve in a short time scale of the H-shell burning phase. Such first-crossing Cepheids spend less time within the IS compared to higher mass Cepheids, thus it is expected to find fewer stars in this part of the IS. On the other hand, evolutionary tracks with $M \gtrsim 4.2M_\odot$ present blue loops that cross the red edge of the IS, therefore the fraction of Cepheids in the second and third crossings, evolving in the longer time scale of the He-core burning

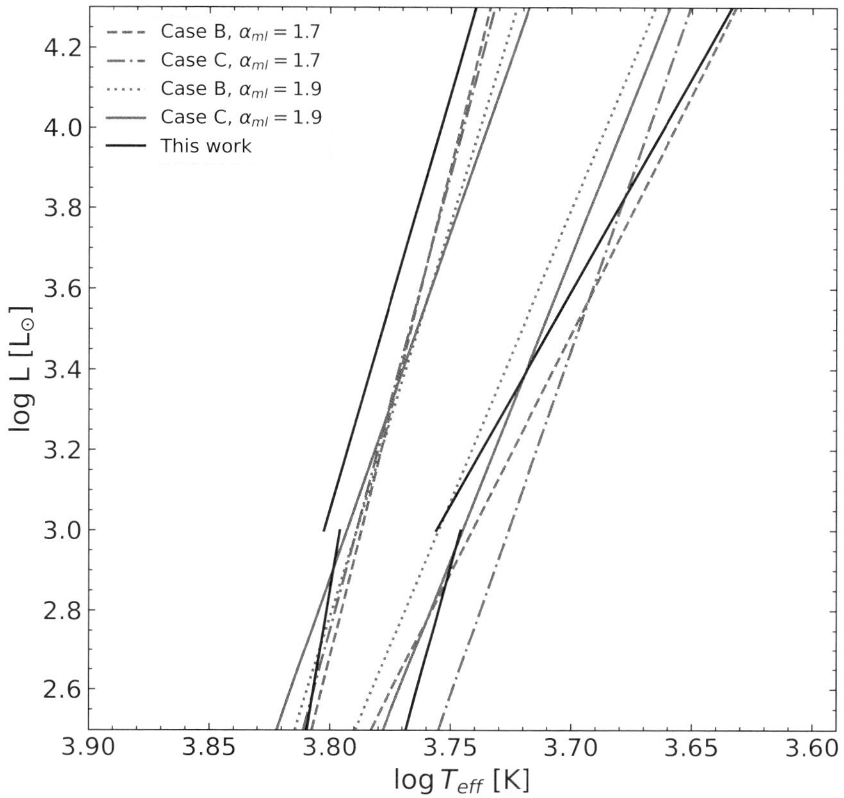

Figure 5. Comparison of theoretical ISs presented in De Somma et al. (2022) (blue and red) and our empirical IS (black) in an HRD. Different line styles represent different cases of increase in the luminosity, over their canonical model and efficiencies of superadiabatic convection α_{ml}.

phase, increases considerably. Since these stars spend more time inside the strip, their total number is much larger than the stars in the lower part of the IS.

4.2. *Comparison with theoretical ISs*

Previous works in the literature have obtained theoretical IS edges and have studied the effects of, for example, rotation, metallicity, and overshooting. By comparing these theoretical results with our empirical IS, it is possible to constrain some of the physical effects studied. Each point of our IS for F Cepheids in the CMD plane was converted to the $\log T_{\rm eff} - \log L$ plane using the color-temperature and bolometric correction calibration presented by Worthey & chul Lee (2011). Then the points were fitted in the same way as for the IS in the color-magnitude diagram. Since our empirical edges are composed of two parts, the conclusions may be different for the lower and upper parts of the IS.

In Fig. 5, we show the theoretical ISs obtained by De Somma et al. (2022). Their models are based on Bag of Stellar Tracks and Isochrones (BaSTI; Hidalgo et al. 2018) evolutionary tracks that take into account core overshooting and mass loss, and were calculated for a metallicity $Z = 0.008$. As far as their pulsation calculations go, De Somma et al. (2022) approximate the effects of overshooting by considering an increase in the luminosity, over their canonical models, of $\Delta(\log L/L_{\odot}) = 0.2$ dex (case B), and $\Delta(\log L/L_{\odot}) = 0.4$ dex (case C), considering in addition 2 values for the superadiabatic convection efficiency, $\alpha_{ml} = 1.7$ and $\alpha_{ml} = 1.9$. Globally, our empirical edges are most consistent with their

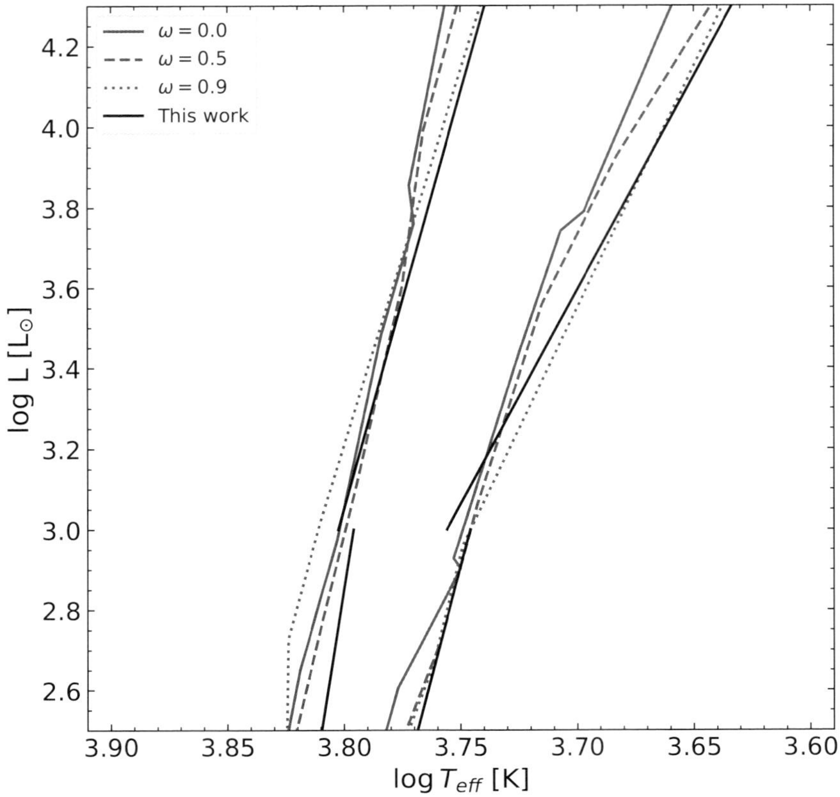

Figure 6. Comparison of theoretical ISs presented in Anderson et al. (2016) (blue and red) and our empirical IS (black) in a HRD. Different line styles represent different rotation rates ω.

models for case B and $\alpha_{\rm ml} = 1.7$. However, there are some noticeable differences – our upper blue edge is systematically hotter, and although the other edges coincide better with the theoretical ones, their slopes are slightly different.

The theoretical instability strips of Anderson et al. (2016) are shown in Fig. 6. They use the Geneva code of stellar evolution (Eggenberger et al. 2008), varying metallicity, and rate of rotation. We compare our results with their ISs for $Z = 0.006$, and three rotation rates $\omega = 0.0$, 0.5, and 0.9. Their edges agree remarkably well with our empirical edges, both in regard to the zero points and the slopes, with only our blue edge being systematically cooler. Globally their models with high rotation rates describe best our determined edges but the lower part seems to favor slightly better those with moderate rotation rates.

Paxton et al. (2019) introduce Radial Stellar Pulsation `RSP`, a functionality in the MESA code to model high-amplitude, self-excited, non-linear pulsations that the star develops when it crosses the IS. The `RSP` model depends on equations describing time-dependent convection, described in Smolec & Moskalik (2008), which also depend on free parameters. Pulsation periods depend weakly on these parameters. However, period growth rates and light curves are sensitive to the choice of these convective variables. Different sets for these parameters are shown in table 4 of Paxton et al. (2019). In Fig. 7, the theoretical ISs borders presented by Paxton et al. (2019, priv. communication) considering set B and D, for $Z = 0.008$ are shown. Their ISs were computed using second and third crossing Cepheid models. Set B of convective parameters adds radiative cooling effects to the convective model, while the parameter set D simultaneously includes

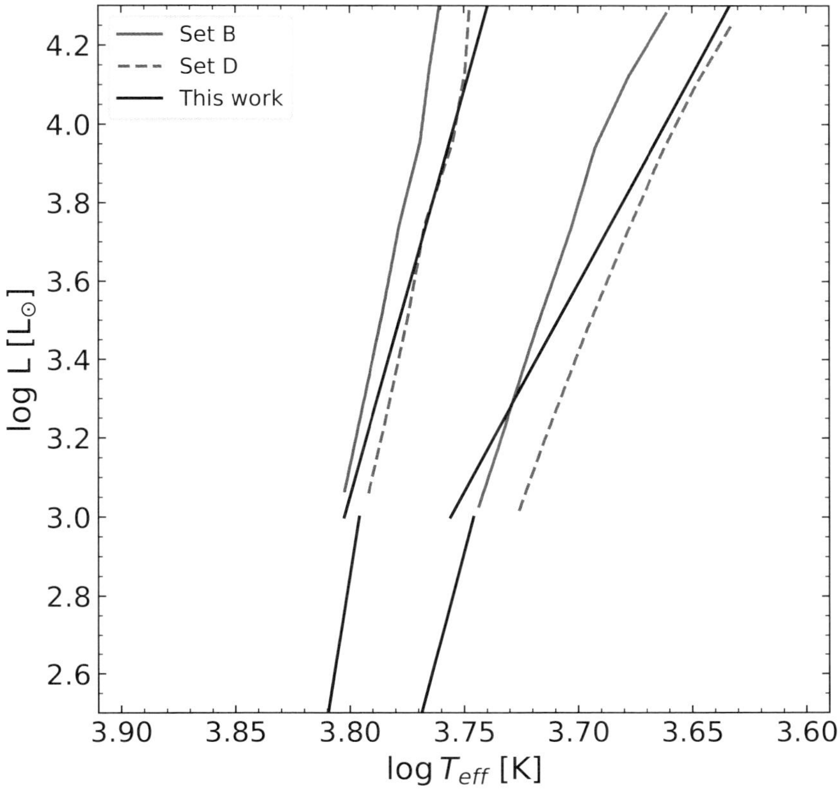

Figure 7. Comparison of theoretical ISs presented in Paxton et al. (2019) (blue and red) and our empirical IS (black) in an HRD. Different line styles represent different sets of convective parameters of RSP.

the effects of turbulent pressure and turbulent flow in addition to radiative cooling. As mentioned in Paxton et al. (2019), the IS edges produced by sets A and C overlap those for convective sets B and D, respectively. We can see that our blue edge is in closer agreement with that of set D, while our red edge lies somewhere in between sets B and D. The red edge is actually described better by set D for luminous long-period Cepheids, and by set B for short-period Cepheids, also if we extrapolate their red edge down to fainter Cepheids.

5. Conclusions

We used a sample of LMC Cepheids from the OGLE-IV catalog to compute the empirical, intrinsic instability strip. We determined the IS edges for F and 1O Cepheids, together and separately, using $V-$ and I-bands. In all cases, a break in the IS was observed at $\log P \sim 3$ days, which is most noticeable in the red edges and for separate 1O and F samples.

Using a set of evolutionary tracks computed with the stellar evolution code MESA, we investigated the scenario in which the origin of this break is explained by depopulation in the faint part of the IS. Such depopulation would be caused by too low an extension of blue loops for lower-mass stars, which would not evolve blueward enough to enter the instability strip. Our results confirmed the conclusions from other studies in the literature on that subject (Bauer et al. 1999; Ripepi et al. 2022).

Consequently, Cepheids with masses below $\sim 4M_\odot$ would be found mostly during the first crossing of the IS, while above this limit there is a much higher contribution of Cepheids at the second and third crossing, for which the time scales are much longer. This change from the domination of the first to the domination of subsequent crossings coincides with the observed break in the IS edges. At the same time at different crossings, we may expect slightly different pulsational properties as the Cepheids are found at different evolutionary stages, i.e. the subgiant and blue loop phases of evolution. These two populations of Cepheids with different properties dominating at a given mass (or period) range, may then affect the position of the IS edges.

We compared our results with theoretical instability strips published in the literature. Our empirical edges are best described by the models of Anderson et al. (2016) with high rotation rates, for periods above the break, and by models with moderate rotation rates below the break. However, regarding the red edge only, models with $\omega = 0.9$ describe well the whole range of pulsation periods.

Regarding the instability strips that result from the models of De Somma et al. (2022) and Paxton et al. (2019), some fine-tuning of parameters would be needed to obtain consistent results. At the moment, no single set of their parameters produces an IS that would correspond well with our empirical edges – they either fall between the different sets or have different slopes. Such fine-tuning is beyond the scope of this work. At the moment for that purpose, we suggest using the models of Anderson et al. (2016) with either high rotation rates or a mixture of moderate and high rotation rates depending on the pulsation period.

Putting constraints on the models, apart from finding the one that describes the observations best, at the same time we can learn about the physical properties of classical Cepheids, their pulsational properties, and the evolution of intermediate-mass stars.

References

Anderson, R. I., Saio, H., Ekström, S., Georgy, C., & Meynet, G. 2016, A&A, 591, A8
Bauer, F., Afonso, C., Albert, J. N., et al. 1999, A&A, 348, 175
Bhardwaj, A., Kanbur, S. M., Macri, L. M., et al. 2016, MNRAS, 457, 1644
Bono, G., Caputo, F., Cassisi, S., et al. 2000a, ApJ, 543, 955
Bono, G., Castellani, V., & Marconi, M. 2000b, ApJ, 529, 293
Caldwell, J. A. R. & Laney, C. D. 1991, in The Magellanic Clouds, ed. R. Haynes & D. Milne, Vol. 148, 249
Chen, B. Q., Guo, H. L., Gao, J., et al. 2022, MNRAS, 511, 1317
De Somma, G., Marconi, M., Molinaro, R., et al. 2022, ApJs, 262, 25
Eggenberger, P., Meynet, G., Maeder, A., et al. 2008, Ap&SS, 316, 43
Fernie, J. D. 1990, ApJ, 354, 295
Fiorentino, G., Caputo, F., Marconi, M., & Musella, I. 2002, ApJ, 576, 402
Fiorentino, G., Marconi, M., Musella, I., & Caputo, F. 2007, A&A, 476, 863
Górski, M., Zgirski, B., Pietrzyński, G., et al. 2020, ApJ, 889, 179
Hidalgo, S. L., Pietrinferni, A., Cassisi, S., et al. 2018, ApJ, 856, 125
Inno, L., Bono, G., Matsunaga, N., et al. 2016, ApJ, 832, 176
Jacyszyn-Dobrzeniecka, A. M., Skowron, D. M., Mróz, P., et al. 2016, Acta Astron.,, 66, 149
Madore, B. F., Freedman, W. L., & Moak, S. 2017, ApJ, 842, 42
Martin, W. L., Warren, P. R., & Feast, M. W. 1979, MNRAS, 188, 139
Musella, I. 2022, Universe, 8, 335
Narloch, W., Pietrzyński, G., Kołaczkowski, Z., et al. 2019, MNRAS, 489, 3285
Paxton, B., Bildsten, L., Dotter, A., et al. 2011, ApJs, 192, 3
Paxton, B., Smolec, R., Schwab, J., et al. 2019, ApJs, 243, 10
Pel, J. W. & Lub, J. 1978, in The HR Diagram - The 100th Anniversary of Henry Norris Russell, ed. A. G. D. Philip & D. S. Hayes, Vol. 80, 229

Petroni, S., Bono, G., Marconi, M., & Stellingwerf, R. F. 2003, ApJ, 599, 522
Pietrzyński, G., Graczyk, D., Gallenne, A., et al. 2019, Nature, 567, 200
Pilecki, B. 2022, in XL Polish Astronomical Society Meeting, ed. E. Szuszkiewicz, A. Majczyna, K. Małek, M. Ratajczak, E. Niemczura, U. B"kak-St"keślicka, R. Poleski, M. Bilicki, & Ł. Wyrzykowski, Vol. 12, 174–177
Pilecki, B., Pietrzyński, G., Anderson, R. I., et al. 2021, ApJ, 910, 118
Pilecki, B., Thompson, I. B., Espinoza-Arancibia, F., et al. 2022, ApJl, 940, L48
Ripepi, V., Chemin, L., Molinaro, R., et al. 2022, MNRAS, 512, 563
Sandage, A., Tammann, G. A., & Reindl, B. 2004, A&A, 424, 43
Sandage, A., Tammann, G. A., & Reindl, B. 2009, A&A, 493, 471
Sharpee, B., Stark, M., Pritzl, B., et al. 2002, AJ, 123, 3216
Skowron, D. M., Skowron, J., Udalski, A., et al. 2021, ApJs, 252, 23
Smolec, R. & Moskalik, P. 2008, Acta Astron.,, 58, 193
Soszyński, I., Udalski, A., Szymański, M. K., et al. 2015, Acta Astron.,, 65, 297
Tammann, G. A., Sandage, A., & Reindl, B. 2003, A&A, 404, 423
Turner, D. G. 2001, Odessa Astronomical Publications, 14, 166
Ulaczyk, K., Szymański, M. K., Udalski, A., et al. 2013, Acta Astron.,, 63, 159
Walmswell, J. J., Tout, C. A., & Eldridge, J. J. 2015, MNRAS, 447, 2951
Worthey, G. & chul Lee, H. 2011, The Astrophysical Journal Supplement Series, 193, 1
Xu, H. Y. & Li, Y. 2004a, A&A, 418, 213
Xu, H. Y. & Li, Y. 2004b, A&A, 418, 225

At the crossroads of astrophysics and cosmology: Period–luminosity relations in the 2020s
Proceedings IAU Symposium No. 376, 2024
R. de Grijs, P. A. Whitelock & M. Catelan, eds.
doi:10.1017/S1743921323003356

The period-luminosity relations of red supergiants

Biwei Jiang[1,2], Yi Ren[3] and Ming Yang[4]

[1]Institute for Frontiers in Astronomy and Astrophysics, Beijing Normal University, Beijing 102206, China
email: bjiang@bnu.edu.cn

[2]Department of Astronomy, Beijing Normal University, Beijing 100875, China

[3]College of Physics and Electronic Engineering, Qilu Normal University, Jinan 250200, China

[4]National Astronomical Observatories, Chinese Academy of Sciences, Beijing 100101, China

Abstract. Red supergiants (RSGs) are the brightest stars in the near-infrared. The existence of their period-luminosity relation (PLR) will be very helpful in determining cosmological distances. This review discusses the development in identification of RSGs, calculation of their light variation period, determination of the PLR, and the uncertainties associated. It is found that the PLR of RSGs exhibits the smallest scatter in the near-infrared, in particular within the $P - M_K$ relation. The PLRs in the LMC, SMC, M31, and M33 show no obvious dependence on the galactic metallicity, which follows approximately the relation as $M_K \approx -3.1 \log P - 1.9$.

Keywords. red supergiant, period-luminosity relation, Magellanic Clouds, M31, M33

1. Introduction

Red supergiants (RSGs) are evolved Population I stars with helium-burning in the core. They are massive with a mass range between about 8-30 $M_\odot$, but the lower and upper mass limits may vary depending on the metallicity and stellar models. With radii up to $1000R_\odot$ or more, luminosities of RSGs can reach to 10^5 $L_\odot$. Meanwhile, their effective temperature $T_{\rm eff}$ ranges from about 3500 K to 4500 K, which makes them the brightest stars in the near-infrared. For example, the brightest RSG can have an absolute magnitude in the K band of $M_K \sim -12$ mag. Even at a cosmological distance of 1Gpc (i.e. $z \sim 0.25$), the apparent K-band magnitude would be $m_K = 28$ mag, neglecting the extinction, which is still visible by the JWST. As a potential cosmological candle, the period-luminosity relation (PLR) of RSGs deserves investigation.

Pulsation is an important process with respect to the structure and evolution of RSGs, which is often considered to play a crucial role for driving the stellar wind (Yoon & Cantiello 2010). Through the wind, pulsation changes stellar structure, and consequently influences seriously the final fate of RSGs to be type II-P supernova or not, e.g., evolves bluewards to become blue supergiants/Wolf-Rayet stars or directly collapses to black hole (Kochanek et al. 2008; Ekström et al. 2012).

Early in 1942, Wilson & Merril (1942) studied the PLR of a handful of RSGs in the Milky Way which were identified by spectroscopy. They concluded that there was an anti-correlation between the light variation period and luminosity. The mistake came mainly from the measurement of the absolute magnitude; the proper motion was used to derive the distance and then used to convert the apparent magnitude to the absolute magnitude. In addition to the distance uncertainty, they failed to correct for interstellar extinction, when such extinction was very poorly understood at the early stage of discovery in 1930 (Trumpler 1930). Because RSGs are generally young stars with an age of about 8-20 Myr, they are supposedly located quite close to the Galactic plane where the extinction is very severe and inhomogeneous. The brightest RSGs may be the most distant and suffer the highest extinction so that the lack of extinction correction can lead to a large underestimation of their brightness. The problem of interstellar extinction has always been an obstacle in determining the PLR of RSGs in the Milky Way (and in other galaxies though to a much less extent).

By assuming that the variability of RSGs is due to radial pulsation, Stothers (1969) calculated the theoretical Q-values of variable RSGs in adiabatic equilibrium and compared with the observational ones. He found that, "With our adopted periods, no P-L or Period-spectrum relation emerges from the data". He suspected that some periods were incorrect and continued with "a suitable readjustment" of these periods, and then concluded that "pulsation seems to be suggested by ... (2) a period-luminosity relation......". Taking α Sco as an example, Stothers (1969) adjusted the observational period, 1733 days to 180 day with a Q-value of 0.06 day. However, modern studies derived its period to be 1650 ± 640 day by Kiss et al. (2006), which confirmed the early observation.

What complicates the determination of the PLR of RSGs in the above works are mainly the ignorance of Galactic interstellar extinction and the complex variation of RSGs. The inhomogeneous extinction leads to the large uncertainty in distance and luminosity calculation. The light variation of RSGs very often contains some irregular components, which may be caused by convection, hot spot, high-order overtone mode, binary, etc. Consequently, the light curve may appear "irregular" and the determination of period becomes difficult and uncertain.

Most RSGs show some degree of variability. The typical amplitude of variation in the V band is about 1.0 mag, while in the near-infrared the amplitude is smaller, about 0.25 mag in the K band. According to the characteristics of light variation, the RSG variability is divided into three classes: irregular that is too complex to be delineated by any period, semi-regular with a short period (several hundred days), and the other with a long secondary period (LSP, with a period of thousand of days). Figure 1 shows the typical light curves of the three types.

2. The period-luminosity relation of red supergiants in the Magellanic Clouds

The success in determining the PLR of RSGs came when astronomers turned their observations to nearby galaxies, i.e. the Large and Small Magellanic Clouds (LMC and SMC), which is very like the case of the Cepheid variables. The same distance in a nearby galaxy means that the relative absolute luminosity can be expressed by the relative apparent one, which makes the PLR more precise.

Glass (1979) observed 38 late-type supergiants in the MCs in the near-infrared $JHKL$ bands. He mentioned the usefulness of RSGs as distance indicators. Afterwards, Feast et al. (1980) used 24 red supergiants in the LMC to determine the PLR. They derived the bolometric magnitude from the $JHKL$ photometry by using a kind of blackbody spectral energy distribution (SED) fitting. Based on the result, they justified the existence of the PLR of red supergiants. Their PLR has the form of $M_{\rm bol} = -8.6 \log P + 16.4$, which roughly coincides with the semi-theoretical relation.

In addition to the same distance, the RSGs in the LMC are not located deep inside the interstellar dust, which reduces the extinction effect. Moreover, the near-infrared extinction is only about a tenth of the visual bands (e.g., the V-band), further reducing the effect of interstellar extinction. One year later, Catchpole & Feast (1981) published the PLR of 22 RSGs in the SMC and concluded that "the SMC red supergiant variables fit

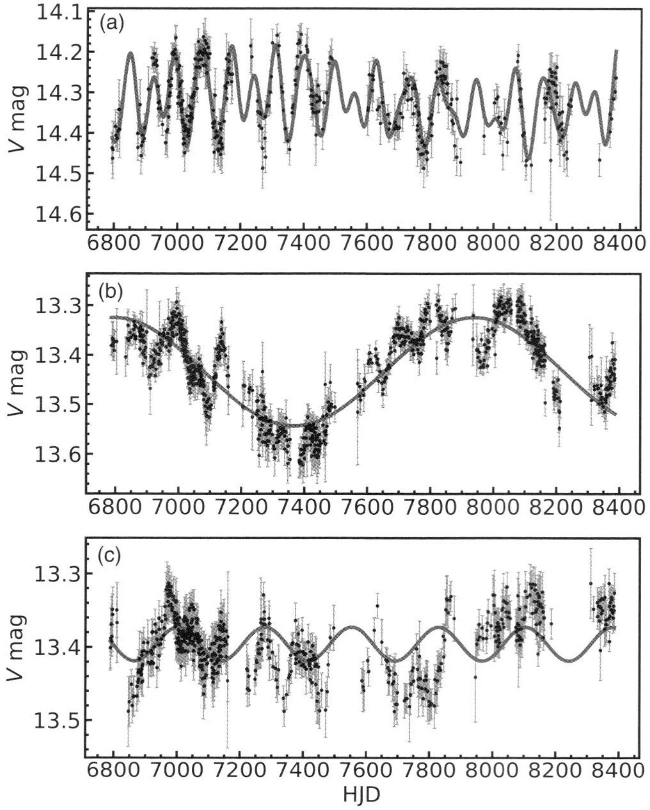

Figure 1. Three types of light variation of red supergiants. Top, middle, and bottom panels show a semi-regular RSG, a RSG with long secondary period, and a RSG with irregular variation, respectively. The red lines show the fitting by periodic functions.

the P-L relation previously derived for the LMC stars". This may be the first suggestion that the PLR of RSGs in the near-infrared is independent of metallicity, since the metallicity of the SMC and LMC differs by a factor of two. Later on, Wood et al. (1983) confirmed the PLR of RSGs in the MCs.

Typically, RSGs have periods about several hundred days with a semi-regular and complex variability. Consequently, a sufficiently long baseline is needed for studying their PLR, which is facilitated by the large-scale time-domain surveys developed quickly since 1990s with the advancement of detectors and driven by various sciences. In order to search for the Massive Halo Compact Objects (MACHOs), which was the candidates for dark matter, several projects like MACHO, OGLE, EROS, etc., were conducted to detect variable objects in densely populated star fields, such as the Galactic bulge and the Magellanic Clouds. In recent decades, more long-term surveys have been developed to monitor supernovae explosions which are important to cosmology. For example, the ASAS and its successor ASAS-SN are survey projects dedicated to relatively bright stars in the whole sky with small telescopes. PTF and its successor ZTF can detect objects as faint as 20 mag in the R- and g-band. These surveys naturally monitor the light variation of RSGs in the field, which greatly increases the sample size of RSGs for which the period can be determined.

Yang & Jiang (2011) and Yang & Jiang (2012) analyzed the light variation of 191 and 126 RSGs in LMC and SMC, respectively. With the nearly a decade long time-series photometric observation by ASAS and MACHO, they calculated the characteristics of light variation to classify the RSGs into three variable classes: irregular, semi-regular, and LSP. For the semi-regular and the LSP variable with distinguishable short period, they determined the PLR in the visual and near-infrared bands which include the 2MASS/JHK_S, Spitzer/IRAC1-4 and Spitzer/MIPS24. It is found that the PLR is tight in the infrared bands such as the 2MASS JHK_S bands and the Spitzer/IRAC bands. Meanwhile, the PLR is relatively sparse in the V band because the visual extinction is larger with larger dispersion than in the infrared bands. There is also relatively large scatter in the MIPS24 band possibly due to the influence of circumstellar dust emission of RSGs.

3. The period-luminosity relation of red supergiants in M31 and M33

The first application of the PLR of RSGs in M33 was by Kinman et al. (1987). They used the PLR derived from the LMC RSGs to calculate the distance according to the periods and apparent K-band magnitudes of RSGs in M33. The derived distance modulus was 24.64 which agrees very well with modern results. This successful application indicates that the PLRs of RSGs in M33 and LMC are consistent with each other.

Soraisam et al. (2018) analyzed the light variation of 255 spectroscopically identified RSGs in M31 based on the PTF data and found that 63 of them showed significant pulsation periods. They derived the PLR which was consistent with those in LMC, SMC and M33 over various metallicities.

Ren et al. (2019) used the time-series data from the iPTF survey spanning nearly 2000 days. For 84 and 56 RSGs classified as semi-regular variables in M33 and M31, respectively, they determined the periods with four popularly used methods, i.e. phase dispersion minimization (PDM), Generalized Lomb-Scargle Periodogram (GLS), Weighted Wavelet Z-transform (WWZ), and Discrete Fourier Transform (DFT). A period was accepted only when the periods derived by at least two methods were consistent with each other. The uncertainty of period was estimated to be about $\sim$ 10% of the period. Such accuracy cannot compete with other types of regular variables like Cepheids, RR Lyraes or Miras. Nevertheless, the PL relations based on the derived periods are tight in the near-infrared, i.e. in the JHK_S bands and the Spitzer/IRAC1 and IRAC2, which is the same as the case of the MCs.

4. Discussion

4.1. *The sample, complete or incomplete*

The early identification of RSGs in an external galaxy relied on spectroscopy of high-luminosity red targets, which showed the spectral features of a RSG (e.g., the broad absorption bands of TiO) and the radial velocity of an extragalactic object. The RSGs in the Magellanic Clouds were mostly identified in this way. Unfortunately, spectroscopy is time-consuming and applicable only to apparently bright objects, i.e. the RSGs in the MCs or the brightest RSGs in M31 and M33 etc. Spectroscopic identification may suffer severe contamination of red dwarfs and giants in the Milky Way, as well as asymptotic giant branch stars (AGBs) in the target galaxy. Later, Massey (1998) developed a method to identify extragalactic RSGs by photometry. They used the $B-V/V-R$ color-color diagram (in Johnson system) to distinguish extragalactic supergiants from foreground dwarf stars on the basis that the B band covers many metallic lines so that $B-V$ is sensitive to stellar surface gravity in addition to effective temperature. In comparison with spectroscopy, photometry is much more efficient and able to detect much fainter RSGs. By using this method, they identified 437 and 776 RSG candidates in M31 and M33 respectively, and the later spectroscopic observations of some of these confirmed that the radial velocities and spectral types are consistent with RSGs. In comparison with a few tens of RSGs in previous studies, this sample size was increased by an order

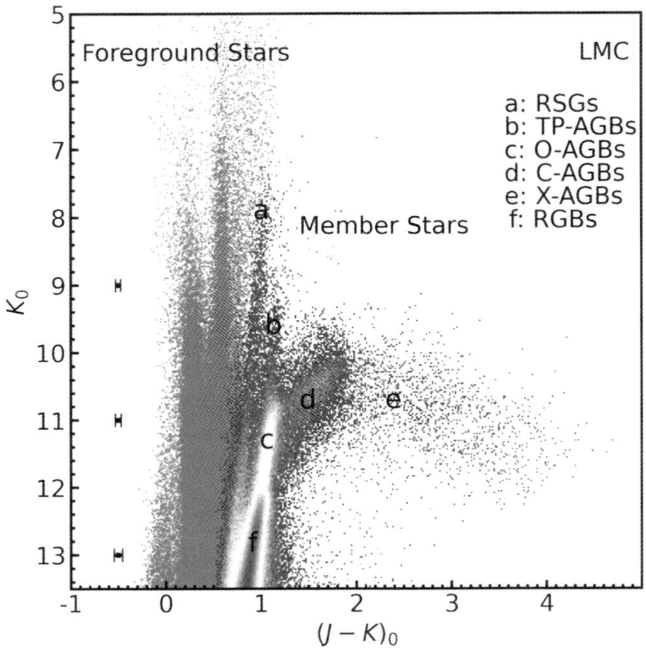

Figure 2. $(J-K)_0/K_0$ of member stars (color density map) in LMC. Removed foreground stars are gray dots. The error bars show the mean error of $J-K$ at different K magnitudes. Member stars include RSGs, thermally pulsing AGBs (TP-AGBs), oxygen-rich AGBs (O-AGBs), carbon-rich AGBs (C-AGBs), extreme AGBs (X-AGBs), and red giant branch stars (RGBs).

of magnitude. It was based on this sample that Ren et al. (2019) analyzed the PLR of RSGs in M31 and M33 in the near-infrared bands.

The photometric method is much more efficient than spectroscopy. However, such an optical color-color diagram is not practical at the faint end because the difference between red supergiant and dwarf stars is small in $B-V$. A photometric accuracy of about 0.01 mag is needed for clear discrimination. Moreover, the interstellar extinction mixes up the two classes. Finally, the modern filter sets have different effective wavelengths and coverages from the Johnson system.

4.1.1. The Magellanic Clouds

The Gaia space telescope brings about astrometric information to identify member stars in addition to providing astrophysical information. Yang et al. (2019) identified the SMC members by using stellar proper motions and parallaxes in the *Gaia*/*DR*2 datasets. The stars within a 5σ range of the proper motion distributions are taken as the candidates for membership. Together with other information such as parallax, radial velocity and the locations in the color-magnitude diagram, they identified 1,405 red supergiants in SMC. In more or less the same way, they further identified 2,974 RSGs in LMC (Yang et al. 2021). On this basis, Ren et al. (2021b) adopted the *Gaia*/*DR*3 datasets which included many more member stars of the MCs than *Gaia*/*DR*2 and refined the method by using the near-infrared color-color diagram (to be described below) and correcting the interstellar extinction. They finally identified 4,823 and 2,138 RSGs in LMC and SMC, respectively. Figure 2 displays the extinction corrected $(J-K)_0/K_0$ diagram of LMC stars identified in this way.

Based on new samples of RSGs in the LMC and SMC and time-series data from the Optical Gravitational Lensing Experiment (OGLE; Udalski et al. 2015; Soszyński et al. 2015) and the All-Sky Automated Survey for SuperNovae (ASAS-SN; Shappee et al. 2014; Kochanek et al. 2017; Jayasinghe et al. 2020), we investigated the PLR of RSGs in the LMC and SMC. The overview and processing of time-series data are as follows:
(1) Overview of Time-Series Data. The ASAS-SN, which contains 20 telescopes worldwide, observes variable objects in the LMC and SMC at a cadence of 4-5 days spanning ~ 2000 days with magnitude limits of ~ 18 at g band and ~ 17 at V band. The OGLE is a long-term photometric survey focused on variables, which observes the brightness of sources in the SMC and LMC at a cadence of 1-2 days and accumulates time-series data over 10 years.
(2) Removal of Outliers. For each light curve, the first quartile (Q_1), the third quartile (Q_3), and the corresponding interquartile range (IQR; IQR $= Q_3 - Q_1$) are calculated from the brightness distribution of a light curve. The photometry points beyond the range (i.e. $< Q_1 - 1.5 \times$ IQR or $> Q_3 + 1.5 \times$ IQR) are regarded as outliers.
(3) Removal of White-Noise Sequences. In this work, ASAS-SN performed forced photometry at the coordinates of the RSGs. This photometry strategy still generates light curves for some sources which are fainter than the detection threshold. So we perform white noise test on light curves. Specifically, if over 95% of the autocorrelation functions (ACFs) of a light curve lie in the region expected for a white-noise process, this light curve is regarded as white-noise and removed.

After time-series data preprocessing, for RSGs in the LMC, there are 2665, 2123, and 2656 light curves in the g, V, and I bands, respectively. For RSGs in the SMC, there are 886, 452, and 1337 light curves in the g, V, and I bands, respectively. There are 4471 and 1924 RSGs with at least one band photometry in the LMC and SMC, respectively.

Then the Lomb-Scargle (LS) periodogram method is introduced to determine the periods of RSGs. The periods of all three bands are determined. If one source has been observed at g, V, and I bands, we prefer the period of the V band, followed by the g band, and finally the I band. Because compared with the I band, the light curve has a higher amplitude in the V band, and there are more measurements than in the g band. To further eliminate the impact of fake periods determined by the LS method, we calculate the Pearson correlation coefficient (R) between the light curves of different bands (i.e. R_{gV}, R_{gI}, and R_{VI}). The RSGs in the LMC (SMC) used to obtain the PLR are required to have at least one Pearson correlation coefficient greater than 0.6 (0.3). In addition, RSGs with period greater than 1000 days are excluded from further analysis.

Due to the increase in the sample size, we can simultaneously observe RSGs in different pulsation modes in the PLR diagram as shown in Figure 3. Therefore, it is necessary to first classify the pulsation modes of RSGs, and we select three lines to fit three different pulsation modes. The slope of the three lines is selected as Yang & Jiang (2011) (i.e. -3.75) and fixed, and the three intercepts are free parameters. During the fitting process, the point is classified to one line to which Euclidean distance from the observation point is smallest. Finally, optimize the overall χ^2 to obtain the intercept of the three lines, and obtain the classification of the RSGs pulsation modes as shown in Figure 3.

For RSGs in fundamental pulsation mode, we calculate their multi-band PLRs, the V band brightness is taken from ASAS-SN, while the JHK_S bands brightness comes from 2MASS. The multi-band PLRs are presented in Figure 4.

4.1.2. The M31 and M33

The application of the Gaia measurements to the MCs greatly increased the sample size of RSGs by about one order of magnitude from previous samples. On the other

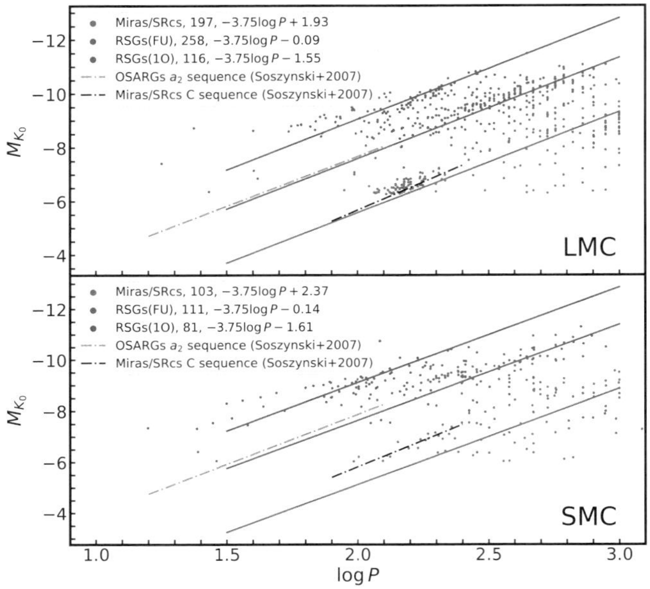

Figure 3. The PLR of the new samples of RSGs in the MCs. The distance moduli of LMC and SMC are 18.26 and 18.72, respectively (Ren et al. 2021b). Blue dots are RSGs in the first overtone pulsation mode. Red dots are RSGs pulsating in the fundamental mode, gray dash-dotted line is OGLE Small Amplitude Red Giants (OSARGs) a_2 PLR sequence (Soszyński et al. 2007) for comparison. Green sources are Mira variables and RSGs with long secondary period, black dash-dotted line is Miras/SRcs C PLR sequence (Soszyński et al. 2007) for comparison.

hand, this also indicates that the RSG samples of 437 and 776 objects (Ren et al. 2019) are far from complete in M31 and M33 and need to be replenished. With a distance modulus (d.m.) of about 24.5 mag, M31 and M33 are much further than the MCs at a d.m. of about 19 mag. Consequently, Gaia measured only few percent of the member stars. The identification by proper motion or parallax of member stars then becomes very inefficient. Ren et al. (2021a) developed a new method based on the near-infrared color-color diagram, i.e. $J-H/H-K$. In methodology, this is similar to the $B-V/V-R$ method. This near-infrared color-color diagram uses the sensitivity of the H-band, that includes many molecular bands, to surface gravity to distinguish the red (super)giants from the foreground dwarfs. As an example, the $J-H/H-K$ diagram of point sources is displayed in Figure 5 for M33, where the intrinsic color indexes of dwarfs, giants and supergiants are plotted for reference (Bessell & Brett 1988). Obviously, the foreground dwarf stars constitute an inverse "V"-shaped branch, while the member red giants, AGBs, and RSGs are well above the dwarf branch. After removing the foreground stars, the color-magnitude diagram becomes very clean where the RSGs can be easily recognized by their high brightness and relatively bluer colors than AGBs. Ren et al. (2021a) identified 5,498 and 3,055 RSGs in M31 and M33 respectively by this method. The spatial distribution of RSGs in M31 and M33 is presented in Figure 6, which coincide with the ring in M31 and spiral arms in M33 as expected for massive stars. Due to the photometric error in the crowded field, the sample may suffer an uncertainty of about 10% for M33 and 20% for M31, but this still is the most complete and pure sample of RSGs in these two galaxies.

4.2. *Universality*

Whether the PLR of RSGs is universal or not is very important for its application to cosmology. It is also important to the physics of RSGs. The first sign of universality

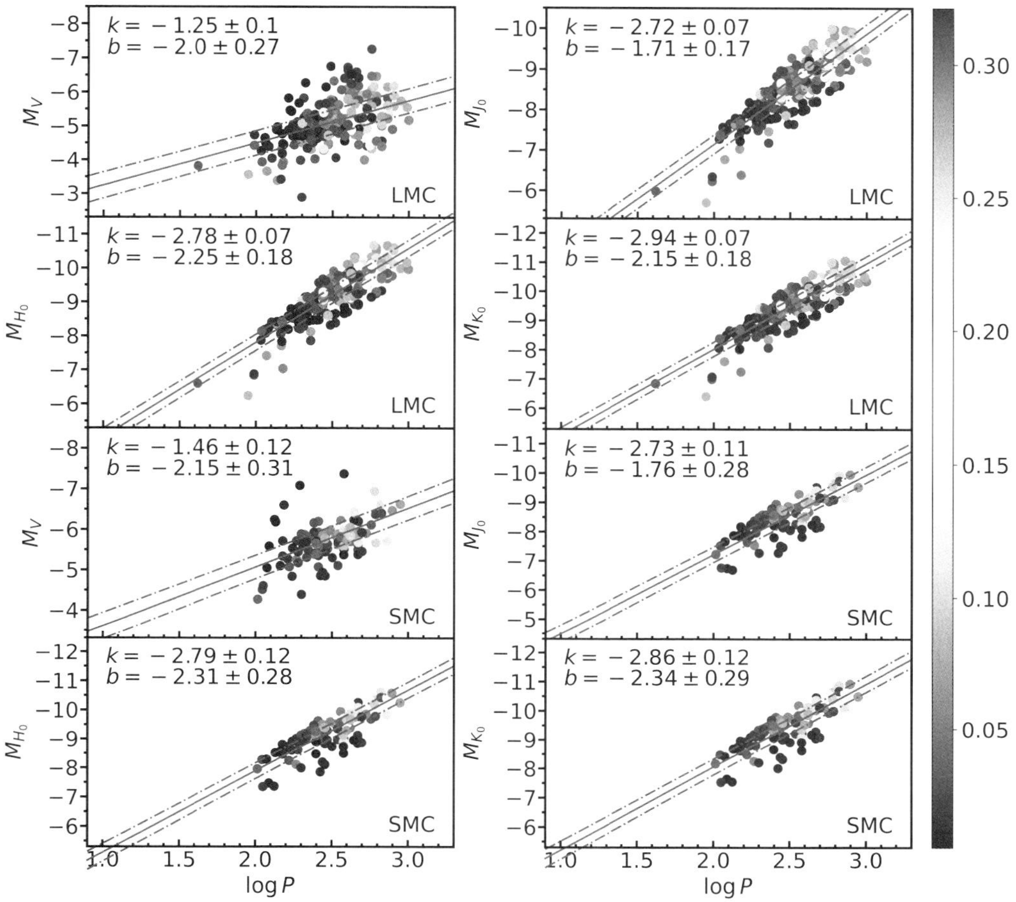

Figure 4. PLR in the $VJHK$ bands for RSGs in fundamental pulsation mode in MCs. The red solid line is a linear fit between the multi-band absolute magnitude and the period. The blue dash-dotted lines indicate the prediction interval at 0.35 significance level. The color of dots decodes the V band amplitude in accord with the color bar (in the unit of magnitude).

appeared when Catchpole & Feast (1981) published the PLR of 22 RSGs in the SMC and concluded that "the SMC red supergiant variables fit the P-L relation previously derived for the LMC", since the metallicity of the SMC is only about half that of the LMC. Later confirmation came from Pierce et al. (2000) who studied the RSGs in the Per OB1 association, the LMC and M33. They estimated the distances to the Per OB1 and M33 by using the PLR of RSGs in the LMC, and obtained results consistent with other methods. This indicated that the PLR derived from the LMC can be applied to the RSGs in much more metal-rich environments.

Ren et al. (2019) compared the $\log P - M_K$ relation of RSGs in the SMC, the LMC, the Milky Way, M33, and M31. They found that the PLR obeyed similar law in these galaxies except the Milky Way, i.e. the slopes are all consistent within the uncertainty. Specifically, $M_{K_S} \sim -3.1 \log P - 1.9$, with a dispersion of about 0.2 mag. The PLR of RSGs in the Milky Way will be more uncertain due to the large dispersion of the estimated distances caused by heterogeneous interstellar extinction. This suggests that the $\log P - M_K$ relations in the K band is universal in these nearby galaxies. In that sense, RSGs can be used as a distance indicator. It must be emphasized that, the universality of PLR in the infrared bands does not imply the PLR is universal in the optical bands.

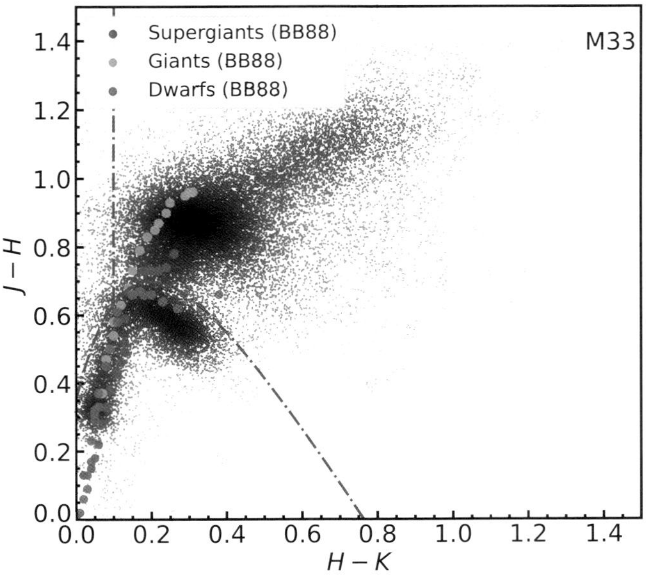

Figure 5. The NIR CCD of the point sources in the field of M33. The blue dash-dotted line is the dividing line between foreground dwarfs and member (super)giants. For comparison, the intrinsic color indexes of dwarfs (green dots), giants (yellow dots), and supergiants (red dots) are taken from Bessell & Brett (1988, BB88 for short).

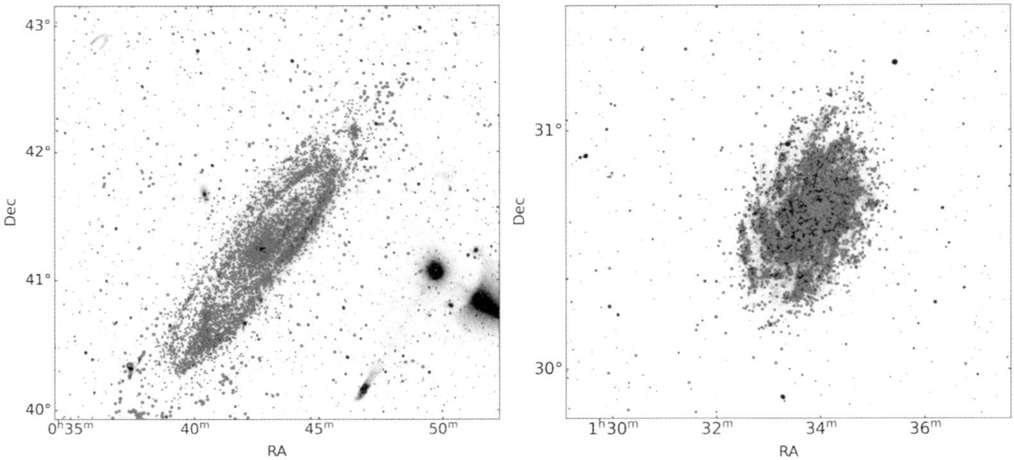

Figure 6. Spatial distribution of RSGs in M31 (left panel) and M33 (right panel). The background is the GALEX ultraviolet observation.

Ramírez & Meléndez (2005) pointed out that metallicity has much weaker influence on stellar infrared colors than optical color. Moreover, the much higher visual extinction with larger dispersion makes the PLR in the optical band very scattered, so that the uncertainty of the $\log P - M_V$ relation is too large to discuss its universality.

Figure 7 shows the PLR in K band. The PLR of RSGs in M31 is derived using the sample from Ren et al. (2021a) and time-series data from ZTF DR16. The periods of RSGs in M33 are taken from Ren et al. (2019), while the K_0 magnitudes come from Ren et al. (2021b).

As can be seen from Figure 7, the PLRs of RSGs in the SMC, LMC, and M31 are consistent within the uncertainty. The slope of the PLR of RSGs in M33 is rather large.

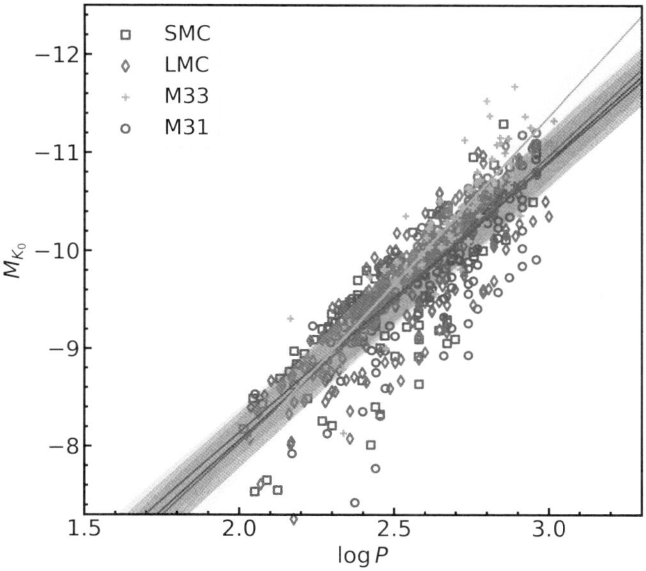

Figure 7. PLRs in K band for RSGs in SMC, LMC, M33, and M31, where only the fundamental mode period is adopted. The distance moduli of SMC, LMC, M33, and M31 are taken from Ren et al. (2021b). The K band magnitudes are extinction corrected.

However, the sample of RSGs in M33 is incomplete, the PLR of RSGs in M33 need to be recalibrated.

4.3. *Pulsation mode*

In the theoretical calculation of Stothers (1969), he assumed the fundamental mode of radial pulsation for red supergiants. Li & Gong (1994) made a comprehensive study on the pulsation of RSGs in the LMC with a linear nonadiabatic analysis. They found that the fundamental mode, and sometimes the first overtone mode, were unstable and that the observed PLR could be well understood by the fundamental-mode hypothesis. Guo & Li (2002) made further linear non-adiabatic calculation of RSGs at different metallicity and predicted the PLR for the fundamental, first-overtone and second-overtone modes.

Yang & Jiang (2011) and Yang & Jiang (2012) compared the PLR derived from the observational data with the theoretical calculation of Guo & Li (2002). They indicated that the majority of RSGs in LMC and SMC might pulsate in the first-overtone mode, while the Galactic RSGs mainly pulsate in the fundamental mode. On the other hand, Ren et al. (2019) identified the pulsation mode of RSGs in M31 and M33 to be mainly in the fundamental mode, also by comparing with the theoretical predictions of Guo & Li (2002). They further studied the ratio of fundamental-to-first overtone mode with respect to the metallicities for the SMC, the LMC, the Milky Way, M33, and M31. It is found that the ratio of RSGs pulsating in the fundamental mode to the first overtone mode increases with the metallicity. This is explained by the metallicity effect on convection. Convection becomes stronger with increasing metallicity, which suppresses the first-overtone mode in the shallower atmosphere than the fundamental mode. Such phenomena were later confirmed by Cepheids. Zhang et al. (2022) calculated the Cepheids ratio of FU/(FU+1O) to be 0.59, 0.60, 0.69, 0.83, and 0.85 for the SMC, the LMC, the Milky Way, M33, and M31, respectively, in order of metallicity, which confirms that the ratio of FU/(FU+1O) increases with metallicity. It should be mentioned that Soraisam et al. (2018) compared their observational results of RSGs in M31 with the MESA model and concluded that the bulk of the pulsating RSGs in their sample are consistent with the fundamental mode. Thus, in spite of different theoretical models and observational analysis, the same conclusion is drawn that the fundamental mode dominates in the relatively metal-rich galaxies.

It can be concluded that RSGs pulsate both in the fundamental and first-overtone mode, while the second-overtone mode has not yet been reported. Although the PLR in the near-infrared appears to be independent of metallicity, the pulsation mode indeed depends on metallicity.

4.4. *Irregular light variation of RSGs*

The period determination of RSGs has always been a troublesome thing. Although quite a few methods are developed to solve the periodicity of light variation, the uncertainty of period can still be a few tens of days. This is due to the complex irregular components in the light curves of RSGs.

The irregular variation actually presents in all three variable types of RSG. The widely accepted mechanism for irregular variation is the large convection cells in the photosphere, since RSGs have a convective envelope due to the high opacity in the low-temperature extended atmosphere. Studies suggested that the surface of RSGs should be dominated by a few large convective cells (Schwarzschild 1975), which was later confirmed by the 3D simulation (Freytag et al. 2002).

Ren & Jiang (2020) investigated the characteristic time scale and amplitude of irregular variation of RSGs in the SMC, the LMC and M31 based on the granulation model. With the timeseries data from the ASAS-SN and iPTF survey, it is found that the granulations in most of the RSGs evolve at a timescale of several days to a year with a characteristic amplitude of 10 to 1000 mmag. The comparison between the SMC, the LMC and M31 indicates that the timescale and amplitude of granulation seem to increase with metallicity.

The irregular variation and periodic variation are generally tangled together. The PLR could be more accurate if the irregular component can be removed or considered together.

In addition, the binary fraction of unevolved massive stars is thought to be 70%-100% (Sana et al. 2012), while studies show that the binary fraction of RSGs is about 20% (Neugent et al. 2020; Patrick et al. 2022). The RSG in a binary system would present another highly different light variation component. This surely makes the light curve more complicated.

4.5. *In the Galaxy*

The PLR of RSGs in our Galaxy suffers large uncertainty. The difficulty within the Galaxy is caused by the interstellar extinction as exhibited in the very first work by Stothers (1969). Because RSGs are mostly located in the Galactic plane, the very serious extinction leads to non-detection of RSGs at relatively large distance, and to uncertainty in the measurement of brightness and temperature. Thus the identification of RSGs in the Milky Way is difficult and uncertain. Except in a few nearby open clusters, only a small number of RSGs are identified. In a recent work by Messineo & Brown (2019), they collected 889 nearby candidate RSGs with reliable parallaxes from Gaia/DR2. With estimation of stellar luminosity and effective temperature, only 43 of them are thought to be highly-probably red supergiants with $M_{\rm bol} < -7.1$ mag. In comparison with the number of RSGs identified in M31 and M33 which are spiral galaxy like our own, this number is only about one percent of the expected. Gehrz (1989) also predicted $\sim$5,000 RSGs in our Galaxy based on the stellar distribution model.

Chatys et al. (2019) analyzed the $\log P - M_K$ relation of 48 RSGs in the Galaxy with the distances from Gaia/DR2. Still, the dispersion is large though the distances may be more accurate than previous measurements. They found that the $\log P - M_K$ relation agreed well with that in the LMC or M33 (Yang & Jiang 2011; Soraisam et al. 2018). A better determination of the RSG PLR in our galaxy can be achieved in a few aspects. A large and more complete sample will reduce the dispersion statistically. A more accurate calculation of interstellar extinction will reduce the error of luminosity. Observation in the near-infrared will detect more distant RSGs as well.

5. Summary

Red supergiants are the near-infrared brightest stars and potential cosmological distance indicators. Their period-luminosity relation has been studied in the Galactic associations and fields, the Magellanic Clouds, M31 and M33. With the large-scale long-term surveys, many more RSGs are identified and their periodicities are measured. It is suggested by various works that the period-M_K relation presents no obvious difference in these systems, which may be attributed to the small influence of metallicity on their infrared brightness and small extinction in the infrared. The dispersion may be further reduced with better determination of periods by taking the irregular variation and binary effect into account. On the other hand, the sample of RSGs in our Galaxy is far from complete, and the measurement of the period-luminosity relation needs more sources and more accurate measurement of distance and extinction.

This work is supported by the NSFC project 12133002, National Key R&D Program of China No. 2019YFA0405503, and CMS-CSST-2021-A09.

References

Bessell, M. S. & Brett, J. M. 1988, *PASP*, 100, 1134
Bouvier, J. 2013, EAS Publications Series, 143
Catchpole, R. M. & Feast, M. W. 1981, *MNRAS*, 197, 385
Chatys, Filip W., Bedding, Timothy R., Murphy, S. et al. 2019, *MNRAS*, 487, 4832
Collier Cameron, A. 1999, Solar and Stellar Activity: Similarities and Differences, 146
Donati, J.-F., Brown, S. F., Semel, M., et al. 1992, *A&A*, 265, 682
Ekström, S., Georgy, C., Eggenberger, P., et al. 2012, *A&A*, 537, A146.
Feast, M. W., Catchpole, R. M, Carter, B. S, et al. 1980, *MNRAS*, 193, 377
Freytag, B., Steffen, M. & Dorch, B. 2002, *AN*, 323, 213
Gehrz, R. 1989, *IAUS*, 135, 445
Glass, I. S. 1979, *MNRAS*, 186, 317
Guo, J. H. & Li, Y. 2002, *A&A*, 565, 559
Kinman, T. D., Mould, J. R. & Wood, P. R. 1987, *AJ*, 93, 833
Kiss, L. L., Szabó, Gy. M., Bedding, T. R. 2006, *MNRAS*, 372, 1721
Kochanek, C. S., Beacom, J. F., Kistler, M. D., et al. 2008, *ApJ*, 684, 1336.
Li, Y. & Gong, Z. G. 1994, *A&A*, 289, 449
Massey, P. 1998, *ApJ*, 501, 153.
Massey, Philip, Olsen, K. A. G., Hodge, Paul W. et al. 2006, *AJ*, 131, 2478
Messineo, M. & Brown, A. G. A. 2019, *AJ*, 158, 20
Neugent, Kathryn F., Levesque, Emily M., Massey, Philip et al. 2020, *ApJ*, 900, 118
Patrick, L. R., Thilker, D., Lennon, D. J.et al. 2022, *MNRAS*, 513, 5847
Pierce, M. J., Jurcevic, J. S. & Crabtree, D. 2000, *MNRAS*, 313, 271
Ramírez, Iván & Meléndez, Jorge 2005, *ApJ*, 626, 465
Ren, Yi & Jiang, Biwei 2020, *ApJ*, 898, 24
Ren, Yi, Jiang, Biwei, Yang, Ming et al. 2019, *ApJS*, 241, 35
Ren, Yi, Jiang, Biwei, Yang, Ming et al. 2021a, *ApJ*, 907, 18
Ren, Yi, Jiang, Biwei, Yang, Ming et al. 2021b, *ApJ*, 923, 232

Sana, H., de Mink, S. E., de Koter, A., et al. 2012, Science, 337, 444
Schwarzschild, M. 1975, *ApJ*, 195, 137
Soraisam, Monika D., Bildsten, Lars, Drout, Maria R. et al. 2018, *ApJ*, 859, 73
Stothers, Richard 1969, *ApJ*, 156, 541
Trumpler, Robert Julius 1930, *Lick Observatory Bulletin*, 420, 154
Yoon, S., & Cantiello, M. 2010, *ApJ*, 717, 62
Wilson, Ralph E. and Merrill, Paul W. 1942, *ApJ*, 95, 248
Wood, P. R., Bessell, M. S. & Fox, M. W. 1983, *ApJ*, 272, 99
Yang, Ming & Jiang, B. W. 2011, *ApJ*, 727, 53
Yang, Ming & Jiang, B. W. 2012, *ApJ*, 754, 35
Yang, Ming, Bonanos, Alceste Z., & Jiang, Biwei et al. 2019, *A&A*, 629, 91
Yang, Ming, Bonanos, Alceste Z., & Jiang, Biwei et al. 2021, *A&A*, 646, 141
Zhang, Zehao, Jiang, Biwei, Ren, Yi et al. 2022, *ApJ*, 928, 139
Bouvier, J. 2013, EAS Publications Series, 143
Collier Cameron, A. 1999, Solar and Stellar Activity: Similarities and Differences, 146
Donati, J.-F., Brown, S. F., Semel, M., et al. 1992, *A&A*, 265, 682
Shappee, B. J., Prieto, J. L., Grupe, D., et al. 2014, *ApJ*, 788, 48
Kochanek, C. S., Shappee, B. J., Stanek, K. Z., et al. 2017, *PASP*, 129, 104502
Jayasinghe, T., Stanek, K. Z., Kochanek, C. S., et al. 2020, *MNRAS*, 491, 13
Udalski, A., Szymański, M. K., & Szymański, G. 2015, *AcA*, 65, 1
Soszyński, I., Udalski, A., Szymański, M. K., et al. 2015, *AcA*, 65, 297
Soszyński, I., Dziembowski, W. A., Udalski, A., et al. 2007, *AcA*, 57, 201

Discussion

Question (Whitelock): I suspect that in your colour-magnitude diagram those objects that you described as thermally pulsing AGB stars are actually hot bottom burning AGB stars. That's not directly relevant to the red supergiants, but I think that's what is going on.

Answer: Maybe yes. We here only plotted the checks from the models. They are consistent with TP-AGB branch of stellar models. So that's why we assigned them TP-AGB stars.

Question (Whitelock): Can you say anything about very long secondary periods in supergiant stars?

Answer: That's a difficult question, there are many models about the long secondary period. I don't think any of these explanations is perfect, like binary or spot or whatever. One of our collaborators [Yi Ren] is very interested in this topic. We will see if we will have any new ideas on this.

Question (Baruch): On an early slide you showed some distances; you tried to generate distances and compared with Gaia. Do you get anything by giving a standard error from the Gaia measurements? Does that help? The SHOES group worked out those 24 microarcseconds, and if they took that away from the Gaia measurements they got really good answers. Can you do the same with your supergiants?

Answer: I have no idea actually now how to compare with this one. SHOES is for much distant galaxies while we are working on the Local Group Galaxies. Well, we derived the distance modulus for these galaxies, but based on the tip of the RGB population, so it has nothing to do with red supergiants.

Question (Kervella): I just wanted to ask if you tried (or maybe you didn't) the subtraction of the infrared excess that may be visible in the spectral energy distribution, to construct a dust-free period-luminosity relation with circumstellar emission removed?

Answer: This is a difficult problem, because the interstellar extinction and circumstellar extinction are actually mixed together. We do have a work to derive the mass-loss rate from the SED of red supergiants (Wang et al, 2021ApJ...912..112W). And then, if we increase the interstellar extinction, the mass-loss rate becomes smaller for sure, because the circumstellar extinction becomes smaller. But we did not decompose these two components. We just have to assume that the interstellar extinction derived in other ways independently are correct.

At the crossroads of astrophysics and cosmology: Period–luminosity relations in the 2020s
Proceedings IAU Symposium No. 376, 2024
R. de Grijs, P. A. Whitelock & M. Catelan, eds.
doi:10.1017/S174392132300265X

Long-Period Variables as distance and age indicators in the era of *Gaia* and LSST

Michele Trabucchi[1,2]

[1]Dipartimento di Fisica e Astronomia Galileo Galilei, Università degli studi di Padova, Vicolo dell'Osservatorio 3, I-35122 Padova, Italy. email: michele.trabucchi@unipd.it

[2]Department of Astronomy, University of Geneva, Ch. Pegasi 51, 1920 Versoix, Switzerland

Abstract. Long-period variables are bright, evolved red giant stars showing periodic photometric changes due to stellar pulsation. They follow one or more period-luminosity and period-age relations, which make them highly promising distance indicators and tracers of young and intermediate-age stellar populations. Such a potential is especially interesting in view of the massive amount of data delivered by modern large-scale variability surveys. Crucially, these applications require a clear theoretical understanding of pulsation physics in connection with stellar evolution. Here, I describe an ongoing effort from our collaboration dedicated to the modelling of stellar pulsation in evolved stars, and how this work is impacting our capability of investigating long-period variables and exploiting them for other astrophysical studies. Furthermore, I present our ongoing work aimed at assessing the potential of semi-regular variables, an often neglected sub-type of long-period variables, to be distance indicators complementary to their better-known, more evolved counterparts, the Mira variables.

Keywords. stars: AGB and post-AGB stars - stars: oscillations - stars: variables: general - stars: distances

1. Introduction

Low- and intermediate-mass stars ($0.8 \lesssim M/M_{\odot} \lesssim 8$) approach the end of their lives through the Asymptotic Giant Branch (AGB) evolutionary phase, during which their envelope becomes unstable to stellar pulsation in low-order oscillation modes. This process is observed in the form of periodic stellar variability characterized by photometric amplitudes up to several magnitudes in visual bands and by periods of order of months up to a few years, hence the name Long-Period Variables (LPVs). This broad class of variable stars does not have a strict definition, and often extends to more massive or less evolved objects, such as core-He-burning red supergiants (RSGs) or stars approaching the tip of the red giant branch (RGB).

Whether or not an object is considered an LPV often depends on the purpose of an observing program, or on its instrumental capabilities. In particular, LPV catalogs from large surveys often include all giant stars with red colors that display photometric changes above the instrumental variability detection threshold. In the 1990s, this kind of surveys have revolutionized our understanding of LPVs (see, e.g., the review contribution by Igor Soszyski in these proceedings), and have revealed a striking degree of complexity in the behaviour of these stars that display multi-periodic variability caused both by pulsation in multiple distinct modes (possibly simultaneously) and by other processes, typically associated with the presence of a binary companion (see Michał Pawlak's contribution in these proceedings).

As many other pulsating stars, LPVs follow period-luminosity relations (see Patryk Iwanek's contribution in these proceedings) that make them excellent candidates to be distance indicators (see Miora Andriantsaralaza's contribution in these proceedings). The Mira variables are the sub-type of LPVs that has enjoyed the most attention in this sense: they are intrinsically very bright, especially at infrared (IR) wavelengths, and show rather regular variations with large photometric amplitudes, so that they are easily observable (even at relatively large distances) and identifiable (see Caroline Huang's contribution in these proceedings).

They also obey a long-known period-age relation by which they can be exploited as stellar chronometers and tracers of young stellar populations. Yet, these applications can in principle be extended to the smaller-amplitude progenitors of Miras, a sub-type of LPVs known as Semi-Regular Variables (SRVs). Often neglected because of their less-regular pulsation behaviour, which is challenging both on observational and theoretical grounds, they carry a wealth of additional physical information that is well within reach of modern time-domain astrophysical surveys, and is ready to be exploited.

More generally, the variability features (in particular the period) of LPVs are tightly connected with their physical properties, and provide a means for probing their structure and inferring their global stellar parameters. Therefore they are valuable tools for advancing our knowledge of the evolution of stars and stellar populations. Understanding the complexity of LPVs and characterizing the connection between their variability and their underlying properties is among the modern challenges for stellar modelling research, which has called for an effort to provide the solid theoretical grounds necessary for supporting the full exploitation of LPVs as tools for the exploration of our cosmic neighbourhood.

2. Stellar pulsation models of LPVs: connecting theory with observations

Stellar pulsation carries valuable information for constraining stellar structure and evolution models, provided that detailed knowledge is available about how the stellar structure determines the pulsation periods (as well as other variability-related observables) and how the pulsation behaviour is affected by evolutionary changes. For this reason, the modelling of stellar pulsation has long been an active field of research.

With a few exceptions (e.g. Xiong et al. 2018) the theoretical investigation of pulsation in AGB stars has been rather dormant in the past few decades, whereas more recently it has seen an enormous progress owing to the development of bleeding-edge three-dimensional radiation-hydrodynamic codes that provide the most realistic picture of convection and its interaction with pulsation (Ahmad et al. 2023; see also the Arief Ahmad's contribution in these proceedings). An inherent shortcoming of these models stems from their extremely time-consuming nature, that makes them currently unfeasible for the exploration of a vast space of stellar parameters necessary for characterizing the AGB phase.

Instead, such an exploration is conveniently performed by adopting linear stellar pulsation codes, that sacrifice a detailed description of physical processes such as the pulsation-convection interaction in favor of computational efficiency. They can be used to construct grids of models to establish a connection between global stellar parameters and pulsation observables, which in turn can be used to link evolutionary calculations with variability in interpolation in the grid nodes. As an alternative, the relation between stellar structure and variability derived from such grids can be is conveniently encapsulated into easy-to-use analytic prescriptions, typically in the form of period-mass-radius relations (as the period is the main variability-related observable and mass and radius are the parameters affecting it the most).

The incomplete coverage of the space of stellar parameters, especially in terms of chemistry, has been the main shortcoming of the various AGB pulsation prescriptions available in literature since the widespread appearance of pulsation codes. For instance, Marigo & Girardi (2007) included variability information into AGB stellar evolutionary tracks by combining several distinct prescriptions, and resorting to a number of approximations to account for a space parameter coverage that was limited in mass, poor in metallicity, and null in terms of the carbon-to-oxygen number abundance ratio (C/O), the very parameter that characterizes the dramatic chemical evolution of these stars.

2.1. *A new grid of LPV pulsation models*

Motivated by the need of incorporating a description of long-period variability into evolutionary models of the Thermally-Pulsing AGB phase, with the aim of attaining a sound physical calibration of the latter (Marigo et al. 2017; Pastorelli et al. 2019, 2020), our group devoted an effort to the production of a new grid of AGB pulsation models with a wide coverage of the space of stellar parameters, including chemistry (Trabucchi et al. 2019).

This was achieved with the linear, radial, non-adiabatic, one-dimensional pulsation code described in Wood & Olivier (2014, and references therein), upgraded to include up-to-date low-temperature gas opacities, with the inclusion of molecular contributions. Special care was paid to ensure the full consistency between the metal mixture adopted for the chemically-homogeneous envelope and that employed to compute the opacity tables (in contrast with a number of previous studies that simply assumed a scaled-solar metal mixture).

The final grid densely covers the mass range $0.6 \leq M/\mathrm{M}_{\odot} \leq 7.0$, the luminosity range $2.5 \leq \log(L/\mathrm{L}_{\odot}) \leq 5.0$, the range of initial metallicity (i.e. at the beginning of the AGB) $0.001 \leq Z_{\mathrm{init.}} \leq 0.017$, C/O between 0.3 (suitable for Hot-Bottom-Burning stars) to 5.0 (for C-rich stars), and various hydrogen abundances ($0.6 \leq X \leq 0.8$). Moreover, each model was computed for two different values of the core mass and three distinct values of the mixing length parameter (thereby offering appropriate coverage in terms of effective temperature). For each combination of such parameters the grid provides the period and growth rate (i.e. the degree of excitation) for five distinct pulsation modes, from the fundamental mode (FM) to the fourth overtone mode (4OM).

As a benchmark for checking the inter-operability between evolutionary tracks and pulsation models, as well as for testing the accuracy of the latter, we combined the pulsation grid with a synthetic model of the population of evolved stars in the Large Magellanic Cloud (LMC), obtaining a simulation of the LPV content of the LMC (Trabucchi et al. 2017). A direct comparison between the near-IR period-luminosity diagram (PLD, see Fig. 1) of the LPVs observed in the LMC by the OGLE-III program (Soszyński et al. 2009) and its simulated counterpart confirmed the accuracy of the models, and led to a novel interpretation of the origin of the period-luminosity relations of LPVs. Indeed, the five† period-luminosity sequences appearing in the observed PLD have been traditionally attributed to five distinct pulsation modes, although there has been disagreement on the exact modal assignment. Surprisingly, our simulated PLD displayed only four such sequences, with one of them (corresponding to 1OM periods) seamlessly covering the area occupied by the observed sequences B and C′ combined.

† We consider here the main period-luminosity sequences, labelled A, B, C, C′, and D according to the nomenclature derived from Wood (2000). However, it is worth recalling that the PLD has a rich fine structure, that various sequence split into sub-ridges, and that additional sequence appear depending on the adopted photometric system (see Wood 2015, for a summary).

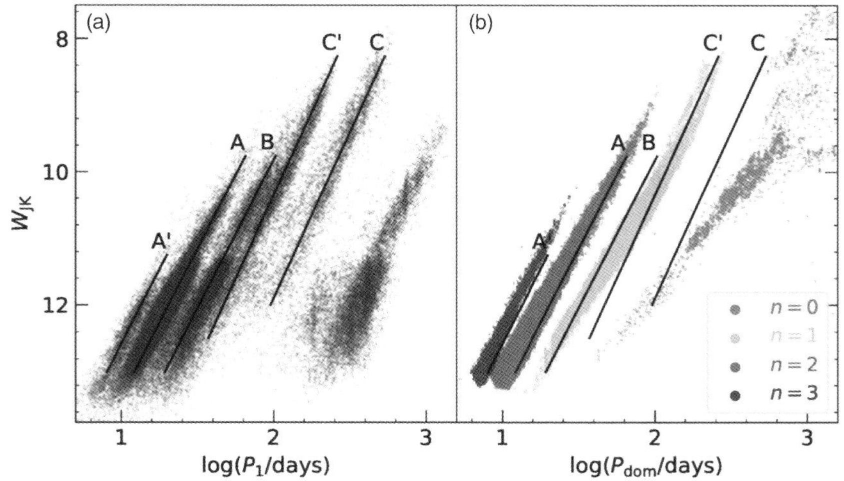

Figure 1. Observed (left, using OGLE-III periods and 2MASS NIR magnitudes) and simulated (right) period-luminosity diagram of LPVs in the LMC. Lines and labels indicate the approximate location of the period-luminosity sequences (using the nomenclature of Wood (2015)). Theoretical dominant periods (color-coded by pulsation modes) are obtained by combining a synthetic stellar population model with linear pulsation models of red giant envelopes. Adapted from Trabucchi et al. (2017).

The reason why the observed distribution of 1OM periods appears split into two sequences is caused by a selection bias linked with the way the PLD is customarily plotted. Indeed, only one period per star is shown in the PLD even in the case of multi-periodic variables such as the LPVs. This approach relies on the reasonable assumption that the 'primary' period of each star (the one with the largest photometric amplitude in the observed light curve, or the strongest signature in the periodogram) is representative of the overall pulsation behaviour of that star. In turn, this relies on the implicit assumption that all observed periods are caused by pulsation. This is actually not the case for many LPV catalogs, among which is OGLE-III, that include Long Secondary Periods (LSPs). This kind of variability is caused by a cloud of dust, dragged by a sub-stellar companion, that periodically eclipses the red giant (Soszyński et al. 2021), and can lead to large enough photometric variations to be identified as primary period. When this happens, the LSPs are drawn on the PLD leading to the appearance of the period-luminosity sequence C, whereas the corresponding pulsation periods are effectively ignored. Hence, this has the subtle side effect of systematically removing information from the PLD. More precisely, pulsation periods are selectively removed from the area of the PLD in the middle of the 1OM period distribution, creating an artificial gap along it and giving rise to the two distinct period-luminosity relations B and C′.

This explanation brings into alignment the two main interpretations, in disagreement with each other, that have emerged since the discovery of the multiple period-luminosity relations of LPVs, and has been confirmed by asteroseismic studies (Yu et al. 2020). Moreover, it opes the way for a better understanding of the nature of LSP-like variability and the relation between the period-luminosity relations and mass-loss on the AGB (McDonald & Trabucchi 2019).

These results are instructive in two ways. On the one hand, they clearly show the degree of complexity of the PLD of LPVs, resulting from their multiperiodic variability caused by the concurrence of multi-mode pulsation and binary-induced variations. On the other hand, it demonstrate the power of the approach based on the combination of

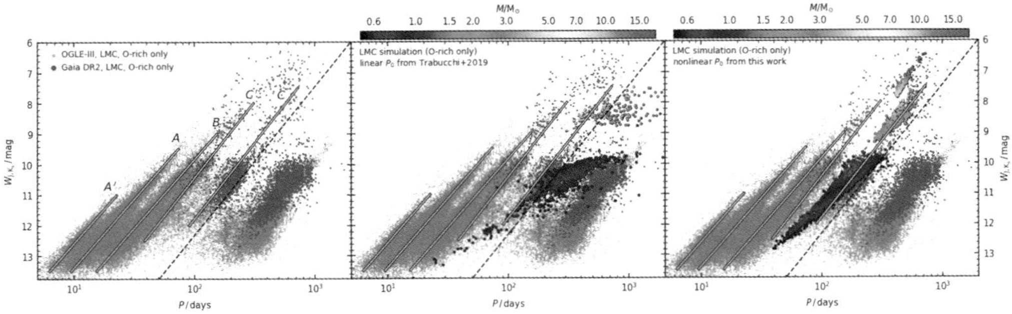

Figure 2. Similar to Fig. 2, but limited to O-rich stars, and comparing observations from OGLE-III and *Gaia* (left-most panel and gray dots in all panels) with inadequate FM periods predicted from linear pulsation models (central panel), and accurate FM periods from 1D hydrodynamic simulations (right-most panel). Simulated data points are color coded by the current stellar mass. Adapted from Trabucchi et al. (2021b).

evolutionary and/or synthetic population models with a suitably wide grid of pulsation models in order to improve our understanding of the period-luminosity relations of LPVs.

2.2. *The fundamental mode and the non-linear regime*

Another result emerging from the study by Trabucchi et al. (2017) is the clear indication that linear pulsation models are not suitable for describing LPVs pulsating predominantly in the FM, such as Mira variables. Indeed, the linear approximation breaks down for these large-amplitude pulsators, a more accurate description of which requires non-linear hydrodynamic calculations. To address this issue, we used the hydrodynamic code described in Keller & Wood (2006) (based on the work by Wood 1974) after applying the same opacity-related upgrades as the linear pulsation code, and performed the first systematic theoretical analysis of the behaviour of LPVs in the non-linear regime of pulsation (Trabucchi et al. 2021b).

The results from hydrodynamic calculations show that the period does not increase indefinitely with radius, as predicted by linear models. Instead, the sensitivity of the period to changes in radius decreases as the star expands, until eventually the period becomes independent of the radius. This confirmed previous indications (Ya'Ari & Tuchman 1996; Lebzelter & Wood 2005; Kamath et al. 2010) that large-amplitude pulsation affects the structure of the pulsating layers of the envelopes, causing them to readjust to a condition of hydrodynamic equilibrium in which their (time-averaged) density is slightly larger than that of an identical model in hydrostatic equilibrium (i.e., static or pulsating with a small amplitude). In turn, the local increase in density causes the pulsation period to decrease (as can be understood by a naive application of the period-mean density relation), hence demonstrating the strongly non-linear nature of FM pulsation in LPVs.

It is worth pointing out that the non-linear period-mass-radius relation provided in Trabucchi et al. (2021b) is significantly more accurate than any such prescription based on linear pulsation models (see, e.g., Trabucchi et al. 2020), as demonstrated by the improved agreement with observational data (see Fig. 2). This can be roughly quantified by considering one of the possible applications of such relations, that is the estimate of the stellar mass based on a knowledge of the FM period and of the stellar radius. By comparing Eq. 1 of Trabucchi et al. (2021b) with Eq. 12 of Marigo & Girardi (2007) one finds that the latter would cause the mass to be underestimated by 50-to-100% in the case of a massive AGB star, whereas it would be overestimated by a factor 2-to-4 in the case of a short-period (~ 100 days), relatively low-mass AGB star, and by twice that much for a similar star with a long period (~ 500 days).

As the hydrodynamic code is computationally more demanding than its linear counterpart, the non-linear grid was limited to a single chemical composition ($Z = 0.006$, O-rich). An extension to cover at least the same nodes of the linear grid is in progress as part of a work preparatory for the exploitation of LPV data from the LSST survey of the upcoming Vera Rubin Observatory (see Dal Tio et al. 2022). The results of the grid are planned to be publicly available and to include several products beyond the pulsation periods, such as template light curves and physical parameters (surface luminosity, radius and temperature) as a function of the phase throughout the pulsation cycle.

3. The period-age relation of LPVs

LPVs have long been known to obey a period-age relation. Between the 1920s and the 1940s it was realized that Galactic Miras with relatively short periods display hotter kinematics compared with Miras with longer periods, and therefore must belong to comparatively older stellar populations (Merrill 1923; Wilson & Merrill 1942; Feast 1963). A crude but convincing theoretical explanation of this anti-correlation between period and stellar age stars from the period-luminosity relation. As longer-period Miras are intrinsically brighter than their shorter-period counterparts, they must be more massive, and hence they must be younger (at least on the average). Whereas in principle valid for all pulsation modes usually active in LPVs, the period-age relation is most useful in the case of the FM. Clearly, it does not only concern the Mira variables, but all LPVs pulsating in the FM, including many SRVs.

Throughout the decades, there have been multiple spikes of interest for the promising applications of the period-age relation, including the recent appearance of several studies focused on mapping the central regions of the Milky Way galaxy, a task for which Mira variables are especially suited (see e.g. the contribution by Megan Lewis in these proceedings). In contrast, there have been comparatively few theoretical studies of the period-age relation (e.g., Wyatt & Cahn 1983; Feast & Whitelock 1987; Eggen 1998). Undoubtedly, the lack of an accurate way of computing FM periods as a function of global stellar parameters must have contributed to hampering such studies. This situation has changed owing to the non-linear prescription derived by Trabucchi et al. 2021a, that we adopted in the attempt of deriving a model-based period-age relation for FM-pulsating LPVs (Trabucchi & Mowlavi 2022).

To do so, we combined the results from hydrodynamic pulsation models with state-of-the-art stellar isochrones in order to simulate the distribution of these age-tracing stars in the period-age plane. In order to compare the result with observations, we took inspiration from the approach of Grady et al. (2019), who calibrated the period-age relation on Miras whose age is known from their membership in star clusters. In addition to the clusters used by Grady et al. (2019) we also adopted a few Magellanic Clouds (MCs) clusters containing a large number of LPVs with measured periods (both Miras and SRVs), and carefully identified the most-likely pulsation modes associated with each period (retaining only FM modes) and the chemical type of each LPV.

Models are in good agreement with observations (Fig. 3), both in terms of the best-fit period-age relation (fully compatible with one derived by Grady et al. 2019) and of the spread of the relation, which turned out to be very large (for instance there is a $\sim$3 Gyr range for the age of a ~ 350 days-period Mira). The spread in period at a given age (and vice versa) originates from the fact that the AGB phase, despite being almost instantaneous compared with previous evolutionary stages, is characterized by important structural changes, most importantly a substantial expansion that leads to large changes in periods. We note that the expansion of the envelope experiences an acceleration during

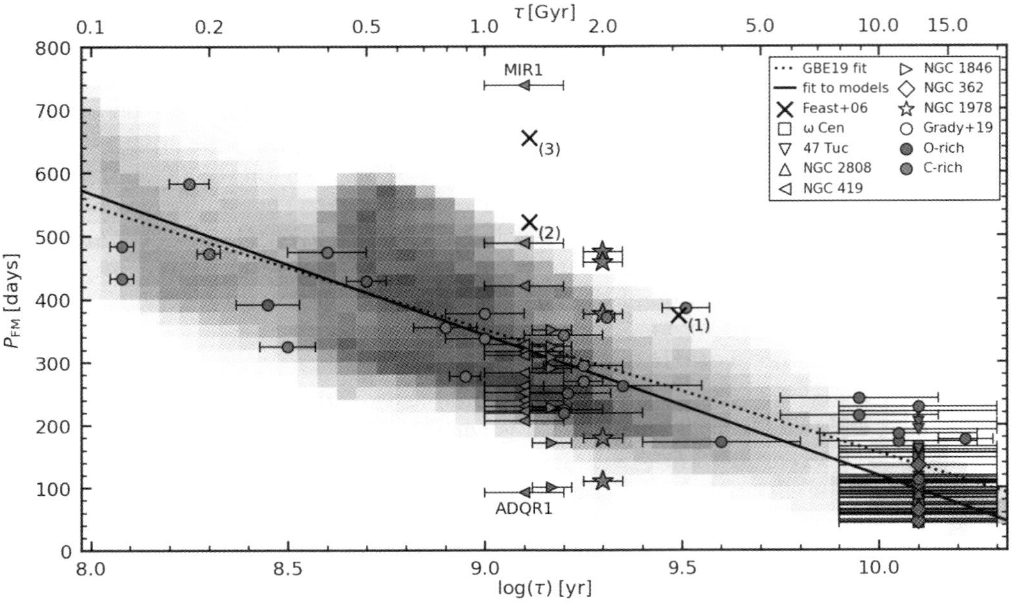

Figure 3. Theoretical FM period-age relation (density map in the background and solid best-fit line) compared with observed periods of LPVs whose age is known from their membership in clusters. Stars belonging to different clusters are drawn with different symbol shapes, whereas the colors indicate different chemical type (blue for O-rich, red for C-rich). The observational best-fit by Grady et al. (2019) is represented by the dashed line. Adapted from Trabucchi & Mowlavi (2022).

the TP-AGB, whereas the period becomes progressively less sensitive to changes in radius as the latter increases. As a result, the theoretical period distribution at a given age is asymmetric, with a peak near its short-period end, in agreement with observations.

Such a spread in the period-age relation is far from ideal for using LPVs as age tracers unless properly characterized. In this respect, we note that advocates of this application traditionally consider only Miras as age indicators. Indeed, the Mira stage is so short compared to the duration of the AGB that it can effectively be treated as its endpoint, thereby selecting a much narrower range of periods at any given age. However, this approach has the downside of excluding a substantial number of other potential age tracers, namely the SRVs that pulsate in the FM and that are often very similar to Miras (see Sect. 4). Since the distinction between the two LPV sub-types is based only on their photometric amplitude, it is hard to apply to LPV models. Indeed, it requires them to be combined with a detailed (time-consuming) treatment of dust formation and radiation-hydrodynamics in the stellar atmosphere (e.g., Bladh et al. 2019), whereas only an approximated description is currently used in the model.

It is also important to keep in mind that Trabucchi & Mowlavi (2022) examined the period-age relation using pulsation results that, strictly speaking, only apply to O-rich chemistry and LMC-like metallicity, while there is some indication (Trabucchi et al. 2019, 2021b) that the chemical composition affects the instability of pulsation, i.e. the instability strip of the various modes, and thus the edges of the period-luminosity and period-age relations. The photometric amplitude of variability is also expected to be metallicity-dependent (especially at optical wavelengths) as it is determined by molecular absorption bands (Reid & Goldston 2002) whose formation is favoured at higher metallicity. This poorly understood dependence might introduce a bias in the way Miras are distinguished from SRVs.

Our incomplete understanding of how the period-luminosity and period-age relations are affected by metallicity, and more in general by specific properties of any given stellar population (such as its star-formation history) is to be kept in mind when LPVs are adopted to map the Galaxy and its neighbourhood. An instructive example of this is offered by the results of Sanders et al. (2022), who used observations of Miras to characterize the properties of the Milky Way's nuclear stellar disc. The age distribution they derive is consistent with a number of previous studies but is not in agreement with the model-based period-age relation or with the relation derived by Grady et al. (2019) from star clusters. Hopefully the extension of our grid of non-linear models to cover various chemical composition will help shedding light on these discrepancies, and enhance our capabilities of using LPVs as age indicators.

4. SRVs as distance indicators

The large-scale optical microlensing surveys initiated at the end of last century (such as MACHO, EROS, MOA, OGLE) have been exceptionally beneficial to the study of variable stars. Among these surveys, the Optical Gravitational Lensing Experiment Udalski et al. (1992) has been especially valuable to understand LPVs. In particular, owing to its high cadence and photometric sensitivity, it allowed for the detailed characterization of the variability and the period-luminosity relations of SRVs and other low-amplitude LPVs. This is clearly illustrated by how OGLE enabled the identification of the OGLE Small-Amplitude Red Giants (OSARGs, Wray et al. 2004), a sub-type of LPVs displaying a rich pattern of radial and non-radial oscillations, possibly bridging the gap between LPVs and solar-like oscillators on the RGB, that would otherwise be classified as irregular variables or even constant stars.

Current and upcoming time-domain astronomical surveys are expected to be comparable to OGLE for characterizing small-amplitude LPVs, whereas they can improve on OGLE in terms of sky coverage (e.g. *Gaia*) or photometric depth (e.g. LSST). It is therefore interesting to assess whether SRVs† have the potential of being distance indicators complementary to Miras. This investigation is motivated by a number of arguments:

- they obey the same PL relation of Miras and/or an additional relation,
- they are longer-lived than Miras, compared to which they are more numerous in any given stellar population (by roughly a factor 10 in the MCs) and trace older stellar populations (thereby representing a potential brighter alternative to RR Lyr variables),
- they are less dusty than Miras, and thus less biased by dust absorption (in the optical) and emission (in the IR),
- their photometric amplitudes are not too small to go undetected at optical wavelengths (where variability is observed), but small enough in the IR (where the PL relations are characterized) to reduce the scatter of the PL relation when it has to rely on single-epoch observations.

It is also instructive to note that, while SRVs are typically fainter than classical Cepheids in blue and visual filters, their optical brightness is comparable in the optical, and SRVs are brighter in the IR (Fig. 4). Moreover, SRVs can be more numerous than Cepheids (e.g. in the LMC). What makes SRVs less appealing than Miras as distance indicators, besides their smaller amplitude (that is becoming less of a problems owing to the improvements in observational instrumentation and data reduction pipelines) is their lesser degree of regularity associated with their multi-periodic nature. While this makes them more challenging to deal with, it should be noted that any period beyond the first carries additional physical information that is worth exploiting.

† We adhere to the classification from the OGLE-III catalog, according to which the SRVs can approximately be identified to the non-Mira LPVs whose primary period lie on sequences C' or C.

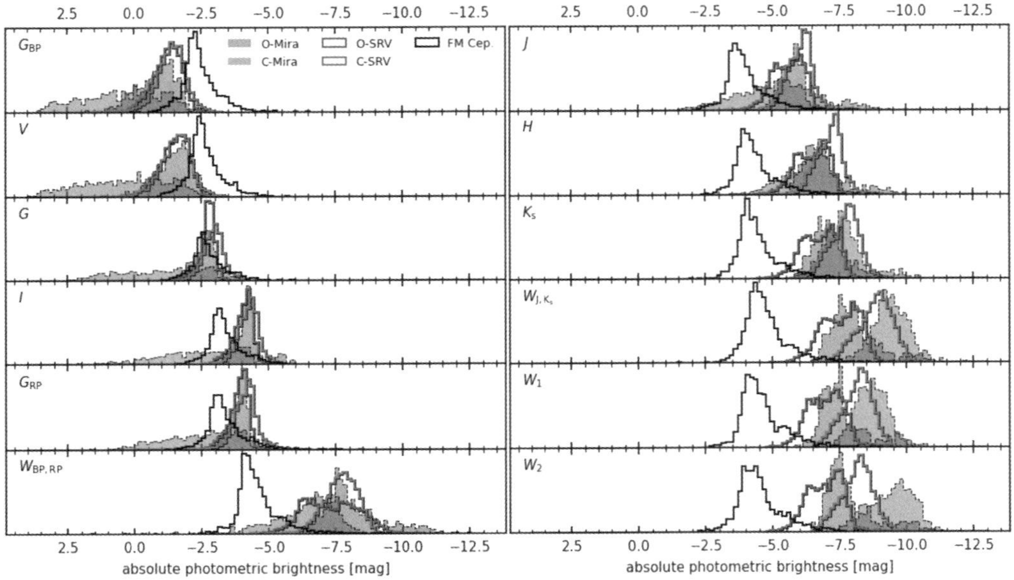

Figure 4. Luminosity functions of LPVs and classical Cepheids (from OGLE-III and OGLE-IV, respectively) in different bands from blue to the mid-IR. A distinction is made between O-rich (blue) and C-rich (red) LPVs, as well as between Miras (filled histograms) and SRVs, while the solid black curves correspond to Cepheids. Each panel correspond to a distinct photometric filter: G, $G_{\rm BP}$ and $G_{\rm RP}$ from *Gaia*, V and I from OGLE, J, H and $K_{\rm s}$ from 2MASS, W_1 and W_2 from AllWISE. The *Gaia* and 2MASS Wesenheit indices, $W_{\rm BP,RP}$ and W_{J,K_s} are defined as in Lebzelter et al. (2018). Panels are sorted approximately by increasing central wavelength of the filters.

In view of such considerations, we undertook a series of exploratory studies aimed at assessing the potential of SRVs, starting with a characterization of these stars in relation with Miras (Trabucchi et al. 2021a) based on OGLE-III observations of LPVs in the MCs. This analysis resulted in the identification of four different sub-groups of SRVs, depending on whether they are mono- or bi-periodic, and whether they pulsate predominantly in the FM or 1OM. The mono-periodic FM-pulsating SRVs are especially similar to Miras, except they display a smaller photometric amplitude, confirming previous findings (e.g., Lebzelter & Hinkle 2002, and references therein). In particular, a sub-set of these stars follow the same sequences as Miras in the PLD as well as in the period-amplitude diagram, without discontinuity. This suggests that SRVs pulsating only in the FM should be treated as the same type of object from a physical perspective, and possibly for the purpose of distance measurements.

A direct benefit of adopting this criterion when selecting distance indicators among LPVs is that the pool from which they are drawn is roughly twice as large. Clearly, this is a rough estimate that holds in a statistical sense and, strictly speaking, only for a population of LPVs very similar to that of the MCs, and for observations obtained with an instrumental setup very close to that of the OGLE program. Nonetheless, there is a net advantage in terms of number of sources that can be observed, and it can be enhanced if the criterion is made more inclusive by accepting also SRVs with smaller amplitudes, bi-periodic, or pulsating in the 1OM.

However, the expansion of the set of LPVs statistically available to be used as distance indicators is not necessarily beneficial in terms of the performance of the PLR as a standard candle. For the purpose of our study, that is a comparative analysis of a large number of PLRs defined in different ways, we quantify their performance in terms of their capacity of correctly predicting the brightness of the sample of LPVs from which they were derived. In other words, we define a sample of LPVs (for instance, a reference set including only Miras, or a set consisting of Miras and a certain group of SRVs, and so on), derive the best-fit PLR of the sample, and compute the brightness residuals, i.e. the differences between the true brightness and the one predicted by the best-fit PLR. We then characterize the distribution of the residuals in terms of its width (that is related to the precision of the PLR) and its shape (i.e. how symmetric and well-centered it is, that is related to the accuracy of the PLR).

In general, the performance of the PLR becomes degraded if the sample is blindly extended to include all SRVs of any given group. Indeed, the SRVs that pulsate in the FM suffer from a larger scatter than Miras around the PLR. However, a more selective addition of sources can is indeed beneficial. For instance, SRVs can be added to the sample by using a lower amplitude threshold than that traditionally used in the definition of Miras. There exists an optimal level of the threshold that maximizes both the precision and accuracy of the PLR. The optimal threshold depends on several factors, such as the photometric band the PLR is observed in, the SRV groups included in the sample, the chemical type of the adopted LPVs, and the pulsation mode (FM or 1OM). In the case of some photometric filters, this approach only leads to marginal improvements (or no improvement at all), and does not justify the use of the extended PLR, whereas the improvement is evident, for instance, in the NIR and in some MIR bands.

This analysis shows that the traditional separation between Miras and SRVs, based purely on amplitude, is not necessarily physically meaningful, nor it is optimal in terms of the performance of the PLR as a distance indicator. Instead, for the purpose of distance (and possibly age) determinations, one could consider a more fluid, efficiency-driven selection criterion. It is worth recalling that the amplitude-based definition is also prone to a number of biases. A relevant example, discussed by Lebzelter et al. (2022), concerns a comparison between sources in the *Gaia* DR3 catalog of LPV candidates and their best matches in the ASAS-SN catalog (Jayasinghe et al. 2021). There exists a discrepancy between the photometric amplitude published in the two catalogs, with the largest differences localized near the Galactic Bulge and plane. In this area, many sources with a relatively small V-band amplitude according to ASAS-SN are shown to have a large G-band amplitude by *Gaia*. The reason behind these differences is that, owing to its higher spatial resolution, *Gaia* light curves are less affected by the compression effect caused by crowding (e.g. Riess et al. 2020). As a result of this subtle effect, a number of Miras are incorrectly classified as SRVs in the ASAS-SN catalog.

At the same time, photometric amplitudes (as well as periods) are effectively independent of distance and extinction, and carry valuable information provided that effects of crowding are properly accounted for. One way of doing so is based on the observation that the amplitude of variability increases with period and luminosity, and in particular it increases *across* the PLRs (e.g. figure 1 of Soszyński et al. 2013). Each PL sequence can thus be treated as a family of lines having slope and zero-point that vary with amplitude, thereby spanning the sequence's entire width (Fig. 5). In other words, the coefficients of the PLR can be thought as amplitude-dependent. This perspective turns the amplitude into a useful second parameter of the PLR, in a fashion that is similar to the introduction of a color term. A similar operation can be done by using the secondary period of bi-periodic SRVs (or the ratio of their period) as the second parameter.

By deriving the best-fit PLR in different amplitude bins for a given sample of LPVs, and then fitting the dependence of the PLR coefficients on the amplitude, we derived various expressions for the period-luminosity-amplitude relation (PALR) whose performance we can characterization of the PLR performance with the same framework described above. The results suggest that this approach effectively accounts for the scatter of the SRVs

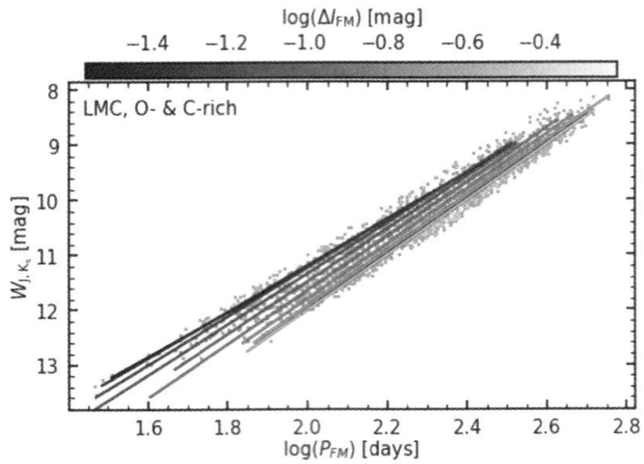

Figure 5. Example best-fit analytic NIR period-luminosity relations for a sample of SRVs in the LMC in different amplitude bins. Each line represent a different relation, and is color-coded by the mean FM amplitude $\Delta I_{\rm FM}$ in the OGLE I band of the corresponding bin. The relations, with varying slope and zero-point, span the full width of the observed period-luminosity sequence.

around the PLR, canceling (at least partially) its negative effect on the performance of the PLR. Compared to the PLR in its traditional form, the PALR displays a distribution of residuals that is both narrower and better-centered, indicating a higher degree of precision and accuracy. The performance is comparable to that obtained when using a PLR limited to Miras or to LPVs with amplitude above an optimized threshold, but can be defined to include a substantially larger set of sources.

5. Conclusions

Long-period variables are powerful astrophysical tools, whose applications range from inferring stellar parameters from variability observations, constraining stellar evolution models, tracing stellar populations and estimating astronomical distances. In order to fully exploit the potential of these stars, a solid theoretical understanding is needed, as well as pulsation models with accurate predictive power. While promising, highly-realistic 3D hydrodynamic simulations of evolved red giant envelopes are being successfully developed, the grid-based approach involving flexible 1D models has proven ideal to widely explore the space of stellar parameters and establish a sound connection between the pulsation of LPVs, their structure and their evolution. In particular, it allowed for a novel interpretation of their period-luminosity relations, showing a higher degree of complexity than expected from previous studies, and providing new insight on the link between pulsation, mass-loss and long secondary periods (whose nature is still debated).

Research in this direction also showed the inadequacy of widespread prescriptions, based on linear pulsation models, adopted to compute fundamental-mode periods of LPVs from their global stellar parameters. A relation based on the result of 1D hydrodynamic pulsation model is able to overcome these shortcomings, accurately describing the period of large-amplitude Mira variables. This new relation prompted an exploratory, theory-based study of the period-age relation of these stars, that is in agreement with observations of LPVs in clusters, but at odds with evidence from Bulge stars, pointing towards the need of more detailed studies of the connection between pulsation and chemical composition.

Finally, the improvement in the capacity of modelling the pulsation of bright red giants opens new avenues for using semi-regular variables, alongside Miras, as potential distance indicators. SRVs have long been neglected for this application, despite following the same period-luminosity relation of Miras (and other ones) because of their smaller amplitude and lesser regularity. Yet, they have a number of properties that motivate assessing their potential as standard candles, such as their being more numerous than Miras and present in older stellar populations. Preliminary results indicate that some SRVs that are especially similar to Miras can help improve their performance for distance determination, and that similar results can be obtained by including information on photometric amplitude of variability as a second parameter in the period-luminosity relation.

References

Ahmad, A., Freytag, B., & Höfner, S. 2023, Astronomy & Astrophysics, 669, A49

Bladh, S., Liljegren, S., Höfner, S., Aringer, B., & Marigo, P. 2019, Astronomy & Astrophysics, 626, A100

Dal Tio, P., Pastorelli, G., Mazzi, A., et al. 2022, Astrophysical Journal, Supplements, 262, 22

Eggen, O. J. 1998, Astronomical Journal, 115, 2435

Feast, M. W. 1963, Monthly Notices of the Royal Astronomical Society, 125, 367

Feast, M. W. & Whitelock, P. A. 1987, in Late Stages of Stellar Evolution, ed. S. Kwok & S. R. Pottasch, 33

Grady, J., Belokurov, V., & Evans, N. W. 2019, Monthly Notices of the Royal Astronomical Society, 483, 3022

Jayasinghe, T., Kochanek, C. S., Stanek, K. Z., et al. 2021, Monthly Notices of the Royal Astronomical Society, 503, 200

Kamath, D., Wood, P. R., Soszyński, I., & Lebzelter, T. 2010, Monthly Notices of the Royal Astronomical Society, 408, 522

Keller, S. C. & Wood, P. R. 2006, Astrophysical Journal, 642, 834

Lebzelter, T. & Hinkle, K. H. 2002, Astronomy & Astrophysics, 393, 563

Lebzelter, T., Mowlavi, N., Lecoeur-Taibi, I., et al. 2022, arXiv e-prints, arXiv:2206.05745

Lebzelter, T., Mowlavi, N., Marigo, P., et al. 2018, Astronomy & Astrophysics, 616, L13

Lebzelter, T. & Wood, P. R. 2005, Astronomy & Astrophysics, 441, 1117

Marigo, P. & Girardi, L. 2007, Astronomy & Astrophysics, 469, 239

Marigo, P., Girardi, L., Bressan, A., et al. 2017, Astrophysical Journal, 835, 77

McDonald, I. & Trabucchi, M. 2019, Monthly Notices of the Royal Astronomical Society, 484, 4678

Merrill, P. W. 1923, Astrophysical Journal, 58, 215

Pastorelli, G., Marigo, P., Girardi, L., et al. 2020, Monthly Notices of the Royal Astronomical Society, 498, 3283

Pastorelli, G., Marigo, P., Girardi, L., et al. 2019, Monthly Notices of the Royal Astronomical Society, 485, 5666

Reid, M. J. & Goldston, J. E. 2002, Astrophysical Journal, 568, 931

Riess, A. G., Yuan, W., Casertano, S., Macri, L. M., & Scolnic, D. 2020, Astrophysical Journal, Letters, 896, L43

Sanders, J. L., Matsunaga, N., Kawata, D., et al. 2022, Monthly Notices of the Royal Astronomical Society, 517, 257

Soszyński, I., Olechowska, A., Ratajczak, M., et al. 2021, Astrophysical Journal, Letters, 911, L22

Soszyński, I., Udalski, A., Szymański, M. K., et al. 2009, Acta Astronomica, 59, 239

Soszyński, I., Wood, P. R., & Udalski, A. 2013, Astrophysical Journal, 779, 167

Trabucchi, M., Lebzelter, T., Mowlavi, N., et al. 2020, in Stars and their Variability Observed from Space, ed. C. Neiner, W. W. Weiss, D. Baade, R. E. Griffin, C. C. Lovekin, & A. F. J. Moffat, 87–90

Trabucchi, M. & Mowlavi, N. 2022, Astronomy & Astrophysics, 658, L1

Trabucchi, M., Mowlavi, N., & Lebzelter, T. 2021a, Astronomy & Astrophysics, 656, A66

Trabucchi, M., Wood, P. R., Montalbán, J., et al. 2017, Astrophysical Journal, 847, 139

Trabucchi, M., Wood, P. R., Montalbán, J., et al. 2019, Monthly Notices of the Royal Astronomical Society, 482, 929

Trabucchi, M., Wood, P. R., Mowlavi, N., et al. 2021b, Monthly Notices of the Royal

Astronomical Society, 500, 1575
Udalski, A., Szymanski, M., Kaluzny, J., Kubiak, M., & Mateo, M. 1992, Acta Astronomica, 42, 253
Wilson, R. E. & Merrill, P. W. 1942, Astrophysical Journal, 95, 248
Wood, P. R. 1974, Astrophysical Journal, 190, 609
Wood, P. R. 2000, Publications of the Astron. Soc. of Australia, 17, 18
Wood, P. R. 2015, Monthly Notices of the Royal Astronomical Society, 448, 3829
Wood, P. R. & Olivier, E. A. 2014, Monthly Notices of the Royal Astronomical Society, 440, 2576
Wray, J. J., Eyer, L., & Paczyński, B. 2004, Monthly Notices of the Royal Astronomical Society, 349, 1059
Wyatt, S. P. & Cahn, J. H. 1983, Astrophysical Journal, 275, 225
Xiong, D. R., Deng, L., & Zhang, C. 2018, Monthly Notices of the Royal Astronomical Society, 480, 2698
Ya'Ari, A. & Tuchman, Y. 1996, Astrophysical Journal, 456, 350
Yu, J., Bedding, T. R., Stello, D., et al. 2020, Monthly Notices of the Royal Astronomical Society, 493, 1388

Question (Whitelock): I wonder if you are going to be looking in detail at the somewhat more massive HBB LPVs, that fall above the standard fundamental PLR. These could be important as distance indicators in the JWST era.

Answer: The current grid extends up to $7M\odot$ and extrapolation of the non-linear period-mass-radius prescription shows good agreement with observations, which is encouraging. We aim to extend the grid further up in mass in the future version of the grid in order to study HBB and super-AGB stars.

Question (Skowron): Have you tried calculating individual distances for SRVs with these relations? I'm curious what errors on individual distances we can expect and how useful these variables would be as, e.g., Galactic structure tracers.

Answer: Currently our work is only exploratory and a differential/comparative analysis. We derive the period-luminosity relations of Miras, then repeat after including SRVs in the sample, we look at the differences and see if there are improvements. We have not applied this to measure distances yet.

Question (Eyer): A question about the use of radial velocities time series of LPVs incoming from Gaia.

Answer: RVs from Gaia will be very useful for a number of applications. The one I'm most interested in is the calibration of the tricky free parameters in the pulsation models. Achieving such a calibration will further improve the predictive power of the models.

At the crossroads of astrophysics and cosmology: Period–luminosity relations in the 2020s
Proceedings IAU Symposium No. 376, 2024
R. de Grijs, P. A. Whitelock & M. Catelan, eds.
doi:10.1017/S1743921323003101

China Space Station Telescope and Variable Star Studies

Xiaodian Chen[1,2,3], Shu Wang[1,2,3], Licai Deng[1,2,3], Richard de Grijs[4,5], Xiaoyue Zhou[1], Xiaohan Chen[1] and Jianxing Zhang[1]

[1]CAS Key Laboratory of Optical Astronomy, National Astronomical Observatories, Chinese Academy of Sciences, Beijing, China
email: chenxiaodian@nao.cas.cn

[2]School of Astronomy and Space Science, University of the Chinese Academy of Sciences, Beijing, China

[3]Department of Astronomy, China West Normal University, Nanchong 637009, China

[4]School of Mathematical and Physical Sciences, Macquarie University, Balaclava Road, Sydney, NSW 2109, Australia

[5]Astrophysics and Space Technologies Research Centre, Macquarie University, Balaclava Road, Sydney, NSW 2109, Australia

Abstract. In the last five years, the number of periodic variable stars has increased by two million. We used the ZTF DR2 data to find and build a catalog that includes 780,000 periodic variable stars. These periodic variable stars were classified into 11 types, which greatly complemented the variable stars in Galactic disk. Based on the latest ZTF DR16 data, we found 2 million variable candidates. We trained a machine learner to classify variable stars, and the learner had a prediction accuracy of 94%. Using millions of variable stars, we carried out studies to optimize the period–luminosity relations and the Galactic structure and the extinction law. With the future China Space Station Telescope, millions of variable stars in the Local Group will be discovered. They help to study the structure of our Local Group and also to cross-check the distance ladders based on different variable stars.

Keywords. Periodic variable stars, Cepheid variable stars, RR Lyrae variable stars, Galaxy structure, Distance indicators

1. Introduction

Variable stars have been studied for hundreds of years, and the number of known variable stars has increased rapidly in the last decade. Some variable stars undergo periodic changes in their luminosity and are called periodic variable stars. There are three main categories of periodic variable stars: pulsating stars, eclipsing binaries, and rotational variable stars. Periodic variable stars can be important distance tracers when they satisfy a simple linear relation between period and luminosity, i.e., the period–luminosity (PL) relation. A large number of periodic variable stars can help to optimize the distance estimates of the galaxies, to understand the three-dimensional (3D) structure of the Milky Way and our Local Group, and to optimize the Hubble constant.

Efficient variable star searches through large fields of view began in the 1990s. The Massive Compact Halo Object survey (Alcock et al. 1993) found 20,000 variable stars in the Large Magellanic Cloud (LMC). The Optical Gravitational Lensing Experiment

(OGLE, Udalski et al. 1992, 2015) has discovered about one million variable stars in nearly 30 years of observations. For the first 20 years, OGLE focused on the LMC and Small Magellanic Cloud (SMC), and in the last decade it has shifted to searching for variable stars in the Galactic bulge and the southern disk. The first all-sky variable star survey was performed by the All-Sky Automated Survey (ASAS, Pojmanski et al. 2005), which found nearly 20,000 periodic variable stars. A deeper search for all-sky variable stars was the Catalina survey (Drake et al. 2014, 2017), which discovered about 100,000 periodic variable stars. The RR Lyrae stars in this catalog were very useful for studying the Galactic halo and streams. The VISTA Variables in the Vía Láctea (VVV, Minniti et al. 2010) survey was an infrared survey that looked for variable stars in the inner disk of the Milky Way. A large number of variable star catalogs were published in 2018. In addition to OGLE, there were the Wide-field Infrared Survey Explorer (WISE) catalog of periodic variable stars (Chen et al. 2018), the All-Sky Automated Survey for Supernovae (ASAS-SN, Jayasinghe et al. 2018), the Asteroid Terrestrial-impact Last Alert System (ATLAS, Heinze et al. 2018), and the variable catalog of Gaia DR2 (Mowlavi et al. 2018; Clementini et al. 2019). These variable catalogs made up for the many variables missing from the Galactic disk, leading to the presentation of a 3D intuitive map of the disk (Chen et al. 2019; Skowron et al. 2019).

Subsequently, the Zwicky Transient Facility (ZTF) catalog of periodic variable stars (Chen et al. 2020) was the largest variable star catalog in the northern sky, including 780,000 periodic variable stars. Gaia DR3 released the largest variable star catalog to date (Rimoldini et al. 2023), including three million periodic variable stars. Future telescopes, including the China Space Station Telescope (CSST) and the Vera C. Rubin Observatory Legacy Survey of Space and Time (LSST), will provide opportunities to discover more extragalactic variable stars. In this paper, we introduce the ZTF catalog of periodic variable stars in Section 2. Section 3 describes our update of this catalog using the latest ZTF data. We discuss our recent studies using variable stars in Section 4. Section 5 is a summary.

2. Periodic variable stars in ZTF DR2

ZTF is a 48-inch Samuel Oschin telescope operating at the Palomar Observatory (Bellm et al. 2019). It has a large field of view (47 square degrees) to capture transient objects. Its public survey program began in March 2018. The survey is scheduled to observe the majority of the northern sky every three days and to observe the Galactic northern disk ($|l| < 7°$) once a day. The ZTF observations have a fixed exposure time of 30 seconds and its corresponding limiting magnitudes in the g, r bands are about 20.5 mag. The main science of the ZTF is to search for transient objects, including supernovae, tidal disruption events, and near Earth asteroids or comets.

We used the ZTF DR2 data to search for periodic variable stars (Chen et al. 2020). The time span of the photometric data is 470 days, which is sufficient for a more accurate determination of the period up to ~ 200 days.The Lomb-Scargle periodogram (Lomb 1976; Scargle 1982) was used to analyze the light curves, and we accepted periodic variable stars when the false-alarm probability of the peaks of the periodogram is less than 0.001. Once excluding aliased periods, such as $P = 0.5, 1, 2...$, the most dominant peak on the periodogram is considered to be the period of one candidate. We then used the period to calculate the phases and adopted a Fourier function $f = a_0 + \sum_{i=1}^{4} a_i \cos(2\pi i t/P + \phi_i)$ to fit the phase-folded light curve. By using the coefficient of determination of the fit R^2, we excluded candidates with poorly fitted light curves.

Table 1. All and new periodic variable stars in ZTF DR2.

Type	Total	New (fraction)
Cep-I	1262	565 (44.8%)
Cep-II	358	154 (43.0%)
RRab	32,518	3034 (9.3%)
RRc	13,875	2178 (15.7%)
EW	36,9707	306,375 (82.9%)
EA	49,943	40,201 (80.5%)
δ Scuti	16,709	15,396 (92.1%)
Mira	11,879	4,997 (42.1%)
SR	119,261	97,737 (82.0%)
RS CVn	81,393	70,957 (87.2%)
BY Dra	84,697	80,108 (94.6%)
Total number	781,602	621,702 (79.5%)

We obtained a sample of 780,000 periodic variable stars, 74% of which are located in the Galactic disk, where previous surveys were very incomplete. Compared to the previous variable star catalogs, 80% of periodic variable stars in our sample are newly discovered. We classified these variable stars using the Density-Based Spatial Clustering of Applications with Noise (DBSCAN) method, with parameters including period, phase difference ϕ_{21}, amplitude, amplitude ratio R_{21}, luminosity, and R^2. The final periodic variable stars were classified into 11 types, including classical Cepheids, type-II Cepheids, fundamental RR Lyrae (RRab) stars, first-overtone RR Lyrae (RRc) stars, δ Scuti stars, EW-type eclipsing binaries, EA-type eclipsing binaries, Miras, semi-regular (SR) variable stars, RS CVn variables and BY Dra variables (see Table 1). Pulsating variable stars and eclipsing binaries with large amplitudes were clearly distinguished using period and light curve parameters (phase difference, amplitude, and amplitude ratio). For pulsating variables with smaller amplitudes and rotational variable stars, the luminosity is required to help their classification. The luminosity was calculated using the Gaia parallax.

3. More variable stars in ZTF DR16

What is more for ZTF variable stars? The currently available catalog is ZTF DR16, which has an increased time span from 470 to 1750 days compared to DR2. The number of epochs in r band has increased from 150 to 350. In our preliminary analysis, the number of periodic variable stars will increase to 2 million. We also plan to add the classification of double-mode variable stars and to add the parameters of eclipsing binaries, such as the depth and duration of the eclipse.

By analyzing the periodogram of each object in ZTF DR16, we identified 2 million periodic variable candidates. As for classification, we tried different machine learning methods and found that the bagged tree method was most suitable. To train the classification learner, we used 0.6 million variable stars from the ZTF DR2 catalog, and the classification accuracy is 93.9% by five-fold cross-validation. The confusion matrix is shown in Fig. 1. The positive prediction ratios are above 90% for all variable stars except the type-II Cepheids and rotational variable stars. It should be noted that if the subtype classification is not considered, the accuracy will be about 99%. As for the feature importance in the classification, the most important feature is the period P, followed by the amplitude ratio R_{21}, the phase difference ϕ_{21} and the amplitude (Amp.). The parallax ϖ and absolute magnitude $M_{W_{gr}}$ are also helpful to improve the accuracy of classification. The maximum value of the power spectral density (PSD), the fit parameters R^2 and root mean square error (RMSE), and the amplitude ratio R_{31} are not very important (see Fig. 2).

Compared to ZTF DR2, there is an increase of 1.2 million periodic variable stars based on ZTF DR16. The new sample includes ~ 600 classical Cepheids, $\sim 100,000$ RR Lyrae

True Class \ Predicted Class	BYDra	CEP	CEPII	DSCT	EA	EW	Mira	RR	RRc	RSCVN	SR
BYDra	13191	7	3		269	1352		69	134	1683	2
CEP	12	1116	25		2	20		16	1	35	7
CEPII	9	56	202			45	4	4		39	10
DSCT				14955	2	540			3	293	
EA	126			1	48855	4162	1		1	381	75
EW	952	13	12	316	4216	372317	4	13	450	5176	571
Mira							8051				406
RR	39	2	2		2	35		34937	168	66	
RRc	35	3		4	2	1016		83	13545	102	
RSCVN	1159	25	14	521	482	10304		111	270	33869	66
SR	1	3	3		16	216	297	1		122	22783

Figure 1. The confusion matrix of our trained learner based on 0.6 million variable stars from the ZTF DR2 catalog.

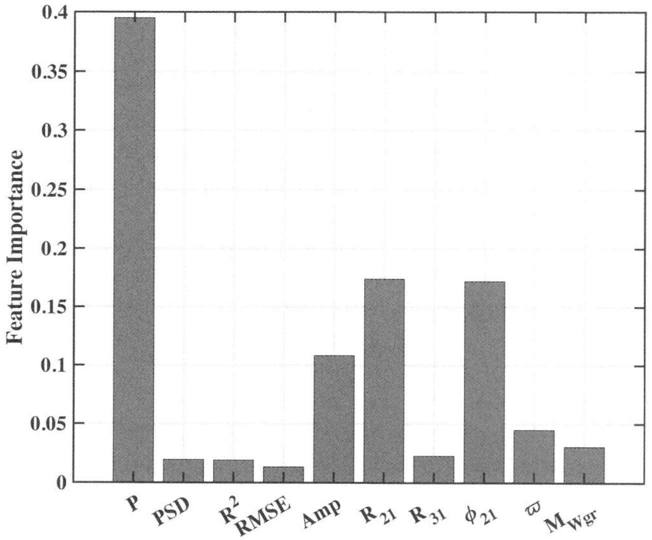

Figure 2. The importance of ten parameters in the classification of periodic variable stars.

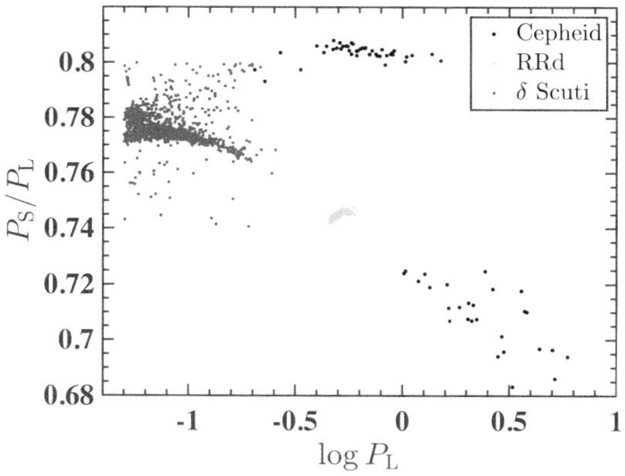

Figure 3. Period–period ratio diagram (Petersen diagram) for δ Scuti stars (red), RR Lyrae stars (green), and classical Cepheids (black). Short and long periods are determined based on ZTF photometry.

stars and $\sim 800,000$ eclipsing binaries. When compared to Gaia DR3, we found that 73% of the sample was in the Gaia variable catalog. This means that Gaia will find the other $20-30\%$ new periodic variable stars in future data releases. The agreement rate for the classification of these two catalogs is close to 99%. For Cepheids, the agreement rate of the classification is slightly lower, about 85%, but the classification of Cepheids can be optimized by visually checking each light curve. For the other variable stars, we planned to check some light curves and thus found some criteria to help the classification.

ZTF DR16 is very suitable for identifying double-mode variable stars. Currently we have classified double-mode variable stars in the DR2 catalog (see Fig. 3). The double-mode variable stars are a very important tracer for metallicity and distance (Chen et al. 2023). In the future, when more dynamical mass measurements are determined to optimize the models, they will become accurate mass tracers. In ZTF DR16, we will find more than 5000 double-mode variable stars.

4. Variable star studies

With millions of periodic variable stars, the PL relation or period–metallicity–luminosity relation of periodic variable stars can be optimized. Moreover, based on more distance tracers, it can help to investigate the Milky Way structure and the extinction law.

We presented a 3D intuitive map of the Milky Way using 1339 high-confidence classical Cepheids selected from the variable star catalogs of WISE, ASAS-SN and ATLAS (Chen et al. 2019). We found that the Galactic disk is warped and can be represented by a linear model of $Z_w = 0.148 \times (R - 9.26)\sin(\phi - 17.4°)$ kpc. Here, Z_w, R and ϕ are the height, onset radius and the angle of the line of nodes (LONs) of the warp. Based on structure and kinematics, we found that the precession of the warp, i.e., the warp LONs are changeable for different galactocentric radii. Recently, based on the proper motion and radial velocity data of ~ 700 classical Cepheids from the Gaia DR3, we analyzed the Milky Way warp from 3D kinematics and determined a warp precession rate of $\omega = 4.4 \pm 2.9$ km s^{-1} kpc^{-1}. Our values are between Poggio et al. (2020) and Chrobáková & López-Corredoira (2021) and support a low warp precession rate.

Open cluster Cepheids (OCCs) are suitable objects for establishing PL relations and period–age relations for classical Cepheids. We performed a census of Galactic OCCs using Gaia early DR3 parallaxes and proper motions, and obtained a total of 33 OCCs,

13 of which are newly discovered (Zhou & Chen 2021). Based on them, we established PL relations of classical Cepheids in Gaia and WISE bands as $M_{G,BP,RP} = -3.32 \log P + (-2.718 \pm 0.049)$ and $M_{W1} = -(3.274 \pm 0.090) \log P + (-2.567 \pm 0.080)$, respectively, which are consistent with the literature (Wang et al. 2018; Ripepi et al. 2019). Based on the cluster ages, we obtained the period–age (PA) relation for fundamental Cepheids as $\log t = -(0.638 \pm 0.063) \log P + (8.569 \pm 0.057)$. The overtone Cepheid is 0.2 dex younger than the fundamental Cepheid at a given period. The error in the PA relation is 17%, and our PA relation is consistent with the model of Anderson et al. (2016), and 0.3 dex older than the model of Bono et al. (2005). It should be noted that the PA relation is better constrained at the older end, but not at the younger end. It is important to search for OCCs with longer periods in younger cluster in the future.

Based on a larger sample, the infrared PL relations of LPVs can be optimized. How the metallicity affects the zero point of the PL relation is still very uncertain. We used the Gaia DR3 LPVs (Lebzelter et al. 2023) and classified them into red giant branch stars, asymptotic giant branch (AGB) stars and red supergiant stars based on the color and absolute magnitude. The majority of LPVs are AGB stars, and we focused on them and classified them as oxygen-rich (O-rich) and carbon-rich (C-rich) AGB stars based on the color. Combining the WISE photometric data, we determined the PL relations for O-rich and C-rich AGB stars in the Milky Way, LMC and SMC, respectively. The PL relations of AGB stars include several sequences, C (fundamental mode), C' (first-overtone mode), B (second-overtone mode) and D (long secondary period) sequences (Wood et al. 1999). Among them, the C sequence is often used for distance measurements. We found that for C-rich AGB stars with different amplitudes, the W_2-band PL relations are different. Large amplitude C-rich AGB stars have a steeper PL relation, which can be explained by the amount of circumstellar dust. During the pulsation, they expand more and produced more circumstellar dust, which leads to a brightening in the mid-infrared magnitudes. Therefore, the near-infrared bands are more suitable for distance measurements using C-rich AGB stars than the mid-infrared bands. We also investigated the metallicity effect based on about 400 AGBs in LMC and SMC with Apache Point Observatory Galactic Evolution Experiment (APOGEE, Majewski et al. 2017) DR17 metallicities (see Fig. 4). The PL relation of C-rich AGB stars is not affected by the metallicity, while the PL relation zero point of O-rich AGB stars becomes brighter with the increasing metallicity. The metallicity effect of O-rich AGB stars is similar to that of the classical Cepheids. The coefficients of the metallicity effect are between $\beta = -0.7$ mag/dex and $\beta = -0.3$ mag/dex for C and C' sequences of the LMC and SMC. This suggests that the metallicity effect should be considered when measuring distances with O-rich AGB stars or O-rich Miras.

The PL relations of variable stars can be used to determine the extinction laws. We studied the classical Cepheids in different regions of LMC and SMC with different extinctions (Wang & Chen 2023). The differences in the apparent magnitudes or colors of classical Cepheids at the same period reflect the extinction and color excesses. Based on them, we estimated the extinction coefficients of the LMC and SMC as $A_{G_{\rm RP}}/E(G_{\rm BP} - G_{\rm RP}) = 1.589 \pm 0.014$ and $A_{G_{\rm RP}}/E(G_{\rm BP} - G_{\rm RP}) = 1.412 \pm 0.041$. Based on the multi-band photometry and APOGEE spectral parameters of red supergiants, the multi-band color excess ratios $\frac{E(\lambda - G_{\rm RP})}{E(G_{\rm BP} - G_{\rm RP})}$ were determined. Combining the extinction coefficients and the color excess ratios, we determined the extinction laws of the LMC and SMC from optical to near-infrared, where the total-to-selective extinction ratio R_V is 3.40 ± 0.07 and 2.53 ± 0.10, respectively. The extinction law of the Milky Way (Wang & Chen 2019) is between those of the LMC and SMC, and we updated the classical extinction laws (Cardelli et al. 1989) by combining our results on these three galaxies. The current extinction laws specific to different environments can be represented by analytical equations (see Equ. 1-5) on wavelength λ. Compared with the classical extinction

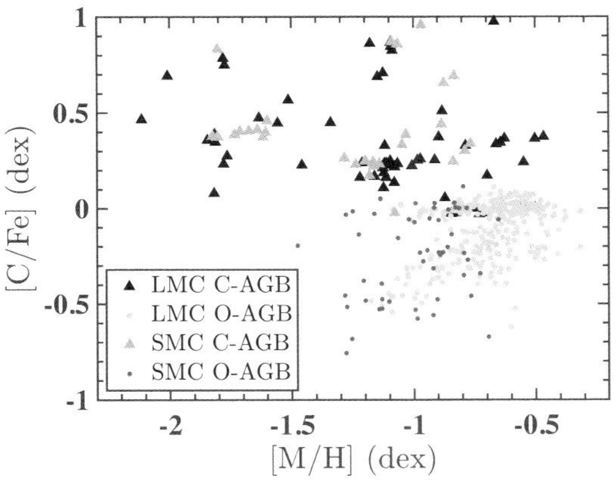

Figure 4. Metallicity and Carbon abundance distributions of C-rich and O-rich AGB stars in LMC and SMC. C-rich AGB stars have a high Carbon abundance. O-rich AGB stars in LMC have a higher metallicity than those in SMC.

law, our results are significantly optimized at wavelengths longer than 600 nm.

$$A_\lambda/A_V = A + B/R_V \quad ; \tag{1}$$

Optical: $0.3\,\mu\text{m} < \lambda < 1.0\,\mu\text{m}$ and $Y = 1/\lambda(\,\mu\text{m}) - 1.82$,

$$\begin{aligned} A = {} & 1.0 + 0.7499Y - 0.1086Y^2 - 0.08909Y^3 \\ & + 0.02905Y^4 + 0.01069Y^5 \\ & + 0.001707Y^6 - 0.001002Y^7 \quad ; \end{aligned} \tag{2}$$

$$\begin{aligned} B = {} & (1.41338Y + 2.28305Y^2 + 1.07233Y^3 \\ & - 5.38434Y^4 - 0.62251Y^5 + 5.30260Y^6 \\ & - 2.09002Y^7) \times (1 - R_V/3.1) \quad . \end{aligned} \tag{3}$$

Near-IR: $1.0\,\mu\text{m} \leq \lambda < 3.33\,\mu\text{m}$,

$$A = (0.3722 \pm 0.0026)\lambda^{-2.07\pm0.03} \quad ; \tag{4}$$

$$B = (-0.5182 \pm 0.0067)\lambda^{-2.07\pm0.03} \times (1 - R_V/3.1) \quad . \tag{5}$$

5. Searching for variable stars with CSST

The CSST is a 2-meter space telescope in the same orbit as the China Manned Space Station and is scheduled to be launched late in 2024. The telescope is equipped with five instruments, of which the Survey Camera is the primary instrument. The Survey Camera has a field of view of 1.1 square degrees and a spatial resolution of 0.15 arc seconds. It will occupy 70% of the CSST's operating time to observe a 17,500 square degree area of the sky in 10 bands. These bands include seven photometric bands (NUV, u, g, r, i, z, y) and three slitless spectra bands covering wavelengths from 255 nm to 1000 nm. The exposure time of the survey is 150 seconds, corresponding to a 5σ limit magnitude of $g = 26.2$ mag. The main science goals of the CSST involve galaxies and cosmology, so the survey avoids the Galactic disk.

The CSST main survey will only observe an object twice in each band, which is not enough for discovering periodic variable stars. We have proposed to perform 20-50 observations of nearby galaxies and dwarf galaxies in the g and r bands during the test phase. Based on these data, it will be possible to detect periodic variable stars from bright to faint in LMC and SMC, including Cepheids, RR Lyrae stars, eclipsing binaries, rotational variable stars, δ Scuti, Miras and SRs. We will also search for Cepheids and RR Lyrae stars in ~ 100 nearby galaxies and dwarf galaxies. We expect to find 1 to 2 million periodic variable stars that will be very helpful to check the consistency of different distance tracers. Based on different distance tracers, we plan to obtain high-precision ($\sigma < 2\%$) distances for dwarf galaxies or galaxies and to present a 3D intuitive map of the Local Group.

6. Summary

In order to search for variable stars more sensitively in the northern sky, we used ZTF DR2 light curves to find and build a catalog including 780,000 periodic variable stars. Based on parameters such as period, light curve parameters, and luminosity, these periodic variable stars were classified into 11 types. 74% of these variable stars are located in the Galactic disk, and 80% are new discovered. This catalog provides an opportunity to study periodic variable stars by a large sample size. Based on the latest ZTF DR16 data, we found 2 million variable star candidates. We trained a learner for variable star classification using the sample of ZTF periodic variable star catalog, and the prediction accuracy of the learner is 94%. By comparing with Gaia DR3, we found an agreement rate of 99% for the classification. Gaia can discover another 20%-30% of variable stars in future data releases.

Millions of new periodic variable stars help to study PL relations, Galactic structure and extinction laws. We used more than one thousand classical Cepheids to establish the 3D map of the Milky Way disk and found the warp is in precession. Based on kinematic parameters of classical Cepheids, we obtained a low precession rate of the warp $\omega = 4.4 \pm 2.9$ km s^{-1} kpc^{-1}. We performed a census of OCCs and obtained 33 OCCs, based on which we determined a PA relation as $\log t = -(0.638 \pm 0.063) \log P + (8.569 \pm 0.057)$. We obtained the extinction laws of the LMC and SMC using classical Cepheids and red supergiants. Combining the Galactic extinction laws, we updated the analytic equations of the extinction laws for different environments R_V. We investigated the PL relations of LPVs and found that the mid-infrared PL relations of C-rich AGB stars are affected by circumstellar dust. For O-rich AGB stars, their PL relation is affected by the metallicity, and the metallicity effect needs to be considered in the distance measurement.

With future CSST, million of variable stars in the Local Group will be found. They help to study the structure of our Local Group, as well as to cross-check the distance ladders of different variable stars.

Acknowledgments: We acknowledge support from the National Natural Science Foundation of China through grants 12173047, 12003046, 12233007, 12133002 and 11903045.

References

Alcock, C., Akerlof, C.W., Allsman, R.A., et al. 1993, *Nature*, 365, 621
Anderson, R.I., Saio, H., Ekström, S., Georgy, C., & Meynet, G. 2016, *A&A*, 591, A8
Bellm, E.C., Kulkarni, S.R., Graham, M.J., et al. 2019, *PASP*, 131, 018002
Bono, G., Marconi, M., Cassisi, S., et al. 2005, *ApJ*, 621, 966
Cardelli, J.A., Clayton, G.C., & Mathis, J.S.: 1989, *ApJ*, 345, 245
Chen, X., Wang, S., Deng, L., et al. 2018, *ApJS*, 237, 28
Chen, X., Wang, S., Deng, L., et al. 2019, *NatAs*, 3, 320
Chen, X., Wang, S., Deng, L., et al. 2020, *ApJS*, 249, 18
Chen, X., Zhang, J., Wang, S., & Deng, L. 2023, *NatAs*, arXiv:2306.10708
Chrobáková Ž., López-Corredoira M. 2021, *ApJ*, 912, 130
Clementini, G., Ripepi, V., Molinaro, R., et al. 2019, *A&A*, 622, A60

Drake, A.J., Graham, M.J., Djorgovski, S.G., et al. 2014, *ApJS*, 213, 9
Drake, A.J., Djorgovski, S.G., Catelan, M., et al. 2017, *MNRAS*, 469, 3688
Heinze, A.N., Tonry, J.L., Denneau, L., et al. 2018, *ApJ*, 156, 241
Jayasinghe, T., Kochanek, C.S., Stanek, K.Z., et al. 2018, *MNRAS*, 477, 3145
Lebzelter, T., Mowlavi, N., Lecoeur-Taibi, I., et al. 2023, *A&A*, 674, A15
Lomb, N.R. 1976, *Ap&SS*, 39, 447
Majewski, S.R., Schiavon, R.P., Frinchaboy, P.M., et al. 2017, *ApJ*, 154, 94
Minniti, D., Lucas, P.W., Emerson, J.P., et al. 2010, *NewA*, 15, 433.
Mowlavi, N., Lecoeur-Taïbi, I., Lebzelter, T., et al. 2018, *A&A*, 618, A58
Poggio E., Drimmel R., Andrae R., et al. 2020, *NatAs*, 4, 590
Pojmanski, G., Pilecki, B., and Szczygiel, D. 2005, *AcA*, 55, 275
Rimoldini, L., Holl, B., Gavras, P., et al. 2023, *A&A*, 674, A14
Ripepi, V., Molinaro, R., Musella, I., et al. 2019, *A&A*, 625, A14
Scargle, J.D. 1982, *ApJ*, 263, 835
Skowron, D. M., Skowron, J., Mróz, P., et al. 2019, *Science*, 365, 478
Wang, S., Chen, X., de Grijs, R., et al. 2018, *ApJ*, 852, 78
Wang, S., & Chen, X. 2019, *ApJ*, 877, 116
Wang, S., & Chen, X. 2023, *ApJ*, 946, 43
Wood P. R., Alcock C., Allsman R. A., et al., 1999, *IAUS*, 191, 151
Udalski, A., Szymanski, M., Kaluzny, J., et al. 1992, *AcA*, 42, 253
Udalski, A., Szymański, M.K., and Szymański, G. 2015, *AcA*, 65, 1
Zhou, X. & Chen, X. 2021, *MNRAS*, 504, 4768

7. Discussion

Question (Whitelock): I may have missed it, but where did your metallicities come from for the AGB stars?

Answer: The metallicities are from APOGEE.

Question (Abdollahi): Which algorithm (model) do you use to classify these variable stars (the machine learning method)?

Answer: Random forest. This method gives an initial classification of variable stars, and then we optimized the classification for different types of variable stars by manually checking the light curve or establishing parameter-based criteria.

Question (Hajdu): Do you have an idea about the saturation limits of the CSST during its normal operation?

Answer: No. I'm not the person who knows. They might test the scanning mode to observe bright stars.

At the crossroads of astrophysics and cosmology: Period–luminosity relations in the 2020s
Proceedings IAU Symposium No. 376, 2024
R. de Grijs, P. A. Whitelock & M. Catelan, eds.
doi:10.1017/S1743921323002983

OH/IR stars and the Period-Luminosity Relation of Mira variables

D. Engels[1], S. Etoka[2], F. Jiménez-Esteban[3], W. Herrmann[4], B. López-Martí[5]

[1]Hamburger Sternwarte, Universität Hamburg, Gojenbergsweg 112, 21029 Hamburg, Germany.
email: dengels@hs.uni-hamburg.de

[2]JBCA, University of Manchester, M13 9PL, Manchester, UK

[3]Centro de Astrobiología (CAB), CSIC-INTA, Camino Bajo del Castillo s/n, 28692, Villanueva de la Cañada, Madrid, Spain

[4]Astropeiler Stockert e.V., Astropeiler Stockert 1-4, 53902 Bad Münstereifel, Germany

[5]Universidad San Pablo-CEU, Campus de Montepríncipe, 28668 Boadilla del Monte, Madrid, Spain

Abstract. The OH/IR stars evolving on the Asymptotic Giant Branch are large-amplitude variable stars with periods in the range of ∼400 to 2500 days, significantly longer than those of the related Mira variables. We use preliminary results from a monitoring program of the 1612-MHz OH maser variations of a sample of >70 OH/IR stars to study a possible extension of Mira Period-Luminosity Relations to longer periods. The period distribution of the sample is split around $P \sim 1100$ days. Using WISE $W3$ absolute magnitudes as proxies for the luminosity and the best available distances, we found no convincing relation between periods and absolute magnitudes. A cause could be rapid evolution of the 'extreme OH/IR stars' (the group with $P > 1100$ days) close to the tip of the AGB, where no increase of luminosity is expected on the short time-scales involved.

Keywords. masers, stars: AGB and post-AGB, circumstellar matter

1. Introduction

Stars evolving on the Asymptotic Giant Branch (AGB) can achieve mass-loss rates in the range $10^{-8} - 10^{-4}\ M_\odot\,\mathrm{yr}^{-1}$ (De Beck et al. 2010, and references therein), which lead to the formation of circumstellar envelopes composed of gas and dust. For the highest mass-loss rates, the dust becomes optically thick and the stars become obscured at visual light and, for the most extreme cases, also in the near-infrared (NIR: 1–5μm). Variable OH/IR stars with high mass-loss rates ($\geq 10^{-6}\ M_\odot\,\mathrm{yr}^{-1}$) are examples of such obscured stars in their final stage of AGB evolution. After most of their mass is lost during the AGB phase, the mass-loss rates drop on short timescales of a few thousand years from late AGB values of $10^{-5} - 10^{-4}$ to post-AGB values of $10^{-8} - 10^{-7}\ M_\odot\,\mathrm{yr}^{-1}$ (Miller Bertolami 2016).

In general, evolved AGB stars are observed as large-amplitude variables with periods >1 year, while post-AGB stars have lost the large-amplitude pulsations, and are almost non-variable. The obscured OH/IR stars are in part large-amplitude periodic variables (hereafter L-AGB stars) with periods ~ 400– ~ 2500 days (fundamental mode pulsators) and in part small-amplitude irregular or almost non-variable stars (hereafter S-pAGB stars). The L-AGB stars can be considered as related to Mira variables. The designation

'OH/IR stars' stems from the way they were discovered (Wilson & Barrett 1970). Because of their optical obscuration, they were found only in the 1970s by searches among bright infrared sources or in blind OH maser surveys with counterparts identified subsequently in the near infrared. In the following, the latter objetcs were confirmed as long-period variable stars at infrared and radio wavelengths (Engels et al. 1983; Herman & Habing 1985).

The extension of the Period-Luminosity Relation (PLR) for Mira variables to the significantly longer periods of the OH/IR stars has been discussed since the first long OH/IR star periods (up to >5 year), were determined. An extended PLR would have been a comfortable tool to determine distances to OH/IR stars. Such a PLR was presented by Engels et al. (1983), but the relation was considered as tentative. As most of the energy is emitted in the mid-IR ($\lambda > 5\mu$m), and no mid-IR observations were available, the bolometric flux determination was rather uncertain. Also the kinematic distances used to determine the OH/IR star luminosities have caveats, because the movement of the OH/IR stars has peculiar velocity components on top of their participation to the Galactic rotation. Jones et al. (1983) argued for a different explanation for the long periods observed. The OH/IR stars in their view would have been former Mira variables, which had undergone a switch in pulsation mode from first overtone to fundamental mode without changing their luminosity.

Because of the large time and efforts to obtain periods for larger samples of OH/IR stars via NIR and radio monitoring programs, progress has been slow. Also the dearth of OH/IR stars in the Large Magellanic Cloud (LMC) (Goldman et al. 2017), where most of the PLR studies for Mira variables were focused on (see Iwanek et al. 2021) and references therein), inhibited progress.

To remedy this, we conducted an NIR monitoring program of the 'Arecibo sample of OH/IR stars' (Jimenéz-Esteban et al. 2021) and are conducting an ongoing monitoring program of the 1612-MHz OH maser emission of the 'Bright OH/IR star sample'†. Also, improved distances were determined for a couple of stars (Engels et al. 2015), using the phase-lag method (Etoka & Engels 2022). We will use the results of these observational programs to re-address the relation between periods and luminosties of OH/IR stars and to compare them with the most recent Mira PLR relations.

2. Variability properties of the 'Bright OH/IR star sample'

The 'Bright OH/IR star sample' comprises 115 stars, originally compiled by Baud et al. (1981) and updated by Engels & Jiménez-Esteban (2007). Almost all of them are located along the Galactic plane at $10 < l < 150°$, and $|b| < 4°$. It is quite complete for bright 1612-MHz OH masers ($F_\nu > 4$ Jy). The monitoring program started in 2008 is mainly made at the Nançay Radio Telescope (NRT) with a cadence of two months. After two years an initial classification (L-AGB, S-pAGB candidates) is made. L-AGB candidates are continued to be monitored until their period and amplitude are determined, while the cadence of observations of the S-pAGB candidates is decreased to 1–2 times a year. Since 2017 the program has been complemented with observations by the Stockert Radio Telescope to follow selected L-AGB stars over several variability cycles.

As of end of 2022, 80 OH/IR stars are classified as L-AGB variables (70%), while 30 (26%) are S-pAGB stars showing only irregular brightness variations. The remaining five objects are the red supergiant NML Cyg or are not classified so far. Two examples are shown in Figure 1. The brighter masers of the sample have been monitored before by Herman & Habing (1985), who reported several sub-groups with different amplitudes and

† see https://www.hs.uni-hamburg.de/nrt-monitoring

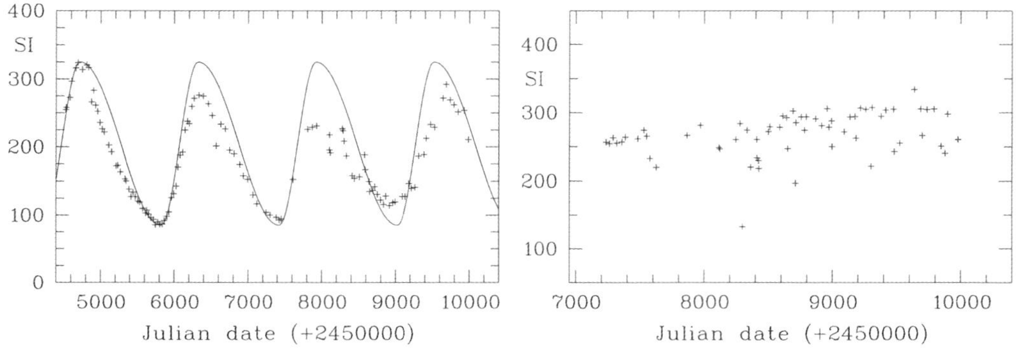

Figure 1. Examples of 1612-MHz OH maser light curves from the 'Bright OH/IR star sample'. Maser brightness SI is the integrated flux density given in units Jy km s^{-1}. *Left:* OH 32.8–0.3, a L-AGB star with a period of ≈4.4 years, observed between 2008 and 2022. *Right:* OH 31.0–0.2, a S-pAGB star, observed 2015–2022. The scatter of the integrated flux is caused by a mix of real fluctuations and calibration uncertainties.

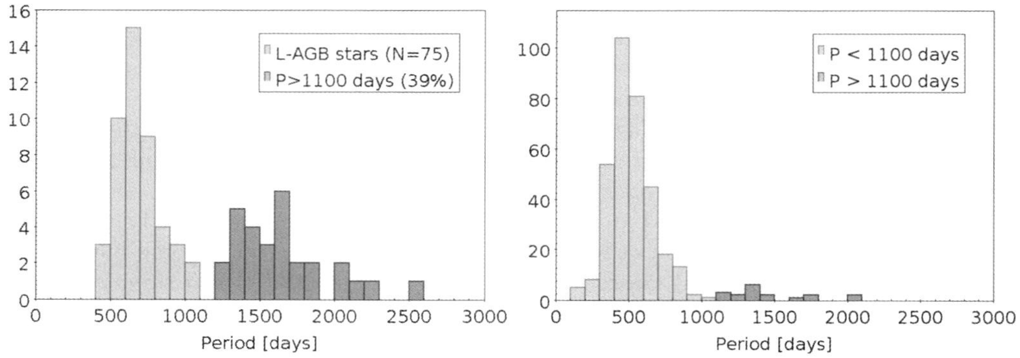

Figure 2. Period distributions. *Left:* 'Bright OH/IR star sample'. Periods are available for 75 OH/IR stars out of the 80 objects classified as L-AGB stars. 29 (39%) objects are 'extreme OH/IR stars' with P > 1100 days. *Right:* 349 OH/IR stars of the 'Arecibo sample of OH/IR stars', 18 with P > 1100 days.

periodicity also among S-pAGB stars. We found no periodicity in any of the S-pAGB stars.

Figure 2 shows the period distribution of the 'Bright OH/IR star sample' in comparison to the 'Arecibo sample'. In both samples two groups separated by period are discernible with the major group having periods $P \leq 1100$ days and a peak in the distribution between 500 and 700 days. The 'Bright OH/IR star sample' shows a prominent separate group of stars with $P > 1100$ days, which we call 'extreme OH/IR stars' hereafter. As the 'Arecibo sample' covers only a small part of the Galactic plane, it contains only few of the 'extreme OH/IR stars'. Of them, 9 (50%) are also in the 'Bright OH/IR star sample'. The confinement of the 'extreme OH/IR stars' to low latitudes suggests that their progenitors were relatively high main-sequence mass stars ($3 - 8\,M_{\odot}$).

3. OH/IR stars and WISE mid-IR Period-Luminosity Relations for Mira variables

Since several photometric surveys have been carried out in the mid-IR wavelength range (e.g. IRAS, MSX, AKARI, and WISE), the spectral energy distributions of OH/IR stars are now much better covered with observations, allowing us to determine more accurate bolometric fluxes than before. The other ingredient for the determination of luminosities is the distances to the OH/IR stars. Mostly kinematic distances are used, which can be obtained from the stellar radial velocities given by the center of the OH maser velocity range, and assuming that the stars follow closely the Milky Way's rotation curve. The latter assumption is probably not generally valid, as there are a fraction of OH/IR stars which have radial velocities incompatible with Galactic rotation. Also for a given radial velocity, in general two (the near- and far-kinematic) distances are possible making the distance determination ambiguous.

More precise distances can be obtained from the phase-lag technique (Etoka & Engels 2022), from radio parallax mesurements (Nakagawa et al. 2016), and also from GAIA (Bailer-Jones et al. 2021) for less obscured stars. Based on phase-lag distances of Van Langevelde et al. (1990) and average kinematic distances, Whitelock et al. (1991) found that OH/IR stars follow the same PLR as Miras in the LMC, albeit with substantial scatter. New phase-lag distances have been determined by Wolak et al. (2013) and Engels et al. (2015) and a few radio parallax distances have been published in recent years (Orosz et al. 2017; Nakagawa 2023), but the number of reliable distance determinations for OH/IR stars is still relatively small.

To determine luminosities for the 'Bright OH/IR star sample', we use preferentially the most recent phase-lag, radio parallax and GAIA distances. For the rest of the sample, we determined kinematic distances using the Bessel distance calculator V2† (Reid et al. 2019) and removed all OH/IR stars without a near-kinematic distance between 1 and 8 kpc. This left all in all 49 OH/IR stars with plausible distances.

To compare the location of the OH/IR stars in a period-luminosity diagram with Mira variables, we used the recently published PLRs for the LMC by Iwanek et al. (2021) and for Galactic Miras by Nakagawa (2023). Both use the absolute magnitude in the WISE $W3$ filter ($\lambda = 11.6\mu$m) as proxy for the luminosity, which is better suited for OH/IR stars, as the commonly used K filter ($\lambda = 2.2\mu$m) is more strongly affected by circumstellar extinction and brightness variations than the wavelength bands in the mid-IR. In Figure 3 we show the location of the OH/IR stars in the period-luminosity diagram together with the proposed WISE $W3$ PLRs. In the left panel only the OH/IR stars with phase-lag, radio astrometric and GAIA distances available are plotted, while in the right panel the remaining OH/IR stars with plausible near-kinematic distances are also included.

Taking into consideration only the OH/IR stars with distances based on (GAIA or radio) parallaxes or phase-lag determinations, we find that the extrapolation of the PLR of Iwanek et al. (2021) to periods >1100 days predicts luminosities higher than observed, while the positions of the OH/IR stars are reasonably consistent with the Nakagawa PLR in this regime. However, if the OH/IR stars with (plausible) kinematic distances are also included, the consistency is lost. The distribution of luminosities is then consistent with an absence of relation between periods and luminosities.

4. Discussion

The absence of a significant PLR for the OH/IR stars in the period range $600 < P < 2500$ days in Figure 3 may be due to uncertainties of the distances used. However, the lack of relation could also be real. The most recent non-linear pulsation models of Trabucchi et al. (2021) do not predict periods ≥ 1000 days for the entire AGB mass range during the evolution on the AGB, so that the periods of the 'extreme OH/IR stars' remain unexplained. These stars may be at the tip of AGB evolution close to the

† http://bessel.vlbi-astrometry.org/

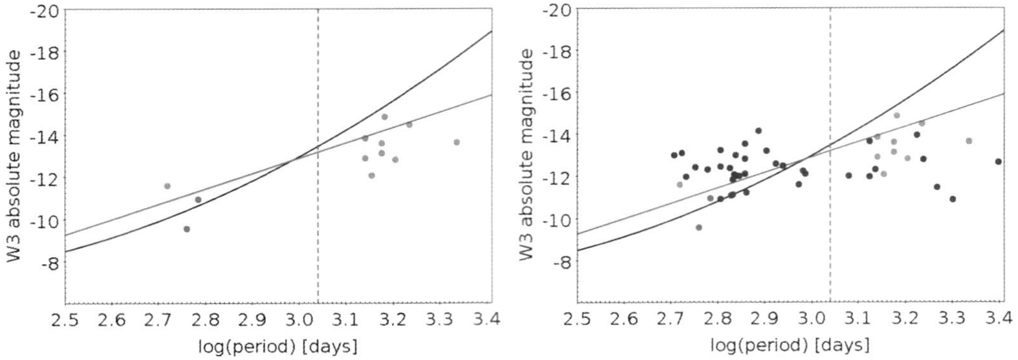

Figure 3. Positions of OH/IR stars from the 'Bright OH/IR star sample' in a period-luminosity diagram based on WISE $W3$ filter photometry and using distances obtained by different methods. Overplotted are the PLRs proposed by Iwanek et al. (2021) (black curve) and Nakagawa (2023) (red line). The vertical dotted line corresponds to a period P=1100 days. *Left:* Only OH/IR stars with distances based on GAIA parallaxes (red), or based on phase-lag determinations or radio parallaxes (green). *Right:* The same, but OH/IR stars added with kinematic distances (blue) between 1–8 kpc.

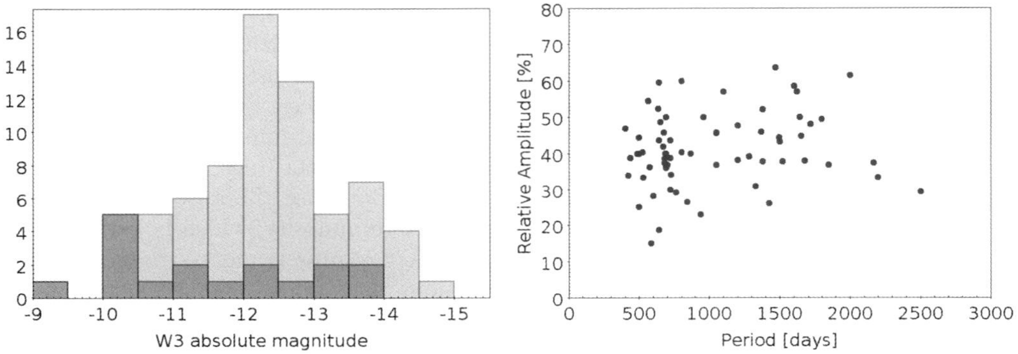

Figure 4. *Left:* WISE $W3$ absolute magnitude distribution of the 'Bright OH/IR star sample' with L-AGB stars shown in red and S-pAGB stars in blue. *Right:* Relative Amplitude (defined as 100 × amplitude / mean integrated flux density SI) of the integrated 1612-MHz OH maser flux density SI of the L-AGB stars in the 'Bright OH/IR star sample'.

transition into the post-AGB phase (Marini et al. 2023). The late AGB evolutionary phase and the early post-AGB phase likely lasts only a couple of thousand years, during which the luminosity does not change significantly. The $W3$ absolute magnitudes of the S-pAGB stars in the 'Bright OH/IR star sample' are similar to those of the L-AGB stars (Figure 4, left panel), corroborating the fact that the AGB maximum luminosity is reached before evolution in the post-AGB phase happens. The group of OH/IR stars with $P < 1100$ days may be ordinary Mira variables, as described by the above mentioned non-linear pulsation models, and may be still evolving on the AGB experiencing thermal pulse cycles increasing their periods and luminosities. On the other hand, the 'extreme OH/IR' stars may already be in their last thermal pulse cycle, with a relatively thin shell above the stellar core, and where the pulsation may appear with different periods and amplitudes for a short period of time (scale hundreds to thousand of years), before the pulsations die out. The star then evolves further as S-pAGB star and the OH masers fade away soon after the transition (Etoka et al. 2023). The cessation of the pulsation might be abrupt, as we have not detected any correlation between amplitudes and periods (Figure

4, right panel), which could be expected if the pulsation would cease slowly over a period of time.

If this scenario is right, then the 'Bright OH/IR star sample' may actually consist of three groups representing different evolutionary phases on the AGB and the early post-AGB phase. These are the Miras (L-AGB stars with $P < 1000$ days), the 'extreme OH/IR stars' stars with longer periods at the border between the AGB and the post-AGB phase, and the post-AGB OH/IR stars (S-pAGB). Maybe, only the first group is expected to follow PLRs derived from shorter-period Mira variables.

References

Bailer-Jones C. A. L., Rybizki J., Fouesneau M., Demleitner M., Andrae R., 2021, AJ, 161, 147

Baud B., Habing H. J., Matthews H. E., Winnberg A., 1981, A&A, 95, 156

De Beck E., Decin L., de Koter A. et al., 2010, A&A, 523, A18

Engels D., Kreysa E., Schultz G. V., Sherwood W. A., 1983, A&A, 124, 123

Engels D. & Jiménez-Esteban F., 2007, A&A, 475, 941

Engels D., Etoka S., Gérard E., Richards A., 2015, in Kerschbaum F., Wing R. F. & J. Hron J., eds., Why Galaxies Care about AGB Stars III: A Closer Look in Space and Time, ASP-CS, 497, 473

Etoka S. & Engels D., 2022, European VLBI Network Mini-Symposium and Users' Meeting 2021, 12–14 July, 2021. Online at https://pos.sissa.it/cgi-bin/reader/conf.cgi?confid=399, id.12

Etoka S., Engels D., Ullrich T., González J. B., López-Martí B., 2023, in Hirota T., Imai H., Menten K. & Pihlstrom Y., eds., Cosmic Masers: Proper Motion toward the Next-Generation Large Projects, Proceedings of the IAU Symposium 380, in press

Goldman S. R., van Loon J. Th., Zijlstra A. A. et al., 2017, MNRAS, 465, 403

Herman J. & Habing H. J., 1985, A&AS, 59, 523

Iwanek P., Soszyński I., Kozłowski S., 2021, ApJ, 919, 99

Jiménez-Esteban F. M., Engels D., Aguado D. S., González J. B., García-Lario P., 2021, MNRAS, 506, 6051

Jones T. J., Hyland A. R., Wood P. R., Gatley I., 1983, ApJ, 273, 669

Marini E., Dell'Agli F., Kamath D. et al., 2023, A&A, 670, A97

Miller Bertolami M. M., 2016, A&A, 588, A25

Nakagawa A., Kurayama T., Matsui M. et al., 2016, PASJ, 68, 78

Nakagawa A., 2023, these proceedings

Orosz G., Imai H., Dodson R. et al., 2017, AJ, 153, 119

Reid M. J., Menten K. M., Brunthaler A. et al., 2019, ApJ, 885, 131

Trabucchi M., Wood P. R., Mowlavi N. et al., 2021, MNRAS, 500, 1575

van Langevelde H.J., van der Heiden R., van Schooneveld C., 1990, A&A, 239, 193

Whitelock P., Feast M., Catchpole R., 1991, MNRAS, 248, 276

Wilson W. J., Barrett A. H., 1970, A&A, 17, 385

Wolak P., Szymczak M., Gérard E., 2013, MNRAS, 430, 2499

Author index